Thomas D. Rossing
Northern Illinois University

The Science of Sound

SECOND EDITION

Addison-Wesley Publishing Company

Reading, Massachusetts • Menlo Park, California
New York • Don Mills, Ontario • Wokingham, England
Amsterdam • Bonn • Sydney • Singapore • Tokyo • Madrid • San Juan

To all those who have taught me
so that I could teach others

Library of Congress Cataloging-in-Publication Data

Rossing, Thomas D., 1929–
    The science of sound/Thomas D. Rossing.—2nd ed.
      p. cm.
    Includes index.
    ISBN 0-201-15727-6
    1. Sound.   2. Music—Acoustics and physics.   I. Title.
QC225.15.R67   1989   89-34321
534—dc20   CIP

ABCDEFGHIJ-HA-943210

Preface to the Second Edition

The past seven years have seen many advances in the areas of acoustics covered in this book. Digital techniques for generating and recording sound have come of age. Compact disc digital audio, for example, is considered by some analysts to be the most successful electronic product ever introduced. Less visible, but no less important, are some very important discoveries about the human auditory system, and the way in which we perceive sounds, including music and speech. New research has led to considerable progress in our understanding of the physics of musical instruments and of concert halls.

The most painful task an author faces is how to incorporate new material into a book without greatly increasing its length. Although the second edition contains a considerable amount of new material, it is only thirty-six pages longer than the first edition. Most chapters can still be covered in two or three class periods. Chapters 5–8 on Perception and Measurement of Sound have been rewritten to take account of recent advances in psychoacoustics and also to make use of the auditory demonstrations now available on a compact disc (from the Acoustical Society of America), which I have found exceedingly useful in my own teaching.

Chapters 28 and 29 discuss digital computers and digital techniques for generating, processing, and recording sound. To make room for this material, discussions of analog synthesizers and of integrated circuits have been shortened. Some topics have been moved to other chapters for pedagogical reasons. I hope that teachers who have used the first edition in their classes will approve of these changes, and I welcome their comments!

T.D.R.

DeKalb, Illinois
September 1989

Preface to the First Edition

━━━━━━━━━━━━━━

The word *sound* is used to describe two different things: (1) an auditory sensation in the ear; (2) the disturbance in a medium, which can cause this sensation. (Making this distinction answers the age-old question, "If a tree falls in a forest and no one is there to hear it, does it make a sound?")

The science of sound, which is called *acoustics,* has become a broad interdisciplinary field encompassing the academic disciplines of physics, engineering, psychology, speech, audiology, music, architecture, physiology, and others. Among the branches of acoustics are architectural acoustics, physical acoustics, musical acoustics, psychoacoustics, electroacoustics, noise control, shock and vibration, underwater acoustics, speech, physiological acoustics, etc.

This book is intended to be an introduction to acoustics, written in nontechnical language, primarily for students without college physics and mathematics. A few basic mathematical ideas (such as logarithms) are introduced, and a brief review of algebra (ninth-grade level) is included in the appendix for those who need it. Chapters vary in the level of sophistication. Chapter 9, for example, presupposes some acquaintance with music; Chapters 28 and 29 describe rather sophisticated technology, in a way that, it is hoped, is understandable to the beginner.

The book is modular in organization. Parts I and II provide an introduction to the rest of the book. A student who has taken a general physics course can begin with Part II. Part V is a prerequisite to Part VII, but apart from that the parts can be studied in almost any order. Thus it can be used as a textbook for a variety of introductory courses in acoustics.

Possible sequences of chapters for various one-semester courses are as follows:

Musical acoustics

I	II	III	IV	VI	
1-4	5-8	9-14	15-17	23	18 chapters

Physics of high-fidelity/electronic sound

I	II	V	VI	VII	
1-4	5-6	18-22	23-25	26-29	18 chapters

Introduction to acoustics (speech, audiology, technology)

I	II	IV	V	VII	VIII	
1-4	5-8	15-16	18-21	23-24	30-32	19 chapters

I have taught courses in musical acoustics for over twenty years. During that period there has been a considerable increase in the number of physics departments that offer such a course not only to music students but to other students with an interest in acoustics and music as well. In recent years there has been a growing interest in high-fidelity sound reproduction, electronic music, and other topics in electro-acoustics. The preliminary version of this book has been used successfully in courses of this type at two universities.

It is highly recommended that laboratory experiments be a part of an introductory acoustics course if at all possible. Musicians readily appreciate that one cannot learn to play the piano by reading a book and listening to lectures, however erudite; the laboratory is our "practice hall." Experiments should be of two types: basic acoustical measurements, and experiments that follow along with the various chapters to illustrate the physical principles. In addition, classroom demonstration experiments should be a regular practice. Many acoustic phenomena can best be demonstrated with a tape-recorded demonstration.

I am indebted to many colleagues who have read various parts of the manuscript and made valuable comments. I especially wish to acknowledge the contributions of William Hartmann, Kenneth Jesse, Garry Kvistad, John Popp, and Carleen Hutchins for their critical reading of various chapters. Over the years I have gained a great deal from discussions with acousticians Uno Ingard, Arthur Benade, Neville Fletcher, Gabriel Weinreich, and musicians too numerous to mention individually. My faithful students Calvin Rose, Craig Anderson, Richard Ross, Mark Gilbert, and Robert Shepherd have been of great assistance in the teaching of our acoustics courses and thereby have contributed in a valuable way.

T.D.R.

DeKalb, Illinois
November 1981

Contents

PART III   169

Acoustics of Musical Instruments

Chapter 9   Musical Scales and Temperament

Chapter 10   String Instruments

Chapter 11   Brass Instruments

Chapter 12   Woodwind Instruments

Chapter 13   Percussion Instruments

PART IV   309

The Human Voice

PART V   367

The Electrical Production
of Sound

PART VI   457
The Acoustics of Rooms

Chapter 23   Auditorium Acoustics

It is the purpose of the first four chapters to discuss briefly the physical principles needed to understand the rest of the book. The level of discussion is comparable to a high school physics course or a college physics course taught for students not majoring in a science. (Such courses are sometimes referred to as ''physics for poets.'')

The reader who has already taken a physics course in high school or college may wish to skip part or all of Part I and begin with Chapter 5. Or Chapters 2–4 may serve as a useful review of the physics of vibrations and waves. On the other hand, the student who has never had a course in physics or physical science is advised to read Chapter 1 very carefully, paying special attention to the definitions of various physical quantities, such as power and energy, that are so easily confused.

The metric system of units (in particular, the Système International, or SI) is used throughout this book. It is the preferred system for scientific work, and in the future (hopefully) it will come into general use in the United States as it has in the rest of the world. Other systems of units are listed in Appendix A.1.

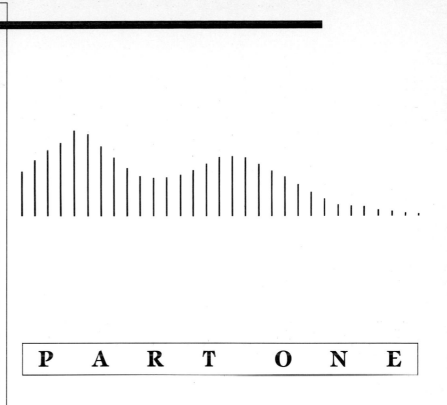

P A R T   O N E

CHAPTER 1
Motion, Force, and Energy

One way to describe the motion of an object, whether it is an air molecule or a spacecraft, is to express its *distance* from a point, its *speed*, and its *acceleration* as functions of *time*. In the following sections we will elaborate on that statement and present definitions of these quantities. In doing so, we will find it convenient to introduce the concept of *coordinates*, for the use of a coordinate system simplifies the description of motion.

1.1 DISTANCE

The position of an object is conveniently expressed as the distance from a certain reference point (e.g., "I am 2½ miles from my office"). However, more information is conveyed if we express direction as well (e.g., "I am 2½ miles northwest of my office"). An alternate way to describe directions makes use of the four cardinal directions of the compass (e.g., "I am 2 miles north and 1½ miles west of my office"). This leaves no doubt as to your position (provided, of course, that the reference point—your office—is well known).

If, instead of using the directions north, south, east, and west, we use the physicist's favorite letters x and y (letting y and $-y$ represent north and south and x and $-x$ represent east and west), we have the

3

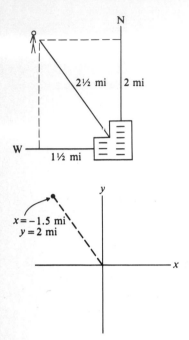

FIG. 1.1
The use of coordinates to express
position and distance.

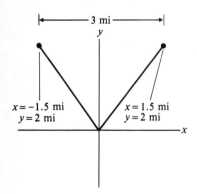

FIG. 1.2
Distance between two points in a
coordinate system.

two-dimensional coordinate system shown in Fig. 1.1. If we agree that the origin of the coordinates (that is, the point at which $x=0$ and $y=0$) is at the office, then the position can be described as $x=-1.5$ mi, $y=2$ mi. It should be fairly obvious that to move from point $x=-1.5$, $y=2$ to point $x=1.5$, $y=2$ one would move a distance of 3 miles, as shown in Fig. 1.2.

To describe position in three-dimensional space, three coordinates are required, so a z-coordinate is added. Much of our discussion, however, will be restricted to objects that move in one direction only; these objects can be described by one coordinate, which we select as y.

Many different units are used to express distance: feet, inches, meters, yards, even furlongs. However, since most of the world uses metric units, and the United States will soon follow this practice, it is prudent to emphasize the use of the metric system. The meter, then, will be our preferred unit of length, with occasional reference to feet and inches. Conversions between various units appear in the appendix.

The metric system uses prefixes to denote various powers of ten as indicated below:

$$
\begin{aligned}
1 \text{ kilometer } &= 1 \text{ km } = 10^3 \text{ meters,} \\
1 \text{ centimeter } &= 1 \text{ cm } = 10^{-2} \text{ meter,} \\
1 \text{ millimeter } &= 1 \text{ mm } = 10^{-3} \text{ meter,} \\
1 \text{ micrometer } &= 1 \text{ } \mu\text{m } = 10^{-6} \text{ meter.}
\end{aligned}
$$

These are the principal prefixes used for distance; others appear in the appendix. (The term "micron" is sometimes used in place of "micrometer.") Note that the first syllable is accented in both "kil′ ometer" and "mic′ rometer."

1.2   SPEED AND VELOCITY

For an object that travels at constant speed, the distance traveled is given by the simple formula:

$$\text{distance} = \text{speed} \times \text{time.}$$

We all realize that if we travel at a steady rate of 50 miles per hour for 2 hours, we will cover 100 miles, as the formula states. The same distance will be covered if our average speed is 50 mph for the two hours, even though our actual speed at different times may vary widely. In fact, average speed can be defined as distance divided by time:

$$\text{speed} = \frac{\text{distance}}{\text{time}}. \tag{1.1}$$

Physicists often speak of *velocity* rather than speed. Velocity specifies the direction of motion as well as the speed, since the components in each direction (x, y, z) are specified. In describing motion in one direction, however, no distinction need be made, and in most of the vibrating systems we wish to discuss, the vibrations are in one direction. Speed is correctly defined as the magnitude of velocity (without regard to direction), and we will use the symbol v for speed, since it is customary to do so; furthermore it reminds us that speed and velocity are closely related. (Note that the speedometer of an automobile indicates speed, since it does not specify the direction of travel.)

To find speed, we must measure both distance and time. This is often done in the laboratory by photographing an object illuminated by a stroboscopic ("strobe") lamp. The photographs in Fig. 1.3 were taken with the lamp flashing ten times per second in a darkened room while the camera shutter remained open. Thus the position of the object was recorded at intervals of 0.1 second. It is easy to see that in Fig. 1.3(a) the speed stays the same, whereas in Fig. 1.3(b) it increases as the object moves from left to right.

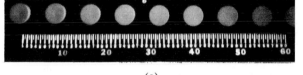

(a)

(b)

We can easily determine the speed of the object in Fig. 1.3(a). During each interval of 0.1 second, it appears to move 7.5 centimeters, so the speed is

$$v = \frac{7.5 \text{ cm}}{0.1 \text{ s}} = 75 \text{ cm/s} = 0.75 \text{ m/s}.$$

FIG. 1.3
Stroboscopic pictures of motion: (a) constant speed; (b) changing speed. Both pictures were made with ten flashes per second.

Since the object moves with constant speed, the average speed and instantaneous speed are the same; this is not the case in Fig. 1.3(b).

Example 1.1   If the speed limit is posted as 30 miles per hour, what is the corresponding limit in meters per second?

Solution: $30 \dfrac{\text{mi}}{\text{hr}} = \dfrac{(30 \text{ mi/hr})(5280 \text{ ft/mi})(0.305 \text{ m/ft})}{3600 \text{ s/hr}}$

$\qquad\qquad = 13.4 \text{ m/s.}$

Example 1.2   A motorist travels 100 miles. The first half of the distance takes one hour; the second half takes $1\frac{1}{2}$ hours. What is the average speed for the journey?

Solution: $v_{av} = \dfrac{d}{t} = \dfrac{100 \text{ mi}}{1 + 1.5 \text{ hr}} = 40 \text{ mi/hr.}$

Example 1.3   A motorist travels 300 miles. During the first half of the journey his average speed is 60 mi/hr, and during the second half his average speed is 30 mi/hr. What is his average speed for the entire journey?

Solution: $t_1 = \dfrac{150 \text{ mi}}{60 \text{ mi/hr}} = 2.5 \text{ hr;}$

$t_2 = \dfrac{150 \text{ mi}}{30 \text{ mi/hr}} = 5.0 \text{ hr.}$

$v_{av} = \dfrac{d}{t} = \dfrac{300 \text{ mi}}{2.5 + 5.0 \text{ hr}} = 40 \text{ mi/hr.}$

(Note that the answer is *not* 45 mi/hr).

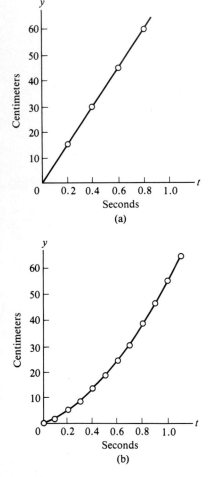

FIG. 1.4

Graphical representation of the motion shown in Fig. 1.3. The object in (a) moved at a constant speed; the object in (b) changed its speed.

1.3   GRAPHICAL REPRESENTATION OF MOTION

If a picture is worth a thousand words, a well-constructed graph must be worth at least five thousand, especially when it comes to describing motion. Suppose we wish to represent the changing position of the objects in Fig. 1.3 using two graphs. One coordinate is the position, represented by y, and the other coordinate is the time t. It does not really matter what point is selected as the zero or starting point (called the origin), but it is convenient to take it as the end of the meter stick; alternatively, it could have been the position of the object at the first flash. The graphs in Fig. 1.4 are the result.

Now we will use these graphs to help us determine average and instantaneous speed. To the graphs we add the useful constructions shown in Fig. 1.5. The distance FE is called Δy, a symbol read as "delta y," which means the "change in y." Similarly, Δt (line DF) represents the "change in t." In Fig. 1.5(a), the average speed during the time interval from $t = 0.2$ s to $t = 0.4$ s is

$$v_{av} = \frac{\Delta y}{\Delta t} = \frac{15 \text{ cm}}{0.2 \text{ s}} = 75 \text{ cm/s.}$$

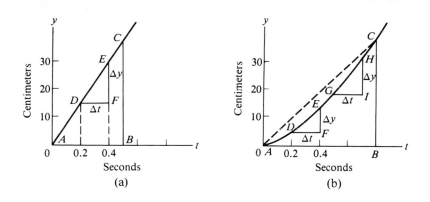

FIG. 1.5
Curves of uniform and changing motion as in Fig. 1.4 with constructions added for determining speed.

In Fig. 1.5(b) it is

$$v_{av} = \frac{\Delta y}{\Delta t} = \frac{8.4 \text{ cm}}{0.2 \text{ s}} = 42 \text{ cm/s}.$$

In Fig. 1.5(a), the speed we calculate should be the same regardless of the time interval Δt selected. In fact, the larger time interval $\Delta t - 0.5$ s (represented by line AB) should be used:

$$v = \frac{\Delta y}{\Delta t} = \frac{37.5 \text{ cm}}{0.5 \text{ s}} - 75 \text{ cm/s}.$$

This is not true in the case of the changing motion in Fig. 1.5(b). Using the time intervals represented by DF, GI, and AB gives three different values of average speed, because the speed is changing:

$$v_1 = \frac{8.4}{0.2} = 42.0 \text{ cm/s}, \quad (DF)$$

$$v_2 = \frac{12.5}{0.2} = 62.5 \text{ cm/s}, \quad (GI)$$

$$v_3 = \frac{37.5}{0.8} = 46.9 \text{ cm/s}. \quad (AB)$$

If the instantaneous speed at a particular time is to be determined, the time interval Δt must be made very small. As Δt becomes smaller and smaller, the line (chord) DE more and more nearly approaches the slope of the curve at point P as shown in Fig. 1.6. Thus, instantaneous speed can be interpreted as the *slope* (steepness) of the curve representing position y as a function of time t.

A graph showing speed as a function of time is also useful in describing motion. If the object photographed in Fig. 1.3 had a speedometer attached to it, we would have a photographic record of speed each time the strobe light flashed. These values of speed could then be plotted on a graph of v versus t. However, we can also deter-

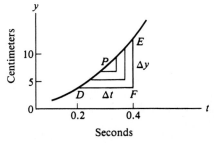

FIG 1.6
Shrinking Δy and Δt in order to obtain instantaneous speed.

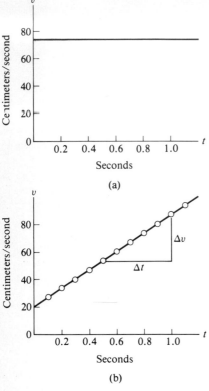

FIG 1.7
Speed as a function of time for the objects shown in Figs. 1.3 and 1.4.

mine speed from the graphs already drawn, since speed at any time is represented by the slope of the curve showing y versus t at that time. Figure 1.7 shows speed as a function of time for each graph in Fig. 1.4.

1.4 ACCELERATION

Acceleration refers to change in speed. When we push down the accelerator pedal in an automobile, we expect the speed to increase. In describing motion, acceleration is defined as the rate of change of speed (just as speed is the rate of change of position). *Average acceleration* is the ratio of change of speed Δv to time interval Δt:

$$a_{av} = \frac{\Delta v}{\Delta t}. \tag{1.2}$$

Instantaneous acceleration at a particular time can be determined by making Δt and Δv very small, just as we determined instantaneous speed v by making Δt and Δy very small.

In Fig. 1.7(a), the speed remains unchanged in time; in Fig. 1.7(b) it increases at a steady rate. The acceleration in Fig. 1.7(a) is therefore zero; in Fig. 1.7(b) it is

$$a = \frac{35 \text{ cm/s}}{0.5 \text{ s}} = 70 \text{ cm/s}^2.$$

Note the units for acceleration; the unit cm appears to the first power, but the unit s appears squared.

An object in *free fall* in the gravitational field of the earth experiences a constant acceleration of 9.8 meters/second². Thus if it begins with no initial speed (up or down), at the end of the first second it will have a speed of 9.8 m/s; at the end of 2 seconds its speed will be 19.6 m/s, and so on.

Note that acceleration does not always increase speed. If an object is thrown upward, the acceleration due to gravity acts to slow it down (sometimes this is called "deceleration"). Figure 1.8 shows stroboscopic photographs of two objects in free fall. In Fig. 1.8(a), the object is dropped from rest; in Fig. 1.8(b), it is thrown upward, slows down, and then begins its descent.

Figure 1.9 shows a stroboscopic photograph of an object with an acceleration that changes its direction with time. The object, a mass attached to a spring, is executing a type of vibratory motion called "simple harmonic" motion, which will be discussed in Chapter 2. The camera has been turned during the exposure, so that multiple images of the object create a photographic record of position as a function of

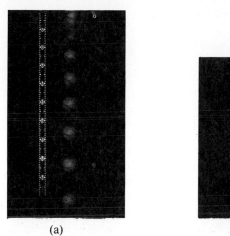
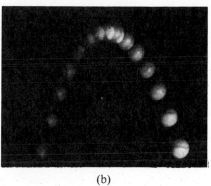

(a)   (b)

FIG. 1.8
Stroboscopic photographs of two
objects in free fall: (a) object dropped
from rest; (b) object thrown upward.
(Photographs by Christopher
Chiaverina.)

time. Note that position y, speed v, and acceleration a all change with
time.

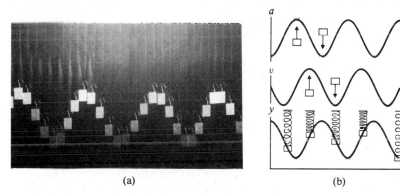

(a)   (b)

FIG. 1.9
Vibratory motion in which y, v, and a
all change with time.

Example 1.4   A bicyclist accelerates from 0 to 10 m/s in
20 s. He then applies the brakes and comes to a stop in 5 s.
What is his average acceleration in each case?

Solution: $a_{av} = \dfrac{\Delta v}{\Delta t} = \dfrac{10-0 \text{ m/s}}{20 \text{ s}} = 0.5 \text{ m/s}^2;$

$\qquad a_{av} = \dfrac{0-10 \text{ m/s}}{5 \text{ s}} = -2 \text{ m/s}^2.$

Example 1.5   A ball is thrown upward with a velocity of
15 m/s. How long does it take to reach its maximum height?

How long does it take to fall back to the ground (neglect air resistance)?

Solution: $a = \dfrac{\Delta v}{\Delta t}$, so $\Delta t = \dfrac{\Delta v}{a} = \dfrac{0 - 15 \text{ m/s}}{-9.8 \text{ m/s}^2} = 1.53 \text{ s.}$

1.5   FORCE, MASS, AND NEWTON'S LAWS

Force is a quantity with which we are all familiar; it can be described as a "push" or a "pull." Practically all human activity involves forces: running, lifting, eating, writing, even standing.

Applying a force to an object may result in either a *distortion* or an *acceleration*. Forces can stretch a spring or bend a wire. Forces can make things move or stop them if they are in motion. If several forces act on an object, it is the *net* force that determines its acceleration. Many years ago, it was discovered that the acceleration of an object is proportional to the net force acting on it. This may be written

$$a \propto F$$

(the symbol $\propto$ indicates proportion). If the net force on a given object is doubled, the acceleration doubles, and so forth.

A statement of proportion can be rewritten as an equation by inserting a constant of proportionality. In the case of force and acceleration, the constant of proportionality needed is the *mass* of the object, and the resulting equation is called *Newton's second law of motion*:

$$F = ma. \tag{1.3}$$

The mass of an object is thus a measure of its opposition to acceleration.

Every object on or near the earth is acted on by gravity. The force of gravity on an object is called its *weight*. Mass and weight are certainly not the same, although they are frequently confused with each other. We measure the mass of an object by comparing it to a known mass, and a convenient way to make this comparison is by "weighing" the object on a scale. Weight w can be defined by Newton's second law,

$$w = mg, \tag{1.4}$$

where $g = 9.8$ meters/second^2 is the acceleration due to gravity of an object in free fall.

In order to express force and mass, they must be assigned units. In the preferred SI (metric) system,

mass is measured in kilograms (kg),
force is measured in newtons (N).

A net force of one newton acting on a mass of one kilogram gives it an acceleration of one meter/second². A mass of one kilogram has a weight of 9.8 newtons.

Suppose that an astronaut on the surface of the moon holds a heavy object. It is much easier to hold this object on the moon, because its weight is only ⅙ as great as it would be on Earth. But an attempt by the astronaut to move the object quickly to a new position would be just as difficult to do as it would be on Earth, because the object's resistance to acceleration is determined by its mass, which is unchanged by the lunar environment.

Example 1.6   How much does a 5-kg mass weigh on the earth and on the moon?

Solution: $w = mg = 5 \text{ kg } (9.8 \text{ m/s}^2) = 49 \text{ N}$ on earth;

$$w = \frac{1}{6}(49) = 8.17 \text{ N on the moon.}$$

Example 1.7   A force of 10 newtons is applied to a mass of 4 kilograms for 5 seconds. Find its acceleration and its speed at the end of 5 seconds.

Solution: $F = ma$, so $a = \dfrac{F}{m} = \dfrac{10 \text{ N}}{4 \text{ kg}} = 2.5 \text{ m/s}^2.$

$a = \dfrac{\Delta v}{\Delta t}$, so $\Delta v = a\Delta t = (2.5 \text{ m/s}^2)(5 \text{ s}) = 12.5 \text{ m/s.}$

1.6   PRESSURE

Newton's second law of motion describes the way in which the net force acting on an object will set that object into motion. So far as that law is concerned, it does not matter whether the net force is due to a single force acting at a point or many forces distributed around the object. In weighing an object, we can consider the entire force of gravity acting at one point, which we appropriately refer to as the *center of gravity* or *center of mass*, even though gravity acts on every part of the object.

There are other times, however, when the distribution of forces is important. For example, you can walk on snow without sinking if you wear snowshoes; on the other hand, if you were to cross a wooden floor in spike heels, you would severely damage the floor. The difference in this case is the *area* over which the same total force is distributed.

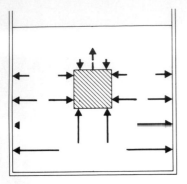

FIG. 1.10
Pressure in a container of fluid (a) acts on all surfaces; (b) is proportional to depth. Buoyant force (dashed arrow) on the immersed object is due to the excess upward force.

It is useful to define a quantity called *pressure* as the force divided by the area over which it is distributed. To be more specific, it is the force acting perpendicular to a surface divided by the area of that surface:

$$p = F_\perp/A. \tag{1.5}$$

Since force is measured in newtons, pressure is measured in newtons/meter^2 (N/m^2) or pounds/square inch in the British system. A 50-kg (110-lb) person standing on spike heels with an area of 20 mm^2 would exert a pressure of nearly 25 million N/m^2 (about 1.8 tons per square inch!) on the floor.

Fluids (liquids and gases) exert forces on the walls of their containers and anything immersed in them. One of the important properties of all fluids at rest, in fact, is that the pressure acts perpendicular to all surfaces (walls of the container as well as immersed objects). The pressure at any point in an open container of fluid (liquid or gas) is determined by the weight of the fluid above that point. For example, the weight of the atmosphere above us results in a pressure of about 10^5 N/m^2 (15 lb/in^2) at sea level; at an altitude of about 5.5 km (3.4 miles), the pressure is only one half as great. The *buoyancy* of an immersed object is due to the excess upward pressure on its bottom surface, as shown in Fig. 1.10.

Slow variations in atmospheric pressure (as measured by a barometer) are indicative of changing weather. Sound waves consist of very small but rapid variations in pressure.

Example 1.8   What is the total inward force due to atmospheric pressure on a sphere one meter in diameter?

Solution: $A = 4\pi r^2 = 4\pi(0.5)^2 = 3.14$ m^2.

$F = pA = (10^5$ N/m$^2)(3.14$ m$^2) = 3.14 \times 10^5$ N.

1.7   WORK AND ENERGY

The terms "work" and "energy" have various meanings in everyday life, but in the language of physics they have very definite meanings. *Work* is done when a force is applied to an object that moves. The work that is done is the product of the force times the distance moved parallel to the force:

$$W = Fd. \tag{1.6}$$

Work is expressed in newton-meters, or *joules* (abbreviated J). If a force of one newton causes an object to move one meter, then one joule of work has been done.

The force of gravity on an object (its weight) is given by the formula *mg*. Thus if an object falls a vertical distance *h*, the work done by gravity is:

$$W = mgh. \tag{1.7}$$

By the same token, raising an object of mass *m* to a height *h* requires an amount of work $W = mgh$.

Energy is perhaps the central idea underlying all branches of science. There are many forms of energy (e.g., mechanical, electrical, thermal, chemical, radiant, nuclear). A great deal of modern technology has as its goal the more efficient conversion of one of these forms of energy into another. For example, our future appears to depend on the development of the technology to convert solar (radiant) energy and the nuclear energy contained in sea water into electrical energy to replace our dwindling supply of oil and gas (chemical energy).

In our study of acoustics, we are concerned mainly with mechanical energy (and, to a lesser extent, electrical energy). Mechanical energy is closely related to work; systems with mechanical energy have the potential to do work. Vibrating systems have mechanical energy; mechanical energy is carried by the moving molecules in a sound wave. Energy, like work, is measured in joules. Sometimes a distinction is made between energy of motion, called *kinetic energy*, and stored energy, called *potential energy*.

Without going into a detailed discussion of energy, let us describe the energy of five completely different systems:

1. A baseball flying through the air has energy of motion, or *kinetic energy*. It is obvious that if it strikes another object (a bat, perhaps?) it can do work by virtue of its kinetic energy. The amount of kinetic energy is given by the formula

$$KE = \tfrac{1}{2}mv^2, \tag{1.8}$$

where *m* is mass and *v* is its speed.

2. A block of wood lifted to a height *h* above the floor has *potential energy*, because if it were allowed to fall, it could do work. The amount of potential energy is given by the formula

$$PE = mgh, \tag{1.9}$$

where *h* is its height above the floor.

3. A spring that has been stretched (or compressed) has potential energy, because if allowed to relax it can do work. If we use the

constant K to denote its spring constant ("stiffness"), the potential energy when it is stretched an amount y from its relaxed length is given by the formula

$$PE = \tfrac{1}{2}Ky^2. \tag{1.10}$$

4. A bottle of gas of volume V whose pressure exceeds atmospheric pressure P_0 by a small amount p has potential energy

$$PE = \frac{1}{2}\frac{V}{P_0}p^2. \tag{1.11}$$

5. A violin string displaced a small distance y at its midpoint has potential energy

$$PE = \frac{2T}{L}y^2, \tag{1.12}$$

where T is the tension in the string and L is its length. When the string is released, this energy is changed into kinetic energy, and thereafter the energy changes back and forth between kinetic and potential. The energy of vibrating systems will be discussed in Chapter 2.

Often the analysis of motion is facilitated by considering the way in which one form of energy is converted into another. For example, if we lift a heavy object we do work, and we give the object potential energy ($PE = mgh$). If we allow the object to free fall, it acquires kinetic energy ($KE = \tfrac{1}{2}mv^2$). The velocity it acquires in falling a distance h can easily be calculated by equating gain of KE to loss of PE without the need to calculate the time of fall:

$$\tfrac{1}{2}mv^2 = mgh$$

$$v^2 = 2gh$$

$$v = \sqrt{2gh}. \tag{1.13}$$

In describing vibrating systems in Chapter 2, we will make use of the fact that such systems continually convert potential energy to kinetic energy and vice versa. Twice during each cycle of oscillation the energy is all kinetic, and twice it is all potential; at other times, the total energy is shared between potential and kinetic forms. As an oscillator slows down due to friction, the total energy decreases, and some of the mechanical energy is converted to another familiar form of energy: *heat*.

1.8 POWER

Note that the definition of work (force × distance) says nothing about the time during which work is done. Raising a 2-kg mass to a height of 1 m requires 19.6 J of work whether the task is done in one second or ten. If the task were done in 4 s, for example, the average rate at which work is done would be 4.9 joules/second (J/s). The rate at which work is done is called *power*. Power is simply work divided by time:

$$\mathscr{P} = \frac{\mathscr{W}}{t}. \tag{1.14}$$

Power can be expressed in J/s, but this unit is used so frequently that it is given a special name, the *watt* (abbreviated W). Electrical equipment is rated according to the number of watts of electrical power it requires. A 100-watt lamp converts electrical energy (to heat and light) at the rate of 100 joules per second. The electric company, which sells electrical energy, installs meters that indicate how much electrical energy has been consumed. Instead of using joules (watt-seconds), however, the billing unit is the kilowatt-hour, the amount of work done by one kilowatt of power in one hour. One kilowatt-hour (kWh) equals $1000 \times 60 \times 60 = 3.6 \times 10^6$ joules.

In the British system of units, work and energy are measured in *foot-pounds* and power in *horsepower*. One horsepower equals 550 foot-pounds per second, which is roughly the rate at which a horse can do work. One horsepower equals 745.7 watts.

Example 1.9   A guitar string 65 cm long and having a tension of 55 N is displaced 8 mm at its midpoint. How much potential energy does it have?

Solution: $\text{PE} = \dfrac{2(55 \text{ N})}{(0.65 \text{ m})} (8 \times 10^{-3}\text{m})^2 = 0.011 \text{ J.}$

Example 1.10   How much power is needed to raise a 2-kg mass to a height of 3 m in 15 s?

Solution: $\mathscr{P} = \dfrac{\mathscr{W}}{t} = \dfrac{mgh}{t} = \dfrac{(2 \text{ kg})(9.8 \text{ m/s}^2)(3 \text{ m})}{15 \text{ s}}$

$= 3.92 \text{ W.}$

1.9   UNITS

The preferred system of units for expressing physical quantities is the SI (Système International) or mks (meter-kilogram-second) system. Besides these three basic units, the system uses such units as newtons, joules, and watts, which are derived in a logical manner from the basic units. (For example, a newton equals kilograms × meters/seconds².) The SI or mks system is described in Appendix A.1.

Another metric system in use, but not as commonly as the mks system, is the cgs system, based on the centimeter, gram, and second. The cgs system is also described in Appendix A.1.

The fps or British system of units is still very much in use, of course, although its popularity is declining as the United States aims toward a conversion to metric units. Besides the basic units of foot, pound, and second, the fps system uses slugs, foot-pounds, horsepower, etc., as described in Appendix A.1.

1.10 SUMMARY

The motion of an object can be described by expressing its distance, speed, and acceleration as functions of time. A graphical representation of the motion consists of a plot of distance, speed, or acceleration as a function of time. Two other basic quantities, force and mass, are related to acceleration through Newton's second law of motion: $F = ma$. Pressure is force per unit area, and is especially important in describing the behavior of objects immersed in a liquid or gas.

Another useful description of an object expresses its kinetic energy (energy of motion) and potential energy (stored energy) as functions of time. Power is the rate at which energy is expanded or the rate at which work is done. The preferred system of units is the mks (SI) system, which uses meters, kilograms, and seconds as its basic units, but also includes derived units such as newtons, joules, watts, and so forth.

References and Suggested Readings

Hewitt, P. G. (1971). *Conceptual Physics*. Boston: Little, Brown.

Physical Science Study Committee (1960). *Physics*. Boston: D.C. Heath.

Project Physics Text (1970). New York: Holt, Rinehart and Winston.

Glossary

acceleration The rate of change of speed or velocity.

coordinates A set of numbers used to locate a point along a line or in space.

force An influence that can deform an object or cause it to change its motion.

gravity The force exerted by the earth on all objects on or near it.

joule A unit of energy or work; one joule is equal to one newton-meter, also one watt-second.

kinetic energy Energy of motion; the capacity to do work by virtue of that motion; equal to one half mass times velocity (or speed) squared.

mass A measure of resistance to change in motion; equal to force divided by acceleration.

newton A unit of force.

potential energy Stored energy; the capacity to do work by virtue of position.

power The rate of doing work; equal to work or energy divided by time.

pressure Force divided by area.

speed The rate at which distance is covered; equal to distance divided by time.

stroboscope A light that flashes at a regular rate, making possible a photographic record of motion.

watt A unit of power; equal to one joule per second.

work The net force on an object times the distance through which the object moves.

Δ The Greek letter delta, denoting change in some quantity.

Questions for Discussion

1. At the same time a rifle is fired in an exactly horizontal position over level ground, a bullet is dropped from the same height. Both bullets strike the ground at the same time. Can you explain why?

2. What are some advantages of using the metric (SI) system of units rather than the English system?

3. In the sixteenth century, Galileo is said to have dropped objects of various weights from the Leaning Tower of Pisa. Since all objects in *free fall* accelerate at 9.8 m/s², one would expect them to reach the ground at the same time. Careful observation, however, indicates that an iron ball will strike the ground sooner than a baseball of the same diameter. Can you explain why? Would the same be true on the moon? (The Apollo astronauts actually photographed a free-fall experiment on the moon using a hammer and a feather.)

4. Think of an object comparable in size to each of the following:
   a) 10^7 m; b) 10^3 m; c) 1 m; d) 10^{-3} m; e) 10^{-10} m.

5. Does shifting to a lower gear increase the power of an automobile? Explain.

6. Draw a diagram, similar to Fig. 1.10, showing how pressure acts on a floating object.

Problems

1. Letting your classroom serve as the "origin" ($x=0$, $y=0$), express the approximate coordinates (x, y) of your place of residence. Let x = the distance east and y = the distance north, as on a map. Use any convenient unit of distance.

2. The speed of a bicycle increases from 5 mph (miles per hour) to 10 mph in the same time that a car increases its speed from 50 mph to 55 mph. Compare their accelerations.

3. The density of water is 1.00 g/cm³ and that of ice is 0.92 g/cm³. What are the corresponding densities in SI units (kg/m³)?

4. If the speed limit is posted as 55 mph, express this in km/hr and in m/s (1 mile = 1.61 km).

5. A car accelerates from rest to 50 mph in 12 s. Calculate its average acceleration in m/s². Compare this to the acceleration of an object in free fall (1 mph = 0.447 m/s).

6. An object weighing one pound (English units) has a mass of 0.455 kg. Express its weight in newtons and thereby express a conversion factor for pounds to newtons.

7. Express your own mass in kilograms and your weight in newtons.

8. Calculate average speed in each of the following cases:
   a) An object moves a distance of 25 meters in 3 seconds.
   b) A train travels two kilometers, the first at an average speed of 50 km/hr and the second at an average speed of 100 km/hr. (*Note:* The average speed is not 75 km/hr.)
   c) A runner runs one kilometer in 3 minutes and a second kilometer in 4 minutes.
   d) An object dropped from a height of 75 meters strikes the ground in 4 seconds.

9. Estimate the total force on the surface of your body due to the pressure of the atmosphere.

10. Calculate the kinetic energy of a 1500-kg automobile with a speed of 30 m/s. If it accelerates to this speed in 20 s, what average power has been developed?

11. An electric motor, rated at ½ horsepower, requires 450 watts of electrical power. Calculate its efficiency (power out divided by power in). What happens to the rest of the power?

12. Calculate the potential energy of:

   a) A 3-kg block of iron held 2 meters above the ground;

   b) A spring with a spring constant $K = 10^3$ N/m stretched 10 cm from its equilibrium length;

   c) A one-liter bottle ($V = 1000$ cm^3) with a pressure 10^4 N/m^2 above atmospheric pressure ($P = 10^5$ N/m^2).

CHAPTER 2

Vibrating Systems

Nature provides many examples of vibrating systems: trees swaying in the wind, atoms in a molecule of water, the motion of the tides, electric current in a flash of lightning, and so on. Although the motion is vastly different in each of these systems, they have several things in common. For one thing, the motion repeats in each regular time interval, which we call the *period* of the vibration; second, some type of *force* constantly acts to restore the system toward its point of equilibrium.

2.1 SIMPLE HARMONIC MOTION

Consider the very simple vibrating system in Fig. 2.1 consisting of a mass *m* attached to the end of a spring. We assume that the amount of stretch in the spring is proportional to the stretching force (which is true of most springs if they are not stretched too far), so that in order to stretch it a length *l* a force *Kl* is required. The symbol *K* is the *spring constant* or "stiffness" of the spring.

Since the spring is vertical, the force of gravity on the mass stretches the spring by an amount that remains constant. In the equilibrium position shown in Fig. 2.1(b), the downward force of gravity on the mass (its weight) is just balanced by the upward force exerted by the spring; therefore the system is in equilibrium. The description of the motion is simplified if we specify the displacement

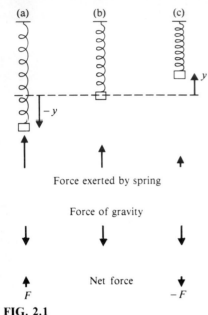

(a)   (b)   (c)

Force exerted by spring

Force of gravity

Net force

F $-F$

FIG. 2.1
A simple vibrator consisting of mass and spring. In (b) the upward force exerted by the spring and the force of gravity balance each other, and the net force F on the mass is zero.

of y of the mass from the equilibrium position and the net force F that acts on the mass. The relationship between F and y is easily shown to be

$$F = -Ky. \tag{2.1}$$

The minus sign in the above equation reminds us that when the mass is below its equilibrium position (y is negative), the net force will be upward (F is positive) as shown in Fig. 2.1(a). Thus the force F could be called a *restoring force*, which always acts in a direction to restore m to its equilibrium position. When the restoring force is proportional to the displacement, as it is in most vibrating systems we study, the motion is given a special name: *simple harmonic motion*. For a system in simple harmonic motion, the period is independent of the amplitude (size) of the vibration. The *frequency f* of vibration is the number of oscillations per second, which is obviously the reciprocal of the period T of one vibration:

$$f = 1/T . \tag{2.2}$$

It is customary to use a unit called the *hertz* (abbreviated Hz) to denote cycles per second. In the case of the vibrating mass-spring system, the frequency of vibration is given by the formula

$$f = \frac{1}{2\pi} \sqrt{\frac{K}{m}}. \tag{2.3}$$

Note that to double the frequency of vibration, the mass m may be reduced to one-fourth its original size, or the spring constant K may be made four times larger.

Example 2.1   Suppose a certain spring stretches 0.10 m when loaded with 2 kg. What is its spring constant? At what frequency will it vibrate when loaded with 2 kg? 0.5 kg?

Solution: At rest $Kl = mg$, so

$$K = \frac{mg}{l} = \frac{2(9.8)}{0.10} = 196 \text{ N/m (newtons per meter)}.$$

Its frequency of vibration when $m = 2$ kg is

$$f = \frac{1}{2\pi} \sqrt{\frac{K}{m}} = \frac{1}{(2)(3.14)} \sqrt{\frac{196}{2}} = \frac{\sqrt{98}}{6.28} = 1.6 \text{ Hz}.$$

When loaded with 0.5 kg, the frequency is

$$f = \frac{1}{2\pi} \sqrt{\frac{K}{m}} = \frac{1}{(2)(3.14)} \sqrt{\frac{196}{0.5}} = \frac{\sqrt{392}}{6.28} = 3.2 \text{ Hz}.$$

A strobe photograph of a mass-spring system was shown in Fig. 1.9. Also shown was a graph of position y as a function of time. A graph of speed v versus time can be made by taking the slope of that graph at every time t. Graphs of y and v as functions of time are shown in Fig. 2.2. Mathematicians refer to curves shaped like these as *sinusoidal* or *sine* curves. The maximum value of y is called the *amplitude*.

2.2 ENERGY AND DAMPING

The formula for the kinetic energy KE of a moving mass was given in Section 1.7:

$$\text{KE} = \tfrac{1}{2}mv^2, \tag{2.4}$$

where m is mass and v is speed. Similarly, the potential energy PE of a spring, stretched or compressed a distance y from its equilibrium length, was given as

$$\text{PE} = \tfrac{1}{2}Ky^2. \tag{2.5}$$

From the graphs in Fig. 2.2, it is clear that v^2 reaches its maximum value when y^2 is zero and vice versa. Thus, the total mechanical energy is constantly changing from kinetic to potential to kinetic. At times t_1 and t_3, potential energy is a maximum, and at t_2 and t_4, kinetic energy is a maximum.

Any real vibrating system tends to lose mechanical energy as a result of friction and other loss mechanisms. Unless the energy is

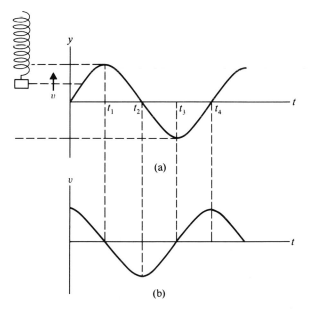

FIG. 2.2
Graphs of simple harmonic motion: (a) displacement vs. time; (b) speed vs. time. Note that speed reaches its maximum when displacement is zero and vice versa.

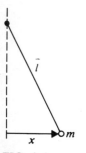

FIG. 2.3
Displacement of a damped vibrator
whose amplitude decreases with time.

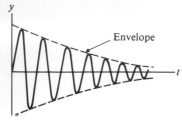

FIG. 2.4
A simple pendulum.

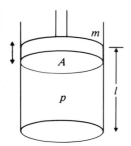

FIG. 2.5
A piston free to vibrate in a cylinder.

renewed in some way, the amplitude of the vibrations will decrease with time, as shown in Fig. 2.3. In many vibrating systems, a certain fraction (usually small) of the energy is lost during each cycle of vibration; the result is a curve that decreases in amplitude in the manner shown in Fig. 2.3. The dotted curve, which indicates the change in amplitude with time, is called the *envelope*, or decay curve. A vibrating system whose amplitude decreases in this way is said to be *damped*, and the rate of decrease is the *damping constant*.

2.3 SIMPLE VIBRATING SYSTEMS

Besides the mass-spring system already described, the following are examples of systems that vibrate in simple harmonic motion:

1. *Pendulum (small angle).* A simple pendulum, consisting of a mass *m* attached to a string of length *l* (see Fig. 2.4), vibrates in simple harmonic motion, provided that $x \ll l$. Assuming that the mass of the string is much less than *m*, the frequency of vibration is

$$f = \frac{1}{2\pi} \sqrt{\frac{g}{l}}, \tag{2.6}$$

where *g* is the acceleration due to gravity. Note that the frequency does not depend on the mass.

2. *A "spring" of air.* A piston of mass *m*, free to move in a cylinder of area *A* and length *l*, vibrates in much the same manner as a mass attached to a spring (see Fig. 2.5). The spring constant of the air in the cylinder is determined by its compressibility and turns out to be $K = \gamma pA/l$, so the frequency is

$$f = \frac{1}{2\pi} \sqrt{\frac{\gamma pA}{ml}}, \tag{2.7}$$

where *p* is the gas pressure, *A* is the area, *m* is the mass of the piston, and γ is a constant that is 1.4 for air.

3. *A Helmholtz resonator.* Another common type of air vibrator, illustrated in Fig. 2.6, is often called a Helmholtz resonator, after H. von Helmholtz (1821–1894), who used it to analyze musical sounds. The mass of air in the neck now serves as the "piston," and the air in the larger volume *V* as the "spring." The frequency of vibration is

$$f = \frac{v}{2\pi} \sqrt{\frac{a}{Vl}}, \tag{2.8}$$

where *a* is the area of the neck, *l* is its length, *V* is the volume of the resonator, and *v* is the speed of sound ($v \approx 344$ m/s).

The Helmholtz resonator can be thought of as having a mass m and a spring constant K that are:

$$m = \rho a l \quad \text{and} \quad K = \frac{\rho \, a^2 v^2}{V}, \qquad (2.9)$$

where ρ is the density of air.

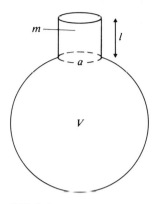

FIG 2.6
A Helmholtz resonator.

Helmholtz resonators can have a variety of shapes and sizes. For example, blowing air across an empty pop bottle causes the air in its neck to vibrate at a fairly low frequency. Note that the smaller the neck area a, the lower the frequency of vibration, which may seem a little surprising at first glance.

Example 2.2 A small flask consists of a sphere 9.8 cm in diameter plus a neck 3 cm in diameter and 10 cm long. At what frequency will it resonate?

Solution: $V = \dfrac{4}{3}\pi r^3 = \dfrac{4}{3}(3.14)(0.049 \text{ m})^3$

$$= 4.93 \times 10^{-4} \text{m}^3;$$

$$a = \pi r^2 = 3.14(0.015)^2 = 7.07 \times 10^{-4} \text{m}^2.$$

$$f = \frac{v}{2\pi}\sqrt{\frac{a}{Vl}} = \frac{344 \text{ m/s}}{2(3.14)}\sqrt{\frac{7.07 \times 10^{-4}\text{m}^2}{(4.93 \times 10^{-4}\text{m}^3)(0.10 \text{ m})}}$$

$$= 207 \text{ Hz}.$$

2.4 SYSTEMS WITH TWO OR THREE MASSES

The vibrating systems considered in the preceding section have one thing in common: A single coordinate is sufficient to describe their motion. In other words, they have one *degree of freedom*. In this section, we will consider vibrators with two or more degrees of freedom. Such systems have more than one *mode* of vibration, and the different modes will generally have different frequencies.

Consider the system consisting of two masses and three springs shown in Fig. 2.7. The system has two "normal" or independent modes, as shown in Figs. 2.7(a) and 2.7(b). In one mode, the masses move in the same direction; in the other, they move in opposite directions. Assuming equal masses and springs with the same stiffness K,

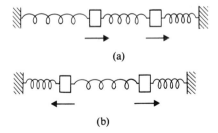

FIG. 2.7
Modes of vibration of a two-mass vibrator. The mode shown in (a), in which the masses move in the same direction, will have the lower frequency.

the frequencies of the two modes are

$$f_a = \frac{1}{2\pi}\sqrt{\frac{K}{m}}, \qquad f_b = \frac{1}{2\pi}\sqrt{\frac{3K}{m}} . \qquad (2.10)$$

Note that mode (a) has the same frequency as the simple mass-spring system shown in Fig. 2.1, whereas mode (b) has a frequency that is 1.7 times that of mode (a).

Modes (a) and (b) are virtually independent of each other. That is, the system can vibrate in mode (a) with minimal excitation of mode (b), and vice versa. If one sets the system into oscillation by giving the two masses a push or pull, the resulting motion will nearly always be a combination of modes (a) and (b). There are many recipes for combining these two modes in different proportions, and thus many ways in which the system can vibrate.

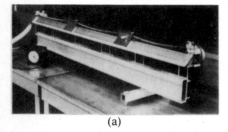

(a)   (b)   (c)

FIG. 2.8
Two-mass vibrators using (a) a linear air track; (b) an air table; (c) masses and springs hung from an overhead rod.

A great deal about the physics of vibration can be learned from watching the motion of a two-mass vibrator. Many physics laboratories have linear air tracks or air tables, on which objects move on a film of air with negligible friction. These are ideal for studying two-mass oscillators. Another convenient arrangement is to hang the masses on long cords from the ceiling or an overhead rod. The cords must be as long as possible to minimize the tendency of the masses to swing like pendulums. These three arrangements are shown in Fig. 2.8.

On the linear air track shown in Fig. 2.8(a), the masses are constrained to move in one direction only. In the systems shown in Figs. 2.8(b) and 2.8(c), however, the masses can move at right angles to the springs as well. Vibrations in this direction are called *transverse* vibrations, whereas vibrations in the direction of the springs are called *longitudinal* vibrations. Some systems vibrate only in transverse modes, some only in longitudinal. The air column of a musical wind instrument, for example, vibrates longitudinally, whereas the membrane of a drum vibrates transversely. A violin string normally vibrates transversely, although longitudinal vibrations (which sound like squeaks or squeals) are occasionally excited by the bowing of unskilled players.

In addition to their two modes of longitudinal vibration, the two-mass systems shown in Figs. 2.8(b) and 2.8(c) have two modes of transverse vibration, which are shown in Fig. 2.9. In the mode of lower frequency, the masses move in the same direction; in the mode of higher frequency, they move in opposite directions; this is similar to the longitudinal modes shown in Fig. 2.7.

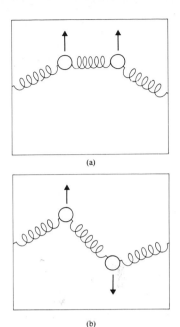

Adding a third mass to the systems of Fig. 2.7 adds additional modes of vibration. In the case of the linear vibrator in Fig. 2.7(a), which vibrates only longitudinally, a third mode of longitudinal vibration appears. The three independent modes of vibration are those shown in Fig. 2.9. The systems in Figs. 2.7(b) and 2.7(c) can vibrate transversely as well; in addition to the three modes of longitudinal vibration in Fig. 2.10, they will have the three independent modes of transverse vibration shown in Fig. 2.11.

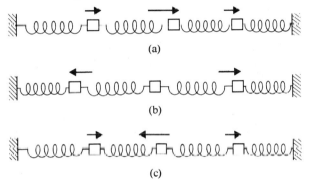

FIG. 2.10
Independent modes of longitudinal vibrations of a three-mass vibrator.

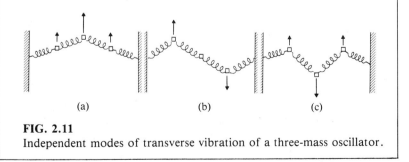

FIG. 2.11
Independent modes of transverse vibration of a three-mass oscillator.

FIG. 2.9
Modes of transverse vibration of a two-mass system. (a) In the mode of lower frequency, masses move in the same direction; (b) in the mode of higher frequency, masses move in opposite directions.

The independent modes shown in Figs. 2.7, 2.9, 2.10 and 2.11 are often called the *normal* modes of the vibrating systems. Getting the system to vibrate in a single normal mode requires special care. It is perhaps best done by driving the system at the frequency of the desired mode (this phenomenon, called *resonance*, will be discussed in Chapter 4). Carefully displacing the masses by the proper amounts and releasing them will also cause the system to vibrate in a single mode.

Example 2.3   The vibrating system in Fig. 2.7 consists of two 0.5-kg masses and three springs having spring constants of 50 N/m. Find the frequencies of its vibrational modes.

Solution: $f_a = \dfrac{1}{2\pi} \sqrt{\dfrac{K}{m}} = \dfrac{1}{2\pi} \sqrt{\dfrac{50 \text{ N/m}}{0.5 \text{ kg}}} = 1.59$ Hz;

$$f_b = \dfrac{1}{2\pi} \sqrt{\dfrac{3(50)}{0.5}} = 2.76 \text{ Hz.}$$

2.5   SYSTEMS WITH MANY MODES OF VIBRATION

In the case of the mass-spring vibrating systems, each new mass added one new mode of longitudinal vibration and one new mode of transverse vibration, if the system was able to vibrate transversely. In general, a system of N masses of the type shown in Fig. 2.8(b) or 2.8(c) will have N longitudinal and N transverse modes of vibration. If the masses were free to move in all three coordinate directions, there would be $2N$ transverse modes of vibration and N longitudinal modes,

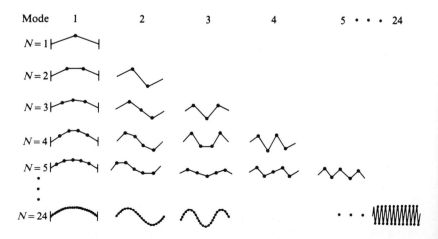

FIG. 2.12
Modes of transverse vibration for mass-spring systems with different numbers of masses. A system with N masses has N modes.

where N is the number of masses. The number of frequencies associated with the transverse modes may be only N, however, since corresponding modes in two directions usually have the same frequency.

The transverse modes of vibration for 1, 2, 3, 4, 5, and 24 masses are sketched in Fig. 2.12. Note that in each case the number of transverse modes equals the number of masses. There are an equal number of longitudinal modes, but they are more difficult to represent in a diagram. In each case, the mode of highest frequency is the one in which adjacent masses move in opposite directions.

Note that as the number of masses increases, the system takes on a wavelike appearance. In fact, a vibrating guitar string can be thought of as a mass-spring system with very large N. The propagation of waves on a string will be considered in Chapter 3.

2.6   VIBRATIONS IN MUSICAL INSTRUMENTS

All musical sound is generated by some type of vibrating system, whether it is a string on a violin, the air column of a trumpet, the head of a drum, or the voice coil of a loudspeaker. Often the vibrating system consists of two or more vibrators that work together, such as the reed and air column of a clarinet, the strings and sounding board of a piano, or the strings and body of a guitar. The acoustics of musical instruments is the subject of Part 3, but a brief description of several common musical vibrators will be made in closing this chapter on vibrating systems.

1. *Vibrating string.* The vibrating string can be thought of as the limit of the mass-spring system (see Fig. 2.11) when the number of masses becomes very large. The string itself has mass and elasticity or "springiness." There will be many modes of vibration, and their frequencies turn out to be very nearly multiples of the frequency of the lowest or *fundamental* mode. When the higher modes have frequencies that are multiples of the fundamental frequency, we call them *harmonics*. Several modes of a vibrating string are illustrated in Fig. 2.13. The guitar (Fig. 2.14), for example, uses vibrating strings.

2. *Vibrating membrane.* Drumheads are membranes of leather or plastic stretched across some type of tensioning hoop or frame. A membrane can be thought of as a two-dimensional string in that its restoring force is due to tension applied from the edge. A membrane, like a string, can be tuned by changing the tension. Membranes, being two-dimensional, can vibrate in many modes that are not normally harmonic. Four modes of vibration of a circular membrane are

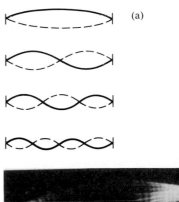

FIG. 2.13
a) Modes of a vibrating string.
b) Strobe picture of a string vibrating in its lowest two modes.

FIG. 2.14
A guitar. Coupling between strings, wood plates, and enclosed air leads to many modes of vibration, which will be discussed in Chapter 10.

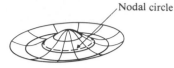
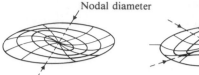

FIG. 2.15
Modes of circular membrane. The first two modes have circular symmetry; the second two do not. (From *Theoretical Acoustics* by P. M. Morse and K. U. Ingard. Copyright © 1968, McGraw-Hill. Used with the permission of McGraw-Hill Book Company.)

illustrated in Fig. 2.15. The first two have circular symmetry; the second two have *nodal* lines (indicated by the arrows), which act as pivots for a rocking motion. Two familiar examples of drums that use vibrating membranes to produce sound are shown in Fig. 2.16.

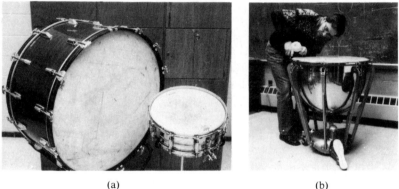

(a)   (b)

FIG. 2.16
(a) Bass drum and snare drum;
(b) timpani.

3. *Vibrating bar.* Many percussion instruments use vibrating bars as sound sources. The stiffness of a bar provides the restoring force when it bends, so no tension need be applied. Thus the ends may be free, as they are in most percussion instruments, or clamped (see Fig. 2.17). The frequencies of the vibrational modes of a uniform bar with free

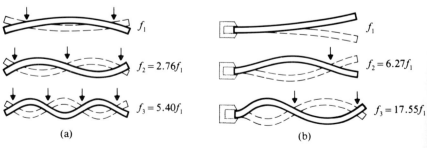

(a)   (b)

FIG. 2.17
Modes of vibrating bars: (a) both ends free; (b) one end clamped.
Arrows locate the modes.

ends (as the bars of a glockenspiel, for example) have the ratios
1:2.76:5.40:8.93, etc., which are nowhere near harmonic. The bars
of marimbas, xylophones (see Fig. 2.18), and other instruments, however, have been shaped to have a quite different set of mode frequency
ratios. Bars can also vibrate longitudinally, but longitudinal modes
are not normally used in musical instruments.

4. *Vibrating plate.* Vibrating plates, like vibrating bars, depend on
their own stiffness for the necessary restoring force. Plates have many
modes of vibration, some exhibiting great complexity.

An interesting way to study the modes of vibration of plates is
through the use of Chladni patterns, first described by E. F. F.
Chladni in 1787. Particles of salt or sand are sprinkled on a vibrating
plate, which is then excited to vibrate in one of its normal modes. The
particles, agitated by the vibrations, tend to collect along nodal lines,
where the vibrations are minimal. Chladni patterns of a circular plate
are shown in Fig. 2.19.

FIG. 2.18
A xylophone. (Courtesy of J. C.
Deagan Co.)

FIG. 2.19
Chladni patterns of a circular plate.
The first four have two, three, four,
and five nodal lines but no nodal
circles: the second four have one or
two nodal circles. (Courtesy of
American Assoc. of Physics Teachers)

FIG. 2.20
(a) Cymbals; (b) gong (left) and tam
tam (right).

Three musical instruments that use vibrating plates are shown in
Fig. 2.20.

5. *Tuning fork.* A tuning fork consists of two bars joined together at
one end. Thus the modes of vibration will resemble those of a bar
clamped at one end, as illustrated in Fig. 2.17(b). Tuning forks are
very convenient standards of frequency; once adjusted, they maintain
their frequency for a long time. The frequency of a tuning fork may be
raised by shortening its length or by removing material near the ends
of the prongs. The frequency can be lowered by removing material
near the base of the prongs, which decreases the stiffness.

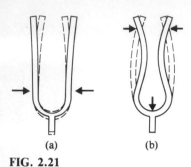

(a)   (b)

FIG. 2.21
Vibrations of a tuning fork:
(a) principal mode; (b) "clang" mode,
which occurs at a higher frequency
than the principal mode.

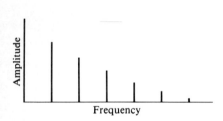

FIG. 2.22
The vibration spectrum of a plucked
string. The spectrum is a recipe that
tells us the frequency and amplitude of
each mode of vibration that is excited.
In this case, the frequencies are
harmonics (multiples) of the
fundamental, but this will not be so in
some vibrating systems.

As shown in Fig. 2.21, two modes of vibration of the tuning fork are the principal mode and the "clang" mode, which occurs at a much higher frequency (nearly three octaves higher in a typical fork). In their normal motion, the bars pivot about two nodes marked by arrows, causing the handle to move up and down. Thus, if the handle is pressed against another object (e.g., a table top), it may cause that object to act as a sounding board. (In a noisy environment, the handle may be touched to one's forehead in order to conduct sound directly to the inner ear.)

6. *Air-filled pipes.* The vibrational behavior of a column of air, as found in an organ pipe or the bore of a trumpet, can be compared to that of the air "spring" we discussed in Section 2.3.2, but is better understood by considering the sound waves within it. Thus we leave the discussion of this type of musical vibrator to Chapter 4.

2.7 COMPLEX VIBRATIONS: VIBRATION SPECTRA

In the preceding three sections, we have considered vibrating systems that can vibrate in several different modes. Each of these modes has a different frequency,* and hence it can be excited individually by some type of driving force at that frequency, as shown in Figs. 2.9–2.19.

More commonly, however, when a vibrating system is excited, it vibrates in several modes at once. A description of its vibrational motion therefore requires a "recipe," which tells us the amplitude and frequency of each of the modes that have been excited. Such a recipe is called the *spectrum* of the vibration. A vibration spectrum of a plucked string is shown in Fig. 2.22. Spectra of this type will appear frequently throughout this book.

When we observe a vibrating system with several modes, we often wish to determine its vibration spectrum. Electronic instruments called *spectrum analyzers* enable us to do this in the laboratory. Spectrum analysis is also called Fourier analysis in honor of the mathematician Joseph Fourier (pronounced "four-yay"), who pioneered in the mathematics of spectrum analysis. Some laboratory spectrum analyzers are described in Chapter 33, and the Fourier analysis of sound waves will be discussed in Chapter 7.

2.8 SUMMARY

Vibrating motion repeats itself in a regular interval of time called the period. Vibrating systems have some type of force acting to restore the

---

* Occasionally two different modes of vibration will have the same natural frequency; they are then called *degenerate* modes. These are rare in musical instruments, however.

system toward its point of equilibrium. In the case of simple harmonic motion, this force is proportional to the displacement.

In a vibrating system, the total mechanical energy changes from kinetic to potential to kinetic during each cycle of vibration. The rate at which the total energy decreases depends on the damping forces. Some systems can vibrate in several independent modes. The actual vibratory motion may be a combination of these modes.

All musical sound is generated by some type of vibrating system. Common vibrators include strings, membranes, bars, plates, and air columns. The familiar tuning fork is a combination of two bars vibrating in opposite directions, for example.

References and Suggested Readings

Chladni, E. F. F. (1787). "Entdeckungen über die Theorie des Klanges," translated excerpts in R. B. Lindsay, *Acoustics: Historical and Philosophical Development*. Stroudsburg, Pa.: Dowden, Hutchinson and Ross, 1973.

Morse, P. M. and K. U. Ingard (1968). *Theoretical Acoustics*. New York: McGraw-Hill.

Rossing, T. D. (1976, 1977). "Acoustics of Percussion Instruments," *The Physics Teacher* **14**: 546, and **15**: 278.

Waller, M. D. (1961). *Chladni Patterns, A Study in Symmetry*. London: Bell and Sons.

Glossary

damping Loss of energy of a vibrator, usually through friction.

envelope Time variation of the amplitude (or energy) of a vibration.

frequency The number of vibrations per second; expressed in hertz (Hz).

fundamental mode The mode of lowest frequency.

harmonics Modes of vibration whose frequencies are multiples of the frequency of the fundamental mode.

Helmholtz resonator A vibrator consisting of a volume of enclosed air with an open neck or port.

longitudinal vibration Vibration in which the principal motion is in the direction of the longest dimension.

node, or nodal line A point or line where minimal motion takes place.

normal modes Independent ways in which a system can vibrate.

period The time duration of one vibration; the minimum time necessary for the motion to repeat.

simple harmonic motion Smooth, regular vibrational motion at a single frequency such as that of a mass supported by a spring.

spectrum A "recipe" that gives the frequency and amplitude of each component of a complex vibration.

spring constant ("stiffness") The strength of a spring; restoring force divided by displacement.

transverse vibration Vibration in which the principal motion is at right angles to the longest dimension.

Questions for Discussion

1. Present an argument to show that the maximum kinetic energy of a mass-spring vibrator is equal to the maximum potential energy. Does the total mechanical energy remain constant throughout a cycle?

2. A damped vibrator is found to decrease its amplitude by one-half every thirty seconds. What is its amplitude 'at the end of five minutes? In theory will it ever stop vibrating? Will it in practice? Explain. (*Hint.* $(\frac{1}{2})^{10} = \frac{1}{1024} \simeq 0.001$.)

3. With the help of Figs. 2.10 and 2.12, make a diagram of the four independent longitudinal modes of vibration for a four-mass vibrator.

4. To excite a tuning fork in its principal mode of vibration with a minimum of "clang" sound, where should you strike it? Of the four microphone positions A, B, C, and D in Fig. 2.23, which will best pick up the sound of the fork? Why?

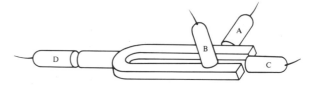

FIG. 2.23

Problems

1. Hanging a mass of one kilogram on a certain spring causes its length to increase 0.2 m.

   a) What is the spring constant K of that spring?

   b) At what frequency will this mass-spring system oscillate?

2. Copy the graphs of displacement and velocity shown in Fig. 2.2, and draw graphs of kinetic energy and potential to the same scale of time.

3. Most grandfather clocks have a pendulum that ticks (makes half a vibration) each second. What length of pendulum is required? (The value of g was given in Chapter 1 as 9.8 m/s².)

4. A bass-reflex loudspeaker enclosure (see Fig. 20.19) is essentially a Helmholtz resonator. Given the following parameters, what resonance frequency might be expected? $V = 0.5$ m³, $a = 0.02$ m², $l = 0.05$ m, speed of sound $v = 343$ m/s at $T = 20°C$.

5. Calculate the maximum potential energy of the mass-spring system described in Problem 1 if its maximum displacement is 5 cm.

6. In the two-mass system shown in Fig. 2.7, each mass is 2 kg and each spring constant $K = 100$ N/m. Calculate the frequencies of modes (a) and (b).

7. Equation 2.3 for the frequency of a simple mass-spring vibrator assumes that the mass of the spring is much smaller than that of the load and thus can be neglected. This will not always be the case. The formula can be refined by letting m be the mass of the load plus one-third the mass of the spring. Suppose that the spring in the example in Section 2.1 has a mass of 100 g (K was found to be 196 N/m). Calculate the vibration frequencies with loads of 0.5 kg and 2 kg, and compare them to those given in the example.

CHAPTER 3
Waves

The world is full of waves: sound waves, light waves, water waves, shock waves, radio waves, X-rays, and others. The room in which you are sitting is being crisscrossed by light waves, radio waves, and sound waves of many different frequencies; vibrational waves of low frequency are propagating through the walls and ceiling. Practically all communication depends on waves of some type. Although sound waves are vastly different from radio waves or ocean waves, all waves possess certain common properties. In this chapter, some of these common properties will be discussed, along with some particular properties of sound waves.

3.1 WHAT IS A WAVE?

One of the first properties noted about waves is that they can transport energy and information from one place to another through a medium, but the medium itself is not transported. A disturbance, or change in some physical quantity, is passed along from point to point as the wave propagates. In the case of light waves or radio waves, the disturbance is a changing electric and magnetic field; in the case of sound waves, it is a change in pressure and density. But in either case, the medium reverts to its undisturbed state after the wave has passed.

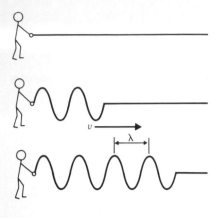

FIG. 3.1
A traveling wave generated by moving
the end of a rope.

All waves have certain things in common. For example, they can be reflected, refracted, or diffracted, as we shall see later in this chapter. All waves have energy, and they transport energy from one point to another. Waves of different types propagate with widely varying speeds, however. Light waves and radio waves travel 3×10^8 meters (186,000 miles) in one second, for example, whereas sound waves travel only 344 meters per second. Water waves are still slower, traveling only a few feet in a second. Light waves and radio waves can travel millions of miles through empty space, whereas sound waves require some material medium (gas, liquid, or solid) for propagation.

3.2   PROGRESSIVE WAVES

Suppose that one end of a rope is tied to a wall and the other end is held, as shown in Fig. 3.1. If the end being held is moved up and down f times per second, a wave with a frequency f will propagate down the rope, as shown. (When it reaches the tied end, a reflected wave will return, but we will ignore this for the moment.) The wave travels at a speed v that is determined by the mass of the rope and the tension applied to it.

If we were to observe the wave carefully (a photograph might help), we would note that the "crests" or "troughs" of the wave are spaced equally; we call this spacing the wavelength λ (the Greek letter lambda).

It is not difficult to see that the wave velocity is the frequency times the wavelength:

$$v = f\lambda. \tag{3.1}$$

That is, if f waves pass a certain point each second and the crests are λ meters apart, they must be traveling at a speed of $f\lambda$ meters per second.

It is possible to propagate either transverse or longitudinal waves in solids, but in general only longitudinal waves propagate through gases and liquids. Figure 3.2 illustrates the propagation of longitudinal and transverse waves in a mass-spring system. This system is also a large-scale model (in one dimension) of a solid crystal, and illustrates ways in which vibrations may propagate in a solid.

In a solid, longitudinal waves travel at a speed represented by the formula

$$v = \sqrt{\frac{E}{\rho}}, \tag{3.2}$$

where ρ is the density of the solid and E is called the elastic modulus (*Young's modulus*). Note that the speed of longitudinal waves in a solid bar is independent of its dimensions. This is not so for transverse

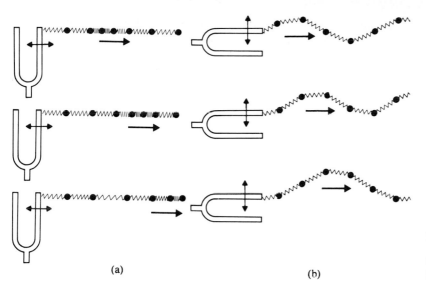

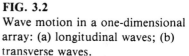

(a)   (b)

FIG. 3.2
Wave motion in a one-dimensional array: (a) longitudinal waves; (b) transverse waves.

waves whose speed is dependent on the dimensions. For a wire or string, the transverse wave velocity (speed) is

$$v = \sqrt{\frac{T}{\mu}}, \tag{3.3}$$

where T is the tension and μ is the mass per unit of length. In a stiff rod or bar, the velocity of transverse waves varies with frequency, and so a simple formula cannot be written. In general, longitudinal waves travel much faster than transverse waves do in solids. The speed of longitudinal (sound) waves in aluminum, for example, is 5000 m/s (about three miles per second).

Example 3.1   The density of steel is 7700 kg/m³ and Young's elastic modulus is 19.5×10^{10} N/m². What is the speed of longitudinal waves in a steel glockenspiel bar? How does this compare with the speed of longitudinal vibrations (sound waves) in air?

Solution: $v = \dfrac{E}{\rho} = \dfrac{19.5 \times 10^{10} \text{ N/m}^2}{7.7 \times 10^3 \text{ kg/m}^3} = 5032$ m/s.

(This is $\dfrac{5032}{343} = 14.7$ times the speed of longitudinal (sound) waves in air.)

Example 3.2   What tension would a steel wire 1 mm in diameter require in order that the transverse and longitudinal wave speeds are equal?

Solution: $v = \sqrt{\dfrac{T}{\mu}}$, so $T = \mu v^2$

$$\rho(\pi r^2)v^2 = 7700(3.14)(5 \times 10^{-4})^2(5032)^2$$

$$= 1.53 \times 10^5 \text{ N}$$

(far greater than the breaking force).

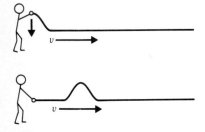

FIG. 3.3
An impulsive wave generated by moving the end of a rope.

3.3   IMPULSIVE WAVES; REFLECTION

Suppose that the rope in Fig. 3.1 is given a single impulse by quickly moving the end up and down. The impulse will travel at the wave speed v and will retain its shape fairly well as it moves down the rope, as illustrated in Fig. 3.3.

The question arises as to what happens when the pulse reaches the end of the rope. Careful observation shows that a pulse reflects back toward the sender. This reflected pulse is very much like the original pulse, except that it is upside down. If the end of the rope were left free to "flop" like the end of a whip, the reflected pulse would be right side up, as illustrated in Fig. 3.4(b). Photographs of impulsive waves on a long string with fixed and "free" ends are shown in Figs. 3.5 and 3.6. Note that the reflected pulse in Fig. 3.5 is upside down. This is called a *reversal of phase*. In Fig. 3.6 the *phase* of the reflected wave remains the same as that of the original wave.

It is instructive to tie the rope at the base of a mirror (see Fig. 3.7). Then the mirror image of the pulse (generated by your image) travels at the same speed and arrives at the end of the rope at the same time as the actual pulse, and appears to continue on as the reflected pulse. (To

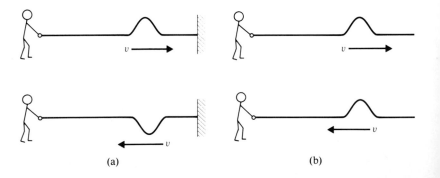

FIG. 3.4
Reflection of an impulsive wave (a) at a fixed end; (b) at a free end.

(a)   (b)

make the sense of the pulse correct, two mirrors can be used to form a corner reflector, but this is really not necessary to achieve the sensation of the reflected pulse coming from a virtual source and meeting the original pulse at the point of reflection.)

One can think of the reflected wave on the rope as coming from an imaginary source, whether or not a mirror is there to show a reflected image. If the rope is tied to a solid object at its far end, the deflection must be zero at all times, even when the pulse arrives; this requires that the pulse be met by a pulse of opposite sense, as shown in Fig. 3.4(a). If the end is free, it snaps like a whip, momentarily doubling its displacement when the pulse arrives. This is equivalent to the arrival of a pulse with the same sense, which then continues as the reflected pulse shown in Fig. 3.4(b).

Several interesting properties of waves can be studied with a wave machine developed at the Bell Laboratories, which consists of a long

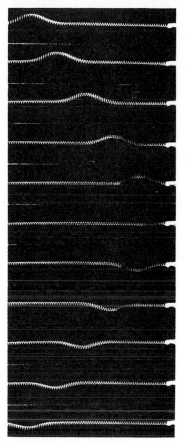

FIG. 3.5
An impulsive wave in a long spring. The pulse travels left to right and reflects back to the left as in Fig. 3.4(a). (From *PSSC Physics*, 2nd ed., 1965, D.C. Heath & Co. with Education Development Center, Newton, Mass.)

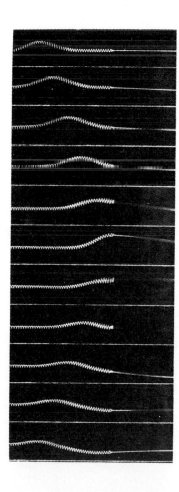

FIG. 3.6
An impulsive wave in a spring showing reflection at a "free" end (actually a very light thread). Compare the reflected pulse to that of Fig. 3.5. (From *PSSC Physics*, 2nd ed., 1965, D. C. Heath & Co. with Education Development Center, Newton, Mass.)

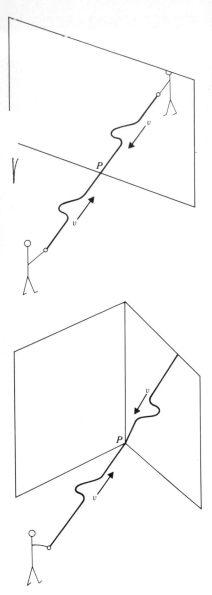

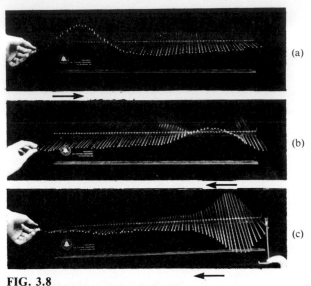

FIG. 3.8
Wave propagation on a "wave machine": (a) incident pulse; (b) reflection at a free end; (c) reflection at a fixed end. (Photographs by Craig Anderson.)

array of rods attached to a wire. Waves travel slowly on this machine; hence they can be observed rather easily. Reflection of a pulse at free and fixed ends is illustrated in the photographs of the wave machine in Fig. 3.8.

3.4   SUPERPOSITION AND INTERFERENCE

An interesting feature of waves is that two of them, traveling in opposite directions, can pass right through each other and emerge with their original identities. The *principle of linear superposition* describes this behavior. For wave pulses on a rope or spring, for example, the displacement at any point is the sum of the displacements due to each pulse by itself. The wave pulses shown in Fig. 3.9 illustrate the principle of superposition. If the pulses have the same sense, they add; if they have opposite sense, they subtract when they meet. These are examples of *interference* of pulses. The addition of two similar pulses is called *constructive interference*; the subtraction of opposing pulses is called *destructive interference*.

Suppose that both ends of a rope (or the wave machine shown in Fig. 3.8) are shaken up and down at the same frequency, so that continuous waves travel in both directions. Continuous waves interfere in much the same manner as the impulsive waves we have just con-

FIG. 3.7
The mirror image of an impulsive wave approaching a point of reflection *P*. In a plane mirror, the two pulses have the same sense, but in a corner mirror the two pulses have opposite sense (just as the incident and reflected pulses on the rope with a fixed end).

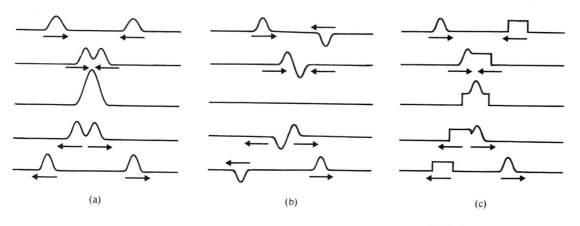

(a)	(b)	(c)

sidered. If two waves arrive at a point when they have opposite sense, they will interfere destructively; if they arrive with the same sense, they will interfere constructively. Under these conditions, the waves do not appear to move in either direction, and we have what is called a *standing wave.*

In the case of two identical waves (same frequency and amplitude) traveling in opposite directions on a rope or spring, there will be alternating regions of constructive and destructive interference, as shown in Fig. 3.10. The points of destructive interference that always have zero displacement are called nodes; they are denoted by *N* in Fig. 3.10. Between the nodes are points of constructive interference, where displacement is a maximum, these are called antinodes. At the antinodes, the displacement oscillates at the same frequency as in the individual waves; the amplitude is the sum of the individual wave amplitudes.

Note that the antinodes in Fig. 3.10, formed by the interference of two identical waves, are one-half wavelength apart. Because these

FIG. 3.9
The superposition of wave pulses that travel in opposite directions: (a) pulses in the same direction; (b) pulses in opposite directions; (c) pulses with different shapes.

FIG. 3.10
Interference of two identical waves in a one-dimensional medium. At times t_1 and t_5 there is constructive interference, and at t_3 there is destructive interference. Note that at points marked *N*, the displacement is always zero.

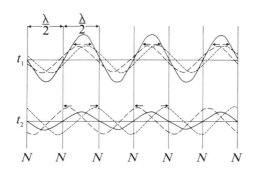

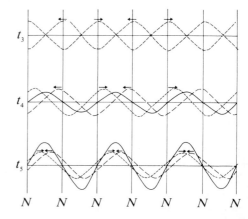

points of maximum displacement do not move through the medium, the configuration is called a standing wave. Standing waves result whenever waves are reflected back to meet the oncoming waves. The case illustrated in Fig. 3.10, in which the forward and backward waves have the same amplitude, is a special case that leads to total interference. If the two incident waves do not have the same amplitude, the nodes will still be points of minimum but not zero displacement.

3.5 SOUND WAVES

Sound waves are longitudinal waves that travel in a solid, liquid, or gas. To aid in your understanding of sound waves, consider a large pipe or tube with a loudspeaker at one end. Although sound waves in this tube are similar in many respects to waves on a rope, they are more difficult to visualize, because we cannot see the displacement of the air molecules as the sound wave propagates down the tube.

Suppose we consider first a single pulse, as we did in the case of the rope. An electrical impulse to the loudspeaker causes the speaker cone to move forward suddenly, compressing the air directly in front of it *very* slightly (even a very loud sound results in a pressure increase of less than 1/10,000 atmospheric pressure). This pulse of air pressure travels down the tube at a speed of about 340 m/s (over 700 miles per hour). It may be absorbed at the far end of the tube, or it may reflect back toward the loudspeaker (as a positive pulse of pressure or a negative one), depending on what is at the far end of the tube.

Reflection of a sound pulse for three different end conditions is illustrated in Fig. 3.11. If the end is open, the excess pressure drops to zero, and the pulse reflects back as a negative pulse of pressure, as shown in Fig. 3.11(b); this is analogous to the "fixed end" condition illustrated in Figs. 3.5 and 3.8(b).* If the end is closed, however, the pressure builds up to twice its value, and the pulse reflects back as a positive pulse of pressure; this condition, shown in Fig. 3.11(c), is analogous to the "free end" reflection of Figs. 3.6 and 3.8(c). If the end is terminated with a sound absorber, there is virtually no reflected pulse. Such a termination is called *anechoic*, which means "no echo."

The speed of sound waves in a gas is given by the formula

$$v = \sqrt{\frac{\gamma R T}{M}}, \qquad (3.4)$$

where T is absolute temperature, M is the molecular weight of the gas, and γ and R are constants for the gas. For air, $M = 2.88 \times 10^{-2}$,

FIG. 3.11
Reflection of a sound pulse in a pipe: (a) incident pulse; (b) reflection at an open end; (c) reflection at a closed end; (d) no reflection from absorbing end.

---

*   In an actual tube with an open end, a little of the sound will be radiated; most of it, however, will be reflected as shown.

TABLE 3.1 Speed of sound in various materials

Substance	Temperature (°C)	Speed (m/s)	Speed (ft/s)
Air	0	331.3	1087
Air	20	343	1127
Helium	0	970	3180
Carbon dioxide	0	258	846
Water	0	1410	4626
Methyl alcohol	0	1130	3710
Aluminum	—	5150	16900
Steel	—	5100	16700
Brass	—	3480	11420
Lead	—	1210	3970
Glass	—	3700–5000	12–16,000

$R = 8.31$, and $\gamma = 1.4$, so $v = 20.1\sqrt{T}$. The absolute temperature T is found by adding 273 to the temperature on the Celsius scale. At $t = 21°C$, for example, $T = 294$ K, so $v = 344$ m/s. At Celsius zero, $v = 332$ m/s. Over the range of temperature we normally encounter, the speed of sound increases by about 0.6 m/s for each Celsius degree, and an approximate formula for the speed of sound is sufficiently accurate:

$$v - 331.3 + 0.6t \text{ m/s,} \tag{3.5}$$

Where t is the temperature in degrees on the Celsius scale.

Sound waves travel much faster in liquids and solids than they do in gases. The speed of sound in several materials is given in Table 3.1.

3.6 WAVE PROPAGATION IN TWO AND THREE DIMENSIONS

Thus far we have considered only waves that travel in a single direction (along a rope or in a pipe, for example). One-dimensional waves of this type are a rather special case of wave motion. More often, waves travel outward in two or three dimensions from a source.

Water waves are a familiar example of two-dimensional waves. Many wave phenomena, in fact, can be studied conveniently by means of a ripple tank in the laboratory. A ripple tank uses a glass-bottom tray filled with water; light projected through the tray forms an image of the waves on a large sheet of paper, as shown in Fig. 3.12.

Three-dimensional waves are difficult to make visible. An ingenious technique has been used to photograph three-dimensional wave patterns (though not the actual waves) at the Bell Laboratories and elsewhere. A tiny microphone and a neon lamp together scan the

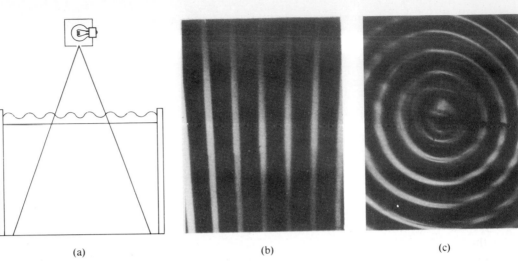

(a)   (b)   (c)

FIG 3.12
(a) A ripple tank for projecting an image of water waves; (b) straight waves on a ripple tank; (c) circular waves on a ripple tank. (Photographs by Christopher Chiaverina.)

FIG 3.13
The pattern of sound waves from a loudspeaker produced by scanning with a microphone and neon lamp. (From Kock, 1971.)

sound field in a dark room while a camera lens remains open in a time exposure. The brightness of the neon lamp is controlled by an amplifier, so that bright streaks appear at wave crests, as shown in Fig. 3.13.

Different types of sources radiate different kinds of patterns. A point source or a source that is spherically symmetric radiates spherical waves. A line source or a source with cylindrical symmetry radiates cylindrical waves. A large flat source radiates plane waves. Real sound sources are never true point sources, line sources, or flat sources, however; what we may have in real life are sources that approximate one of these geometries.

A source that is very small compared to a wavelength of sound approximates a point source and emits nearly spherical waves. A small enclosed loudspeaker will radiate nearly spherical waves at low frequency, as shown in Fig. 3.14(a). A column of small loudspeakers may resemble a line source at low frequency and emit cylindrical waves, as shown in Fig. 3.14(b). Symmetrical radiation patterns of this type can be observed outside, away from reflecting objects, or in an anechoic (echo-free) room.

3.7 THE DOPPLER EFFECT

Ordinarily, the frequency of the sound waves that reach the observer is the same as the frequency of vibration of the source. There is a notable exception, however, if either the source or the observer is in motion. If they are moving toward each other, the observed frequency increases; if they are moving apart, the observed frequency decreases. This apparent frequency shift is called the *Doppler effect*.

The Doppler effect is explained quite simply with the aid of Fig. 3.15. Suppose that a source *S* emits 100 waves per second. An observer at rest *O* will count 100 waves per second passing him. However, an observer *O′* moving toward the source will count more waves since he "meets" them as he moves, just as the driver of an automobile meets oncoming traffic. The apparent frequency (the rate at which the observer meets waves) will be

$$f' = f_s \frac{v + v_o}{v},\qquad (3.6)$$

where f_s is the frequency of the source, v_o the speed of the observer, and v the speed of sound. Note that after the observer passes the sound source, v_o must be subtracted from v. Thus the frequency drops abruptly as the observer passes the source.

There is also a Doppler effect if the source is in motion. You have probably observed a drop in pitch or frequency of the noise as a truck or car passes by while you are standing at the side of the road. The case of the moving source is shown in Fig. 3.15(b). The source emitted the wave numbered 1 when it was at position S_1, number 2 when at S_2, etc. The wave fronts resemble spheres with centers constantly shifting to the right as the source moves. Thus the observer O receives waves at a greater rate than he or she would from a stationary source. If the

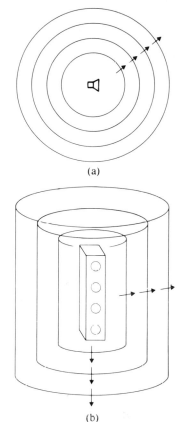

(a)

(b)

FIG 3.14
Sound wave patterns from (a) a single small loudspeaker ("point" source); (b) a column of loudspeakers ("cylindrical" source).

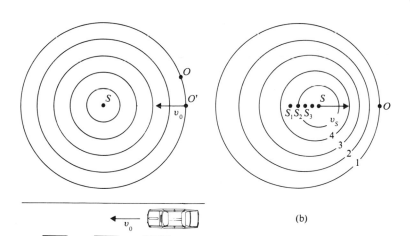

(a)

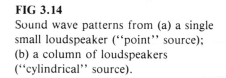

(b)

FIG. 3.15
The Doppler effect: (a) observer moving toward the sound source; (b) source moving toward the observer.

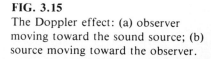

speed of the source is v_s, the apparent frequency will be

$$f = f_s \frac{v}{v - v_s} .\qquad(3.7)$$

Note that if the source moves directly toward the observer, the frequency will drop abruptly, not gradually, as the source passes by.

Example 3.3   An automobile horn emits a tone with a frequency of 440 Hz. What is the apparent frequency when the automobile approaches an observer at 55 mph (25 m/s) and what is the apparent frequency when it recedes at this same speed?

Solution: $f' = f_s \dfrac{v}{v - v_s} = 440 \dfrac{343}{343 - 25} = 475$ Hz;

$$f' = 440 \frac{343}{343 + 25} = 410 \text{ Hz}.$$

Note that the pitch drops by 14% (more than two semitones on the musical scale).

Example 3.4   A police radar "speed gun" transmits microwaves having a frequency of 9600 MHz. What is the upward shift in frequency for waves reflected from an automobile traveling at 55 mph (25 m/s)?

Solution: View the automobile as a mirror moving at 25 m/s; the "image" of the source appears to move at twice this speed or 50 m/s.

$$\Delta f = f' - f_s = f_s \frac{v}{v - v_s} - f_s$$

$$= 9.6 \times 10^9 \frac{3 \times 10^8}{3 \times 10^8 - 50} - 9.6 \times 10^9$$

$$\cong 9.6 \times 10^9 \left(1 + \frac{50}{3 \times 10^8} - 1 \right)$$

$$= 1600 \text{ Hz}$$

(Although this frequency shift is only one part in six million, it can readily be measured—as many speeding motorists know—by mixing together the transmitted and reflected microwaves).

3.8   REFLECTION

The reflection of wave pulses of one dimension on a rope or in a tube was discussed in Sections 3.3 and 3.5. Waves of two or three dimensions undergo similar reflections when they reach a barrier. The reflection of light waves by a mirror is a phenomenon familiar to all of us, as is the echo that results from clapping one's hands some distance away from a large wall, which reflects the sound waves back to the source. Figure 3.16(a) shows the reflection of water waves from a straight barrier. Note that the spherical reflected waves appear to come from a point behind the barrier. This point, which is called the *image* is denoted by S' in Fig. 3.16(b). It is the same distance from the reflector as the source S is.

Reflection of waves from a curved barrier can lead to the *focusing* of energy at a point, as shown in Fig. 3.17. A curious case of sound focusing, which occurs in "whispering galleries," is shown in Fig. 3.17(b). Sound originating from a source S is reflected by a curved barrier, "beamed" to a second curved barrier, and focused at O.

Well-known examples of whispering galleries are found in the Museum of Science and Industry in Chicago and in the National Capitol in Washington. The curved ceilings of certain auditoriums, such as the Mormon Tabernacle in Salt Lake City, make it possible to

(a)

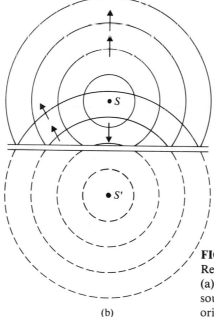

(b)

FIG. 3.16
Reflection of waves from a barrier:
(a) waves on a ripple tank from a point source; (b) reflected waves appear to originate from image S'.

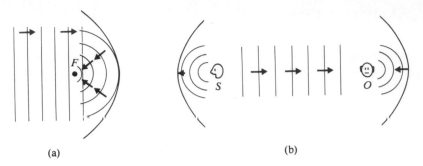

(a)   (b)

FIG. 3.17
Reflection of waves by a curved barrier: (a) incoming waves are focused at
F by a curved reflector; (b) "whispering gallery" in which two curved
reflectors beam sound from source *S* to observer *O* with great efficiency.

transmit whispers between selected spots. However, the focusing of
sound by curved walls is frequently detrimental to the acoustics of
auditoria, as we will discuss in Chapter 20.

3.9   REFRACTION

When the speed of waves changes, a phenomenon called *refraction*
occurs, which can result in a change in the direction of propagation or
a "bending" of the waves. The change of speed may occur abruptly as
the wave passes from one medium to another, or it may change
gradually within the medium. These two situations are illustrated in
Fig. 3.18.

   The situation illustrated in Fig. 3.18(b), which sometimes occurs
during the cool evening hours, causes sounds to be heard over great
distances. Since the speed of sound increases with temperature (Section 3.5), the sound travels faster some distance above the ground
where the temperature is greater. This results in a bending of sound

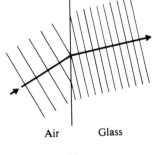

Air   Glass

(a)

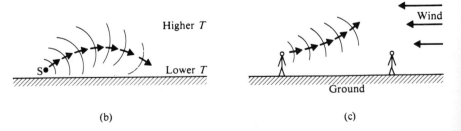

(b)   (c)

FIG 3.18
Refraction of waves: (a) light waves passing from air to glass; (b) sound
waves in the atmosphere when temperature varies with height; (c) sound
traveling against the wind.

downward as shown. Sound that would ordinarily be lost to the upper atmosphere is refracted back toward the ground.

Figure 3.18(c) shows why it is difficult to be heard when yelling against the wind. (It is *not* because the wind blows the sound waves back; even a strong wind has a speed much less than that of sound). Refraction results because the wind speed is less near the ground than it is some distance above it. Since the speed of sound with respect to the air (in this case, moving air) remains the same, the ground speed of the sound changes with altitude. The resulting refraction causes some of the sound to miss its target.

3.10 DIFFRACTION

When waves encounter an obstacle, they tend to bend around the obstacle. This is an example of a phenomenon known as *diffraction*. Diffraction is also apparent when waves pass through a narrow opening and spread out beyond it. Examples of the diffraction of water waves, light waves, and sound waves are shown in Figs. 3.19, 3.20, and 3.21.

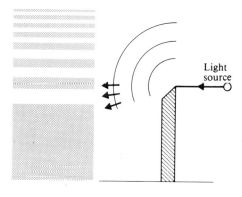

(a)

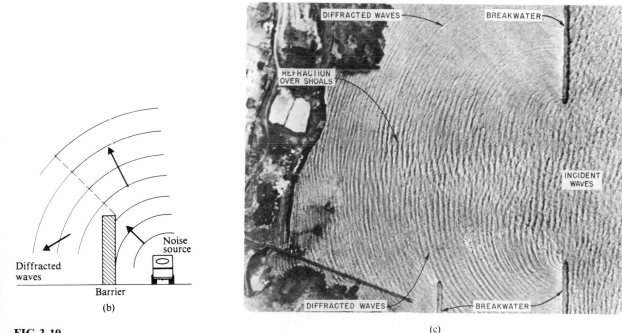

(b)

(c)

FIG 3.19

Diffraction of waves by a barrier: (a) shadow of a straight edge magnified to show diffraction of light; (b) diffraction of sound waves allows noise to "leak" around a wall; (c) ocean waves in a harbor. (Photograph (c) courtesy of University of California at Berkeley.)

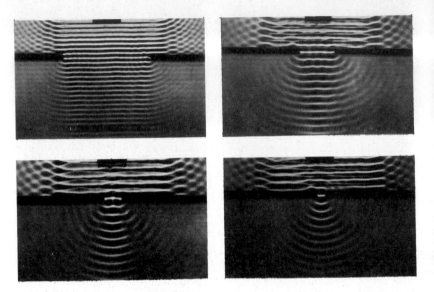

FIG 3.20

Diffraction of water waves passing through openings of various sizes. The narrower the opening (compared to the wavelength), the greater the diffraction. (Courtesy of Film Studio, Educational Development Center.)

An important point to remember is that it is the size of the opening in relation to the wavelength that determines the amount of diffraction. A loudspeaker 0.2 m (8 in.) in diameter, for example, will distribute sound waves of 100 Hz ($\lambda = 3.4$ m) in all directions, but waves of 2000 Hz ($\lambda = 0.2$ m) will be much louder directly in front of the speaker than at the sides, since diffraction will be minimal.

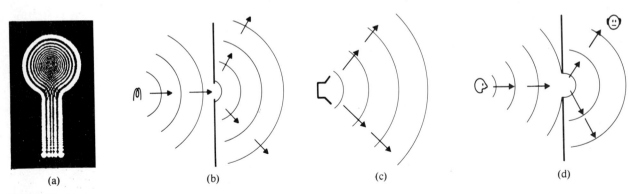

(a)   (b)   (c)   (d)

FIG. 3.21

Diffraction of waves passing through narrow openings: (a) light waves passing through a keyhole; (b) light waves through a very narrow slit; (c) sound waves from a loudspeaker; (d) diffraction allows sound to be heard "behind" a doorway.

3.11   INTERFERENCE

In Section 3.4, we pointed out that interference between incident and reflected waves leads to standing waves. Standing waves exist in a room due to interference between waves reflected from the ceiling, walls, and other surfaces; these can be observed by moving one's head around while a pure tone is played through a loudspeaker.

Waves from two identical sources provide another example of interference. Constructive and destructive interference lead to minima and maxima in certain directions, as shown in Fig. 3.22. The interference patterns are determined by the spacing of the two sources compared to a wavelength.

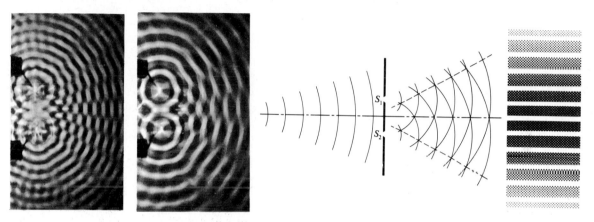

(a)   (b)

FIG 3.22
Interference of waves from two identical sources: (a) water waves in a ripple tank; (b) light waves from two slits illuminated by the same light source. (From *PSSC Physics*, 2nd ed., 1965, D. C. Heath & Co. with Education Development Center, Newton, Mass.)

3.12   SUMMARY

We are surrounded by waves of many types (light waves, radio waves, sound waves, water waves, etc.). These quite different types of waves have many properties in common. All carry energy; all can be reflected, refracted, and diffracted; interference leads to regions of minimum and maximum amplitude. However, the speeds at which these waves travel varies widely. Waves can be classed as transverse or longitudinal depending on the direction of the vibrations. Sound waves are longitudinal vibrations of molecules that result in pressure fluctuations. The speed of sound waves in air increases with temperature.

Some wave phenomena can be understood best by considering wave propagation in one dimension and extending the ideas to two- and three-dimensional waves.

References and Suggested Readings

Baez, A. V. (1967). *The New College Physics.* San Francisco: Freeman.

Kock, W. E. (1965). *Sound Waves and Light Waves.* New York: Doubleday.

Kock, W.E. (1971). *Seeing Sound.* New York: Wiley.

Physical Science Study Committee (1965). *Physics*, 2nd ed. Boston: D. C. Heath.

Project Physics—Text (1975). New York: Holt, Rinehart and Winston.

Glossary

absolute temperature The temperature (in Kelvins) on a scale that has its zero at the lowest attainable temperature ($-273°C$); absolute temperature is found by adding 273 to the Celsius temperature.

anechoic Echo free; an anechoic room is one whose walls, ceiling, and floor are covered with sound-absorbing material, usually in the shape of wedges.

diffraction The spreading out of waves when they encounter a barrier or pass through a narrow opening.

Doppler effect The shift in apparent frequency when the source or observer is in motion.

impulsive wave A brief disturbance or pressure change that travels as a wave.

interference The interaction of two or more identical waves, which may support (constructive interference) or cancel (destructive interference) each other.

longitudinal wave A wave in which the vibrations are in the direction of propagation of the wave; *example*: sound waves in air.

reflection An abrupt change in the direction of wave propagation at a change of medium (by waves that remain in the first medium).

refraction A bending of waves when the speed of propagation changes, either abruptly (at a change of medium) or gradually (e.g., sound waves in a wind of varying speed).

standing wave A wavelike pattern that results from the interference of two or more waves; a standing wave has regions of minimum and maximum amplitude called nodes and antinodes.

superposition The motion at one point in a medium is the sum of the individual motions that would occur if each wave were present by itself without the others.

transverse wave A wave in which the vibrations are at right angles to the direction of propagation of the wave; *example*: waves on a rope.

wavelength The distance between corresponding points on two successive waves.

Young's modulus An elastic modulus of a solid; the ratio of force per unit area to the stretch it produces.

Questions for Discussion

1. Although ocean waves are often described as transverse waves, the motion of a small bit of water is actually in a circle. Why could strictly transverse waves not exist on a water surface?

2. Mine operators carefully select the right atmospheric conditions for blasting operations in order to minimize community disturbance. What atmospheric conditions would be optimum?

*Problems*_____

1. Electromagnetic waves travel through space at a speed of 3×10^8 m/s. Find the frequency of the following. (1 nm $= 10^{-9}$ m.)

   a) Radio waves with $\lambda = 100$ m

   b) Waves of red light ($\lambda = 750$ nm)

   c) Waves of violet light ($\lambda = 500$ nm)

   d) Microwaves with $\lambda = 3$ cm (used in police radar).

2. Two trumpet players tune their instruments to exactly 440 Hz. Find the difference in the apparent frequencies due to the Doppler effect if one plays his or her instrument while marching away from an observer and the other plays while marching toward the observer. Is this enough to make them sound out of tune? (Assume 1 m/s as a reasonable marching speed.)

3. How much will the velocity of sound in a trumpet change as it warms up (from room temperature to body temperature, for example)? If the wavelength remains essentially the same (the expansion in length will be very small), by what percentage will the frequency change?

4. At what frequency does the wavelength of sound equal the diameter of the following? (1 in. $= 0.0254$ m)

   a) A 15-in. woofer

   b) a 3-in. tweeter

5. A nylon guitar string has a mass of 8.3×10^{-4} kg/m and the tension is 56 newtons. Find the speed of transverse waves on the string.

6. The audible range of frequencies extends from approximately 50–15,000 Hz. Determine the range of wavelengths of audible sound.

7. The distance from the bridge to the nut on a certain guitar is 63 cm. If the string is plucked at the center, how long will it take the pulse to propagate to either end and return to the center? (Use the speed calculated in Problem 5.)

8. Find the speed of sound in miles per hour at 0°C. This is called Mach 1. A supersonic airplane flying at Mach 1.5 is flying at 1.5 times this speed. Find its speed in miles per hour.

9. A thunder clap is heard 3 s after a lightning flash is seen. Assuming that they occurred simultaneously, how far away did they originate?

10. The density of aluminum is 2700 kg/m^3 and Young's elastic modulus is 7.1×10^{10} N/m^3. Compare the speed of longitudinal waves in aluminum to those in steel (see Example 3.1).

11. Compare the speed of sound calculated from equations (3.4) and (3.5) when $t = 30°C$.

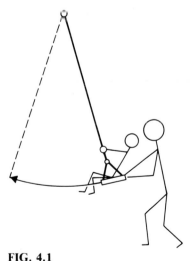

CHAPTER 4

Resonance

Consider a simple mechanical system: a child in a swing (Fig. 4.1). The swing has a natural frequency that is determined by its length (as the pendulum described in Section 2.3). If the swing is given a small push at the right time in each cycle, its amplitude gradually increases. This is an example of *resonance*. The swing receives only a small amount of energy during each push, but provided this amount is larger than the energy lost due to friction during each cycle, the *amplitude* of swing increases.

4.1 RESONANCE OF A MASS-SPRING VIBRATOR

Consider a mass-spring system similar to the one discussed in Section 2.1. Suppose that the spring is attached to a crank, as shown in Fig. 4.2. Let the crankshaft revolve at a frequency f and let the natural frequency of the mass-spring system be f_0. If f is varied slowly, the amplitude A of the mass is observed to change, reaching its maximum A_{max} when $f = f_0$. The mass is forced to vibrate at the frequency f of the crank, but when f matches f_0, the natural frequency of the system, resonance occurs. At resonance the maximum transfer of energy occurs, and the amplitude builds up to a value A_{max} determined by the friction in the system. The graph of amplitude A as a function of fre-

FIG. 4.1
An example of resonance: a child in a swing.

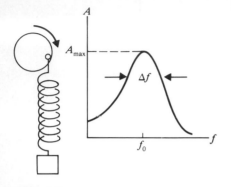

FIG. 4.2
Resonance of mass-spring vibrator
driven at a frequency f.

quency f, shown in Fig. 4.2, is a nearly symmetrical curve, with a width Δf often called the *linewidth*. The linewidth is usually measured at an amplitude of 71% of A_{max} ($A_{max}/\sqrt{2}$).

Just as A_{max} depends on the rate of energy loss (due to friction or *damping*), so Δf also depends on energy loss. For a heavily damped system, Δf is large, and A_{max} is small. For a system with little loss, a "sharp" resonance with small Δf and large A_{max} occurs. Engineers define a quantity $Q = f_0/\Delta f$ to characterize the sharpness of a resonance. (The use of the letter Q comes from the term "quality factor" used to describe electrical circuits.) A high Q circuit is one with a sharp resonance; a low Q circuit has a broad resonance curve.

The linewidth Δf and the Q associated with a resonance are intimately related to the damping constant and the decay curve of a vibrator described in Section 2.2. A vibrator that loses its energy slowly will have a sharp resonance, and a vibrator that loses its energy rapidly will have a broad resonance. If the vibrator is set into motion and left to vibrate freely, its decay time is directly proportional to the Q of its resonance.

4.2   PHASE OF DRIVEN VIBRATIONS

If we carefully observe the direction of motion of the crank and the mass, we note an interesting phenomenon. At low frequencies, far below resonance, the two move in the same direction. At frequencies far above resonance, however, they move in opposite directions.

We describe this phenomenon by using the term *phase*, which may be thought of as a specification of the "starting point" of a vibration. At low frequencies, the entire system follows the motion of the crank, and the spring hardly stretches at all. As the frequency of the crank increases, however, it is more difficult to move the mass, and thus it begins to lag behind the driving force supplied by the crank. At resonance, the mass is one-fourth cycle behind the crank, although its amplitude builds up to its maximum value. As the crank frequency is increased still further, the phase difference becomes greater until finally the mass is one-half cycle behind the crank; that is, the mass and the crank move in opposite directions, as shown at the right in Fig. 4.3. The higher the Q, the more abrupt is this transition from "in phase" to "opposite phase." A vibrator with a lot of damping, on the other hand, exhibits a gradual change of phase, as shown in Fig. 4.4.

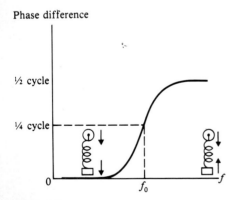

FIG. 4.3
Phase difference between crank and
mass in a driven mass-spring system.
At resonance they differ in phase by
one-fourth of a cycle.

4.3   STANDING WAVES ON A STRING

In Section 3.4, we learned how interference between two waves traveling in opposite directions leads to standing waves. We also learned how reflection occurs when waves or pulses reach the boundary of the

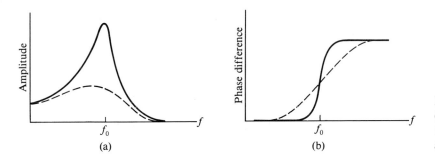

FIG. 4.4
(a) Response and (b) phase difference for vibrators with more (dotted curve) and less (solid curve) damping.

medium in which they propagate. We now combine these two ideas, and show how the modes of vibration or resonances of acoustical systems can be interpreted as waves propagating back and forth between the boundaries.

A simple and familiar example of such a system is a string of length L with both ends fixed as shown in Fig. 4.5. In its fundamental mode (that is, the standing wave with the lowest frequency and the longest wavelength), the string vibrates as shown in Fig. 4.5(a). The wavelength λ can be seen to be twice the string length, so the frequency is $f_1 = v/2L$. In the second mode, shown in Fig. 4.5(b), the wavelength λ equals the string length, so $f_2 = v/L = 2f_1$. Continuing to higher modes, we find that they have frequencies $3f_1$, $4f_1$, etc. The frequency of the nth mode will be

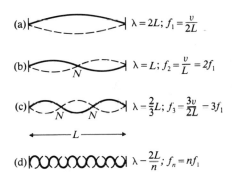

$$f_n = n \frac{v}{2L} = nf_1. \tag{4.1}$$

FIG. 4.5
Modes of vibration of resonances of a vibrating string as standing waves. The nodes are denoted by N. Note that the frequencies are harmonics of the fundamental frequency f_1.

Substituting the expression for wave speed given in Section 3.2 gives the following expression for the modes of a vibrating string:

$$f_n = \frac{n}{2L} \sqrt{\frac{T}{\mu}}, \tag{4.2}$$

where T is tension and μ is mass per unit length.

If a string is driven at the frequency of one of its natural modes, resonance can occur. There are many ways in which to apply the driving force, three of which are illustrated in Fig. 4.6. The magnetic drive shown in Fig. 4.6(b) works only for a string of steel or other magnetic material. The moving violin bow applies a rather complicated driving force (to be described in Chapter 10), which has components at several different frequencies. The string may also be driven by *electromagnetic force,* even when the string is made of a nonmagnetic metal, by placing a permanent magnet near the string and passing an alternating current of the desired frequency through the string.

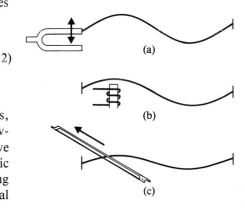

FIG. 4.6
Three ways to drive a string at one of its resonances: (a) a tuning fork; (b) an electromagnet; (c) a violin bow.

Example 4.1 A steel guitar string with a diameter of 0.3 mm and 65 cm long has a tension of 100 N. Find the frequencies of its first three modes of vibration. The density of steel is 7700 kg/m³.

Solution: $\mu = \pi r^2 \rho = \pi (1.5 \times 10^{-4})^2 (7700)$

$$= 5.44 \times 10^{-4} \text{ kg/m};$$

$$f_1 = \frac{1}{2\pi} \sqrt{\frac{T}{\mu}} = \frac{1}{2(0.65)} \sqrt{\frac{100}{5.44 \times 10^{-4}}} = 330 \text{ Hz};$$

$$f_2 = \frac{2}{2\pi} \sqrt{\frac{T}{\mu}} = \frac{2}{2(0.65)} \sqrt{\frac{100}{5.44 \times 10^{-4}}} = 660 \text{ Hz } (= 2f_1);$$

$$f_3 = 3f_1 = 990 \text{ Hz.}$$

4.4 PARTIALS, HARMONICS, AND OVERTONES

It is appropriate at this point to clarify nomenclature in order to avoid confusion. We will use the term *harmonics* to refer to modes of vibration of a system that are whole-number multiples of the fundamental mode, and also to the sounds that they generate. (It is customary to stretch the definition a bit so that it includes modes that are *nearly* whole-number multiples of the fundamental: 2.005 times the fundamental rather than 2, for example.) Thus we say that the modes of an ideal vibrating string are harmonics of the fundamental, but the modes of a real string are usually so close to being whole-number multiples that we also speak of them as harmonics. Note that the term "first harmonic" refers to the fundamental.

Many vibrators do not have modes that are whole-number multiples of the fundamental frequency, however, and the term *overtones* is used to denote their higher modes of vibration. Harmonics are therefore described as overtones whose frequencies are whole-number multiples of the fundamental frequency. Minor confusion arises, however, from the fact that the term harmonics includes the fundamental, but the term overtones does not. Thus the second harmonic is the first overtone, the third harmonic is the second overtone, and so forth.

There is another term in common use that refers to modes of vibration of a system or the components of a sound: *partials*. Partials include all the modes or components, the fundamental plus all the overtones, whether they are harmonics or not. The term *upper partials* excludes the fundamental and thus is a synonym of overtones, but use of the former will be avoided in this book.

The actual motion of a vibrating string is a combination of the various modes of vibration. The way in which these modes or partials combine is given by the *spectrum* of the vibration. A vibration spectrum is like a recipe that specifies the relative amplitudes of the partials. Similarly, the spectrum of a sound specifies the amplitudes of its partials, as we will discuss in Chapter 7.

4.5 OPEN AND CLOSED PIPES

The reflection of sound pulses at open and closed ends of pipes was described in Section 3.5. At an open end, a pulse of positive pressure reflects back as a negative pulse; at a closed end, it reflects as a positive pulse. These two end conditions can be used to arrive at the resonances for open and closed pipes.

The motion of the vibrating air in a pipe is a little harder to visualize than the transverse vibrations of a string, since the motion of the air is longitudinal. The displacement of the air is greatest at an *open* end, but the pressure variation is maximum at a *closed end*. A pressure-sensitive microphone inserted into the tube will pick up the most sound at the points where the pressure variations above and below atmospheric pressure are maximum. Thus in Figs. 4.7 and 4.8, both the air motion and the pressure variations are shown.

In an actual pipe, the pressure variations do not drop to zero right at the open end of the pipe, but rather a small distance beyond. Thus

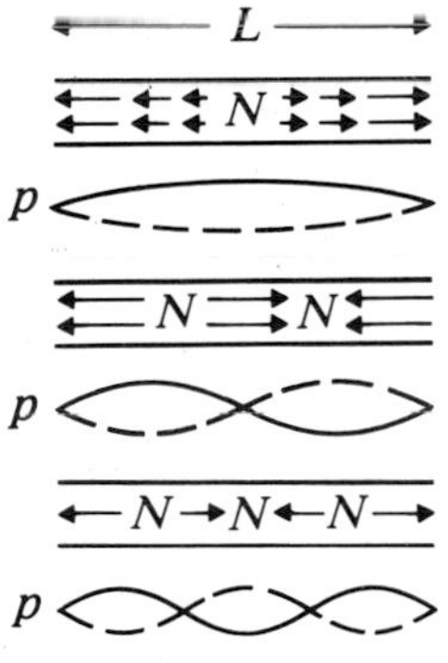

$$\lambda = 2L; \quad f_1 = \frac{v}{2L}$$

$$\lambda = L; \quad f_2 = \frac{v}{L} = 2f_1$$

$$\lambda = \frac{2}{3}L; \quad f_3 = \frac{3v}{2L} = 3f_1$$

$$f_n = nf_1 \qquad (4.3)$$
$$(n = 1, 2, 3 \ldots)$$

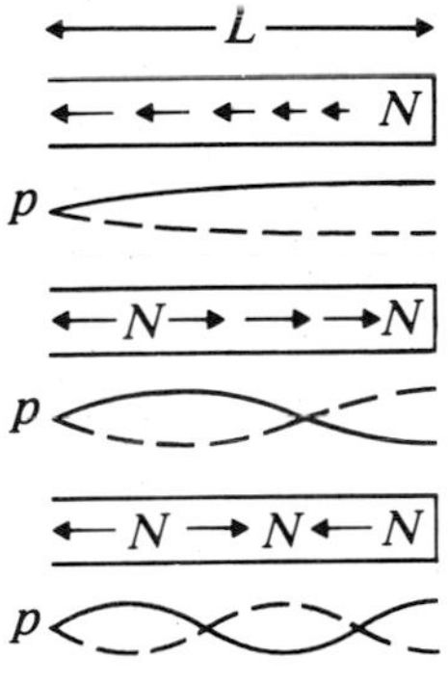

$$\lambda = 4L; \quad f_1 = \frac{v}{4L}$$

$$\lambda = \frac{4}{3}L; \quad f_3 = \frac{3v}{4L} = 3f_1$$

$$\lambda = \frac{4}{5}L; \quad f_5 = \frac{5v}{4L} = 5f_1$$

$$f_n = nf_1 \qquad (4.4)$$
$$(n = 1, 3, 5 \ldots)$$

FIG. 4.7
Modes of vibration or resonances of an open pipe. At the open ends the pressure is equal to atmospheric pressure. The resulting modes include both odd-numbered and even-numbered harmonics. Minimum displacement occurs at the nodes denoted by N.

FIG. 4.8
Modes of vibration or resonances of a closed pipe. At the closed end, the air motion is minimum but the pressure is maximum. The resulting modes include odd-numbered harmonics only. Minimum displacement occurs at the nodes denoted by N

the pipe appears to have an acoustic length that is slightly greater than its physical length. For a cylindrical pipe of radius r, the additional length, called the *end correction*, is $0.61r$. Twice this amount should be added to the length of a pipe with two open ends to obtain its acoustic length.

Resonance of a tube can be demonstrated by placing a tuning fork near one end of it, as shown in Fig. 4.9(a). A piston at the closed end makes it possible to change the resonance frequency. Blowing through a closed ("stopped") organ pipe may excite several of its resonances. Gentle blowing excites the lowest mode, but blowing much harder causes the pipe to vibrate in its first overtone, which for a closed pipe is the third harmonic ($f_3 = 3f_1$), a musical twelfth above the lowest mode.

In Chapter 15, which deals with speech production, we will be interested in resonances of the human vocal tract that allow us to enunciate various vowel sounds. There, we will consider not only reflections from open and closed ends but from constrictions and changes in pipe diameter as well. At every such discontinuity, a portion of the sound wave is reflected, thus leading to standing waves and resonances. If reflections occur at several places along the pipe, the resonances (in the vocal tract, they are called *formants*) can become rather complex.

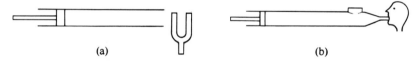

(a)   (b)

FIG. 4.9
Resonance of a closed tube demonstrated by (a) holding a tuning fork near one end; (b) blowing on an organ pipe.

Example 4.2 Find the first three modes of vibration of a pipe 0.75 meter long with open ends (neglect end corrections).

Solution: $f_1 = \dfrac{v}{2L} = \dfrac{343}{2(0.75)} = 229$ Hz;

$$f_2 = \frac{2v}{2L} = \frac{2(343)}{2(0.75)} = 457 \text{ Hz};$$

$$f_3 = \frac{3(343)}{2(0.75)} = 658 \text{ Hz}.$$

Example 4.3 Find the first three modes of vibration of a pipe 0.75 meter long with one open end and one closed end (neglect end correction).

Solution: $f_1 = \dfrac{v}{4L} = \dfrac{343}{4(0.75)} = 114$ Hz;

$$f_3 = \frac{3(343)}{4(0.75)} = 343 \text{ Hz};$$

$$f_5 = \frac{5(343)}{4(0.75)} = 572 \text{ Hz}.$$

4.6 ACOUSTIC IMPEDANCE

A quantity that acoustical engineers find very useful is *acoustic impedance* Z_A. It is defined as the ratio of sound pressure p to volume velocity U and is measured in acoustic ohms:

$$Z_A = p/U. \tag{4.5}$$

The volume velocity U is the amount of air that flows through a specified area per second due to passage of a sound wave. In the case of sound propagating in a tube, the specified area would be the cross-sectional area of the tube.

In the case of plane waves propagating in a tube, the acoustic impedance can be found by using the formula $Z_A = \rho v/S$, where ρ is the density of the air (1.15 kg/m^3 at room temperature), v the speed of sound, and S the cross-sectional area of the tube. Hence $Z_A \simeq 400/S$, with S measured in square meters. We will find the numerical value of Z_A much less important than the fact that it varies inversely with area S. Thus when there is a constriction, a change in diameter, or a sidebranch in a tube, the impedance change at that point leads to reflection of sound waves.

Acoustic impedance is analogous to electrical impedance, which is the ratio of voltage to electrical current (see Chapter 15). In this case, the voltage is the forcing function that causes current to flow. In the case of a sound wave, sound pressure is the forcing function that causes air flow with volume velocity U. In the chapters on wind instruments, use will be made of acoustic input impedance, which is the ratio of pressure to volume velocity at the input (mouthpiece) of a wind instrument.

55

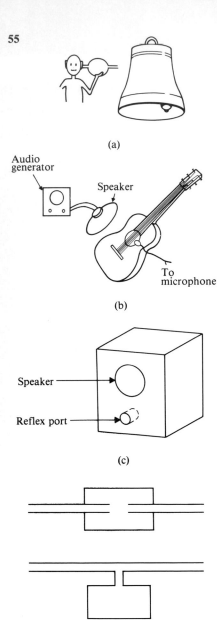

(a)

(b)

(c)

(d)

FIG. 4.10
Some examples of Helmholtz resonators: (a) analysis of vibrations of a bell; (b) air resonance of a guitar body; (c) bass reflex loudspeaker cabinet; (d) two types of side-branch mufflers.

4.7 HELMHOLTZ RESONATOR

The Helmholtz resonator was described as a vibrating system in Section 2.3. As a resonator, it has many applications in acoustics. Only a few of them will be described.

1. Before the invention of microphones, amplifiers, and spectrum analyzers, Helmholtz resonators were used to study vibrating objects and analyze complex sounds. In Fig. 4.10(a) the modes of vibration of a bell are being probed by a Helmholtz resonator having a second opening through which resonant sound can be heard by the investigator.

2. The main air resonance of a violin or guitar is essentially a Helmholtz resonance. The frequency can be determined by blowing across the *f*-holes of a violin or the sound hold of a guitar and listening for the pitch of the resonance. A more precise method is to insert a microphone inside the instrument and generate sound outside by means of a loudspeaker and audio generator.

3. Bass reflex loudspeaker cabinets are designed so that radiation from the back of the speaker cone excites the Helmholtz resonance of the cabinet and appears at the reflex port in phase with the front of the speaker.

4. Some automobile mufflers make use of side branches that absorb sound at their resonance frequency, as shown in Fig. 4.10(d).

4.8 SYMPATHETIC VIBRATIONS: SOUNDBOARDS

The amount of sound radiated by a vibrating system depends on the amount of air it displaces as it moves (the volume velocity defined in Section 4.6). A vibrating string or the narrow prongs of a tuning fork displace very little air as they vibrate; thus they radiate a small amount of sound. The moving cone of a loudspeaker and the vibrating membrane of a drum, on the other hand, radiate sound rather efficiently.

It is possible to increase the sound radiation from a tuning fork by pressing its handle against a wood plate or table top, so that the tuning fork forces the large wood area to vibrate. The vibrations of the wood, called sympathetic vibrations, may or may not occur at a frequency near a resonance of the wood plate, but nevertheless they amplify the sound because of the large surface set into vibration.

Violins, guitars, cellos, lutes, and other string instruments depend almost completely on sympathetic vibrations of the wood sounding box for radiation of their sound. Most of the sound radiation in these instruments comes from sympathetic vibrations of the top plate, which is driven by the vibrating strings through the bridge. The top plate has many resonances of its own distributed throughout the playing range, and these resonances, to a large part, determine the quality of the instrument. Sympathetic vibration of the wood also sets the air inside the instrument into vibration, so that sound is radiated from the *f*-holes (violin) or sound hole (guitar). String instruments will be discussed in Chapter 10.

Pianos and harpsichords have large soundboards with many resonances of their own closely spaced throughout the playing range. Vibrations are transmitted from the string to the soundboard through the bridge as in the violin or guitar. We will discuss pianos and harpsichords in Chapter 14.

4.9 SUMMARY

Resonance occurs when a force applied to a vibrating system varies with a frequency at or near the natural frequency of the system. Linewidth, Q, and maximum response are ways to describe the sharpness of the resonance, which in turn depends on the amount of damping in the system. The phase difference between the vibrating object and the driving force changes near the frequency resonance.

Normal modes of vibration of strings, pipes, and similar systems may be described as standing waves. Standing waves, which consist of waves propagating back and forth between the boundaries, lead to resonances in vibrating systems. Acoustic impedance, which is the ratio of sound pressure to volume velocity, is a quantity useful in the analysis of acoustical systems. Since sound radiation depends on air displacement, radiation from a vibrating string is greatly enhanced by the sympathetic vibration of a soundboard.

References and Suggested Readings

Backus, J. (1969). *The Acoustical Foundations of Music.* New York: Norton. (See Chapter 4.)

Benade, A. H. (1976). *Fundamentals of Musical Acoustics.* New York: Oxford. (See Chapter 10.)

French, A. P. (1966). *Vibrations and Waves.* New York: Norton. (See Chapter 4.)

Sears, F. W., and M. W. Zemansky (1980). *College Physics,* 5th ed. Reading, Mass.: Addison-Wesley.

Glossary

acoustic impedance A measure of the difficulty of generating flow (in a tube, for example); it is the ratio of the sound pressure to the volume velocity due to a sound wave.

amplitude The height of a wave; the maximum displacement of a vibrating system from equilibrium.

damping Energy loss in a system that slows it down or leads to a decrease in amplitude.

electromagnetic force The force that results from the interaction of an alternating electric current with a magnetic field.

fundamental The mode of vibration (or component of sound) with the lowest frequency.

harmonic A mode of vibration (or a component of a sound) whose frequency is a whole-number multiple of the fundamental frequency.

Helmholtz resonator A vibrator consisting of a volume of enclosed air with an open neck or port.

linewidth The width Δf of a resonance curve, usually measured at 71 percent of its maximum height; a measure of the sharpness of a resonance (a sharp resonance is characterized by a small linewidth).

overtone A mode of vibration (or component of a sound) with a frequency greater than the fundamental frequency.

partial A mode of vibration (or component of a sound); includes the fundamental plus the overtones.

phase difference A measure of the relative positions of two vibrating objects at a given time; also the relative positions, in a vibration cycle, of a vibrating object and a driving force.

Q A parameter that denotes the sharpness of a resonance; $Q = f_0 / \Delta f$, where f_0 is the resonance frequency and Δf is the linewidth.

resonance When a vibrator is driven by a force that varies at a frequency at or near the natural frequency of the vibrator, a relatively large amplitude results.

soundboard A sheet of wood or other material that radiates a substantial amount of sound when it is driven in sympathetic vibration by a vibrating string or in some other manner.

spectrum A recipe for vibratory motion (or sound) that specifies the relative amplitudes of the partials.

Questions for Discussion

1. If a child in a swing is pushed with the same impulsive force in each cycle, will the amplitude increase by the same amount in each cycle?

2. List as many examples of Helmholtz resonators as you can other than those given in Section 4.7. Are the resonances sharp or broad?

3. Attach a mass to a spring, as in Fig. 2.1 or 4.2, and determine the approximate resonance frequency by moving the top of the spring up and down by hand.

Then move it at frequencies below and above resonance, and carefully describe the force exerted on your hand in each case.

4. Does the end correction given in Section 4.5 lower all harmonics of a pipe proportionally, or does it result in the overtones going out of tune? An exact expression for the end correction shows that it varies slightly with wavelength. Does that change your answer?

Problems

1. A particular vibrator has a resonance frequency of 440 Hz and a Q of 30. What is the linewidth of its resonance curve?

2. Sketch a waveform that represents the displacement of the mass in Fig. 4.2 as a function of time. Then carefully sketch a second wave one-fourth cycle in advance of the first to represent the driving force at resonance. Label each curve correctly.

3. Determine the frequencies of the fundamental and first overtone (second partial) for the following. Neglect end corrections.

   a) A 16-foot open organ pipe

   b) a 16-foot stopped organ pipe (one open end, one closed end)

4. Extend Figs. 4.7 and 4.8 to include two more modes each.

5. Find the difference in the fundamental frequency, calculated with and without the end correction, of an open organ pipe 2 m long and 10 cm in diameter.

6. A nylon guitar string 65-cm long has a mass of 8.3×10^{-4} kg/m and the tension is 56 newtons. Find the frequencies of the first four partials.

7. A steel bar one meter long is held at the center and tapped on one end. Since its ends are free to move, its modes of longitudinal vibration will be similar to those of the air in a pipe open at both ends. Using the speed of sound given in Table 3.1, calculate the frequencies of the first three longitudinal modes.

8. Determine the frequencies of the pipes in Problem 3 if helium is substituted for air. (The speed of sound in helium is given in Table 3.1.)

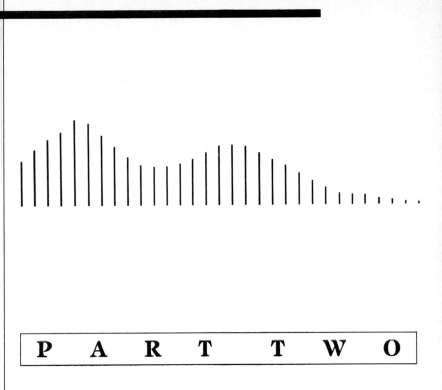

Psychoacoustics is the science that deals with the perception of sound. This interdisciplinary field overlaps the academic disciplines of physics, biology, psychology, music, audiology, and engineering, and utilizes principles from each of them. Our understanding of sound perception has increased substantially in recent years.

Loudness, pitch, timbre, and duration are four attributes used to describe sound, especially musical sound. These attributes depend in a rather complex way on measurable quantities such as sound pressure, frequency, spectrum of partials, duration, and envelope. The relationship of the subjective attributes of sound to physical quantities is the central problem of psychoacoustics and it has received a great deal of attention in recent years. Another interesting subject is the method of using rather subtle clues to localize the direction of a sound source and also draw surprisingly accurate conclusions about the nature of the acoustic environment. (In a dark unfamiliar room, for example, we could probably point to a speaker and also conclude whether the room is large or small.)

The next four chapters introduce some of the important topics of psychoacoustics. Chapter 5 briefly describes the human auditory system and the hearing process. Chapters 6 and 7 discuss three important attri-

P A R T   T W O

butes of sound: loudness, pitch, and timbre. Finally, Chapter 8 discusses several phenomena having to do with combination tones and how they relate to musical sound.

CHAPTER 5
Hearing

The human auditory system is complex in structure and remarkable in function. Not only does it respond to a wide range of stimuli, but it precisely identifies the pitch and the timbre (quality) of a sound and even the direction of the source. Much of the hearing function is performed by the organ we call the ear, but recent research has emphasized how much hearing depends on the data processing that occurs in the central nervous system as well.

5.1 THE HEARING FUNCTION

The range of pressure stimuli to which the ear responds represents a variation of over a million times. The energy content of an extremely loud sound is about a million million (10^{12}) times greater than that of the weakest sound that can be heard. At some sound frequencies, the vibrations of the eardrum may be as small as 10^{-8} mm, about one-tenth the diameter of the hydrogen atom. It is estimated that the vibrations of the very fine membrane in the inner ear that transmits this stimulus to the auditory nerve are nearly 100 times smaller yet in amplitude (Békésy, 1960).

The frequency range of hearing varies greatly among individuals; a person who can hear over the entire audible range of 20–20,000 Hz is unusual. The ear is relatively insensitive to sounds of low frequency; for example, its sensitivity at 100 Hz is roughly 1000 times less than its sensitivity at 1000 Hz. Sensitivity to sounds of high frequency is greatest in early childhood and decreases gradually throughout life, so that an adult may have difficulty hearing sounds beyond 10,000 or 12,000 Hz. (This deterioration of perception of high frequencies, termed *presbycusis*, is compared in Chapter 31 to noise-induced hearing loss.)

Another remarkable quality of the auditory system is its selectivity. From the blended sounds of a symphony orchestra, a listener can pick out the sound of a solo instrument. In a noisy room crowded with people, it is possible to pick out a single speaker. Even during sleep the conditioned ear of a mother can respond to the cry to an infant. We can train ourselves to sleep through the noise of city traffic but to awaken at the sound of an alarm clock or unusual noise.

5.2   STRUCTURE OF THE EAR

For convenience of description it is usual to divide the ear into three sections: the outer ear, the middle ear, and the inner ear (see Fig. 5.1). The *outer ear* consists of the external *pinna* and the *auditory canal* (meatus), which is terminated by the *eardrum* (tympanum). The pinna

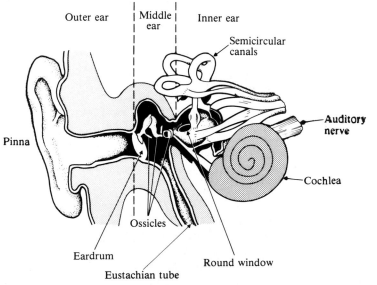

FIG. 5.1
A schematic diagram of the ear, showing outer, middle, and inner regions. This drawing is not to scale; for purposes of illustration, the middle ear and inner ear have been enlarged.

helps, to some extent, in collecting sound and contributes to our ability to determine the direction of origin of sounds of high frequency. The auditory canal acts as a pipe resonator that boosts hearing sensitivity in the range of 2000 to 5000 Hz.

The *middle ear* begins with the eardrum, to which are attached three small bones (shaped like a hammer, an anvil, and a stirrup) called *ossicles*. The eardrum, which is composed of circular and radial fibers, is kept taut by the tensor tympani muscle. The eardrum changes the pressure variations of incoming sound waves into mechanical vibrations to be transmitted via the ossicles to the inner ear.

The ossicles perform a very important function in the hearing process. Together they act as a lever, which changes the very small pressure exerted by a sound wave on the eardrum into a much greater pressure (up to 30 times) on the oval window of the inner ear. This function, which an engineer might call a mechanical transformer, is illustrated in Fig. 5.2. The lever action of the ossicles provides a factor of about 1.5 in force multiplication, whereas the remaining factor of about 20 in pressure comes from the difference in the areas of the eardrum and round window (the same force distributed over a smaller area results in a greater pressure, as explained in Section 1.6).

Another function of the small bones is to protect the inner ear from very loud noises and sudden pressure changes. Loud noise triggers two sets of muscles; one tightens the eardrum and the other pulls the stirrup away from the oval window of the inner ear. This response to loud sounds, called the *acoustic reflex*, will be discussed in Chapter 6.

Since the eardrum makes an airtight seal between the middle and outer parts of the ear, it is necessary to provide some means of pressure equalization. The *Eustachian tube*, which connects the middle ear to the oral cavity, is such a safety device. If the Eustachian tube is slow to open, a "popping" may be heard in the ears when the outside air pressure changes, for example, during a rapid change in altitude. It is remarkable that all these middle ear functions take place in a space approximately the size of an ordinary sugar cube!

The marvelously complex *inner ear* contains the *semicircular canals* and the *cochlea*. The semicircular canals contribute little or nothing to hearing; they are the body's horizontal-vertical detectors necessary for balance. The spiral cochlea, a masterpiece of miniaturization, contains all the mechanism for transforming pressure variations into properly coded neural impulses.

The cross-section of the cochlea in Fig. 5.3 shows three distinct chambers that run the entire length: the *scala vestibuli*, the *scala tympani*, and the *cochlear duct*.

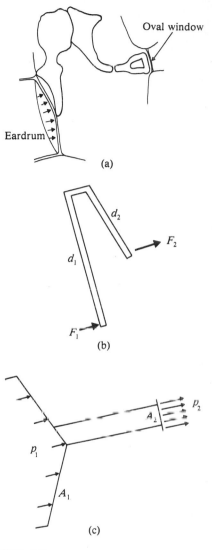

FIG. 5.2
Pressure amplification by the ossicles. (a) Three bones link the eardrum to the inner ear. (b) Lever action: A smaller force acts through a larger distance, resulting in a larger force acting through a smaller distance. (c) Pressure multiplication by piston action: A small pressure on a large area produces the same force as a large pressure on a small area.

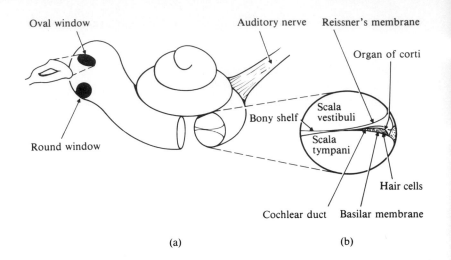

Oval window

Auditory nerve   Reissner's membrane

Organ of corti

Bony shelf

Scala
vestibuli

Round window

Scala
tympani

Hair cells

Cochlear duct   Basilar membrane

(a)

(b)

FIG. 5.3
A schematic diagram of (a) the cochlea; (b) a section cut from the cochlea.

The cochlea is filled with liquid and surrounded by rigid bony walls. Actually there are two different liquids, called perilymph (in the canals or scala) and endolymph (in the cochlear duct); the total capacity of the cochlea is only a fraction of a drop. Perilymph is similar to spinal fluid, whereas endolymph is similar to the fluid within cells. The two liquids are kept separate by two membranes: *Reissner's membrane* and the *basilar membrane*. Reissner's membrane is exceedingly thin, approximately two cells thick.

Resting on the basilar membrane is the delicate and complex *organ of Corti*, a gelatinous mass about 1½ inches long. This "seat of hearing" contains several rows of tiny *hair cells* to which are attached nerve fibers. A single row of inner hair cells contains about 7000 cells, whereas about 24,000 outer hair cells occur in several rows. Each hair cell has many hairs or *cilia* that are bent when the basilar membrane responds to a sound. The bending of the cilia is thought to stimulate the hair cells, which in turn excite neurons in the auditory nerve.

In order to understand how the basilar membrane vibrates, we can see the cochlea uncoiled and simplified in Fig. 5.4. The cochlea then appears as a tapered cylinder divided into two sections by the basilar membrane. (Because the cochlear duct is quite thin, we can ignore it—as a first approximation—and consider the two sections separated by a single membrane.) At the larger end of the cylinder are the oval and round windows, each closed by a thin membrane, and near the far end of the basilar membrane is a small hole called the *helicotrema* connecting the two sections. The basilar membrane terminates just short

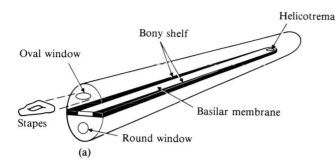

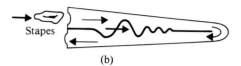

(b)

(a)

of the smaller end of the cylinder, so that fluid can transmit pressure waves around the end of the membrane.

When the stapes (stirrup) vibrates against the oval window, hydraulic pressure waves are transmitted rapidly down the scala vestibuli, inducing ripples in the basilar membrane. High tones create their greatest amplitude in the region near the oval window where the basilar membrane is narrow and stiff. On the other hand, low tones create ripples of greatest amplitude where the membrane is slack at the far end, (see Fig. 5.5). Thus the initial frequency analysis takes place in the cochlea, although we will see in Chapter 6 that much of the sense of pitch is determined in the central nervous system, where the data from the auditory nerve are processed.

The conversion of the mechanical vibrations of the basilar membrane into electrical impulses in the auditory nerve is accomplished in the organ of Corti. When the basilar membrane vibrates, the "hairs" of the hair cells are bent, thus generating nerve impulses that travel to the brain. The impulse rate on the auditory nerve depends on both the intensity and the frequency of the sound.

The overall hearing mechanism is illustrated in Fig. 5.6. Sound waves propagate through the ear canal, excite the eardrum, and cause mechanical vibrations in the middle ear. The stapes vibrating against the oval window causes pressure variations in the cochlea, which in turn excite mechanical vibrations in the basilar membrane. These vibrations of the basilar membrane cause the hair cells to transmit electrical impulses to the brain via the auditory nerve.

FIG. 5.4

(a) A schematic diagram of uncoiled cochlea showing the basilar membrane and oval and round windows.
(b) When the stapes (stirrup) presses against the oval window, a pressure pulse propagates through the cochlear fluid toward the round window, causing ripples to occur in the basilar membrane.

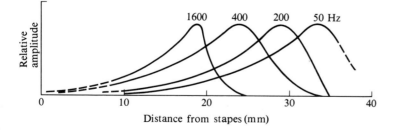

FIG. 5.5

Basilar membrane displacement amplitude as a function of distance for several different frequencies. (After von Békésy, 1960).

FIG. 5.6
A schematic representation of the ear, illustrating the overall hearing mechanism. Sound waves in the outer ear cause mechanical vibrations in the middle ear, and eventually nerve impulses that travel to the brain to be interpreted as sound.

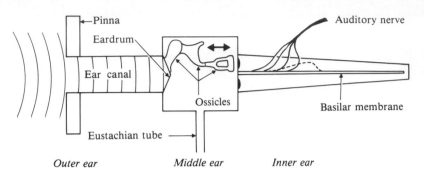

Some sounds are heard through vibrations of the skull that reach the inner ear. Hearing by bone conduction plays an important role in speaking. The sounds of humming or clicking one's teeth are heard almost entirely by bone conduction. (If you stop your ears with your fingers, thus interfering with the air path, the humming may actually sound louder.) During speaking or singing, two different sounds are heard, one by bone conduction and one by air conduction. The recorded sound of your own voice sounds very unnatural to you because only the airborne sound is received by the microphone, whereas you are used to hearing both components in your own voice.

Many researchers have contributed to our understanding of the hearing process, but two scientists deserve special mention: Hermann von Helmholtz and Georg von Békésy.

Hermann Ludwig Ferdinand von Helmholtz (1821–1894) was a physician and a man of many sciences. He did pioneering work in the field of physiology, mathematics, thermodynamics, optics, and acoustics. He invented the ophthalmoscope used to study the interior of the eye and formulated an important theory of color perception. In 1862 he published his monumental book *On the Sensations of Tone as a Physiological Basis for the Theory of Music*, which has been reprinted many times and is useful even today to researchers in psychoacoustics. Helmholtz envisioned the fibers of the basilar membrane as selective resonators tuned, like the strings of a piano, to different frequencies. Thus a complex sound would be analyzed into its various components by selectively exciting fibers tuned to the frequency of one of the components. It turned out that Helmholtz was nearly, but not quite, correct in this assumption, as we shall learn in Chapter 7.

Georg von Békésy, a communications engineer in Budapest, Hungary, became interested in the mechanism of hearing while studying ways to improve telephones. In order to carry out his studies, Békésy carefully removed cochleas from the ears of animal and human cadavers. For his careful and extensive research, he was awarded a Nobel prize in 1961.

In order to illustrate vibrations of the basilar membrane, Békésy built several mechanical models of the cochlea, one of which is shown in Fig. 5.7. A brass tube with a slit at the top is covered by a plastic of varying thickness with a raised ridge. The tube is closed with a piston at one end and a fixed plate at the other, and filled with water. The elasticity of the plastic varies along its length in much the same manner as the basilar membrane. Thus when the piston is driven at various frequencies, the point of maximum excitation moves up and down the tube, which can be felt by placing one's forearm in gentle contact with the ridge of the plastic (Békésy, 1960, 1970). The cochlear model, as well as other instruments used by Békésy, can be seen and operated at a small museum at the University of Hawaii.

Much of Békésy's success was due to the careful techniques he developed for removing the cochleas of fresh cadavers. Working under a microscope with microtools of his

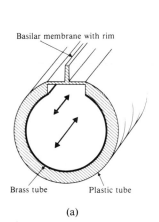

Basilar membrane with rim

Brass tube   Plastic tube

(a)

(b)

FIG. 5.7
Cochlear model constructed by Békésy. The observer notes that the point of maximum sensation moves up and down the forearm as the frequency of the sound changes.

own design, he was able to lay open a part of the basilar membrane. The cochlear fluid was drained and replaced by a salt solution with a suspension of powdered aluminum and coal. By observing light scattered from the suspended powder, he discovered an undulation in the basilar membrane when the cochlea was excited by sound.

Békésy studied the ears of many different mammals. An amusing story is told about his excitement in learning that an elephant had died in the Budapest zoo. He traced the carcass to a local glue factory where he was able to recover the elephant's cochleas. To Békésy's delight, traveling waves were observable also in the basilar membrane of the elephant (Stevens and Warshofsky, 1965).

5.3   SIGNAL PROCESSING IN THE AUDITORY SYSTEM

Signal processing in the auditory system can be divided into two parts: that done in the peripheral auditory system (ears themselves), and that done in the auditory nervous system (brain). The ears process an acoustic pressure signal by first transforming it into a mechanical vibration pattern on the basilar membrane, as shown in Figs. 5.4 and 5.5, and then representing this pattern by a series of pulses to be transmitted by the auditory nerve. Perceptual information is extracted at various stages of the auditory nervous system.

It is possible, by inserting a tiny electrode into the auditory nerve, to pick up the electrical signals traveling in a single fiber of the auditory nerve from the cochlea to the brain (Tasaki, 1954). The signal consists of a series of voltage spikes, each spike corresponding to the stimulation of a hair cell attached to the basilar membrane. The spikes are found to be closely correlated to the mechanical vibration pattern on the basilar membrane up to frequencies of about 4000 or 5000 Hz.

Each auditory nerve fiber responds over a certain range of frequency and sound pressure. The "tuning curves" in Fig. 5.8, for example, show response thresholds for six different fibers in the auditory nerve of a cat (Kiang and Moxon, 1974). Each nerve fiber has a characteristic frequency (CF) at which it has maximum sensitivity. Fibers with a high CF show a rapid rolloff in sensitivity above their CF but a long "tail" below it. A 60-dB stimulus at 200 Hz, for example, causes spikes to appear on all six fibers.

By sophisticated techniques such as probing with laser light (Khanna and Leonard, 1982) and using the Mössbauer effect (Johnstone and Boyle, 1967), it has been found that basilar membrane displacements in live animals show a much sharper frequency response than those of Fig. 5.5 in the cochlea of a dead animal. Rhode and

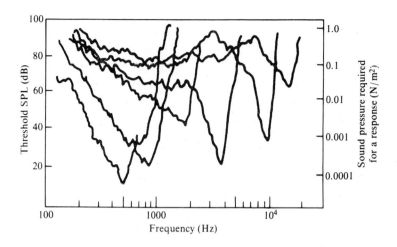

FIG. 5.8
Tuning curves for six different fibers in the auditory nerve of a cat showing response thresholds as functions of frequency (adapted from Kiang and Moxon, 1974). The relationship of sound pressure level to sound pressure is discussed in Chapter 6.

Robles (1974) found that within several hours after death, the basilar membrane response decreases 10–15 dB, the frequency of maximum response shifts downward, and the response curve broadens. In fact, the mechanical frequency response of the basilar membrane in live cochleas is quite comparable to the tuning curves observed in nerve fibers (Fig. 5.8, for example). There is some evidence for sharpening of neural tuning curves further along the neurological pathway, however.

If we were to observe the spikes on a nerve fiber when the stimulus is a tone of a single frequency, we would note that the time between spikes almost always corresponds to one or two or more periods of the tone. Although the nerve fiber does not fire at the peak of every vibration cycle in the basilar membrane, it rarely fires at any other time. The situation is a little more complicated when the stimulus is a complex tone, but still we find that the pattern of spikes on the auditory nerve carries accurate information about the frequency spectrum of the stimulus tone.

Consider a stimulus consisting of the pure tones C_4 (523 Hz) and C_5 (1046 Hz), spaced one octave apart. Their neural tuning curves (or frequency response curves) shown in Fig. 5.9(a) show very little overlap, so very few hair cells respond to both frequencies. Processing of the one component in the brain is only slightly affected by the presence of the other one.

As the interval between the two components decreases, the situation changes. Their amplitude envelopes show more and more overlap, as in Fig. 5.9(b) and (c), so an increasing number of hair cells are stimulated by both components. This leads to many interesting auditory phenomena, some of which will be discussed in Chapters 6, 7, and 8.

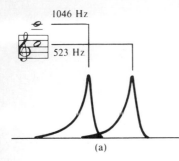

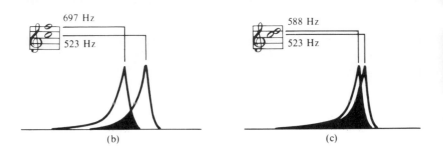

(a)   (b)   (c)

FIG. 5.9
Frequency response curves for pairs of
pure tones. As the interval between
them decreases, their response curves
show increasing overlap.

5.4   CRITICAL BANDS

When two pure tones are so close in frequency that there is considerable overlap in their amplitude envelopes on the basilar membrane,
they are said to lie within the same *critical band*. Critical bands are of
great importance in understanding many auditory phenomena, such
as loudness, pitch, and timbre. They have been defined and measured
in a variety of ways (Fletcher, 1940; Plomp, 1976; Zwicker, et al.,
1957).

Each critical band may be regarded as a data collection unit on the
basilar membrane. About 24 critical bands span the audible frequency
range, and the regions on the basilar membrane to which each of these
corresponds is about 1.3 mm long and embraces about 1300 neurons
(Scharf, 1970). The *critical bandwidth* varies with center frequency,
as shown in Fig. 5.10, having nearly a constant value at low frequency
and being roughly proportional to frequency at high frequency. Bandwidths are found to vary substantially, depending upon the type of
experiment.

FIG. 5.10
Critical bandwidth as a function of the
critical band center frequency.
Bandwidths are typical of those
reported in various experiments.

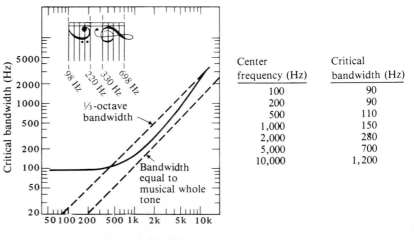

Center frequency (Hz)	Critical bandwidth (Hz)
100	90
200	90
500	110
1,000	150
2,000	280
5,000	700
10,000	1,200

Center frequency (Hz)

> **Critical Bands as Musical Intervals**
>
> Over much of the audible range, critical bands are slightly less than $\frac{1}{3}$-octave in width, as indicated in Fig. 5.10. An octave is the musical interval between two tones whose frequencies are in the ratio 2:1. The ratio of frequencies of two tones that are $\frac{1}{3}$-octave apart is $\sqrt[3]{2} = 1.26$. In musical language, $\frac{1}{3}$-octave equals four semitones or a major third. Sound analyzers that measure sound pressure in each of about 30 $\frac{1}{3}$-octave bands are quite common (30 such bands are required to span the audible range as compared to only 24 critical bands because critical bands are substantially greater than $\frac{1}{3}$-octave at low frequency; see Fig. 5.10); one such analyzer is shown in Fig. 33.9.

5.5   BINAURAL HEARING AND LOCALIZATION

"Nature," said the ancient Greek philosopher Zeno, "has given us one tongue, but two ears, that we might hear twice as much as we speak." Excellent advice.

The most important benefit we derive from binaural hearing is the sense of localization of the sound source. Although some degree of localization is possible in monaural listening, binaural listening greatly enhances our ability to sense the direction of the sound source.

Lord Rayleigh, who contributed so much to our understanding of acoustics, was one of the first to explain binaural localization of sound. In 1876 Rayleigh performed experiments (which, unknown to him, had been performed nearly a century earlier by Giovanni Venturi, an Italian scientist remembered for his work on fluid dynamics) to determine his ability to localize sounds of different frequencies. He found that sounds of low frequency were more difficult to locate than those of high frequency. According to Rayleigh's explanation, a sound coming from one side of the head produces a more intense sound in one ear than in the opposite ear, because the head casts a "sound shadow" for sounds of high frequency, as shown in Fig. 5.11(a). At low frequency, however, the shadow effect is small because sound waves of long wavelength diffract around the head. At 1000 Hz, the sound level is about 8 decibels greater at the ear nearest the source, but at 10,000 Hz the difference could be as great as 30 decibels.

Sounds of low frequency can be localized, although with slightly less accuracy than those of higher frequency. In 1907, Rayleigh offered a second theory of localization to explain low-frequency effects. A sound coming from the side strikes one ear before the other,

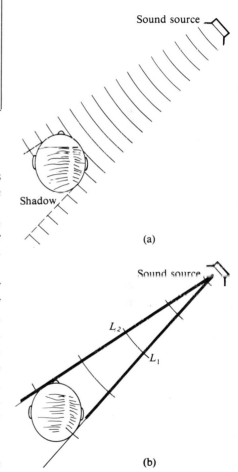

FIG. 5.11
Localization of a sound source. (a) At frequencies above 4000 Hz, localization is due to intensity difference at two ears. (b) At frequencies below 1000 Hz, localization is due to an interaural time difference between sound traveling paths L_1 and L_2.

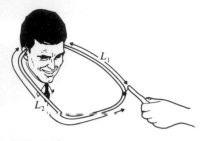

FIG. 5.12
An experiment illustrating the sensitivity of the ear to interaural time difference. Tapping the tube so that $L_1 = L_2$ causes the sound to appear centered. When $L_2 > L_1$ the sound appears to come from the left.

and thus the sounds in the two ears will be slightly out of phase, as shown in Fig. 5.11(b). He confirmed this theory by experiments with two tuning forks tuned to slightly different frequencies, so that their relative phases constantly changed. The sound of the beating tone moved from right to left and back again.

Several experiments have confirmed the fact that for frequencies up to about 1000 Hz, localization occurs mainly through detection of the phase difference at the two ears (for steady sounds) or the difference in arrival time (for clicks), as illustrated in Fig. 5.12. Above 4000 Hz, localization by intensity difference takes over. Between 1000 and 4000 Hz, the accuracy of localization declines, with a high error rate around 3000 Hz demonstrating that the two mechanisms do not overlap appreciably.

At high frequencies (about 5000 Hz and upward), the pinna aids in localization of a sound, particularly in distinguishing between sound coming from the front or the back, because it receives sound with slightly greater efficiency from the front. Some animals have the ability to aim their pinnas toward sounds of interest, but human beings must turn the entire head to change pinna orientation.

An important corollary to sound localization is the so-called *precedence effect* (sometimes referred to as the *Haas effect*), which applies to efforts to localize a sound source in a room. If similar sounds arrive within about 35 ms (0.035 s), the apparent direction of the sound source is the direction from which the first arriving sound comes. The ear automatically assumes this to be the direct sound and successive sounds to have been reflected one or more times. This effect will be discussed further in Chapter 23.

5.6   MEASURING SENSATIONS: PSYCHOPHYSICS

Information about the world around us comes from our senses: vision, taste, smell, touch, and hearing. Each of our sensory organs responds to a particular type of stimulus over a limited range of energies. Our eyes, for example, respond to electromagnetic waves over an extremely narrow range of frequency compared to the wide range of electromagnetic radiations all around us.

Perception involves not only the reception of information by the appropriate sensory organ, but the coding, transmission, and processing of this information by the central nervous system. Our understanding of how this is accomplished has advanced remarkably in recent years, but still remains only fragmentary. (This may be due partly to the fact that research in this area involves several disciplines: physics, psychology, physiology, speech and hearing, engineering, mathematics, etc.) An excellent source of information about perception is a collection of articles from *Scientific American* (1972). It appears that many perceptual abilities are intrinsic; others are acquired or developed through experience and training.

The study of the relationships between stimuli and the subjective sensations they produce is the basis of *psychophysics*, so named by a pioneer in the field, G. T. Fechner. Inspired by earlier work on the subject by Ernst Weber, Fechner spent many years trying to determine quantitative relationships between stimulus and perceived sensation, and in 1860 published many of his findings in a monumental book entitled *Elements of Psychophysics*. He summed up much of this work in a simple mathematical law relating sensation to stimulus, which is often referred to today as *Fechner's law.* It expresses the relationship between stimulus and sensation rather simply: As stimuli are increased by *multiplication,* sensations increase by *addition.* For example, as the intensity of a sound is doubled, its loudness increases by one step on a scale. Mathematicians call such a relationship logarithmic; Fechner's law states that sensation grows as the logarithm of the stimulus.

Fechner argued that the same relationship applies to any stimulus and its corresponding sensation: to light and vision, etc. Recent investigations have pointed out its inadequacies; nevertheless, Fechner's law served as a basis for psychophysical theory for nearly a century thereafter. Fechner answered his early critics by saying "The Tower of Babel was never finished because the workers could not reach an understanding on how they should build it; my psychophysical edifice will stand because the workers will never agree on how to tear it down" (Stevens and Warshofsky, 1965, p. 82). We will return to the subject in the following chapters. At this time it is appropriate to review or introduce (depending on the background of the reader) some properties of logarithms and logarithmic scales.

5.7 LOGARITHMS AND POWERS OF TEN

It is inconvenient to write out numbers such as 1,530,000,000 and 0.000087. These same numbers can be written better as 1.53×10^9 and 8.7×10^{-5}, respectively, since $10^9 = 1,000,000,000$ and $10^{-5} = 0.00001$. Other powers of ten are

$$
\begin{aligned}
10^3 &= 1000, \\
10^2 &= 100, \\
10^1 &= 10, \\
10^0 &= 1, \\
10^{-1} &= 0.1, \\
10^{-2} &= 0.01, \quad \text{etc.}
\end{aligned}
$$

On some electronic calculators, the scientific notation used to display very large and very small numbers expresses 10^3 as E3 and 10^{-3} as E-3, so 1,530,000,000 would be expressed as 1.53 E9 and 0.000087 as 8.7 E-5 (E is an abbreviation for "exponent," which means the power to which ten is raised). Other calculators omit the letter E but leave a space between the first part of the number and its exponent (e.g., 1531 becomes 1.531 03 in scientific notation).

To multiply two numbers in scientific or exponential notation, we *add* the exponents; to divide, we *subtract* exponents. Thus

$$(10^3)(10^4) = 10^7;$$
$$(5 \times 10^2)(3 \times 10^5) = 15 \times 10^7 = 1.5 \times 10^8;$$
$$(3 \times 10^{-3})(2 \times 10^5) = 6 \times 10^2;$$
$$\frac{10^4}{10^2} = 10^2;$$
$$\frac{6 \times 10^5}{3 \times 10^3} = 2 \times 10^2.$$

In general, then, $(10^A)(10^B) = 10^{A+B}$, and $10^A/10^B = 10^{A-B}$.

Closely related to exponents and scientific notation are logarithms. Logarithms are defined as follows: The logarithm to the base 10 of a number x is equal to the power to which 10 must be raised in order to equal x." That is, if $x = 10^y$, then $y = \log x$. For example, $100 = 10^2$, so $2 = \log 100$ (here $x = 100$, $y = 2$); or $1000 = 10^3$, so $3 = \log 1000$ ($x = 1000$, $y = 3$).

Although logarithms to other bases are used in mathematics, we nearly always use base 10 in acoustics. Most calculators use *log x* to denote the logarithm of x to base 10, and *ln x* to denote the logarithm to the base 2.7183. On some calculators, the inverse logarithm is computed by pressing "INV" and then "log"; on others a 10^x key is used.

The following identities are useful for performing calculations with logarithms:

$$\log AB = \log A + \log B;$$
$$\log A/B = \log A - \log B;$$
$$\log A^n = n \log A.$$

The logarithms of some numbers are as follows:

x	$\log x$	x	$\log x$
1	0	6	0.778
2	0.301	7	0.845
3	0.477	8	0.903
4	0.602	9	0.954
5	0.699	10	1.000

Using this table and the identities listed above, we can compute the logarithms of many numbers. For example:

log 400 = log 4 + log 100 = 0.602 + 2 = 2.602 (first identity);
log 2.5 = log 5 − log 2 = 0.699 − 0.301 = 0.398 (second identity);
log 25 = 2 log 5 = (2)(0.699) = 1.398 (third identity).

If one remembers that log 2 = 0.3, many logarithms can be estimated closely. For example:

$$\log 4 = \log 2 \times 2 = 0.3 + 0.3 = 0.6;$$

$$\log 5 = \log \frac{10}{2} = 1 - 0.3 = 0.7;$$

$$\log 8 = \log 2 \times 2 \times 2 = 0.3 + 0.3 + 0.3 = 0.9.$$

A *logarithmic scale* is one on which equal distances represent the same factor anywhere along the scale (in contrast to a *linear scale*, on which equal distances represent equal increments). Logarithmic and linear scales are shown in Fig. 5.13.

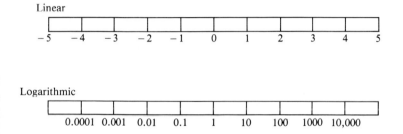

FIG. 5.13
Linear and logarithmic scales. On the linear scale, moving one unit to the right adds an increment of one; on the logarithmic scale, moving one unit to the right multiplies by a factor of ten.

Sound frequencies are usually represented on a logarithmic scale for reasons that will become clear later on. In Fig. 5.14, the distance from 20 to 200 Hz is the same as from 200 to 2000 Hz or from 2000 to 20,000 Hz.

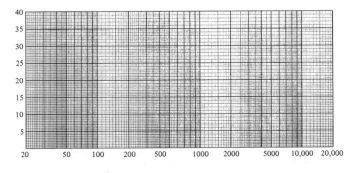

FIG. 5.14
Graph paper with a logarithmic scale of frequencies. Such graph paper is called "semilog," because only one axis is logarithmic. On "log-log" graph paper, both axes are logarithmic.

5.8   SUBJECTIVE ATTRIBUTES OF SOUND

Four attributes are frequently used to describe sound, especially musical sound. They are loudness, pitch, timbre, and duration. Each of these subjective qualities depends on one or more physical parameters that can be measured. Loudness, for example, depends mainly on sound pressure but also on the spectrum of the partials and the physical duration. Pitch depends mainly on frequency, but also shows lesser dependence on sound pressure and envelope. Timbre is a sort of catchall, including all those attributes that serve to distinguish sounds with the same pitch and loudness. Table 5.1 relates subjective qualities to physical parameters, and is presented here as an introduction to the next three chapters, which discuss loudness, pitch, and timbre in more detail.

TABLE 5.1   Dependence of subjective qualities of sound on physical parameters

Physical Parameter	Subjective Quality			
	Loudness	Pitch	Timbre	Duration
Pressure	+++	+	+	+
Frequency	+	+++	++	+
Spectrum	+	+	+++	+
Duration	+	+	+	+++
Envelope	+	+	++	+

+ = weakly dependent; ++ = moderately dependent; +++ = strongly dependent.
Note: Spectrum refers to the frequencies and amplitudes of all the partials (components) in the sound. The physical duration of a sound and its perceived (subjective) duration, though closely related, are not the same. Envelope includes the attack, the release, and variations in amplitude. These parameters will be discussed in Chapters 6, 7, and 8.

5.9   SUMMARY

The human auditory system responds to pressure stimuli over a range of a million times. The frequency range of hearing extends from 20 to 20,000 Hz for some individuals, substantially less for others. Like other sensations, hearing tends to follow *logarithmic* relationships. The *outer ear* boosts hearing sensitivity in the middle frequency range, and aids in determining the direction of a sound. The middle ear contains three small bones, called *ossicles*, which transmit sound pressure from the eardrum to the inner ear. The main part of the inner ear is the *cochlea*, which transforms pressure variations into neural impulses. Much of our ability to determine the direction of a sound source depends on *binaural* hearing, with a different mechanism of localization being dominant at high and low frequencies.

References and Suggested Readings

Békésy, G. von (1960). *Experiments in Hearing*. New York: McGraw-Hill.

Békésy, G. von (1970). "Enlarged Mechanical Model of the Cochlea with Nerve Supply," in *Foundations of Modern Auditory Theory,* Vol. I. Ed., J. V. Tobias. New York: Academic Press.

Denes, P. B., and E. N. Pinson (1973). *The Speech Chain*. Garden City, N.Y.: Anchor.

Flannagan, J. L. (1972). *Speech Analysis, Synthesis and Perception*. New York: Springer-Verlag.

Fletcher, H. (1940). "Auditory Patterns," *Revs. Modern Physics* **12**: 47.

Helmholtz, H. von (1877). *On the Sensation of Tone as a Psychological Basis for the Study of Music,* 4th ed. Trans., A. J. Ellis. New York: Dover, 1954.

Johnstone, D. M., and A. J. F. Boyle (1967). "Basilar Membrane Vibration Examined with Mössbauer Technique," *Science* **158**: 389.

Khanna, S. M., and D. G. B. Leonard (1982). "Basilar Membrane Tuning in the Cat Cochlea," *Science* **215**: 305.

Kiang, N. Y. S., and E. C. Moxon (1974). "Tails of Tuning Curves of Auditory-nerve Fibers," *J. Acoust. Soc. Am.* **55**: 620.

Plomp, R. (1976). *Aspects of Tone Sensation*. London: Academic Press.

Readings from *Scientific American* (1972). *Perception: Mechanisms and Models*. San Francisco: W. H. Freeman. This collection of readings contains reprints of 38 articles published during 1950–1971 with introductions to each section by R. Held and W. Richards. "The Ear" by Békésy (1957) is included.

Rhode, W. S., and L. Robles (1974). "Evidence from Mössbauer Experiments for Nonlinear Vibration in the Cochlea," *J. Acoust. Soc. Am.* **55**: 588.

Scharf, B. (1970). "Critical Bands" in *Foundations of Modern Auditory Theory,* ed. J. Tobias. New York: Academic Press.

Scharf, B., and S. Buus (1986). "Audition I. Stimulus, Physiology, Thresholds" in *Handbook of Perception and Human Performance*. Ed. K. R. Boff, L. Kaufman, and J. P. Thomas. New York: J. Wiley.

Stevens, S. S., and F. Warshofsky (1965). *Sound and Hearing*. New York: Time Life.

Tasaki, I. (1954). "Nerve Impulses in Individual Auditory Nerve Fibres of Guinea Pigs," *J. Neurophysiology* **17**: 97.

Teas, D. C. (1970). "Cochlear Processes," in *Foundations of Modern Auditory Theory*. Ed., J. V. Tobias. New York: Academic Press.

Zwicker, E., G. Flottorp, and S. S. Stevens (1957). "Critical Bandwidth in Loudness Summation," *J. Acoust. Soc. Am.* **29**: 548.

Glossary

auditory canal A tube in the outer ear that transmits sound from the external pinna to the eardrum.

basilar membrane A membrane in the cochlea that separates the cochlear duct from the scala tympani and to which the organ of Corti is attached.

cochlea The spiral organ of the inner ear containing the sound-sensing mechanism.

critical band Frequency band within which two or more tones excite many of the same hair cells on the basilar membrane and thus are difficult to distinguish as separate tones.

eardrum (tympanum) The fibrous membrane that terminates the auditory canal and is caused to vibrate by incoming sound waves.

Eustachian tube A tube connecting the middle ear to the oral cavity that allows the average pressure in the middle ear to equal atmospheric pressure.

exponent The number expressing the power to which 10 or some other number is raised.

Fechner's (Weber's) law An empirical law expressing the way in which sensation varies with stimulus.

hair cells The tiny sensors of sound in the cochlea.

linear scale A scale in which moving a given distance right or left adds or subtracts a given increment.

localization The ability to determine the location or direction of a sound source.

logarithm (of a number) The power to which 10 (or some other base) must be raised to give the desired number.

logarithmic scale A scale on which moving a given distance right or left multiplies or divides by a given factor.

organ of Corti The part of the cochlea containing the haircells; the "seat of hearing."

ossicles Three small bones of the middle ear that transmit vibrations from the eardrum to the cochlea.

pinna The external part of the ear.

psychophysics The study of the relationship between stimuli and the sensations they produce.

Reissner's membrane A membrane in the cochlea that separates the cochlear duct from the scala vestibuli.

scala vestibuli A canal in the ear that transmits pressure variations from the oval window to the cochlear duct.

Questions for Discussion

1. If everyone's hearing sensitivity were reduced by ten decibels, in what ways would our lives probably be different?

2. What advantage is there in having our various senses respond on a (nearly) logarithmic rather than a linear scale?

3. Before the development of radar, a device used to determine the direction of aircraft consisted of two sound-receiving horns, each of which transmitted sound to one ear. Comment on the effectiveness of such a device.

4. Listen to a tape recording of your own voice and compare its sound to what you hear when you speak and sing. Try to describe the difference in terms of relative balance between high and low frequency components, etc.

Problems

1. Assume that the outer ear canal is a cylindrical pipe 3 cm long, closed at one end by the eardrum. Calculate the resonance frequency of this pipe (see Fig. 4.6). Our hearing should be especially sensitive for frequencies near this resonance.

2. At what frequency does the wavelength of sound equal the distance between your ears? What is the significance of this with respect to your ability to localize sound?

3. The effective area of the eardrum is estimated to be approximately 0.55 cm². During normal conversation, the sound pressure variations of about 10^{-2} newtons per square meter reach the eardrum. What force is exerted on the eardrum (force = pressure × area)?

4. Pressure is force per unit area. Calculate the pressure when a force of 500 newtons (approximate weight of a 110-lb person) is supported by:

   a) Spike heels having an area of 10^{-5} m² each;

   b) Standard heels having an area of 10^{-2} m² each.

   Comment on the likelihood of denting the floor in each case.

5. Measure the distance between your ears. Divide this distance by the speed of sound (Table 3.1) to find the

maximum difference in arrival time $\Delta t = (L_2 - L_1)/v$ that occurs when a sound comes directly from the side.

6. Calculate the difference in arrival time at the two ears for a sound that comes from a 45-degree direction (from the northwest, for example, when the listener faces north).

7. Perform the following arithmetic operations.

a) $(1.6 \times 10^{-8})(5.0 \times 10^3)$

b) $\dfrac{4.5 \times 10^{-2}}{1.5 \times 10^{-3}}$

c) $1.3 \times 10^3 + 4.3 \times 10^2$

d) $4.2 \times 10^2 - 5.4 \times 10^{-2}$

8. Find the following logarithms using the logarithms of the numbers 1–10 and the three identities given.

a) log 50

b) log 0.5

c) $\log 3 \times 10^{10}$

d) log 16

9. Given log x, find the number x in each case.

a) $\log x = 0.3$

b) $\log x = 3.0$

c) $\log x = 1.3$

d) $\log x = -0.3$

CHAPTER 6

Sound Pressure, Power and Loudness

In this chapter, we will discuss the quality of loudness and the physical parameters that determine it. The principal such parameter, we learned in Table 5.1, is sound pressure. Related to the sound pressure are the sound power emitted by the source and the sound intensity (the rate of energy flow in a sound wave). The sound pressure can be measured directly, however, and our ears (like most microphones) respond to sound pressure.

6.1 SOUND PRESSURE LEVEL

In a sound wave there are extremely small periodic variations in atmospheric pressure to which our ears respond in a rather complex manner. The minimum pressure fluctuation to which the ear can respond is less than one billionth (10^{-9}) of atmospheric pressure. This threshold of audibility, which varies from person to person, corresponds to a sound pressure amplitude of about 2×10^{-5} N/m^2 (newtons/meter^2) at a frequency of 1000 Hz. The threshold of pain corresponds to a pressure amplitude approximately one million (10^6) times greater, but still less than 1/1000 of atmospheric pressure.

Because of the wide range of pressure stimuli, it is convenient to measure sound pressures on a logarithmic scale, called the *decibel* (dB) *scale*. Although a decibel scale is actually a means for comparing two sounds, we can define a decibel scale of sound level by comparing sounds to a reference sound with a pressure amplitude $p_0 = 2 \times 10^{-5}$ newtons/meter^2, assigned a sound pressure level of 0 dB. Thus, we define *sound pressure level* (SPL or L_p) as

$$L_p = 20 \log p/p_0 \qquad (6.1)$$

$$= 10 \log \frac{p^2}{p_0^2},$$

Expressed in other units, $p_0 = 20$ micropascals $= 2 \times 10^{-4}$ dynes/cm$^2 = 2 \times 10^{-4}$ microbars. (For comparison, atmospheric pressure is 10^5 N/m^2, or 10^6 microbars).

Sound pressure levels are measured by a sound level meter, consisting of a microphone, an amplifier, and a meter that reads in decibels. Sound pressure levels of a number of sounds are given in Table 6.1. If you have access to a sound level meter, it is recommended that you carry it with you to many locations to obtain a feeling for different sound pressure levels.

TABLE 6.1   Typical Sound Levels One Might Encounter.

Jet takeoff (60 m)	120 dB	
Construction site	110 dB	*Intolerable*
Shout (1.5 m)	100 dB	
Heavy truck (15 m)	90 dB	*Very noisy*
Urban street	80 dB	
Automobile interior	70 dB	*Noisy*
Normal conversation (1 m)	60 dB	
Office, classroom	50 dB	*Moderate*
Living room	40 dB	
Bedroom at night	30 dB	*Quiet*
Broadcast studio	20 dB	
Rustling leaves	10 dB	*Barely audible*
	0 dB	

Example 6.1   What sound pressure level corresponds to a sound pressure of 10^{-3} N/m^2?

Solution: $L_p = 20 \log \dfrac{10^{-3}}{2 \times 10^{-5}} = 34.0$ dB.

Example 6.2.   How much force does a sound wave at the pain threshold ($L_p \simeq 120$ dB) exert on an eardrum having a diameter of 7 mm?

Solution: $L_p = 120 = 20 \log \dfrac{p}{p_0}$;

$$p = p_0 \text{ inv} \log \frac{120}{20} = 20 \text{ N/m}^2;$$

$$F = pA = 20\pi(3.5 \times 10^{-3})^2 = 1.54 \times 10^{-3} \text{ N}.$$

6.2 SOUND POWER AND INTENSITY LEVELS

In addition to the sound pressure level, there are other levels expressed in decibels, so one must be careful when reading technical articles about sound or regulations on environmental noise. One such level is the *sound power level* (PWL or L_W), which identifies the total sound power emitted by a source in all directions. Sound power, like electrical power, is measured in watts (one watt equals one joule of energy per second). In the case of sound, the amount of power is very small, so the reference selected for comparison is the picowatt (10^{-12} watt). The sound power level (in decibels) is defined as

$$L_W = 10 \log W/W_0, \tag{6.2}$$

where W is the sound power emitted by the source, and the reference power $W_0 = 10^{-12}$ watt.*

The relationship between sound pressure level and sound power level depends on several factors, including the geometry of the source and the room. If the sound power level of a source is increased by 10 dB, the sound pressure level also increases by 10 dB, provided everything else remains the same. If a source radiates sound equally in all directions and there are no reflecting surfaces nearby (a free field), the sound pressure level decreases by 6 dB each time the distance from the source doubles. Propagation in a free field is described in the box below.

Another quality described by a decibel level is *sound intensity,* which is the rate of energy flow across a unit area. The reference for measuring sound intensity level is $I_0 = 10^{-12}$ watt/m^2, and the *sound intensity level* (IL or L_I) is defined as

$$L_I = 10 \log I/I_0. \tag{6.3}$$

For a free progressive wave in air (e.g., a plane wave traveling down a tube or a spherical wave traveling outward from a source), sound

---

* In Chapter 1 we used the symbol $\mathscr{P}$ for mechanical power (the rate of doing work), also expressed in *watts*. Sound power might logically be given the same symbol, but in using W we follow the prevalent practice.

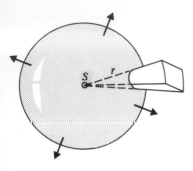

FIG. 6.1
Spherical sound waves in a free field.
The power from source *S* is distributed
over a spherical surface $4\pi r^2$ in area.

pressure level and sound intensity level are nearly equal ($L_p \simeq L_I$).
This is not true in general, however, because sound waves from many
directions contribute to sound pressure at a point. When we speak of
simply sound level, we nearly always mean sound pressure level L_p,
since that is what is indicated by our sound-measuring instruments.

Free Field

When a point source (or any source that radiates equally in all
directions) radiates into free space, the intensity of the sound
varies as $1/r^2$ (and the sound pressure varies as $1/r$), where *r* is
the distance from the source *S*. This may be understood as a
given amount of sound power being distributed over the surface of an expanding sphere with area $4\pi r^2$ (see Fig. 6.1).
Thus the intensity is given by

$$I = W/4\pi r^2, \tag{6.4}$$

where *W* is the power of the source. An environment in which
there are no reflections is called a *free field*. In a free field, the
sound intensity level decreases by 6 dB each time the distance
from the source is doubled. The sound intensity level (or
sound pressure level) at a distance of one meter from a source
in a free field is 11 dB less than the sound power level of the
source. This is easily shown as follows:

$$I = \frac{W}{4\pi r^2} = \frac{W}{4\pi(1)} \, ;$$

$$L_I = 10 \log \frac{I}{10^{-12}} = 10 \log \frac{W}{10^{-12}} - 10 \log 4\pi = L_W - 11 \approx L_p.$$

Similarly, it can be shown that at a distance of two meters, L_I
is 17 dB less than L_W.

Hemispherical Field

More common than a free field is a sound source resting on
a hard, sound-reflecting surface and radiating hemispherical
waves into free space (see Fig. 6.2). Under these conditions,
the sound intensity level L_I and the sound pressure level L_p at
a distance of one meter are 8 dB less than the sound power
level, once again diminishing by 6 dB each time the distance is
doubled. In actual practice, few sound sources radiate sound
equally in all directions, and there are often reflecting surfaces nearby that destroy the symmetry of the spherical or
hemispherical waves.

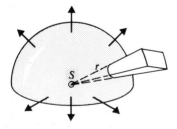

FIG. 6.2
Hemispherical sound waves from a
source *S* on a hard reflecting surface.
The power is distributed over a surface
$2\pi r^2$ in area.

Example 6.3 If a trombone bell has an area of 0.1 m^2 and the power radiated from the bell during a very loud note is 1.5 W, what is the average intensity and sound intensity level at the bell?

Solution: $I = \dfrac{W}{A} = \dfrac{1.5}{0.1} = 15 \text{ W/m}^2$;

$$L_I = 10 \log \frac{I}{I_0} = 10 \log \frac{15}{10^{-12}}$$

$$= 10 \log 15 \times 10^{12} - 132 \text{ dB}.$$

Example 6.4 The sound pressure level 1 m from a noisy motor resting on a concrete floor is measured to be 95 dB. Find the sound power and the sound power level of the source.

Solution 1: $L_I = 10 \log \dfrac{I}{I_0} \simeq L_p = 95$ dB;

$$I = I_0 \text{ inv log } \frac{95}{10} = 3.16 \times 10^{-3} \text{ W/m}^2;$$

$$W = 2\pi r^2 I = 2\pi(1)^2(3.16 \times 10^{-3}) = 1.98 \times 10^{-2} \text{ W}.$$

Solution 2: For a hemispherical field,

$$L_w = L_p(1\text{m}) + 8 = 95 + 8 = 103 \text{ dB};$$

$$W = W_0 \text{ inv log } \frac{103}{10} = 1.98 \times 10^{-2} \text{ W}.$$

6.3 MULTIPLE SOURCES

Very frequently we are concerned with more than one source of sound. The way in which sound levels add may seem a little surprising at first. For example, two uncorrelated sources, each of which would produce a sound level of 80 dB at a certain point, will together give 83 dB at that point. Figure 6.3 gives the increase in sound level due to additional equal sources. It is not difficult to see why this is the case, since doubling the sound power raises the sound power level by 3 dB, and thus raises the sound pressure level 3 dB at our point of interest. Under some conditions, however, there may be interference between waves from the two sources, and this doubling relationship will not hold true.

When two waves of the same frequency reach the same point, they may interfere constructively or destructively. If their amplitudes are

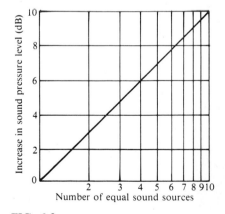

FIG. 6.3
Addition of equal (uncorrelated) sound sources.

both equal to A, the resultant amplitude may thus be anything from zero up to 2 A. The resultant intensity, which is proportional to the amplitude squared, may thus vary from 0 to 4 A^2. If the waves have different frequencies, however, these well-defined points of constructive and destructive interference do not occur. In the case of sound waves from two noise sources (as in the case of light from two light bulbs), the waves include a broad distribution of frequencies (wavelength), and we do not expect interference to occur. In this case, we can add the energy carried by each wave across a surface or, in other words, the intensities.

It is worth reviewing the difference between sound pressure level, sound power level, and sound intensity level, since all three are expressed in decibels. Nearly always it is *sound pressure level* that is measured, and from it the other two are calculated. *Sound power level* expresses the strength of the source, irrespective of the location of source and observer. *Sound intensity level* expresses the rate at which energy flows across a unit area. For a free progressive wave, the intensity is proportional to the square of the sound pressure (actually the average value of p^2). Thus a doubling of the sound pressure results in an increase of 6 dB in sound pressure level L_p , whereas doubling the sound power increases the sound power L_W by only 3 dB. Similarly, a ten-fold increase in sound pressure represents a 20 dB increase in L_p, but a ten-fold increase in sound power increases L_W by 10 dB. Table 6.2 may be helpful.

TABLE 6.2   Sound pressure levels and sound power levels

p/p_0	L_p	W/W_0	L_w
2	6 dB	2	3 dB
4	12 dB	4	6 dB
10	20 dB	10	10 dB
20	26 dB	20	13 dB
50	34 dB	50	17 dB
100	40 dB	100	20 dB
1000	60 dB	1000	30 dB

Combining Sound Levels

In the case of two uncorrelated sound sources, what we really want to add are the mean-square sound pressures (average values of p^2) at a point. If the sound pressure at a certain point due to Source 1 is p_1, and the sound pressure at that same point due to Source 2 is p_2, then the sound pressure level due to both sources is

$$L_p = 10 \log \frac{p_1^2 + p_2^2}{p_0^2}.$$

In the case of two equal sources, $p_1 = p_2$, so that

$$L_p = 10 \log 2 \frac{p_1^2}{p_0^2} = 10 \log \frac{p_1^2}{p_0^2} + 10 \log 2 = 20 \log \frac{p_1}{p_0} + 3.$$

Thus, two sources which by themselves each cause $L_p = 40$ dB at a certain location will cause $L_p = 43$ dB at that same location when sounded together. (This result is also obtained from the graph in Fig. 6.3.)

Example 6.5 With one violin playing, the sound pressure level at a certain place is measured to be 50 dB. If three violins play equally loudly, what will the sound pressure level most likely be at the same location?

Solution: $L_p = 10 \log \dfrac{p_1^2 + p_2^2 + p_3^2}{p_0^2}$

$$= 10 \log \frac{p_1^2}{p_0^2} + 10 \log 3$$

$$= 50 + 4.8 = 54.8 \text{ dB}$$

(This result could also be determined from Fig. 6.3.)

Example 6.6. If two sound sources independently cause sound levels of 50 and 53 dB at a certain point, what is the SPL at that point when both sources contribute at the same time?

Solution: $50 = 20 \log \dfrac{p_1}{p_0}$,

so $p_1 = p_0$ inv log $\dfrac{50}{20} = 6.32 \times 10^{-3}$ N/m²;

likewise $p_2 = 8.93 \times 10^{-3}$ N/m²;

$$L_p = 10 \log \frac{p_1^2 + p_2^2}{p_0^2}$$

$$= 10 \log \frac{3.99 \times 10^{-5} + 7.98 \times 10^{-5}}{4 \times 10^{-10}} = 54.8 \text{ dB}$$

(Note that the answer is *not* $50 + 53 = 103$ dB.)

6.4 LOUDNESS LEVEL

Although sounds with a greater L_I or L_p usually sound louder, this is not always the case. The sensitivity of the ear varies with the frequency and the quality of the sound. Many years ago Fletcher and Munson (1933) determined curves of equal *loudness level* (L_L) for pure tones (that is, tones of a single frequency). The curves shown in Fig. 6.4, recommended by the International Standards Organization, are quite similar to those of Fletcher and Munson. These curves demonstrate the relative insensitivity of the ear to sounds of low frequency at moderate to low intensity levels. Hearing sensitivity reaches a maximum between 3500 and 4000 Hz, which is near the first resonance frequency of the outer ear canal, and again peaks around 13 kHz, the frequency of the second resonance.

The contours of equal loudness level are labeled in units called *phons*, the level in phons being numerically equal to the sound pressure level in decibels at $f = 1000$ Hz. The phon is a rather arbitrary unit, however, and is not widely used in measuring sound. It is important, however, to note the relative insensitivity of the ear to sounds of low frequency, which is one reason why weighting networks are used in sound-measuring equipment.

Sound level meters have one or more weighting networks, which provide the desired frequency responses. Generally three weighting networks are used; they are designated A, B, and C. The C-weighting

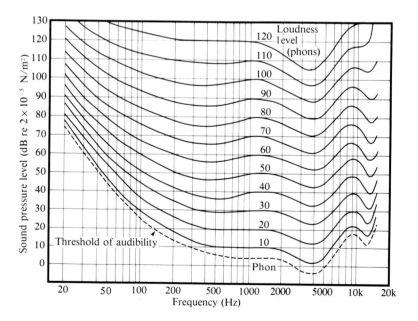

FIG. 6.4
Equal-loudness curves for pure tones (frontal incidence). The loudness levels are expressed in phons.

network has an almost flat frequency response, whereas the A-weighting network introduces a low-frequency rolloff in gain that bears rather close resemblance to the frequency response of the ear at low sound pressure level. A sound level meter is shown in Fig. 6.5, along with the frequency responses of A-, B-, and C-weighting networks.

Measurements of sound level are usually made using the A-weighting network; such measurements are properly designated as $L_p(A)$ or *SPL* (A) in dB, although the unit dBA or dB(A) is often used to denote A-weighted sound level. Many sound level meters have both fast and slow response, the slow response measuring an "average" level. Inside a building, the C-weighted sound level may be substantially higher than the A-weighted sound level, because of low-frequency machinery noise, to which the ear is quite insensitive.

Although it is difficult to describe a sound environment by a single parameter, for many purposes the A-weighted sound level will suffice. At low to medium sound levels, it is reasonably close to the true loudness level so that dBA (easily measured with a sound level meter) may be substituted for phons without too much error.

There are many examples of interesting sound environments to measure. In the classroom, one can ask the entire class to shout loudly, then half the class to do so, one-fourth of the class, etc. The sound level should drop about 3 dB in each step. One can also measure traffic noise, noise near a construction site, sound level at a concert, noise in an automobile, and so on. In each case the A-weighted sound level should be measured, although it may be interesting to measure the C-weighted level (which places more emphasis on sounds of low frequency) as well.

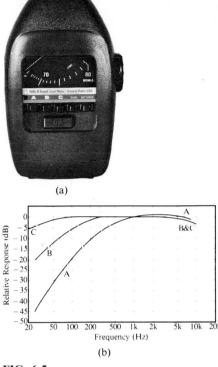

(a)

(b)

FIG. 6.5
Sound-level meter with the frequency response of its A-, B-, and C-weighting networks. (Photography courtesy of GenRad, Inc.).

6.5 LOUDNESS OF PURE TONES: SONES

In Chapter 5, we mentioned Fechner's law, relating sensation to stimulus. The logarithmic relationship in that law was found to provide only a rough approximation to listeners' estimates of their own sensations of loudness. In an effort to obtain a quantity proportional to the loudness sensation, a loudness scale was developed in which the unit of loudness is called the *sone*. The sone is defined as the loudness of a 1000-Hz tone at a sound level of 40 decibels (a loudness level of 40 phons).

For loudness levels of 40 phons or greater, the relationship between loudness S in sones and loudness level L_L in phons recommended by the International Standards Organization (ISO) is

$$S = 2^{(L_L - 40)/10}. \tag{6.5}$$

An equivalent expression for loudness S that avoids the use of L_L is

$$S = Cp^{0.6}, \tag{6.6}$$

where p is the sound pressure and C depends on the frequency

Equations (6.5) and (6.6) are based on the work of S. S. Stevens, which indicated a doubling of loudness for a 10-dB increase in sound pressure level. Some investigators, however, have found a doubling of loudness for a 6-dB increase in sound pressure level (Warren, 1970). This suggests the use of a formula in which loudness is proportional to sound pressure (Howes, 1974):

$$S = K(p - p_0), \tag{6.7}$$

where p is sound pressure and p_0 is the pressure at some threshold level.

One way to represent the loudness graphically is to combine one of the above psychophysical expressions for loudness with the curves of equal loudness given in Fig. 6.4. The solid curves in Fig. 6.6 make use of Eq. (6.5) to give the loudness of pure tones at different frequencies. A slight difference is found between sounds that arrive from straight ahead (frontal incidence) and those that arrive from all directions (random incidence); the curves in Fig. 6.6 are drawn for random incidence, as would be found in a live room. The peaks that occur near 4000 Hz are due to the main resonance of the ear canal; the smaller ones above 10 kHz are due to the second resonance at approximately

FIG. 6.6
Subjective loudness of pure tones (solid curves) and "musical" tones with five harmonics (dashed curves) as a function of frequency and sound level (L_p). These curves represent "average" judgments of subjective loudness, and the variation from person to person will be substantial. (These curves were calculated using Eq. 6.5 and Fig. 6.9.)

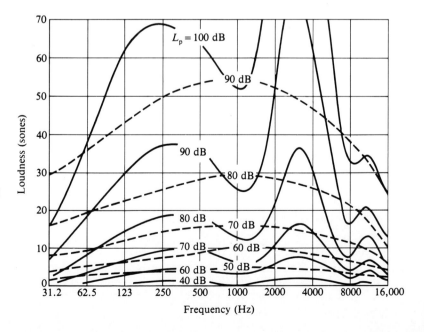

three times the main resonance frequency (the ear canal resembles a pipe closed at one end by the eardrum).

Also shown in Fig. 6.6, as dashed curves, are predictions of the subjective loudness of "musical" tones, which consist of a fundamental plus four harmonic overtones. These were calculated using Fig. 6.9 by giving each of the five harmonics the same amplitude. The sound pressure levels are what would be measured on a sound level meter with C-weighting; the sound level of each harmonic is 7 dB below the total level (see Fig. 6.3). Note that the peaks due to ear canal resonances that are prominent for tones of a single frequency are pretty well averaged out for tones with several harmonics, which is fortunate for the performing musician. The loudness of complex tones is discussed in Sections 6.6 and 6.7.

Example 6.7   Find the loudness level and the loudness of a 500-Hz tone with $L_p = 70$ dB.

Solution: From Fig. 6.4, the loudness level is $L_L = 74$ phons.

The loudness is: $S = 2^{(74-40)/10} = 10.6$ sones.

This result could also be obtained by noting the intersection of the 70-dB (solid) curve with 500 Hz in Fig. 6.6.

6.6   LOUDNESS OF COMPLEX TONES: CRITICAL BANDS

As pointed out in Table 5.1, loudness depends mainly on sound pressure, but it also varies with frequency, spectrum, and duration. We have already seen how loudness depends on frequency; now we will consider its dependence on the spectrum of the sound.

If we were to listen to two pure tones having the same sound pressure level but with increasing frequency separation, we would note that when the frequency separation exceeds the *critical bandwidth*, the total loudness begins to increase. Broadband sounds, such as those of jet aircraft, seem louder than pure tones or narrowband noise having the same sound pressure level. Figure 6.7 illustrates the dependence of loudness on bandwidth with fixed sound pressure level and center frequency. Note that loudness is not affected until the bandwidth exceeds the critical bandwidth, which is about 160 Hz for the 1-kHz center frequency shown.

One way to estimate critical bandwidth is to increase the bandwidth of a noise signal while decreasing the amplitude in order to keep the power constant. When the bandwidth is greater than a critical

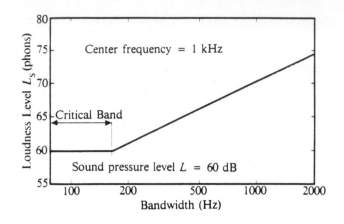

FIG. 6.7
The effect of bandwidth on loudness.

band, the subjective loudness increases above that of a reference noise signal because the stimulus now extends over more than one critical band (Demonstration 3 in Houtsma, Rossing, and Wagenaars, 1988).

The perceived loudness of broadband (*white*) noise is compared to that of a 1000-Hz tone having the same SPL in Fig. 6.8. At a sound pressure level of 55 dB, the white noise is judged to be about twice as loud as the 1000-Hz tone, but at higher and lower levels the difference is substantially less (Scharf and Houtsma, 1986).

The dependence of loudness on stimulus variables, such as sound pressure, frequency, spectrum, duration, etc., appears to be about the same whether the sound is presented to one ear (monaurally) or to both ears (binaurally). However, a sound presented to both ears is judged nearly twice as loud as the same sound presented to one ear only (Scharf and Houtsma, 1986).

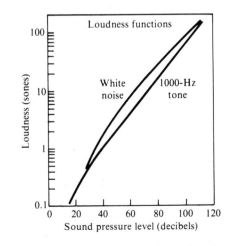

FIG. 6.8
Loudness of white noise compared to that of a 1000-Hz tone at the same sound pressure level. Fifteen subjects judged the noise presented binaurally through headphones (Scharf and Fishken, 1970).

6.7 LOUDNESS OF COMBINED SOUNDS

The loudness of combined sounds is a subject of considerable interest. How many violins must play together, for example, in order to double the loudness? Or, how does the loudness of traffic noise depend on the number of vehicles? We stated in Section 6.3 that the intensities (or mean-square pressures) from two or more uncorrelated sound sources add together to give a total intensity. The loudness is not necessarily additive, however. Accepted methods for combining loudness are given in the following box.

When two or more tones are mixed, the way in which their individual loudnesses combine depends on how close they are to each other in frequency. We can have three different situations:

1. If the frequencies of the tones are the same or fall within the critical bandwidth, the loudness is calculated from the total intensity $I = I_1 + I_2 + I_3 + \cdots$. If the intensities I_1, I_2, I_3, etc., are equal, the increase in sound level is as shown in Fig. 6.5. The loudness may then be determined from the combined sound level by using Fig. 6.6.

2. If the bandwidth exceeds the critical bandwidth, the resulting loudness is greater than that obtained from simple summation of intensities. As the bandwidth increases, the loudness approaches (but remains less than) a value that is the sum of the individual loudnesses:

$$S = S_1 + S_2 + S_3 + \cdots . \tag{6.8}$$

3. If the frequency difference is very large, the summation becomes complicated. Listeners tend to focus primarily on one component (e.g., the loudest or the one of highest pitch) and assign a total loudness nearly equal to the loudness of that component (Roederer, 1975).

To determine the loudness in sones of a complex sound with many components, it is advisable to measure the sound level in each of the ten standard octave bands (or in ⅓-octave bands). Octave bands are frequency bands one octave wide (that is, the maximum frequency is twice the minimum frequency). Octave-band analyzers, available in many acoustics laboratories, usually have a filter that allows convenient measurement of the sound level in standard octave bands with center frequencies at 31, 63, 125, 250, 500, 1000, 2000, 4000,

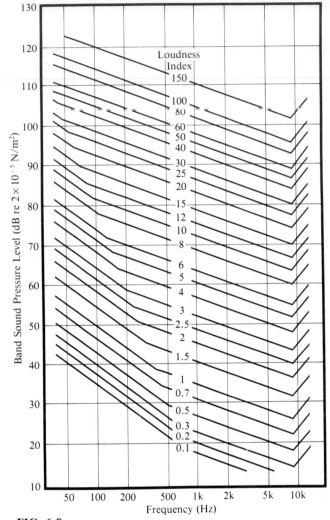

FIG. 6.9
Curves for determining loudness indexes (sometimes called
"Stevens's method").

8000, and 16,000 Hz. Once these levels have been measured,
the chart in Fig. 6.9 (from ISO Recommendation No. 532) can
be used to find the loudness in sones. The loudness index S_i
for each octave band can be determined from the chart. Then
the loudness in sones is determined by summing these S_i's
according to the formula

$$S = S_m + 0.3\Sigma S_i, \tag{6.9}$$

> where S_m is the greatest of the loudness indexes and ΣS_i is the sum of indexes for all the other bands. The formula tells us to add 100 percent of the loudest band plus 30 percent of all the others.

Is this seemingly complicated procedure necessary? For precise determination of loudness, yes. For estimating loudness, no. A pretty fair estimate of loudness can be made by using an ordinary sound level meter to measure the A-weighted sound level. To estimate the number of sones, let 30 dBA correspond to 1.5 sones and double the number of sones for each 10-dBA increase, as shown in Table 6.3. This procedure works quite well at low to moderate levels, because the A-weighting is a reasonable approximation to the frequency response of the ear.

TABLE 6.3   Chart for estimating loudness in sones of complex sounds from A-weighted sound levels

$L_p(A)$	30	40	50	55	60	65	70	75	80	85	90	dB
S	1.5	3	6	8	12	16	24	32	48	64	96	sones

Since the previous paragraphs have dealt with numbers, formulas, and graphs, it is appropriate to make a few comments on how they apply to music, environmental noise, and audiometric measurements. It should be emphasized that loudness is subjective, and its assessment varies from individual to individual. On the average, a sound of four sones sounds twice as loud as a sound of two sones, but some listeners may regard it as three times louder or one and a half times louder.

Interesting examples illustrating the importance of loudness phenomena in music appear throughout the literature. Roederer (1975, p. 86) discusses the selection of combinations of organ stops. Benade (1976, p. 244) describes how a saxophone was made to sound louder at the same sound pressure level by a change in timbre.

6.8   MUSICAL DYNAMICS AND LOUDNESS

Variations in loudness add excitement to music. The range of sound level in musical performance, known as the *dynamic range*, may vary from a few decibels to 40 dB or more, depending on the music (loud peaks and pauses may cause the instantaneous level to exceed this range). The approximate range of sound level and frequency heard by the music listener is shown in Fig. 6.10.

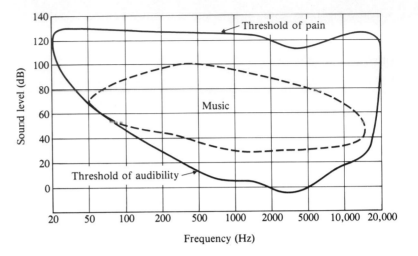

FIG. 6.10
Approximate range of frequency and sound level of music compared to the total range of hearing.

Composers use dynamic symbols to indicate the appropriate loudness to the performer. The six standard levels are shown in Table 6.4.

Measurements of sound intensity of a number of instrumentalists have shown that seldom do musical performers actually play at as many as six distinguishable dynamic levels, however. In one study, the dynamic ranges of 11 professional bassoonists were found to vary from 6 to 17 decibels with an average of 10 dB (Lehman, 1962). A 10-dB increase in sound level, you will recall from Section 6.4, is usually said to double the loudness (expressed in sones). Most listeners would have considerable difficulty identifying six different levels within a dynamic range of 10 dB. Dynamic ranges of several instruments are given in Table 6.5.

The dynamic ranges in Table 6.5 are for single notes played loudly and softly. Several instruments have much more sound power near the top of their playing range than near the bottom. (Fortissimo on a French horn, for example, is found to be nearly 30 dB greater at C_5

TABLE 6.4 Standard levels of musical dynamics

Name	Symbol	Meaning
fortissimo	ff	Very loud
forte	f	Loud
mezzo forte	mf	Moderately loud
mezzo piano	mp	Moderately soft
piano	p	Soft
pianissimo	pp	Very soft

TABLE 6.5   Dynamic ranges of musical instruments

Instrument	Average dynamic range (dB) (Clark and Luce, 1965)	Maximum dynamic range (dB) (Patterson, 1974)
Violin	14	40
Viola	16	
Cello	14	
String bass	14	30
Recorder		10
Flute	7	30
Oboe	7	
English horn	5	
Clarinet	8	45
Bassoon	10	40
Trumpet	9	
Trombone	17	38
French horn	18	
Tuba	13	

than at C_2, although the difference between ff and pp on any note of the scale may be 20 dB or less.)

Measurement of the dynamic ranges of various musical instruments and players is an instructive and relatively easy experiment for the reader to perform. The dynamic ranges of most players we have measured fall close to those reported by Luce and Clark.

6.9   MASKING

When the ear is exposed to two or more different tones, it is a common experience that one may mask the others. *Masking* is probably best explained as an upward shift in the hearing threshold of the weaker tone by the louder tone and depends on the frequencies of the two tones. Pure tones, complex sounds, narrow and broad bands of noise all show differences in their ability to mask other sounds. Masking of one sound can even be caused by another sound that occurs a split second after the masked sound.

Some interesting conclusions can be made from the many masking experiments that have been performed:

1. Pure tones close together in frequency mask each other more than tones widely separated in frequency.

2. A pure tone masks tones of higher frequency more effectively than tones of lower frequency.

3. The greater the intensity of the masking tone, the broader the range of frequencies it can mask.

4. If two tones are widely separated in frequency, little or no masking occurs.

5. Masking by a narrow band of noise shows many of the same features as masking by a pure tone; again tones of higher frequency are masked more effectively than tones of lower frequency than the masking noise.

6. Masking of tones by broadband ("white") noise shows an approximately linear relationship between masking and noise level (that is, increasing the noise level 10 dB raises the hearing threshold by the same amount). Broadband noise masks tones of all frequencies.

7. *Forward masking* refers to the masking of a tone by a sound that ends a short time (up to about 20 or 30 milliseconds) before the tone begins. Forward masking suggests that recently stimulated cells are not as sensitive as fully rested cells.

8. *Backward masking* refers to the masking of a tone by a sound that begins a few milliseconds later. A tone can be masked by noise that begins up to ten milliseconds later, although the amount of masking decreases as the time interval increases (Elliott, 1962). Backward masking apparently occurs at higher centers of processing where the later-occurring stimulus of greater intensity overtakes and interferes with the weaker stimulus.

9. Masking of a tone in one ear can be caused by noise in the other ear, under certain conditions; this is called *central masking*.

Some of the conclusions about masking just stated can be understood by considering the way in which pure tones excite the basilar membrane (see Fig. 5.5). High-frequency tones excite the basilar membrane near the oval window, whereas low-frequency tones create their greatest amplitude at the far end. The excitation due to a pure tone is asymmetrical, however, having a tail that extends toward the high-frequency end as shown in Fig. 6.11. Thus it is easier to mask a tone of higher frequency than one of lower frequency. As the intensity of the masking tone increases, a greater part of its tail has amplitude sufficient to mask tones of higher frequency.

The curves in Fig. 6.12 illustrate the masking effects of pure tones, narrowband noise, and broadband noise. In (a) and (b) the masking (in dB) is the increase in the sound level of a tone that makes it audible in the presence of the masking sound. For example, in the presence of a 400-Hz tone of 60 dB (the center curve in Fig. 6.12a), a 500-Hz tone

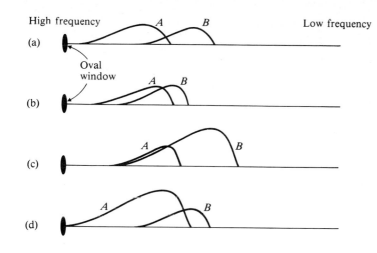

FIG. 6.11
Simplified response of the basilar membrane for two pure tones *A* and *B*. (a) The excitations barely overlap; little masking occurs. (b) There is an appreciable overlap; tone *B* masks tone *A* and somewhat more than the reverse. (c) The more intense tone *B* almost completely masks the higher-frequency tone *A*. (d) The more intense tone *A* does not completely mask the lower-frequency tone *B*.

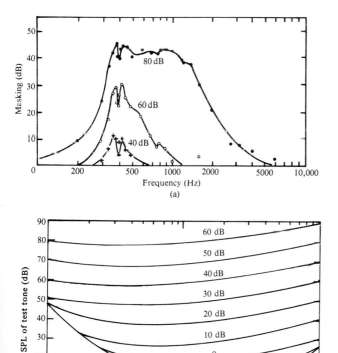

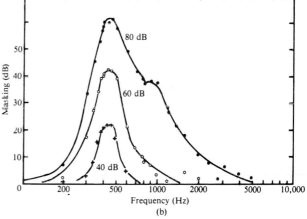

FIG. 6.12
(a) Masking by a pure tone (400 Hz). (b) Masking by narrowband noise (90 Hz wide, centered at 410 Hz). (c) Masking by broadband (white) noise. (Parts (a) and (b) from Egan and Hake, 1950; part (c) from Hawkins and Stevens, 1950. Reprinted by permission of the Am. Inst. of Physics.)

would require a sound level 22 dB above the normal threshold (see Fig. 6.4) to be barely audible, but in the presence of 60 dB of narrow-band noise centered on 410 Hz (the center curve in Fig. 6.12b), the threshold for the same 500-Hz tone is raised 40 dB.

Since the curves in Figs. 6.12 (a) and (b) are for the same subject, a comparison can be made of masking by a pure tone of 400 Hz and by narrowband noise centered on 410 Hz. Note that the masking due to the pure tone dips at 400 Hz due to interference effects or beats (see Section 8.2). Slight dips are also apparent at harmonics of the masking frequency (800 and 1200 Hz).

Masking by Broadband Noise

Masking by broadband noise is a little more complicated. In Fig. 6.12(c), the sound level of the test tone is the vertical axis. White noise has the same amount of energy in each interval of frequency. The noise levels of the masking (white) noise are given in dB per cycle (Hz); the readings on a sound level meter (C-weighting) would be about 40 dB greater for each curve. The sound level required for a test tone at any frequency to be audible is nearly equal to the effective sound level of the noise within a critical bandwidth at that frequency. With a noise level of 40 dB per cycle, for example, a 500-Hz test tone must have a level of about 58 dB to be audible (Fig. 6.12c). The effective sound level within the critical band centering on 500 Hz can be estimated by considering the noise as due to 110 sources, each producing a level of 4 dB, since the critical bandwidth at 500 Hz is 110 Hz (see Fig. 6.8). One hundred ten 40-dB sources within the critical band would give an effective sound level of $40 + 10 \log 110 = 60$ dB, which is close to the level observed. At 10,000 Hz the critical bandwidth is 1200 Hz, so the effective sound level is 71 dB for 40 dB/cycle noise.

6.10   LOUDNESS REDUCTION BY MASKING

Sounds are seldom heard in isolation. The presence of other sounds not only raises the threshold for hearing a given sound but generally reduces its loudness as well (this is sometimes called *partial masking*).

Figure 6.13 shows how white noise reduces the apparent loudness of a 1000-Hz tone. Compared to the tone in quiet (same curve as in Fig. 6.8), the loudness functions in white noise are steeper. Rising from an elevated threshold, the partially masked tone eventually comes to its full unmasked loudness when the noise level is less than 80 dB. In more intense noise, the loudness does not reach its full un-

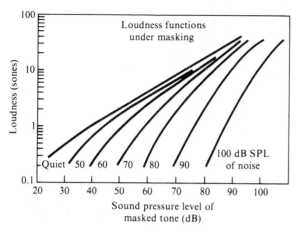

FIG. 6.13
Loudness functions for a 1000-Hz tone
partially masked by white noise at
various sound pressure levels. Subjects
adjusted the level of the tone in quiet
so that it sounded as loud as the tone
with noise (From Scharf, "Loudness"
in *Handbook of Perception,* vol. 4, ed.
E. C. Canteretta & M. P. Friedman,
NY, © 1978 by Academic Press.
Reprinted by permission.)

masked value, but the function approaches the same slope as the function without masking noise (Scharf and Houtsma, 1986).

6.11   LOUDNESS AND DURATION: IMPULSIVE SOUNDS AND ADAPTATION

How does the loudness of an impulsive sound compare to the loudness of a steady sound at the same sound level? Numerous experiments have pretty well established that the ear averages sound energy over about 0.2 s (200 ms), so loudness grows with duration up to this value. Stated another way, loudness level increases by 10 dB when the duration is increased by a factor of 10. The loudness level of broadband noise seems to depend somewhat more strongly on stimulus duration than the loudness level of pure tones, however. Figure 6.14 shows the approximate way in which loudness level changes with duration.

The ear, being a very sensitive receiver of sounds, needs some protection to avoid injury by very loud sounds. Up to 20 dB of effective protection is provided by muscles attached to the eardrum and the ossicles of the middle ear. When the ear is exposed to sounds in excess of 85 dB or so, these muscles tighten the ossicular chain and pull the stapes (stirrup-shaped bone) away from the oval window of the cochlea. This action is termed the *acoustic reflex.*

Unfortunately the reflex does not begin until 30 or 40 ms after the sound overload occurs, and full protection does not occur for another 150 ms or so. In the case of a loud impulsive sound (such as an explosion or gunshot), this is too late to prevent injury to the ear. In fact a tone of 100 dB or so preceding the loud impulse has been proposed as a way of triggering the acoustic reflex to protect the ear (Ward, 1962). It is interesting to speculate what type of protective mechanism, analogous to eyelids, might have developed in the audi-

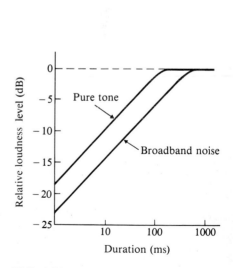

FIG. 6.14
Variation of loudness level with
duration. (After Zwislocki, 1969).

tory system had the loud sounds of the modern world existed for millions of years (earlids, perhaps?).

Like most other sensations, loudness might be expected to decrease with prolonged stimulation. Such a decrease is called *adaptation*. Under most listening conditions, however, loudness adaptation appears to be very small. A steady 1000-Hz tone at 50 dB causes little adaptation, although the loudness of a tone that alternates between 40 and 60 dB appears to decrease in loudness over the first two or three minutes, as do tones within about 30 dB of threshold (Scharf and Houtsma, 1986).

Exposure to a loud sound affects our ability to hear another sound at a later time. This is called *fatigue* and may result in both a temporary loudness shift (TLS) and a temporary threshold shift (TTS). TLS and TTS appear to be greatest at a frequency a half octave higher than that of the fatiguing sound. Noise-induced TTS is discussed in Chapter 31.

6.12   SUMMARY

Each of the quantities sound pressure, sound power, and sound intensity has an appropriate decibel level that expresses the ratio of these quantities in relation to a reference level. Sound pressure can be measured directly by a sound level meter, which may offer one to three different frequency weightings. Loudness level (in phons) expresses the sound pressure level of an equally loud 1000-Hz tone, whereas the loudness (in sones) expresses a subjective rating of loudness. Expressing the loudness of complex tones is fairly subtle, involving critical bandwidth, masking one tone by another, etc. The loudness of impulsive sounds increases with their duration up to about 0.2 s. The dynamic range of music covers about 40 dB, although individual instruments have dynamic ranges considerably less than this. Composers use six standard levels to indicate loudness. The ear is partially protected from loud sounds by the acoustic reflex.

References and Suggested Readings

Benade, A. H. (1976). *Fundamentals of Musical Acoustics*. New York: Oxford.

Clark, M., and D. Luce (1965). "Intensities of Orchestral Instrument Scales Played at Prescribed Dynamic Markings," *J. Audio Eng. Soc.* **13**: 151.

Egan, J. P., and H. W. Hake (1950). "On the Masking Pattern of a Simple Auditory Stimulus," *J. Acoust. Soc. Am.* **22**: 622.

Elliott, L. L. (1962). "Backward and Forward Masking of Probe Tones of Different Frequencies," *J. Acoust. Soc. Am.* **34**: 1116.

Fletcher, H., and W. A. Munson (1933). "Loudness, Definition, Measurement and Calculation," *J. Acoust. Soc. Am.* **6:** 59.

Hawkins, J. E., Jr., and S. S. Stevens (1950). "Masking of Pure Tones and Speech by Noise," *J. Acoust. Soc. Am.* **22:** 6.

Houtsma, A. J. M., T. D. Rossing, and W. M. Wagenaars (1988). *Auditory Demonstrations* (Philips Compact Disc #1126-061 and text). Woodbury, N.Y.: Acoustical Society of America.

Howes, W. L. (1974). "Loudness Function Derived from Data on Electrical Discharge Rates in Auditory Nerve Fibers," *Acustica* **30:** 247.

International Standards Organization, Standards R131, 26, R357 and R454 (International Organization for Standardization, 1 Rue de Narembé, Geneva, Switzerland).

Lehman, P. R. (1962). "The Harmonic Structure of the Tone of the Bassoon," Ph.D. thesis, University of Michigan, Ann Arbor.

Littler, T. S. (1965). *The Physics of the Ear.* New York: Macmillan.

Patterson, B. (1974). "Musical Dynamics," *Sci. Am.* **231**(5): 78.

Plomp, R. (1976). *Aspects of Tone Sensation.* London: Academic.

Roederer, J. G. (1975). *Introduction to the Physics and Psychophysics of Music,* 2nd ed. New York: Springer-Verlag.

Scharf, B. (1978). "Loudness," in *Handbook of Perception,* Vol 4. Ed., E. C. Carterette and M. P. Friedman. New York, Academic Press.

Scharf, B., and D. Fishken (1970). "Binaural Summation of Loudness," *J. Experimental Psych.* **86:** 374–79.

Scharf, B. and A. J. M. Houtsma (1986). "Audition II. Loudness, Pitch, Localization, Aural Distortion, Pathology," in *Handbook of Perception and Human Performance.* Ed., K. R. Boff, L. Kaufman, and J. P. Thomas. New York: J. Wiley.

Stevens, S. S., and G. Stevens (1975). *Psychophysics Introduction to its Perceptual, Neural and Social Prospects.* New York: Wiley.

Ward, W. D. (1962). "Studies on the Aural Reflex III. Reflex Latency as Inferred from Reduction of Temporary Threshold Shift from Impulses," *J. Acoust. Soc. Am.* **34:** 1132.

Warren, R. M. (1970). "Elimination of Biases in Loudness Judgements for Tones," *J. Acoust. Soc. Am.* **48:** 1397.

Zwislocki, J. J. (1969). "Temporal Summation of Loudness: An Analysis," *J. Acoust. Soc. Am.* **46:** 431.

Glossary

acoustic reflex Muscular action that reduces the sensitivity of the ear when a loud sound occurs.

critical bandwidth The frequency bandwidth beyond which subjective loudness increases with bandwidth (see also definition in chapter 5).

decibel A dimensionless unit used to compare the ratio of two quantities (such as sound pressure, power, or intensity), or to express the ratio of one such quantity to an appropriate reference.

intensity Power per unit area; rate of energy flow.

intensity level $L_I = 10 \log I/I_0$, where I is intensity and $I_0 = 10^{-12}$ W/m^3 (abbreviated IL or L_I).

loudness Subjective assessment of the "strength" of a sound, which depends on its pressure, frequency, and timbre; loudness may be expressed in sones.

loudness level Sound pressure level of a 1000-Hz tone that sounds equally loud when compared to the tone in question; loudness level is expressed in phons.

masking The obscuring of one sound by another.

phon A dimensionless unit used to measure loudness level; for a tone of 1000 Hz, the loudness level in phons equals the sound pressure level in decibels.

sone A unit used to express subjective loudness; doubling the number of sones should describe a sound twice as loud.

sound power level $L_w = 10 \log W/W_0$, where W is sound power and $W_0 = 10^{-12}$ W (abbreviated PWL or L_W).

sound pressure level $L_p = 20 \log p/p_0$, where p is sound pressure and $p_0 = 2 \times 10^{-5}$ N/m^2 (or 20 micropascals) (abbreviated SPL or L_p).

white noise Noise whose amplitude is constant throughout the audible frequency range.

Questions for Discussion

1. Which will sound louder, a pure tone of $L_p = 40$ dB, $f = 2000$ Hz, or a pure tone of $L_p = 65$ dB, $f = 50$ Hz?

2. If two identical loudspeakers are driven at the same power level by an amplifier, how will the sound levels due to each combine? Does it make a difference whether the program source is stereophonic or monophonic?

3. How is it possible for one sound to mask a sound that has already occurred (backward masking)? Speculate what might happen in the human nervous system when such a phenomenon occurs.

4. How long must a burst of broadband (white) noise be in order to be half as loud as a continuous noise of the same type?

Problems

1. What sound pressure level is required to produce minimum audible field at 50, 100, 500, 1000, 5000, and 10,000 Hz?

2. What sound pressure level of 100-Hz tone is necessary to match the loudness of a 3000-Hz tone with $L_p = 30$ dB? What is the loudness level (in phons) of each of these tones?

3. With one violin playing, the sound level at a certain place is measured as 50 dB. If four violins play equally loudly, what will the sound level most likely be at this same place?

4. If two sounds differ in level by 46 dB, what is the ratio of their sound pressures? their intensities?

5. A loudspeaker is supplied with five watts of electrical power, and it has an efficiency of 10 percent in converting this to sound power. What is its sound power level? If we assume that the sound radiates equally in all directions, what is the sound pressure level at a distance of one meter? at a distance of four meters?

6. A 60-Hz tone has a sound pressure level of 60 dB measured with C-weighting on a sound level meter. What level would be measured with A-weighting?

7. Using the graphs in Figs. 6.4 and 6.6, calculate the loudness level in phons and loudness in sones for a 100-Hz tone with $L_p = 50$ dB and a 400-Hz tone with $L_p = 80$ dB.

8. What A-weighted sound level would be obtained on a sound level meter for each of the tones in Problem 7 (see Fig. 6.5)? Compare the loudness estimate by using Table 6.3 with the values obtained in Problem 7.

9. Find the sound pressure and the intensity of a sound with $L_p = 50$ dB.

CHAPTER 7
Pitch and Timbre

Pitch has been defined as that characteristic of a sound that makes it sound high or low or that determines its position on a scale. For a pure tone, the pitch is determined mainly by the frequency, although the pitch of a pure tone may also change with sound level. The pitch of complex sounds also depends on the spectrum (timbre) of the sound and its duration. In fact, the pitch of complex sounds has been one of the most interesting objects of study in psychoacoustics for quite a few years.

7.1 PITCH SCALES

The American National Standards Institute (1960) defines *pitch* as "that attribute of auditory sensation in terms of which sounds may be ordered on a scale extending from low to high." This definition probably leads most of us to think of a musical scale. Are there other pitch scales besides musical scales? Is there a subjective scale of pitch similar to the sone scale of loudness discussed in the previous chapter?

Pitch is a subjective sensation. Two persons hearing the same sound may assign it different positions on a pitch scale. In fact, some listeners may assign a different pitch to a sound depending upon whether it is presented to the right or left ear (this is called "binaural diplacusis").

The basic unit in most musical scales is the *octave*. Notes judged an octave apart have frequencies nearly (but not always exactly, as we will see) in the ratio 2:1. As early as the sixth century BC, according to legend, Pythagoras of Athens noted that if one segment of a string is half as long as the other, the pitches produced by plucking the two segments have a special similarity. Errors of one octave are frequently made in judging the pitch of a musical note (if you don't believe this, ask a musician to whistle a note and then to name the octave in which the note lies).

In music, the octave is subdivided in different ways, as we shall see in Chapter 9. Western music normally divides the octave into 12 intervals called *semitones;* these are given note names (A through G with sharps and flats) and designated on musical staves.

Psychophysical Pitch Scales

Various attempts have been made to establish a psychophysical pitch scale. If an average listener were allowed to listen to a tone of 4000 Hz followed by a tone of low frequency and then asked to tune an oscillator to a pitch halfway between, a likely choice would be something around 1000 Hz. On a scale of pitch, then, 1000 Hz is judged as halfway between 0 and 4000 Hz. The unit used for subjective pitch is the *mel;* the scale is arranged so that doubling the number of mels doubles the pitch. From 0 to 2400 mels spans the frequency range

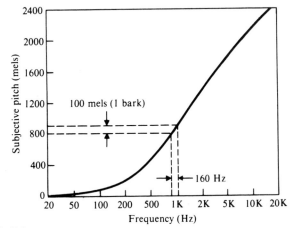

FIG. 7.1
Pitch scale vs. frequency scale. On the pitch scale, 100 mels is close to the width of the critical band, which is 160 Hz at a center frequency of 1000 Hz (dashed lines). (From Wightman and Green, 1974. Reprinted by permission of the Am. Inst. of Physics.)

0 to 16 kHz; the correspondance between mels and hertz is shown in Fig. 7.1.

Another psychophysical scale is based on critical bands of hearing. A critical bandwidth is designated "one *bark*." Interestingly enough, it turns out that one bark is very nearly equal to 100 mels, so the two scales are actually quite similar.

A numerical scale of pitch (in mels) is not nearly so useful as a numerical scale of loudness (in sones), however. Pitch is usually related to a musical scale where the octave, rather than the critical bandwidth, is the "natural" pitch interval.

7.2   PITCH DISCRIMINATION

The ability to distinguish between two nearly equal stimuli is often characterized, in psychophysical studies, by a *difference limen* or *just noticeable difference* (jnd). Two stimuli will be judged "the same" if they differ by less than the jnd.

The jnd for pitch has been found to depend on the frequency, the sound level, the duration of the tone, and the suddenness of the frequency change. It also depends on the musical training of the listener and to some extent on the method of measurement. Figure 7.2 shows the average jnd (of four subjects) for pure tones at a sound level of 80 dB. From 1000 to 4000 Hz, the jnd is approximately 0.5 percent of the pure tone frequency, which is about one-twelfth of a semitone. Sometimes the term *frequency resolution* is used to denote the jnd divided by the frequency ($\Delta f / f$).

By comparing the upper and lower curves in Fig. 7.2, we can see that critical bandwidth is roughly equal to 30 difference limens or jnd's at all center frequencies. This remarkable result suggests that the same mechanism in the ear is responsible for critical bands and for pitch discrimination. It is quite likely related to regions of excitation along the basilar membrane (see Section 5.4).

It is interesting to compare pitch discrimination to color discrimination. Whereas the visible spectrum extends over one octave (violet light has roughly twice the frequency of red) and includes 128 just noticeable differences (distinguishable colors), the auditory spectrum covers about 10 octaves with 5000 jnd's.

7.3   PITCH OF PURE TONES

We have already noted that pitch depends mainly on frequency; pitch scaling with frequency was discussed in Section 7.1. We now consider the pitch dependence of pure tones on other physical quantities such

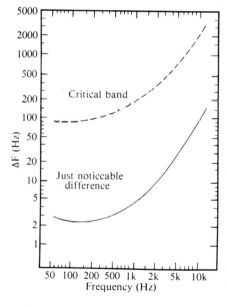

FIG. 7.2
Just-noticeable difference (jnd) in frequency determined by modulating the frequency of a tone at 4 Hz. Note that the jnd at each frequency is nearly a constant percentage of the critical bandwidth. (From Zwicker, Flottorp, and Stevens, 1957).

as sound pressure, duration, envelope, and the presence of other sounds.

Pitch and Sound Level

Early experiments on pitch versus sound level reported substantially larger pitch dependence on sound level than more recent studies do. Early work by Stevens (1935) indicated shifts in pitch as large as two semitones (apparent frequency changes of 12 percent) as the sound level of pure tones increased from 40 to 90 dB. Tones of low frequency were found to fall in pitch with increasing intensity; tones of high frequency rise in pitch with increasing intensity, and tones of middle frequency (1–2 kHz) show little change. (This has sometimes been referred to as "Stevens's rule.") Stevens found the maximum downward shift with sound level at 150 Hz and the largest upward shift with sound level around 8000 Hz.

It now appears that the effect is small, even for pure tones, and varies from observer to observer; in one experiment, for example, five musically trained subjects hear pitch lowerings that varied from 0 to 75 cents (75 cents = ¾ semitone) when a 250-Hz tone increased from 40 to 90 dB (Ward, 1970). Whereas pitch changes for individuals tend to follow Stevens's rule, then, averaging over a group of observers makes the changes insignificant. Figure 7.3 shows the pitch shifts of pure tones with frequencies from 200 Hz to 6000 Hz averaged over 15 subjects.

The small pitch changes shown in Fig. 7.3, as well as the larger changes described by early investigators, are for pure tones. Less is known about the effect for complex tones. Studies with musical instruments have generally shown very small pitch change with intensity (around 17 cents for an increase from 65 to 95 dB, for example). Whether the pitch of a complex tone rises or falls with increasing intensity appears to depend on which partials (above or below 1000 Hz) are predominant (Terhardt, 1979).

FIG 7.3
Pitch shift of pure tones as a function of sound pressure level. Shifts are shown in both percent and cents (100 cents = 1 semitone). The curves are based on data from 15 subjects. (After Terhardt, 1979).

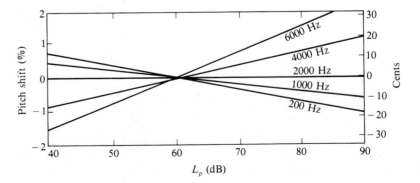

In contrast with the results in Fig. 7.3, however, increasing the amplitude of short tone bursts causes a downward shift in pitch over a wide range of frequency. Similar results are found in experiments using 12-ms bursts (Doughty and Garner, 1948) and 40-ms bursts (Rossing and Houtsma, 1986).

A phenomenon of pitch change that has been observed during reverberant decay may be due in part to pitch change with sound level, although other effects appear to contribute as well. This phenomenon is quite apparent when one is listening to pipe organ music in churches with substantial reverberation; the pitch often appears to rise as the sound level diminishes after a loud chord ends (Parkin, 1974).

It is fortunate for performing musicians and listeners alike that the change in pitch with sound level for complex tones is much less than was reported from early experiments with pure tones. Musical performance would be very difficult if substantial changes of pitch occurred during changes in dynamic level.

Pitch and Duration

How long must a tone be heard in order to have an identifiable pitch? Although early experiments by Savart (1840) indicated that a sense of pitch develops after only two cycles, most subsequent investigations indicated that a longer duration is required (see Fig. 7.4). Very brief tones are described as "clicks," but as the tones lengthen, the clicks take on a weak sensation of pitch, increasing in strength upon further lengthening.

The transition from "click" to "tone" depends on sound level; if the tone does not begin abruptly, but rather with a soft onset, tone recognition times as short as 3 ms are possible, which is shorter than the attack time of most musical instruments (Winckel, 1967).

It has been suggested that the dependence of pitch on duration follows a sort of "acoustical uncertainty principle" $\Delta f \Delta t = K$, where Δf is the uncertainty in frequency and Δt is the duration of a tone burst. Under optimum conditions, K can be less than 0.1 (Majerník and Kalužný, 1979). When the tone duration falls below 25 ms, the pitch may appear to change, although slightly different results are reported by various investigators (Rossing and Houtsma, 1986).

The ear has an especially high sensitivity for detecting frequency changes of pure tones. The jnd for frequency change in pure tones Δt is less than for noise, provided that the amplitude of the pure tone remains constant. Even with a band of noise as narrow as 10 Hz at a center frequency of 1500 Hz (which sounds like a pure tone of varying amplitude), Δt will be six times greater than for a pure tone of 1500 Hz (Zwicker, 1962).

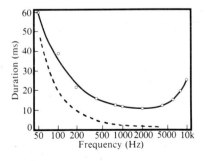

FIG. 7.4

The duration required for a given tone to produce a definite pitch. The solid line is from the data of Bürck et al. (1935); the dashed line is the duration of two cycles (Savart, 1840).

Pitch and Envelope

The perceived pitch of a short exponentially decaying sinusoidal tone is found to be consistently higher than a simply gated sine tone with the same frequency and energy (Hartmann, 1978). Rossing and Houtsma (1986) found the same effect for tones with rising exponential envelopes, and found that the pitch shift depends on the sound pressure level as well as the rate of rise or fall of the tone envelope.

The reason for the envelope dependence of pitch is not clear, but it appears to be related to the pitch shift with intensity discussed earlier in this section. It certainly is an effect that musicians should take into account when dealing with the pitch of percussion instruments.

Effect of Interfering Sounds

Sounds are seldom heard in isolation. Another factor that influences the pitch of a pure tone is the presence of other interfering sounds. Experiments both with a second interfering tone and with interfering noise can be summarized as follows:

1. If the interfering tone has a frequency below that of the test tone, an *upward* shift always occurs.

2. If the interfering tone frequency is above that of the test tone, a *downward* shift is observed at low frequencies.

3. Interfering noise always causes an upward pitch shift if it has a lower frequency than the test tone (but if it has a high frequency, the shift can occur in either direction).

4. The pitch shift increases with the amount by which the interfering tone or noise amplitude exceeds that of the test tone (Terhardt and Fastl, 1971).

7.4   PITCH OF COMPLEX TONES: VIRTUAL PITCH

When the ear is presented with a tone composed of exact harmonics, it is easy to predict what pitch will be heard. It is simply the lowest common factor in these frequencies, which is the fundamental. The ear identifies the pitch of the fundamental, even if the fundamental is very weak or missing altogether. For example, if the ear hears a tone having partials with frequencies of 600, 800, 1000, and 1200 Hz, the pitch will nearly always be identified as that of a 200-Hz tone, the "missing fundamental." This is an example of what is called *virtual pitch,* since the pitch does not correspond to any partial in the complex tone. The ability of the ear to determine a virtual pitch makes it possible for the undersized loudspeaker of a portable radio to produce bass tones, and also forms the basis for certain mixture stops on a pipe organ.

If a strong fundamental is not essential for perceiving the pitch of a musical tone, the question arises as to which harmonics are most important. Experiments have shown that for a complex tone with a fundamental frequency up to about 200 Hz the pitch is mainly determined by the fourth and fifth harmonics. As the fundamental frequency increases, the number of the dominant harmonics decreases, reaching the fundamental itself for $f_0 = 2500$ Hz and above (Plomp, 1976). Consider, for example, a tone A_3 with a frequency $f_0 = 220$ Hz; if the fourth and fifth harmonics were raised in frequency, the pitch of the tone would most likely appear to rise even though the fundamental remained at 220 Hz.

When the partials of the complex tone are not harmonic, however, the determination of virtual pitch is more subtle. According to recent theories of pitch, the ear picks out a series of nearly harmonic partials somewhere near the center of the audible range, and determines the pitch to be the largest near-common factor in the series (Goldstein, 1973). Several demonstrations of virtual pitch are presented by Houtsma, Rossing, and Wagenaars (1987).

Musical examples of the ability of the auditory system to formulate a virtual pitch from "near harmonics" in a complex tone are the sounds of bells and chimes. In each case the pitch of the "strike tone" is determined mainly by three partials that have frequencies almost in the ratio 2:3:4, as we shall see in Chapter 13. In the case of the bell, there is usually another partial with a frequency near that of the strike tone, which reinforces it. In the case of the chime, however, there is none: The pitch is purely subjective.

7.5 SEEBECK'S SIREN AND OHM'S LAW: A HISTORICAL NOTE

About the middle of the eighteenth century, A. Seebeck performed a series of experiments on pitch perception that produced some significant, if surprising, results. As a source of sound, Seebeck used a siren consisting of a rotating disc with periodically spaced holes that created puffs of compressed air at regular intervals, as shown in Fig. 7.5(a). Seebeck noted that this siren produced sound with a very strong pitch corresponding to the time between puffs of air. Doubling the number of holes, as shown in Fig. 7.5(b), raised the pitch exactly an octave, as expected.

However, using a disc with unequal spacing of holes, as shown in Fig. 7.5(c), produced an unexpected result: The pitch now heard matched that heard with the siren in (a). This may be understood by

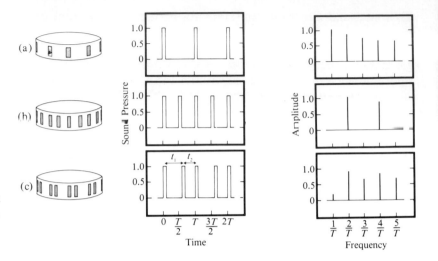

FIG. 7.5

Three different sirens used by Seebeck along with the waveforms and spectra of sound they generate. (After Wightman and Green, 1974.)

studying the corresponding waveforms (amplitude vs. time) and spectra (amplitude vs. frequency) shown in Fig. 7.5. In (a) the spectrum has components at the fundamental frequency $1/T$ (where T is the period) and its harmonics ($2/T$, $3/T$, etc.). In (b) the fundamental frequency is twice as great ($2/T$), and the harmonics occur at $4/T$, $6/T$, etc.; thus, the pitch is an octave higher. In (c) where the spacing between puffs is alternately t_1 and t_2, the period of repetition is $T = t_1 + t_2$; thus, harmonics occur at the same frequencies as in (a), although the fundamental is very much weaker. The pitch therefore matches that of case (a), although the quality or timbre of the sound is quite different.

It is quite easy and instructive to repeat Seebeck's experiment using an electronic pulse generator to generate the waveforms shown in Fig. 7.5. What one hears in the case of waveform (c) is two tones an octave apart, the lower tone becoming softer as $t_2 \to t_1$, disappearing rather abruptly when $t_2 \doteq t_1$, whereas the upper tone remains relatively constant in loudness.

About the time Seebeck was performing his experiment, G. S. Ohm adapted Fourier's theorem on spectrum analysis (see Section 7.10) to acoustics and formulated what is often known as "Ohm's acoustical law" (or "Ohm's second law," his "first" law having dealt with electrical circuits). Ohm believed that a pitch corresponding to a certain frequency could be heard only if the acoustic wave contained power at that frequency. Thus he criticized Seebeck's interpretation of his siren experiment that periodicity, rather than fundamental frequency, determines pitch. In the case of the waveform shown in Fig. 7.5(c), however, the sensation of pitch is far too strong to be explained on the basis of the weak component or partial at the fundamental frequency, and thus Ohm's law is contradicted. Ohm finally suggested that the phenomenon was due to an acoustical illusion (Wightman and Green, 1974).

In his monumental work *On the Sensations of Tone as a Physiological Basis for the Theory of Music,* H. von Helmholtz (1877) supported Ohm's position, adding the important idea of distortion products generated in the ear. For pure tones, these distortion products would be primarily harmonics of the pure tone (harmonic distortion). For the waveforms shown in Fig. 7.5(a) and (c), however, distortion would produce sum and difference tones, resulting in the generation of a strong fundamental, since difference tones between all the adjacent partials would be at this frequency.

Experiments with filtered sound by H. Fletcher (1934) and others appeared to support Helmholtz. When the lower harmonics of a complex tone are filtered out, the pitch remains the same. This phenomenon can be demonstrated by recording the sound of a musical instrument and playing it back through a high-pass filter to remove the fundamental (and even the lower harmonics). The missing fundamental is supplied by the ear of the listener.

7.6 THEORIES OF PITCH: PLACE PITCH vs. PERIODICITY PITCH

Two major theories of pitch perception have gradually developed on the basis of numerous experiments in many different laboratories. They are usually referred to as the place (or frequency) theory and the periodicity (or time) theory. Before discussing these theories, let us briefly review the relationship between frequency and period.

A periodic waveform is one that repeats itself after a certain interval of time, called the period T. The reciprocal of the period is the fundamental frequency f_1. If the waveform is complex, it can be resolved into a spectrum of partials with frequencies $2f_1$, $3f_1$, etc., called the harmonics (see Section 2.7). A periodic waveform need not have energy at its fundamental frequency f_1, as will become apparent later in this chapter. In a pulse waveform, the fundamental *frequency* is not necessarily the same as the pulse *rate*. The waveform in Fig. 7.5(c), for example, has $2/T$ pulses per second, although its fundamental frequency $f_1 = 1/T$ is only half as great. In determining pitch, the ear apparently performs *both a time analysis and a frequency analysis* of the sound wave and reaches its final decision after a considerable amount of computation!

The idea that vibrations of different frequencies excite resonant areas on the basilar membrane is often referred to as the *place theory* of hearing. According to this theory, the cochlea converts a vibration in time into a vibration pattern in space (along the basilar membrane), and this in turn excites a spatial pattern of neural activity. The place theory explains many aspects of auditory perception but fails to explain others.

Helmholtz regarded the basilar membrane as a frequency analyzer, with transverse fibers "tuned" to resonate at frequencies determined by their length, mass, and tension. A complex wave of sound pressure would excite regions of the basilar membrane corresponding to the frequencies of its components or partials, the higher frequencies acting on regions near the oval window, and the lower frequencies acting closer to the far end where the membrane is thick and loose. (Helmholtz was nearly correct; later investigations showed that the individual fibers are not free to resonate, but the membrane as a whole can create the effect of resonances.)

In his experiments with cochleas removed from human cadavers, Békésy provided support for the place theory of pitch perception. By ingenious and careful experiments, he directly observed wavelike motions of the basilar membrane caused by sound stimulation. Just as Helmholtz had suggested, the place of maximum vibration moved up and down the basilar membrane as the frequency of the sound wave changed (see Figs. 5.5 and 5.7).

More recent experiments have pointed to limitations in the place theory of pitch perception, however. One difficulty is in explaining fine frequency discrimination. In order to respond to rapid changes in frequency, a resonator must have considerable damping. But damping decreases selectivity, that is, the ability to discriminate between small differences in frequency. Another difficulty arises in attempting to explain why we hear a complex tone as one entity with a single pitch.

According to the *periodicity theory* of pitch, the ear performs a *time* analysis of the sound wave. Presumably the time distribution of the electrical impulses carried by the auditory nerve has encoded into it information about the time distribution of the sound wave. This information is decoded by a process called *autocorrelation* (to be discussed in Section 8.13) in the central nervous system.

In the late 1930s, J. F. Schouten and his colleagues in the Netherlands performed experiments that supported the periodicity theory of pitch. Schouten studied stimuli, such as those shown in Fig. 7.6, in which the pitch corresponds to the repetition rate of the pulses, 200 Hz. In the waveform shown in Fig. 7.6(b), the fundamental component has been canceled out by addition of an out-of-phase signal of 200 Hz; the pitch remains unchanged at 200 Hz, the frequency of the "missing fundamental." Schouten then added a pure tone of 206 Hz. If a distortion product of 200 Hz were actually present in the ear, as suggested by the hypothesis of Helmholtz, beats should be heard at a frequency of six per second. No beats were heard.

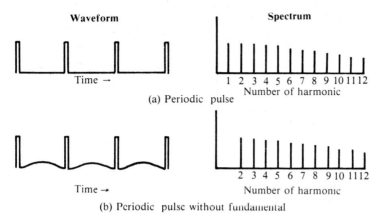

Waveform   **Spectrum**

1 2 3 4 5 6 7 8 9 10 11 12
Number of harmonic

Time →

(a) Periodic pulse

2 3 4 5 6 7 8 9 10 11 12
Number of harmonic

Time →

(b) Periodic pulse without fundamental

FIG. 7.6
Cancellation of the fundamental frequency of a complex signal. Part (a) shows a periodic pulse train and its spectrum. By appropriate adjustment of phase and amplitude, the fundamental may be canceled as shown in (b). In both cases, however, the pitch of the signal corresponds to the fundamental. (After Schouten, 1940).

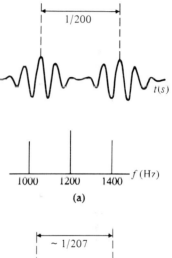

(a)

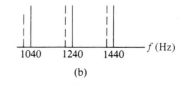

(b)

FIG. 7.7
Waveforms for pitch-shift experiments of the Schouten type: (a) carrier of 1200 Hz modulated at 200 Hz; (b) carrier at 1240 Hz modulated at 200 Hz.

Schouten continued his experiments with a type now called pitch shift experiments. Using amplitude modulation, he produced complex waveforms in which the frequencies of individual components could be shifted by the same amount, thus leaving the spacing between components undisturbed. For instance, a carrier frequency of 1200 Hz modulated by a 200-Hz signal produces components at 1000 Hz and 1400 Hz (called "sidebands") along with the 1200-Hz component. Such a waveform, shown in Fig. 7.7, has a clear pitch of 200 Hz. If the carrier frequency is changed to 1240 Hz, however, the components are shifted to 1040, 1240, and 1440 Hz. The pitch is now found to shift to about 207 Hz, even though the difference frequency remains at 200 Hz. This experiment can be repeated in the laboratory using a generator with provision for amplitude modulation or an electronic music synthesizer. (See Demonstration 21, Houtsma, Rossing, and Wagenaars, 1987).

The virtual pitch can be estimated by dividing the component frequencies by successive integers 5, 6 and 7 to obtain a "nearly common factor." In this case $1040/5 = 208$, $1240/6 = 206.7$ and $1440/7 = 205.7$. Averaging these three factors together gives 207 Hz, which the auditory system accepts as the frequency of the "missing fundamental." Using 4, 5, and 6 or 6, 7, and 8 leads to less consistent trial factors, so the auditory system prefers the 207-Hz factor.

Schouten explained the pitch-shift phenomena as due to synchronous firing in the auditory nerve due to an unresolved "residue" of high-frequency components. These components, too close in frequency to be resolved on the basilar membrane, retain the periodicity of the original tone envelope. Schouten's residue theory of pitch provided a reasonable alternative to the distortion hypothesis of Helmholtz, but subsequent experiments (e.g., Plomp (1967), Ritsma (1967)) showed that the pitch of complex tones is determined by the low-frequency (resolved) components rather than by the high-frequency (unresolved) residue. An excellent historical review of the subject is given by Plomp (1967).

The importance of some sort of *central* pitch processor in the nervous system was illustrated by experiments in which a single harmonic of a missing fundamental was presented to one ear and a different harmonic to the other ear (Houtsma and Goldstein, 1972). The resulting virtual pitch heard this way (*dichotic* presentation) appeared to be as strong as when both harmonics were presented to the same ear (*monotic* presentation). In both monotic and dichotic presentations, the virtual pitch tends to deteriorate with increasing harmonic number.

Modern Theories of Pitch

Modern theories of pitch, given such names as "optimum processor theory" (Goldstein, 1973), "virtual pitch theory" (Terhardt, 1974), and "pattern transformation theory" (Wightman, 1973), describe how the ear-brain processor determines the pitch of complex tones. Each of them has attractive features. A detailed discussion of them is beyond the scope of this book.

Quite a few experiments have been conducted to evaluate the predictions of these theories (references are given in Scharf and Houtsma (1986) and in Houtsma and Rossing (1987)). Some of these experiments compare the observed pitch shifts in the complex tone with those observed in the partials due to masking noise, amplitude envelope change, intensity change, etc. Others compare complex tones made up of high and low partials, partials of unequal amplitude, etc. The general conclusion appears to be that none of the current pitch theories is completely successful in explaining all the experiments.

One might correctly conclude from the foregoing discussion that both the place and periodicity theories of pitch have validity. Clues from both frequency and time analyses of the sound are used to determine pitch, although one or the other may predominate under certain conditions. For low-frequency tones, the time (periodicity) analysis appears to be more important, whereas at high frequencies, the frequency analysis in the basilar membrane ("place clues") plays a more important role. The relative importance of each type of clue and the frequency range over which the clues predominate are still under study.

Repetition Pitch: A Demonstration of Periodicity Pitch

In 1693, astronomer Christiaan Huygens, standing at the foot of a staircase at the castle at Chantilly de la Cour in France, noticed that sound from a nearby fountain produced a certain pitch. He correctly concluded that the pitch was caused by periodic reflections of the sound against the steps of the staircase. *Repetition pitch* due to interference between noise and its delayed repetition is discussed by Bilsen and Ritsma (1969/70), who describe several historical examples, including those shown in Fig. 7.8.

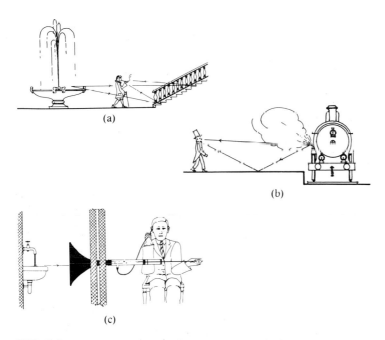

FIG. 7.8
Examples of repetition pitch (from Bilsen and Ritsma, 1969/70: (a) Huygens (1693) observed periodic reflections of the noise of a fountain against the steps of a staircase; (b) Minnaert (1941) observed the interference of the hissing sound from a locomotive with its reflection from a platform; (c) Hermann (1912) observed interference between the noise of running water and its reflection in a tube of adjustable length.

Repetition pitch can be demonstrated in a number of ways, including the two shown in Fig. 7.9. In both cases, broadband noise is combined with identical noise that has traveled a distance L farther, and thus is delayed by a time $T = L/v$, where v is the speed of sound. The perceived pitch corresponds to a frequency $f = 1/T = v/L$ and is easy to identify for time delays of 1 to 7 ms. Some blind persons make use of this phenomenon to locate obstructions by observing the interference between the direct and reflected sound.

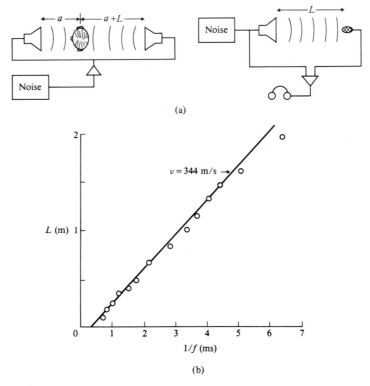

(a)

(b)

FIG. 7.9
(a) Two ways to demonstrate repetition pitch by combining noise with similar noise delayed by time $T = L/v$; (b) Pitch change with time delay using the second arrangement in (a). (After Rossing and Hartmann, 1975).

7.7 ABSOLUTE PITCH

A subject that has held considerable fascination, but also caused no small amount of controversy, is *absolute pitch*. The term refers to the ability to recognize and define (e.g., by naming or singing) the pitch of a tone without the use of a reference tone. This ability is often compared to absolute recognition of color (e.g., green) without any comparison to a standard spectrum. Whereas absolute color recognition is possessed by about 98 percent of the population (only 2 percent being partially or totally colorblind), absolute pitch recognition is rare (less than 0.01 percent of the population appears to have it).

Absolute pitch contrasts with relative pitch, which most persons have to some degree. Nearly all persons can tell whether one tone is higher than another; persons with some musical experience or training can recognize intervals between tones with varying degrees of precision. Someone with a well-trained ear can tell when the frequency of a second tone deviates as little as one percent from the expected interval, although these judgments are not as accurate as they are consistent. For example, the frequency that a person judges, with great consistency, to be an octave above a 1000-Hz tone may actually be 2060 Hz. Relative pitch, in fact, is a remarkable sensory ability that has no counterpart in our other senses. We cannot judge a color that has twice the frequency of a reference color; the only comparable judgment in the visual domain might be selection of a complementary color, and few people develop the ability to do that with great accuracy.

Psychologists have studied absolute pitch for at least 75 years, and during that time there has been considerable discussion and some controversy concerning its origin. In particular, there is less than unanimous agreement as to whether absolute pitch is inherited, acquired, or possibly both. At least four different theories about absolute pitch have developed (Ward, 1963):

1. *Heredity theory.* The faculty for developing absolute pitch is inherited, just as the ability for color identification is (unless one inherits colorblindness). The child, so gifted, learns pitch names in early life just as color names are learned.

2. *Learning theory.* The opposite point of view, that absolute pitch can be acquired by almost anyone by diligent and constant practice, is not too widely held.

3. *Unlearning theory.* The ability to develop absolute pitch is nearly universal, but is simply trained out of most people at an early age (by emphasis on relative pitch, for example).

4. *Imprinting theory.* Imprinting is a term used to describe rapid irreversible learning that takes place at a specific developmental stage (used to explain, for example, why ducklings will follow for the rest of their life the first moving object they see after hatching). Proponents of this theory feel that nearly all children could be taught absolute pitch at the appropriate stage of development.

Bachem (1955) distinguishes between "chroma" and "tone height" as two separate components of pitch. All A's up and down the scale have the same chroma or quality but differ in tone height (possessors of absolute pitch frequently make octave errors in identifying tones). Above about 5000 Hz, chroma tends to become fixed, while tone height continues to increase, so that absolute pitch identification is not possible.

At least one person with absolute pitch has reported a change in his internal pitch standard with time (Vernon, 1977). At age 52 he noted a tendency to identify keys one semitone higher than they should be. He was troubled because he heard the overture to Wagner's "Die Meistersinger" in the "effeminate" key of C♯ rather than the "strong and masculine" key of C. By age 71, however, it had moved still further into the sturdier key of D! The shift of internal pitch standard may have been due to a change in elasticity of the basilar membrane with age; in other words a tone of a given frequency was invoking maximum activity at a different place on the basilar membrane than in earlier years.

However it develops, absolute pitch is a remarkable ability. Absolute pitch (inherited or acquired) may continue to be a controversial subject for some time to come, because of the obvious difficulty of experimenting with human subjects in isolation. If one really wants a child to acquire absolute pitch (it has disadvantages as well as advantages!), one should probably begin as early as possible to play "find the note" games on the piano.

7.8   PITCH STANDARDS

The advantages of a universal pitch standard are so obvious that it is quite remarkable that for so many years there was none. Pipe organs were built with A's tuned all the way from 374 to 567 Hz (Helmholtz, 1877). In 1619, Praetorius suggested a pitch of 424 Hz; Handel's tuning fork reportedly vibrated at 422.5 Hz. This pitch standard prevailed, more or less, for two centuries, and it is the pitch standard for which Hayden, Mozart, Bach, and Beethoven composed.

Early in the nineteenth century pitch began to rise, probably due to an increased use of brass instruments, which were found to sound more brilliant at the higher pitch. In 1859 a commission appointed by the French government (which included Berlioz, Meyerbeer, and Rossini) selected 435 Hz as a standard. Early in the twentieth century a "scientific pitch," with all the C's being powers of 2 (128, 256, 512, and so on), appeared; this leads to about 431 Hz for A. Unfortunately, tuning forks made to this standard are still being distributed by scientific and medical supply houses.

In 1939 an International Conference in London unanimously adopted 440 Hz as the standard frequency for A_4, and this is almost universally used by musicians. A few orchestras have once again begun a ''pitch-raising'' game by tuning to 442 or even 444 Hz for greater brightness. This is unfortunate, however, since instruments designed to play well at one pitch may not retain their tone or intonation at another (this is especially true of woodwinds). Singers of today sing the arias of Mozart and Beethoven about a semitone above the pitch for which they were written; most violins of the old masters have already had to be strengthened by adding stouter bass bars and necks to accommodate the increased string tension of today's pitch standard.

Tuning forks have served as convenient pitch standards since the time of Handel. More recently, quartz crystals have provided us with a more precise standard for measuring frequency as well as time. Electronic frequency counters and stroboscopic tuners have made it possible for every physics laboratory as well as every band or orchestra to have precise and dependable frequency standards. The United States Bureau of Standards broadcasts an exceedingly precise 440-Hz tone on its shortwave radio station WWV for checking local standards.

The frequency of most musical instruments changes with temperature, and those using wood and gut also change with humidity. The velocity of sound increases about 0.6 m/s for each Celsius degree, so the pitch of a wind instrument rises about 3 cents ($\frac{3}{100}$ of a semitone) per degree of temperature rise (the slight lowering of pitch due to expansion in length is negligible). String instruments generally fall in pitch due to relaxing tension as temperature rises.

7.9   TIMBRE OR TONE QUALITY

The word *timbre,* borrowed from French, is used to denote the ''tone quality'' or ''tone color'' of a sound. The American National Standards Institute (1960) defines it: ''Timbre is that attribute of auditory sensation in terms of which a listener can judge two sounds similarly presented and having the same loudness and pitch as dissimilar.'' An explanatory note is added: ''Timbre depends primarily on the spectrum of the stimulus, but it also depends upon the waveform, the sound pressure, the frequency location of the spectrum, and the temporal characteristics of the stimulus.'' This definition suggests that judgment of timbre must take place under conditions of equal loudness and pitch (and probably equal duration as well), and so Pratt and Doak (1976) have suggested an alternative definition: ''Timbre is that attribute of auditory sensation whereby a listener can judge that two sounds are dissimilar using any criteria other than pitch, loudness or duration.''

Timbre may be described as a "multidimensional attribute" of sound (Plomp, 1970); it is impossible to construct a single subjective scale of timbre of the type used for loudness (sones) and pitch (mels), for example. Two recent attempts to construct subjective scales, by asking listeners to rate various verbal attributes of steady sounds, are illustrated in Fig. 7.10. Each investigator found the dull–sharp (brilliant) scale the most significant.

In discussing timbre, and especially in reading about the many experiments on timbre described in the literature, it is important to distinguish between the timbre of *steady* complex tones and those that include *transients* or other variations with time. Plomp (1970) suggested the possibility of using *tone color* to refer to the perceptual differences between steady complex tones; this suggestion has not been widely accepted, however.

A thorough investigation of the timbre of steady tones was carried out by Helmholtz (1877). Helmholtz demonstrated that the sounds of most musical instruments (including the vocal cords) consist of series of harmonics that determine the timbre. Furthermore, he carefully described a way in which the ear could comprehend timbre. On the basis of his experiments, he formulated the following general rules:

1. Simple tones, such as those of tuning forks and widely stopped organ pipes, have very soft, pleasant sound, free from roughness but dull at low frequencies.

2. Musical tones with a moderately loud series of harmonics up to the 6th (such as those produced by the piano, the French horn, and the human voice) sound richer and more musical than simple tones, yet remain sweet and soft if the higher harmonics are absent.

3. Tones consisting of only odd harmonics (narrow stopped organ pipes, clarinet) sound hollow and, if many harmonics are present, nasal. When the fundamental predominates, the quality of tone is rich; when the fundamental is not sufficiently strong, the quality of tone is poor.

4. Complex tones with strong harmonics above the 6th or 7th are very distinct, but the quality of tone is rough and cutting.

FIG. 7.10
Subjective rating scales for timbre: (a) Pratt and Doak, 1976; (b) von Bismarck, 1974.

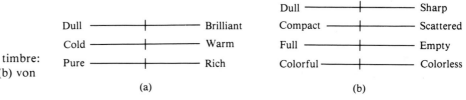

Helmholtz continued with careful experiments to determine the dependence of timbre on the relative phases of the harmonics. Using electrically driven tuning forks and tuned resonators (of the type we now call Helmholtz resonators), he concluded that timbre does not depend on phase differences between the harmonics. Unfortunately, Helmholtz could detect only very slow changes in phase in his experiments (a limitation that he apparently recognized), and thus some interesting dynamic phase effects were overlooked. So thorough were the studies of Helmholtz that until about 1950 very little new information of significance appeared in the literature.

Before continuing the discussion of timbre, we will investigate the Fourier analysis of a tone.

7.10 FOURIER ANALYSIS OF COMPLEX TONES

The determination of the harmonic components of a periodic waveform is called *Fourier analysis,* after the mathematician Joseph Fourier (1768–1830), who formulated an important mathematical theorem: *Any periodic vibration, however complicated, can be built up from a series of simple vibrations, whose frequencies are harmonics of a fundamental frequency, by choosing the proper amplitudes and phases of these harmonics.* Constructing a complex tone from its harmonics (the opposite of Fourier analysis) is called *Fourier synthesis.* The terms *spectrum analysis, harmonic analysis,* and *sound analysis* are sometimes used to describe Fourier analysis applied to sound. A specification of the strengths of the various harmonics (usually in the form of a graph) is called a *spectrum.*

Spectra of four different complex waveforms are shown in Fig. 7.11. Although they sound rather harsh and unmusical (with the exception of the flutelike triangle wave), these waveforms are frequently used to create sound in electronic music synthesizers (see Chapter 27). The square wave, for example, is composed of only odd-numbered harmonics with amplitudes in the ratio $1/n$. Thus if the fundamental has frequency f and amplitude A, the other harmonics in the spectrum will have frequencies of $3f$, $5f$, $7f$ . . . , and amplitudes $A/3$, $A/5$, $A/7$. . . . The triangle wave has odd harmonics with amplitudes in the ratio $1/n^2$ (that is, A, $A/9$, $A/25$ . . .). The sawtooth wave, on the other hand, has both odd-numbered and even-numbered harmonics with amplitudes in the ratio $1/n$ (A, $A/2$, $A/3$ . . .).

Figure 7.12 illustrates how Fourier synthesis works. The first six harmonics of a sawtooth wave are shown individually and collectively. Note that when combined *in the proper phase,* the first six harmonics approximate a sawtooth wave, although "wiggles" occur that diminish with the addition of higher harmonics.

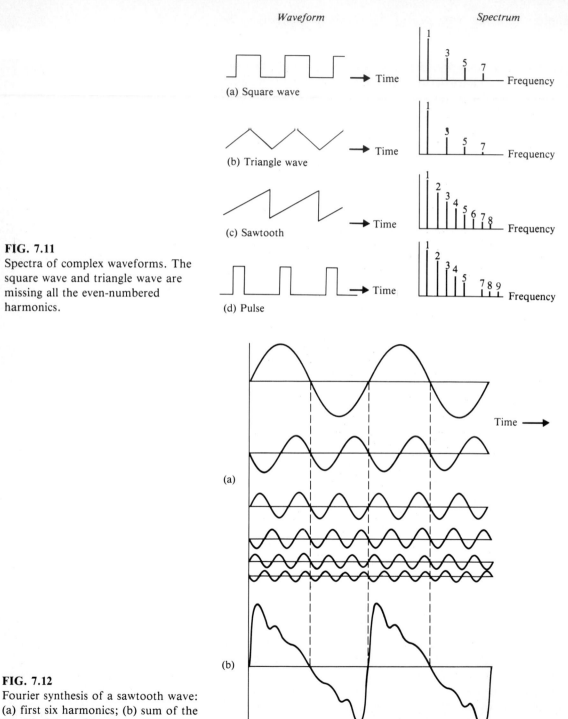

FIG. 7.11
Spectra of complex waveforms. The square wave and triangle wave are missing all the even-numbered harmonics.

FIG. 7.12
Fourier synthesis of a sawtooth wave: (a) first six harmonics; (b) sum of the first six harmonics.

Textbooks present many "typical" spectra of musical instruments. It should be emphasized, however, that sound spectra from a given instrument vary widely according to the way in which the instrument is played (soft, loud, high, low, or mid-range) and how the sound is recorded (near field, far field, reverberant field, direction of microphone from the instrument, etc).

One way to determine the spectrum of harmonics is by direct computation from the recorded waveform. One of the earliest instruments developed for recording waveforms was the *phonodeik* designed by D. C. Miller (1916), which used a vibrating mirror to direct a beam of light onto a moving film. Most of the sound spectra in early publications were calculated from phonodeik recordings.

Modern spectrum analyzers are of two types: digital and analogue. Digital-spectrum analyzers begin by sampling one period of the wave at regular intervals and feeding these samples into a digital computer. The computer then calculates the amplitude and phase of each harmonic.

Analogue spectrum analyzers use filters or other electronic circuits to isolate the harmonics one after another. If this is done very rapidly (in a few milliseconds), the analyzer is called a *real-time* spectrum analyzer, which is very useful for studying changing sounds or spectra during attack and decay of sounds.

Some interesting information about timbre can be obtained by averaging many spectra (from the same instrument, for example). Figure 7.13 shows averages of 512 spectra of a clarinet and a male voice.

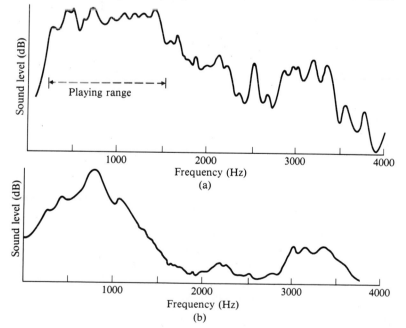

FIG. 7.13

Time-averages of spectra; (a) clarinet; (b) tenor singing "ah."

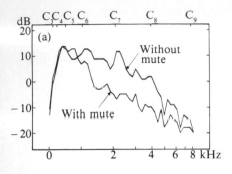

FIG. 7.14
Long-time average spectra of a violin with and without a mute. (From Jansson and Sundberg, 1975).

In each case, the pitch is varied by playing (singing) up and down the scale during the recording. The significance of the various maxima will become clear after reading about woodwind instruments (Chapter 12) and voice formants (Chapter 15).

Long-time-average spectra have been used extensively at the Royal Institute of Technology in Stockholm to study musical instruments and the singing voice. A long-time-average spectrum contains information on the written music, the performance, the musical instrument, and the room in which it is played. The effect of varying any one of these factors can be studied by holding the others constant. Figure 7.14 shows the long-time-average spectra of a violin played with and without a mute, for example.

It should be mentioned that whereas the effects of phase on timbre are small for steady tones, the ear is in fact quite sensitive to *changes* in phase, especially if they take place at a regular rate. This is illustrated by the phenomenon described as "second-order beats," to be discussed in Chapter 8.

7.11   TIMBRE AND DYNAMIC EFFECTS: ENVELOPE AND DURATION

In Sections 7.9 and 7.10, the discussion focused on the timbre of steady complex tones. Transients and other dynamic effects, however, play an important role in determining the timbre of musical and speech sounds, as you can prove to yourself by two simple experiments.

Record the sounds of a number of different musical instruments. In the first experiment, play the tape backward (so that the attack transient occurs at the end). You will hear some curious effects. For example, a piano played backward sounds like a reed organ or harmonium. (This is illustrated in Demonstration 29, Houtsma, Rossing, and Wagenaars, 1987). For a second experiment, cut and splice the tape, so that the attack transient is removed. Without attack transients, a remarkable similarity is noted between dissimilar pairs of instruments such as a French horn and a saxophone and even a trumpet and an oboe.

Berger (1963) performed an experiment in which the sounds of various instruments were presented with the first and last half seconds removed; using 30 band students as a jury of listeners, he obtained the "confusion matrix" shown in Table 7.1. Note that with the transients removed, the sound of an alto saxophone was correctly identified by only four jurists, whereas eleven jurists thought it was a French horn. Also surprising is the confusion of tenor saxophone with clarinet, since the "woody" tone of a clarinet emphasizes odd-numbered harmonics.

TABLE 7.1   Listener judgments of recorded wind-instrument tones presented with first and last half seconds removed (Berger, 1963).

Stimulus	Flute	Oboe	Clarinet	Tenor saxophone	Alto saxophone	Trumpet	Cornet	French horn	Baritone	Trombone	No answer
Flute	1	2		1	6	5	4			4	7
Oboe		28									2
Clarinet	1	1	20	4	3						1
Tenor saxophone			25	2	1						2
Alto saxophone				3	4		1	11	5	5	1
Trumpet	8				6	2	3	4	1	3	3
Cornet		1				12	15				2
French horn	1			2	3			5	6	6	7
Baritone			1	1	2	3	2	4	7	3	7
Trombone	2	1		5	3			1	5	9	4

During attack, the various partials of a musical sound may develop at different rates. Figure 7.15 shows the attack waveform of an organ pipe tone, along with the onset of the first five harmonics. During the attack transient, the waveform is not exactly periodic; note the difference from cycle to cycle. Note that the second harmonic of the organ pipe develops slightly faster than the others; in other wind instruments, the fundamental is often found to lead.

Strong and Clark (1967) performed some interesting experiments in which they interchanged spectra and time envelopes of wind instrument tones. They synthesized many tones, each time using the envelope characteristic of one instrument with the spectrum of another, and asked listeners to identify the instrument. They found that in the cases of the oboe, clarinet, bassoon, tuba, and trumpet, the spectrum is much more important than the envelope; in the case of the flute, the envelope is more important than the spectrum; in the cases of the

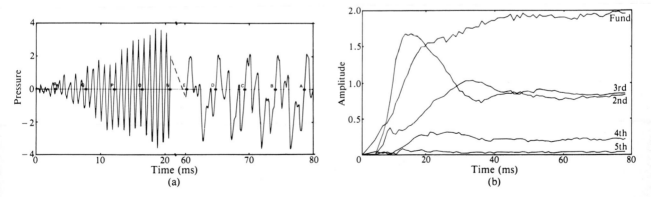

FIG. 7.15

(a) The waveform of an attack transient. (b) Amplitudes of the first five harmonics of the attack transient of a 110-Hz diapason organ pipe. (From Keeler, 1972).

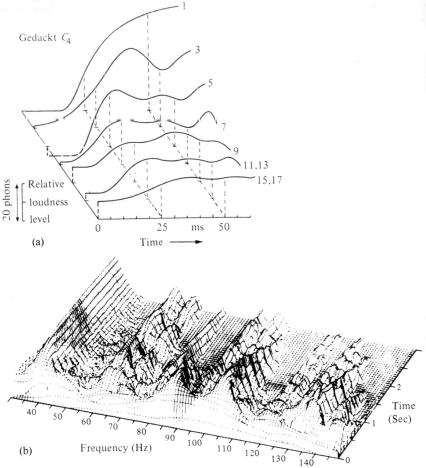

FIG. 7.16
Examples of three dimensional spectral plots that illustrate attack transients in musical instruments: (a) Gedackt 8-foot C_4 organ pipe (Pollard and Jansson, 1982a); (b) Bass drum (Bassett, 1982). The vertical coordinate in each case is sound pressure level or loudness level.

trombone and French horn, spectrum and envelope appear to be of comparable importance. The general principle seems to be that the spectrum takes on greatest importance when it has a maximum in a unique location within its playing range.

Real-time spectrum analyzers (see Chapter 33) and digital computers have made it possible to represent transient features in the spectra of sounds on three-dimensional plots relating sound pressure level, frequency, and time. Figure 7.16 shows attack transients in the sounds of an organ pipe and a bass drum, measured by two different investigators.

Several attempts have been made to apply a technique called "multidimensional scaling" to musical timbre. In multidimensional scaling, the investigator gathers perceptual data consisting of subjective similarities between all pairs in a set of stimuli. From these, a set of "dimensions" are selected that best fit the data. One of these dimensions is usually the spectrum of the sound. The other dimension may be the presence of high-frequency energy during the attack, or the extent to which the higher harmonics rise and fall in synchronism (Grey, 1977).

7.12 TRISTIMULUS DIAGRAMS

An effective way to represent timbre graphically is to use a *tristimulus diagram* analogous to a chromaticity diagram used to represent color. The eye contains three different types of cone receptors: one type responds primarily to red light, one to green light, and one to blue light. The brain is able to judge the spectral distribution of light by comparing the signals from the three types of receptors. A tristimulus diagram represents the three different stimulus variables x, y, and z on a two-dimensional diagram by defining them in such a way that $x + y + z = 1$.

The hearing mechanism, of course, does not have three different types of sensors. However, Pollard and Jansson (1982b) have suggested that salient features of a spectrum envelope can be represented by three appropriate variables representing the effective loudness of three different parts of the spectrum.

In the tristimulus diagrams in Fig. 7.17, the fraction of the total intensity in partials 2, 3, and 4 is plotted along the y-axis and the

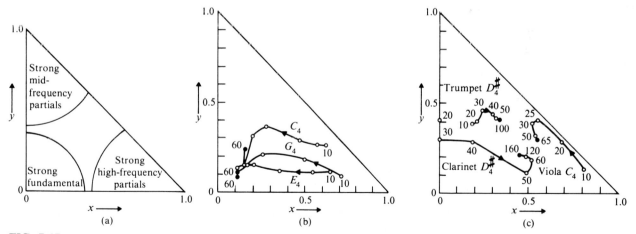

FIG. 7.17
Tristimulus diagrams representing the timbre of musical instruments. The fraction of the intensity in partials 2, 3, and 4 is plotted along the y-axis, and the fraction in higher harmonics along the x-axis. (a) Regions where tones with strong fundamentals, midfrequency partials, and high-frequency partials would be found. (b) Timbre of three Gedackt organ pipes from 10 to 60 ms after attack. (c) Attack transients of trumpet, clarinet, and viola (Pollard and Jansson, 1982b).

fraction in all higher harmonics along the *x*-axis. The third variable is the fractional intensity of the fundamental. Figure 7.17(a) indicates regions where tones with strong fundamentals, mid-frequency partials, and high-frequency partials would be found. A pure tone would lie at the origin. Figure 7.17(b) compares the spectra (timbre) from 10 to 60 ms after attack of three Gedackt organ pipes, including the C$_4$ pipe shown in Fig. 7.16(a). Figure 7.17(c) shows the starting transients of tones on a trumpet, clarinet, and viola. Note that although the steady state tones (black circles) are reasonably close to one another, the paths traversed during the attack transients are completely different.

7.13   VIBRATO

Discussion of vibrato often leads to some lively differences of opinion, much of it centering on definitions, since the terms *vibrato* and *tremolo* are used in different ways (often interchangeably) throughout the literature of music and psychoacoustics.

The definition recommended by the American National Standards Institute (1960) is "The vibrato is a family of tonal effects in music that depend on periodic variations of one or more characteristics in the sound wave." The important note is added: "When the particular characteristics are known, the term 'vibrato' should be modified accordingly: e.g., frequency vibrato, amplitude vibrato, phase vibrato and so forth." In keeping with this recommendation, we use the term *frequency vibrato* to refer to frequency modulation (FM) and *amplitude vibrato* to refer to amplitude modulation (AM). In practice it is virtually impossible to have frequency vibrato without amplitude vibrato because of the effect of room resonances and resonances in the source instrument. Amplitude vibrato without frequency vibrato is possible (in the case of a vibraphone with resonators that open and close periodically, for example), but would be the exception rather than the rule.

Vibrato was studied extensively some 40 to 50 years ago by C. E. Seashore and colleagues at the University of Iowa, and many of their findings are confirmed by more recent experiments (Ward, 1970). Vibrato appears to vary with individual performers, an "average" rate for both singers and instrumentalists being around 7 Hz. Singers seem to use a slightly greater depth of frequency vibrato than instrumentalists do, however.

You can perform interesting experiments on vibrato using an audio generator with provision for frequency modulation (many generators can be frequency modulated by a second oscillator), or with an electronic music synthesizer. Try varying both the *rate* and the *depth* of frequency modulation. You will probably find that with modulation rate in the range of 1 to 5 Hz, you can recognize the periodicity of pitch change (most clearly around 4 Hz). Beginning at about 6 Hz, however, the tone takes on a single average pitch with intensity fluctuations at the frequency of the vibrato. At a still higher rate (around 12 Hz), the sound becomes a rather unpleasant confusion of more than one tone. It is not difficult to see why performers choose a vibrato rate around 7 Hz.

The parameters of a natural vibrato fluctuate slightly during the duration of a tone. Tones from electronic instruments, which have a fixed rate and depth of vibrato, sound artificially rigid. Analyses of the vibrato used by opera singers Maria Callas and Dietrich Fischer-Dieskau show that both singers use deep vibratos (Winckel, 1975). The rates of vibrato and trill used by Callas were the same, and in fact her transition from vibrato to trill was made with no change of phase. When trained singers sing duets, they reportedly adjust their vibratos to have identical rate and phase (but not necessarily depth); the adjustment is most likely subconscious (Winckel, 1975).

Vibrato is said to cover up small errors in frequency. Fletcher, Blackham, and Geertsen (1965) found that the vibrato of many violinists apparently centers 15 to 20 cents above the target pitch. Vibrato makes identification of vowel sounds more difficult, and tends to conceal formant frequencies of singers that may deviate substantially from the corresponding formant frequencies of normal speech (Sundberg, 1975).

7.14 BLEND OF COMPLEX TONES

Our auditory system has the ability to listen to complex sounds in different modes. When we listen *analytically,* we hear the different partials separately; when we listen *synthetically* or holistically, we focus on the whole sound and pay little attention to the partial sounds. Listeners differ in the degree to which they listen analytically or synthetically. If a two-tone complex of 800 and 1000 Hz is followed by one of 750 and 1000 Hz, for example, an analytic listener will hear one partial go down in pitch; a synthetic listener will hear a virtual pitch rising a major third from 200 to 250 Hz (Demonstration 25, Houtsma, Rossing, and Wagenaars, 1987).

A tone with several harmonic partials, whose frequencies and relative amplitudes remain steady, is generally heard as a single tone, even if the total intensity changes. However, when one of the harmonics is turned off and on, it stands out clearly (Demonstration 1, Houtsma, Rossing, and Wagenaars, 1987). The same is true if one of the harmonics is given a "vibrato" (i.e., its frequency, its amplitude, or its phase is modulated at a slow rate).

One of the most remarkable feats of our auditory system is its ability to single out complex tones from a complex background, such as the sounds of different instruments in a symphony orchestra or conversation at a cocktail party, for example. In the former case, the ear interprets certain partial tones as belonging to one particular instrument, other partials as belonging to another instrument. In other words, it looks for familiar or likely sets of partial tones and fuses these together into a single complex tone at the same time it hears a blend of many instrument sounds. As we have seen, the mechanism for this analysis is partially understood, but much research remains to be done.

Erickson (1975) addresses this subject from the standpoint of a composer. In an enlightening chapter, "Some Territory Between Timbre and Pitch," he discusses three ways in which a complex sound can be heard: (1) as a chord; (2) as a pitch (with timbre); (3) as a sound (an unpitched sound without definite pitch or pitches such as the sound of a bass drum). These three concepts can be represented as the apexes of a triangle (see Fig. 7.18) with the "grey areas" between them represented by the sides of the triangle.

Transformation from a chord to the fused condition described as "a sound," for example, is illustrated by the music of Edgard Varese. A pitch (with timbre)-to-chord transformation occurs in the unusual chanting of Tibetan lamas recorded and described by Smith, Stevens, and Tomlinson (1967). The chanting is done in such a way that certain harmonics of the voice become audible as separate pitches, giving the effect of one person singing a continuous chord.

It is well known that the partials in a piano tone are stretched further apart than partials in a true harmonic series (see Chapter 14). Stretching the partials even further apart causes the sounds to become bell-like or chimelike. More surprising, perhaps, is the observation that compression of the partials also produces bell-like timbres (Slaymaker, 1970). Individual partials, in both cases, can be singled out more easily than the harmonic partials of the usual musical tone; the transformation can be described as going from pitch (with timbre) to an inharmonic chord as the partials are stretched or compressed beyond certain limits.

Inharmonicity in the partials of a complex tone appears to be detected in a different way for low and high harmonics. For low harmonics, the inharmonic partial appears to "stand out" when it is mistuned by an amount that varies from 1 to 3% in different subjects. For high harmonics, on the other hand, the mistuning is detected as a kind of "beat" or "roughness," presumably reflecting a sensitivity to changing phase of the mistuned harmonic relative to the other harmonics (Moore, Peters, and Glasberg, 1985).

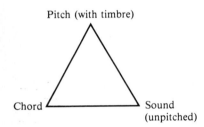

FIG. 7.18
Three ways in which a complex sound can be heard. (From Erickson, 1975).

7.15 SUMMARY

Pitch has been defined as the characteristic of a sound that gives it the sensation of high or low. It is determined mainly by the frequency of a tone, but sound level, spectrum, and duration also influence pitch. Early models for pitch perception regarded the basilar membrane as a frequency analyzer of high resolution (place theory), but more recent studies have shown that much of the determination of pitch is contributed by a temporal analysis in the central nervous system (periodicity pitch). The ear is able to assign a pitch to complex sounds composed of inharmonic partials and even to some presentations of wideband noise. Some persons have the ability to identify pitch independent of a reference pitch (absolute pitch).

Timbre or tone quality depends on the frequency of a tone, its time envelope, its duration, and the sound level at which it is heard. Any complex waveform that is periodic can be constructed from simple tones with the right frequency and phase; determination of the spectrum of simple tones is called *spectrum analysis* or *Fourier analysis*. Under most conditions, the timbre of a complex sound is insensitive to the phase of its components. Periodic variation of the frequency and amplitude, called *vibrato,* lends warmth and blend to musical tones. A vibrato rate of about 7 Hz is common in musical performance.

References and Suggested Readings

American National Standards Institute (1960). "USA Standard Acoustical Terminology." S1.1–1960.

Bachem, A. (1955). "Absolute Pitch," *J. Acoust. Soc. Am.* **27**: 1180.

Bassett, I. G. (1982). "Vibration and Sound of the Bass Drum," *Percussive Notes* **19**(3): 50.

Berger, K. W. (1963). "Some Factors in the Recognition of Timbre," *J. Acoust. Soc. Am.* **36**: 1888.

Bilsen, F. A., and R. J. Ritsma (1969/70). "Repetition Pitch and Its Implication for Hearing Theory," *Acustica* **22**: 63.

Bismarck, G. von (1974). "Timbre of Steady Sounds: A Factorial Investigation of Its Verbal Attributes," *Acustica* **30**: 146.

Burck, W., P. Kotowski, and H. Lichte (1935). "Die Horbarkeit von Laufzeitdifferenzen," *Elektrotechn. Nachr.-Techn.* **12**: 355.

Doughty, J. M., and W. M. Garner (1948). "Pitch Characteristics of Short Tones II: Pitch as a Function of Duration," *J. Exp. Psychol.* **38**: 478.

Erickson, R. (1975). *Sound Structure in Music.* Berkeley: University of California. (See especially Chapter 2.)

Evans, E. F., and J. P. Wilson, eds. (1977). *Psychophysics and Psychology of Hearing.* London: Academic Press.

Fletcher, H. (1934). "Loudness, Pitch and Timbre of Musical Tones and Their Relation to the Intensity, the Frequency, and the Overtone Structure," *J. Acoust. Soc. Am.* **6**: 153. See also *Phys. Rev.* **23**: 427 (1924).

Fletcher, H., E. D. Blackham, and O. N. Geertsen (1965). "Quality of Violin, Viola, Cello and Bass-Viol Tones," *J. Acoust. Soc. Am.* **37**: 851.

Goldstein, J. L. (1973). "An Optimum Processor Theory for the Central Information of Complex Tones," *J. Acoust. Soc. Am.* **54**: 1496.

Grey, J. M. (1977). "Multidimensional Perceptual Scaling of Musical Timbres," *J. Acoust. Soc. Am.* **61**: 1270.

Hartmann, W. M. (1978), "The Effect of Amplitude Envelope on the Pitch of Sinewave Tones," *J. Acoust. Soc. Am.* **63**: 1105.

Helmholtz, H. L. F. von (1877). *On the Sensations of Tone*

as a Physiological Basis for the Theory of Music, 4th ed. Trans. A. J. Ellis. New York: Dover, 1954.

Houtsma, A. J. M., and J. L. Goldstein (1972). "The Central Origin of the Pitch of Complex Tones: Evidence from Musical Interval Recognition," *J. Acoust. Soc. Am.* **51**: 520.

Houtsma, A. J. M., and T. D. Rossing (1987). "Effects of Signal Envelope on the Pitch of Short Complex Tones," *J. Acoust. Soc. Am.* **81**: 439.

Houtsma, A. J. M., T. D. Rossing, and W. M. Wagenaars (1987), *Auditory Demonstrations* (Philips Compact Disc #1126-061 and text). Woodbury, N. Y.: Acoustical Society of America.

Jansson, E. V., and J. Sundberg (1975). "Long-Time-Average Spectra Applied to Analysis of Music. Part I: Method and General Applications," *Acustica* **34**: 15.

Keeler, J. S. (1972). "Piecewise-Periodic Analysis of Almost-Periodic Sounds and Musical Transients," *IEEE Trans. Audio and Electroacoustics* **AU-20**: 338.

Majernik, V., and J. Kalužný (1979). "On the Auditory Uncertainty Relations," *Acustica* **43**: 132.

Miller, D. C. (1916). *Science of Musical Sounds.* New York: Macmillan.

Moore, B. C. J., R. W. Peters, and B. R. Glasberg (1985). "Threshold for the Detection of Inharmonicity in Complex Tones," *J. Acoust. Soc. Am.* **77**: 1861.

Parkin, P. H. (1974). "Pitch Change During Reverberant Decay," *J. Sound and Vibration* **32**: 530.

Plomp, R. (1967). "Pitch of Complex Tones," *J. Acoust. Soc. Am.* **41**: 1526.

Plomp, R. (1970). "Timbre as a Multidimensional Attribute of Complex Tones," in *Frequency Analysis and Periodicity Detection in Hearing.* Eds., R. Plomp and G. Smoorenburg. Leiden: Sijthoff.

Pollard, H. F., and E. V. Jansson (1982a). "Analysis and Assessment of Musical Starting Transients," *Acustica* **51**: 249.

Pollard, H. F., and E. V. Jansson (1982b). "A Tristimulus Method for the Specification of Musical Timbre," *Acustica* **51**: 162.

Pratt, R. L., and P. E. Doak (1976). "A Subjective Rating Scale for Timbre," *J. Sound and Vibration* **45**: 317.

Ritsma, R. (1967). "Frequencies Dominant in the Perception of the Pitch of Complex Sounds," *J. Acoust. Soc. Am.* **42**: 191.

Rossing, T. D., and W. M. Hartmann (1975). "Perception of Pitch from Time-Delayed White Noise," AAPT meeting, Anaheim, California (January 1975).

Rossing, T. D., and A. J. M. Houtsma (1986). "Effects of Signal Envelope on the Pitch of Short Sinusoidal Tones," *J. Acoust. Soc. Am.* **79**: 1926.

Savart, F. (1840). "Uber die Ursachen der Tonhohe," *Ann. Phys. Chem.* **51**: 555.

Scharf, B., and A. J. M. Houtsma (1986). "Audition II: Loudness, Pitch, Localization, Aural Distortion, Pathology," in *Handbook of Perception and Human Performance,* Vol. 1. Eds., K. R. Boff, L. Kaufman, and J. P. Thomas. New York: J. Wiley.

Schouten, J. F. (1940). "The Perception of Pitch," *Philips Tech. Rev.* **5**: 286. (This article summarizes much of the contents of several articles in *Proc. Ned. Acad. Wet.* 41 and 43).

Smith, H., K. N. Stevens, and R. S. Tomlinson (1967). "On an Unusual Mode of Chanting by Tibetan Lamas," *J. Acoust. Soc. Am.* **41**: 1262.

Stevens, S. S. (1935). "The Relation of Pitch to Intensity," *J. Acoust. Soc. Am.* **6**: 150.

Strong, W., and M. Clark, Jr. (1967). "Perturbations of Synthetic Orchestral Wind-Instrument Tones," *J. Acoust. Soc. Am.* **41**: 277.

Sundberg, J. (1975). "Vibrato and Vowel Identification," in Report STL-QPSR 2-3/1975, Speech Technology Laboratory, Royal Institute of Technology, Stockholm.

Terhardt, E. (1974). "Pitch, Consonance, and Harmony," *J. Acoust. Soc. Am.* **55**: 1061.

Terhardt, E. (1979). "Calculating Virtual Pitch," *Hearing Research* **1**: 155.

Terhardt, E., and H. Fastl (1971). "Zum Einfluss von Störtönen und Störgeräuschen auf die Tonhöhe von Sinustönen," *Acustica* **25**: 53.

Vernon, P. E. (1977). "Absolute Pitch: A Case Study." *Br. J. Psychol.* **68**: 485.

Ward, W. D. (1963). "Absolute Pitch," *Sound* **2**(3): 14; **2**(4): 33.

Wightman, F. L. (1973). "The Pattern-Transformation Model of Pitch," *J. Acoust. Soc. Am.* **54**: 407.

Wightman, F. L., and D. M. Green (1974). "The Perception of Pitch," *Am. Scientist* **62**: 208.

Winckel, F. (1967). *Music, Sound and Sensation.* Trans., T. Binkley. New York: Dover.

Winckel, F. (1975). "Measurements of the Acoustic Effectiveness and Quality of Trained Singers' Voices," 90th meeting of Acoustical Society of America, San Francisco.

Zwicker, E. (1962). "Direct Comparisons between the Sensations Produced by Frequency Modulation and Amplitude Modulation," *J. Acoust. Soc. Am.* **34:** 1425.

Zwicker, E., G. Flottorp, and S. S. Stevens (1957). "Critical Bandwidth in Loudness Summation," *J. Acoust. Soc.* **29:** 548.

Glossary

absolute pitch The ability to identify the pitch of any tone without the aid of a reference.

analytic listening Listening to a complex tone in a way that individual components or partial tones are heard as separate entities.

bark An interval of frequency equal to a critical bandwidth.

critical band The frequency bandwidth at which subjective response (to loudness, pitch, etc.) changes rather abruptly (see Chapter 6).

distortion An undesired change in waveform. Two common examples are harmonic distortion and intermodulation distortion. *Harmonic distortion* means that harmonics are generated by altering the waveform in some way ("clipping" the peaks, for example). *Intermodulation distortion* refers to the generation of sum and difference tones.

envelope The amplitude of a tone as a function of time.

Fourier analysis, or **spectrum analysis** The determination of the component tones that make up a complex tone or waveform.

Fourier synthesis The creation of a complex tone or waveform by combining its spectral components.

fundamental The lowest common factor in a series of harmonic partials. The fundamental frequency of a periodic waveform is the reciprocal of its period.

harmonic A partial whose frequency is a multiple of some fundamental frequency.

inharmonic partial A partial that is not a harmonic of the fundamental.

just noticeable difference (jnd) or **difference limen** The minimum change in stimulus that can be detected.

mel The unit of subjective pitch; doubling the number of mels doubles the subjective pitch for most listeners. The critical band is about 100 mels wide.

octave The basic unit in most musical scales. Notes judged an octave apart have frequencies nearly in the ratio 2:1.

partial tone (or partial) One of the components in a complex tone (it may or may not be a harmonic of the fundamental).

period The smallest increment of time over which a waveform repeats itself.

periodic quantity One that repeats itself at regular time intervals.

periodicity pitch Pitch determination on the basis of the period of the waveform of a tone.

phase The fractional part of a period through which a waveform has passed, measured from a reference.

pitch An attribute of auditory sensation by which sounds may be ordered from low to high.

place theory of pitch A view of the basilar membrane as a frequency analyzer of high resolution; pitch is determined by sensing the place on the basilar membrane that has maximum excitation.

repetition pitch Pitch sensation created by the interference of a sound with a time-delayed repetition.

residue theory of pitch A view that components of a tone that cannot be resolved by the basilar membrane (the residue) are analyzed in time by the central nervous system.

semitone One step on a chromatic scale. Normally $\frac{1}{12}$ of an octave.

spectral dominance A view that certain partials dominate in the determination of the pitch of a complex tone.

spectrum The "recipe" for a complex tone that gives the amplitude and frequency of the various partials.

subjective pitch Pitch determined to have a frequency that does not correspond to that of any partial.

synthetic (holistic) listening Listening to a complex tone in a way that focuses on the whole sound rather than the individual partials.

timbre An attribute of auditory sensation by which two sounds with the same loudness and pitch can be judged dissimilar.

transient A sound that does not reoccur, at least on a regular basis.

tristimulus diagram A way of representing timbre graphically in terms of the relative loudness of three different parts of the spectrum.

vibrato Tonal effect in music resulting from periodic variation of amplitude, frequency, and/or phase.

virtual pitch Subjective pitch created by two or more partials in a complex tone (two examples are the "missing fundamental" of a filtered tone and the strike note of a bell).

Questions for Discussion

1. A "tonic" chord in the key of A consists of tones with frequencies of 440, 550, and 660 Hz. When such a chord is played on the piano or by three instruments, why is this not heard as a single tone with a pitch of 110 Hz (the "missing fundamental")?

2. Have you ever experienced the pitch change during reverberation described by Parkin? Would this effect be apparent on a recording with reverberation?

3. Discuss the advantages and disadvantages to a performing musician of possessing absolute pitch.

4. Try to account for the most prevalent "confusions" in Berger's experiment (Table 7.1) in the identification of instrument tones without the transients.

Problems

1. At what point would you divide a 65-cm guitar string (as Pythagoras did) so that the two segments sound pitches one octave apart?

2. From Fig. 7.2, find the jnd at frequencies of 200, 1000, and 5000 Hz.

3. By referring to Fig. 7.2, show that the critical band comprises roughly 30 jnd's. (Compare them at 200, 1000, 5000, and 10,000 Hz, for example.)

4. According to Fig. 7.3, how many cents does the pitch of a 200-Hz tone fall when the sound pressure level is changed from 50 to 90 dB?

5. In Fig. 7.1 let $t_1 = 7$ ms and $t_2 = 3$ ms. Determine $1/T$, $2/T$, and $3/T$. What is the frequency of the pitch that would be heard? What is the pulse rate?

6. From Fig. 7.15(b), determine the approximate rise times of the first and second harmonics of a diapason organ pipe.

7. In the range of 1–4 kHz, the just noticeable difference (jnd) is approximately 0.5 percent of the frequency. Show that this is about one-twelfth of a semitone (a semitone corresponds to a frequency ratio of 1.0595).

8. From Fig. 7.1, determine the number of mels in an octave from

   a) C_3 (131 Hz) to C_4 (262 Hz);

   b) C_4 to C_5 (523 Hz);

   c) C_5 to C_6 (1046 Hz).

9. If a pure tone with a frequency of 800 Hz is modulated at 150 Hz, what sidebands are produced? According to the theory discussed in Section 7.6, what virtual pitch will probably be heard? (Try dividing by various sets of integers such as 4, 5, 6 and 5, 6, 7, etc.)

10. If the steps in Fig. 7.8(a) are 30 cm deep, what pitch would most likely be heard? Compare this to the pitch of an organ pipe 30 cm long (Chap. 4).

11. Compare the tension in a violin string tuned to a standard A (440 Hz) with the tension in the same string tuned to match Handel's tuning fork (422 Hz). (See Section 3.2).

12. What are the frequencies of the first four partials in a 300-Hz square wave?

13. What is the frequency of the maximum sound level in the spectrum of Fig. 7.13(a)?

14. What is the strongest partial in the Gedackt organ pipe sound (Fig. 7.16(a)) at $t = 25$ ms? at $t = 50$ ms?

15. Compare the relative intensities of the fundamental, mid-frequency partials, and high-frequency partials in the C_4 Gedackt organ pipe (in Fig. 7.17(b)) at $t = 10$ ms, $t = 40$ ms and $t = 60$ ms (the fundamental is found by using the relationship $z = 1 - x - y$).

CHAPTER 8

Combination Tones and Harmony

In this chapter we will focus our discussion on the various effects that can occur when two or more tones reach the ear simultaneously. We begin with the simplest case, two pure tones with the same frequency, then proceed to two pure tones of different frequency, and finally to complex tones and chords.

8.1 LINEAR SUPERPOSITION

A *linear* system is one in which doubling the driving force doubles the response. If two driving forces are applied to a linear system simultaneously, the response will be the sum of the responses to the driving forces individually. That is, the response of the system to one driving force is not affected by the presence of the second one.

More than likely, you have heard the terms "linear" or "linearity" applied to components of a high-fidelity sound system. A loudspeaker with a linear response, for example, can reproduce the sound of a clarinet and, at the same time and quite independently, can reproduce a violin sound when presented with (electrical) driving forces characteristic of each instrument. If the loudspeaker were not completely linear in its response, one signal would influence (i.e., modulate) the other, and intermodulation distortion would be the result. This will be discussed in Chapter 22.

To determine the response of a linear system to two simple harmonic (pure tone) driving forces, we simply add together the two indi-

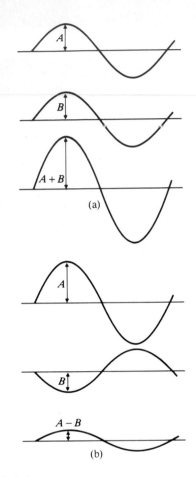

FIG. 8.1
Linear superposition of two simple
harmonic motions at the same
frequency: (a) same phase; (b) opposite
phase.

vidual responses at each point in time. If two simple harmonic motions at the same frequency are superimposed, the resultant will also be a simple harmonic motion at that frequency. The amplitude of the resultant will depend not only on the amplitude of the components but also on the fractional part of the period through which each component has passed. This fraction of the period is known as the *phase*. Figure 8.1 illustrates two cases as examples.

In the first case, the two motions are identical in phase and the displacements add. In the second case, they are opposite in phase and the displacements subtract (if $A = B$, total cancellation is the result). The concept of phase is a useful one in the comparison of two simple harmonic motions as well as in their superposition. It merits discussion in some detail.

8.2 PHASE ANGLE

Simple harmonic motion can be described as the projection of uniform circular motion onto an axis, as shown in Fig. 8.2. As the circle rotates, the projection of point P on the y-axis moves in simple harmonic motion. The angle ϕ indicates how far the circle has turned, so as ϕ increases, the corresponding point P moves to the right on the graph of displacement vs. time. During one complete revolution, ϕ increases by 360°, and the point moves to the right a distance T on the time axis (T is the period of the motion in seconds).

The representation of simple harmonic motion as the projection of circular motion can be demonstrated in several ways, one of which is shown in Fig. 8.3. The shadow of a wheel with a crank is projected alongside a mass vibrating in simple harmonic motion at the end of a spring (see Section 2.1). If the speed of the wheel is adjusted so that the time required for one revolution is the same as the period of the mass-spring vibrator, its shadow will move up and down in synchronism with the mass. Both the mass and the shadow of the crank move up and down in simple harmonic motion.

FIG. 8.2
Simple harmonic motion represented as
the projection on a vertical axis of the
point P moving around a circle at a
uniform rate.

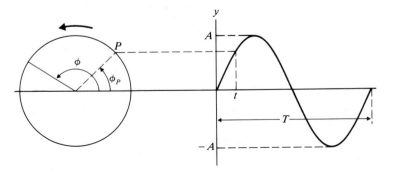

Simple harmonic motion is often referred to as *sinusoidal* motion, because the projection of circular motion on the *y*-axis is given analytically by a trigonometric function called the sine; that is,

$$y = A \sin \phi = A \sin 360 \frac{t}{T},$$

where the amplitude of the vibration *A* also equals the radius of the circle projected. Also, 360 is the number of degrees in a complete circle, t/T is the fraction of a complete circle subtended by angle ϕ in time *t*, *T* is a complete period, and sin is the abbreviation for sine. The language of trigonometry will not be used in this book, however.

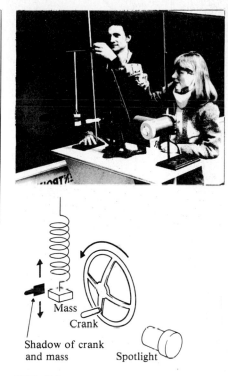

FIG. 8.3
Demonstration of simple harmonic motion as the projection of circular motion. By rotating the wheel at the proper rate, the shadow of a crank can be made to move up and down in synchronism with a mass-spring vibrator. (Photograph by Christopher Chiaverina).

We now add a second point in simple harmonic motion, such as point *Q* in Fig. 8.4. The projection of point *Q* moves with the same period *T* and amplitude *A* as point *P*, but it obviously has a different motion. At any given time *t*, points *P* and *Q* will be at different points on the circle. Thus the projected points reach their maximum and minimum values at different times. At any given time *t*, the positions of *P* and *Q* on the two curves result from different angular positions ϕ_P and ϕ_Q of the rotating points. The difference $\phi_P - \phi_Q$, which remains constant, is called the *phase difference* between the two simple harmonic motions.

The complete description of a given simple harmonic motion requires three parameters: the *period* (or frequency), the *amplitude*, and the initial *phase*. In the case of sound, phase has significance only when it is used to compare two or more waves or vibrations.

8.3   COMBINATION OF TWO SIMPLE HARMONIC MOTIONS

In order to radiate a pure tone, a loudspeaker cone moves in and out with a motion that is essentially simple harmonic motion. How does

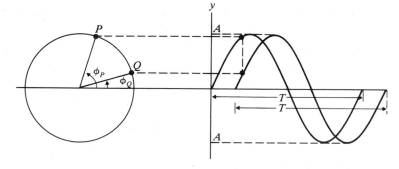

FIG. 8.4
Two points *P* and *Q* move with the same period *T* and amplitude *A*, and maintain a constant phase difference $\phi_P - \phi_Q$.

a loudspeaker cone move when two pure tones are radiated? If its response is linear, its motion at any time will be the linear combination or superposition of two simple harmonic motions. When it radiates a complex tone, its motion will be a combination of all the spectral components of the complex tone (see Section 7.10).

We consider first the combination of two simple harmonic motions having the same period (frequency) but different amplitudes and initial phases. Two cases were illustrated already in Fig. 8.1: same phase ($\phi_B - \phi_A = 0$) and opposite phase ($\phi_B - \phi_A = 180°$). Another important case is the one in which the phase difference $\phi_B - \phi_A = 90°$. In this case the displacement curve of one simple harmonic motion reaches its maximum when the other curve is at zero, as shown in Fig. 8.5. We would expect the resultant curve obtained by linear superposition to reach its maximum value somewhere between the maxima of curves A and B, which indeed it does. The amplitude of the resultant curve can be shown to be $\sqrt{A^2 + B^2}$ in this case.

FIG. 8.5
Linear superposition of simple harmonic motions with the same period (frequency) but with a phase difference $\phi_B - \phi_A = 90°$. (Compare with Fig. 8.1.)

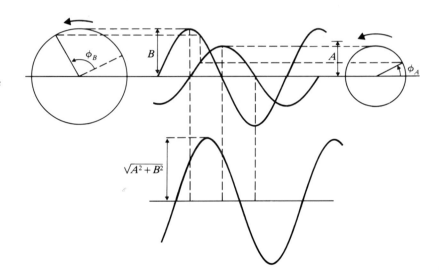

The amplitude of the resultants for the values of phase difference between simple harmonic motions A and B can be calculated. The results are summarized in Table 8.1.

TABLE 8.1   Amplitude of resultant obtained by superposition of two simple harmonic motions with the same frequency.

Phase difference $\phi_B - \phi_A$	0	45°	90°	135°	180°	270°
Amplitude of resultant	$A + B$	$\sqrt{A^2 + B^2 + 1.4AB}$	$\sqrt{A^2 + B^2}$	$\sqrt{A^2 + B^2 - 1.4AB}$	$A - B$	$\sqrt{A^2 + B^2}$

8.4   PURE TONES WITH SLIGHTLY DIFFERENT FREQUENCIES: BEATS

A *pure tone* is a sound wave with a single frequency. The superposition of two pure tones proceeds in the same way as the superposition of two simple harmonic motions, which we discussed in Section 8.3. If the tones have the same frequency, the resultant amplitude will be somewhere between $A + B$ and $A - B$ (or $B - A$, if B is larger than A), depending on their phase difference (see Table 8.1). A pure tone is often referred to as a *sine wave* for the same reason that simple harmonic motion is called sinusoidal (see the box on p. 145).

If two pure tones have slightly different frequencies, f_1 and $f_1 + \Delta f$, the phase difference $\phi_B - \phi_A$ changes continually with time, and so the amplitude of the resultant tone changes also. The amplitude of the resultant varies between $A + B$ and $A - B$ at a frequency Δf. These slow periodic variations in amplitude at frequency Δf are called *beats*. If the amplitudes A and B are equal, the resultant amplitude varies between $2A$ and 0, but this is not necessary for beats to occur.

In the case of two pure tones of slightly different frequency, linear superposition at our ears leads to a sensation of audible beats at the *difference frequency* Δf. These beats are heard as a pulsation in the loudness of a tone at the "average" frequency $f = \frac{1}{2}(f_1 + f_2)$. An example of beats is shown in Fig. 8.6.

So long as the frequency difference Δf is less than about 10 Hz, the beats are easily perceived. When Δf exceeds 15 Hz, the beat sensation disappears, and a characteristic roughness appears. As Δf increases still further, a point is reached at which the "fused" tone at the average frequency gives way to two tones, still with roughness. The respective resonance regions on the basilar membrane are now separated sufficiently to give two distinct pitch signals, but the excitations corresponding to the two pitches still overlap to give an effect of roughness (see Fig. 5.9). When the separation Δf exceeds the width of the critical band (see Section 5.4), the roughness disappears, and the two tones begin to blend. This process is illustrated by the graph (not to scale) in Fig. 8.7.

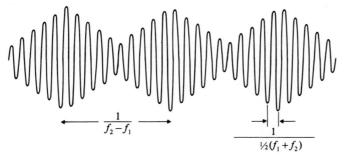

$$\frac{1}{f_2 - f_1}$$

$$\frac{1}{\frac{1}{2}(f_1 + f_2)}$$

FIG. 8.6
Waveform with beats due to pure tones with frequencies f_1 and $f_2 = f_1 + \Delta f$.

FIG. 8.7
Schematic representation of
frequencies heard when pure tones of
frequencies f_1 and f_2 are superimposed.
Note that the disappearance of beats
occurs around $\Delta f = 15$ Hz regardless
of the values of f_1 and f_2; the critical
bandwidth (c.b.) and the "fusion
frequency" (F) increase with f_2 and f_1.
(After Roederer, 1975).

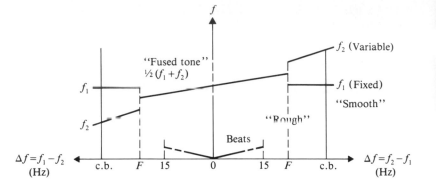

Probably the easiest way to illustrate beats is by connecting two
audio generators to an amplifier and a loudspeaker. A pair of resistors
(about 1000 ohms) can serve as a "mixer." One should keep the frequency of one generator constant at f_1 and vary the other slowly above
and below f_1. Note how the frequency of beats Δf changes as f_2
changes, and note also the frequency f_2 at which the beats disappear
and two separate tones can be distinguished. It is instructive to view
the resultant waveform on an oscilloscope at the same time that it is
heard by using an arrangement like that shown in Fig. 8.8.

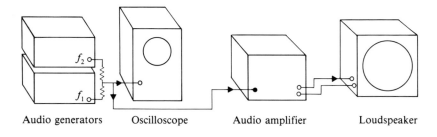

Audio generators   Oscilloscope   Audio amplifier   Loudspeaker

FIG. 8.8
Demonstration of beats with electrical
signals from two audio generators.

The limit of pitch discrimination or "fusion frequency," the point
at which the single fused tone changes to two tones, varies with center
frequency in a manner somewhat like the critical band. It varies from
about a semitone (at 500 Hz) to more than a whole tone* (below 200
Hz and above 4000 Hz) (Plomp, 1964), but is always less than the
critical bandwidth. At the same time it is 7 to 30 times larger than the
just noticeable difference (jnd) for frequency (see Fig. 7.2). In other
words, we can detect a very small change in frequency of a pure tone,
but two tones sounded together may have to differ by a semitone or
more in order to be heard as separate tones.

---

*For a semitone interval $\Delta f/f \approx 0.06$, and for a whole-tone interval $\Delta f/f \approx 0.12$.
These and other musical intervals will be discussed in Chapter 9.

Beats can also be heard when tones of slightly different frequency are presented separately to our two ears; these are called *binaural beats*. They are difficult to detect and are best heard with f_1 around 500 Hz and Δf in the range of 5 to 20 Hz (Perrott and Nelson, 1969). They have been described as a "muffled" sound. When Δf is less than 5 Hz, the beats change to a "rotating tone," a single tone that appears to move around inside the head. Binaural beats can be heard over a wide range of intensity of the stimulus tones; they apparently originate in the central auditory processor, not the individual ears. The rotating tone sensation is probably a result of the continuous change in phase, which makes it appear that the sound source is changing its direction (see Section 5.5).

Beats may be used to tune two audio generators or musical instruments to precisely the same pitch. Although we have discussed beats between pure tones, beats with a slightly different timbre occur between complex tones, and we will discuss beats between tones with various ratios of frequency in Section 8.12. (It is these so-called *second-order beats* that are counted by piano tuners to achieve just the right interval between two notes.)

8.5   THE MUSICAL STAFF: MUSICIANS' GRAPH PAPER

In Section 5.7 we discussed linear and logarithmic scales and the way in which they may be used to represent quantities on graphs. Graph paper that has one logarithmic axis is called semilogarithmic graph paper, and we use it quite frequently for representing quantities that are functions of frequency (see Fig. 5.13). When frequency is represented on a logarithmic scale, octaves (frequency ratio of 2:1) and other familiar musical intervals become equidistant anywhere on the scale (the piano keyboard approximates a logarithmic scale; an octave requires a reach of the same distance anywhere on the keyboard).

Music is written on staves which approximate logarithmic frequency scales. Normally a musical staff consists of five lines, and a clef sign is placed at the beginning of each staff to show the exact location of a particular note (G, F, or C). Piano music is normally written on two staves (connected by a brace) which carry treble and bass clef signs. The five lines of a musical staff are separated either by three or four semitones, as shown in Fig. 8.9.

Although the lines on a normal musical staff are evenly spaced, we have drawn the staves in Fig. 8.9 with two different spacings, so that they exactly fit a logarithmic frequency scale and also indicate clearly whether the lines are separated by an interval of a minor third (three semitones) or a major third (four semitones). Note that the two

FIG. 8.9
The musical staff as semilogarithmic graph paper. Some lines are spaced three semitones (minor third) apart, and some are spaced four semitones (major third) apart. The scale of time on the linear horizontal axis indicates the tempo.

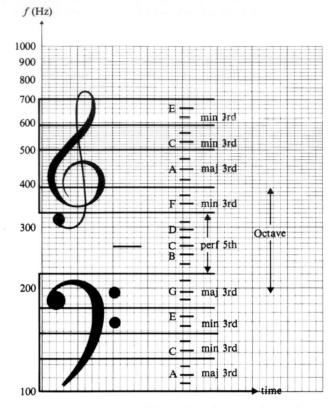

staves in Fig. 8.9 (used in most piano and vocal scores) span a frequency range of just under three octaves from 98 to 698 Hz. This is a small part of the total 7½-octave range of a piano, so ledger lines are added above and below the staves. Also, the notation 8va is used to indicate that notes on the staff are to be played one octave higher than normal (as in Fig. 8.11). Several years ago, we proposed the use of three new clefs to extend written music over a full 10 octaves, the range of audible sound (see Appendix A.7). The musical world has not rushed to adopt our "invention," however.

The horizontal time axis in music is linear. But instead of marking off the axis in seconds, the musical staff uses bar lines (the distance between two bar lines is sometimes called a *measure*). The rate at which the music is to be played is denoted in several ways: by the time signature (e.g., 2/4, 3/4, 6/8, etc.), by the meter (e.g., ♩ = 64, which means 64 quarter notes per minute), and by designations for tempo (e.g., adagio, allegro, vivace, etc.). In 4/4 time at a meter of ♩ = 64, there will be four "beats" per bar or measure, each quarter note receiving one beat with a duration of 1/64 minute, or 60/64 second. Thus each bar of four beats should equal 240/64 = 3.75 s. The two

passages shown in Fig. 8.10 would be represented by straight lines on a graph of frequency (logarithmic scale) vs. time (linear scale). This is because semitone intervals are written as quarter notes (♩) and whole-tone intervals as half notes (♩).

FIG. 8.10
Two passages that would appear as straight lines on a graph of log frequency vs. time.

8.6   COMBINATION TONES

When two tones are sounded together, a lower tone is frequently heard. This undertone is called a *difference tone* or *Tartini tone* after the Italian violinist Tartini, who reportedly discovered it around 1714. If the two tones have frequencies f_1 and f_2, this difference tone, which is an example of a *combination tone,* occurs at a frequency $f_2 - f_1$ (or $f_1 - f_2$).

Difference tones can be demonstrated by using two audio generators, two flutes, or even two soprano voices. The notes shown in Fig. 8.11 should produce a difference tone melody; another such passage is given by Stickney and Englert (1975). I often demonstrate difference tones with a plastic whistle (originally obtained from a box of breakfast cereal) that emits loud tones with frequencies of 1727, 1896, and 2081 Hz. Difference tones at 169, 185, and 374 Hz are clearly audible, and in fact fast beats occur between the 169- and 185-Hz difference tones (see Rossing, 1974). Police in London use whistles of this type.

Other combination tones that can be heard have frequencies given by $2f_1 - f_2$ and $3f_1 - 2f_2$ (Plomp, 1965). The $2f_1 - f_2$ tone, called the cubic difference tone, may be detected at stimulus levels as low as 15 dB, but is heard only in a limited range of frequency below f_1 (Smoorenburg, 1972). Other members of the class given by $f_1 - k(f_2 - f_1)$ can also be detected but with some difficulty ($4f_1 - 3f_2$, for example).

FIG. 8.11
Playing these notes on a piccolo or descant recorder should produce a difference-tone melody about three octaves lower, as shown on the lower clef.

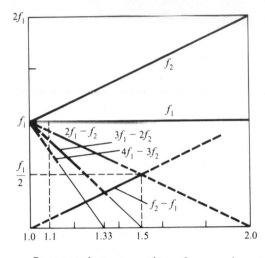

FIG. 8.12
Most prominent combination tones for a two-tone presentation consisting of one tone of fixed frequency f_1 and one of variable frequency f_2. Stronger and weaker tones are indicated by solid and dashed lines, respectively. (After Plomp, 1965).

Suppose that we continue the experiment with two tones described in Section 8.4 by increasing f_2 until it reaches twice the value of f_1. The difference tone $f_2 - f_1$ would be heard over most of the range, and the cubic difference tone $2f_1 - f_2$ would be heard over much of the range as well. These two tones are shown in Fig. 8.12, as well as the primary tones f_2 and f_1 and the less prominent difference tones $3f_1 - 2f_2$ and $4f_1 - 3f_2$. The cubic and higher order difference tones are best heard in the region just above $f_2/f_1 = 1.1$. The two most prominent difference tones are also shown on musical staves in Fig. 8.13 along with the primary tones.

Combination tones have many applications in music, several of which are discussed by Hindemith (1937) in his classic book on musical composition. For example, very low tones can be produced on a small organ by the combination tone from two smaller pipes instead of requiring a large pipe tuned to the desired low pitch. (The reader should be warned, however, that not all of Hindemith's ideas are consistent with more recent experimental results in perception and psychoacoustics).

FIG. 8.13
Combination tones on a musical staff. As in Fig. 8.12, f_1 is fixed and f_2 is variable. Note that the difference tone $(f_2 - f_1)$ and the cubic difference tone $(2f_1 - f_2)$ cross at $f_2 = 1.5f_1$, just as they do in Fig. 8.12. This ratio of frequencies corresponds to an interval called a "perfect fifth."

8.7 MODULATION OF ONE TONE BY ANOTHER

Closely akin to combination tones is the phenomenon of *amplitude modulation* (AM), which will be further discussed in Chapter 22 on components for high-fidelity sound reproduction. When vibrations

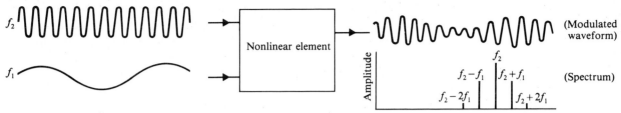

FIG. 8.14
Amplitude modulation of one
vibration by another in a nonlinear
element. The spectrum of the
modulated wave shows
sidebands at $f_2 - f_1$ and $f_2 + f_1$.

occur at two different frequencies in a nonlinear system, various *side-bands* are generated, having frequencies equal to the various differ-ence tones and summation tones as well. If one of the frequencies is considerably less than the other, we describe the process as "modula-tion of the higher frequency by the lower one." That is, the amplitude of the high-frequency component changes at the frequency of the low-frequency component, as shown in Fig. 8.14.

Note that the principal sidebands occur at $f_2 + f_1$ and $f_2 - f_1$ with much smaller sidebands at $f_2 \pm 2f_1$, etc. In the case of pure tones, they are not usually heard as separate tones but as components of a complex tone. One example of audible sidebands occurs in the pitch-shift experiments described in Section 7.6. Intermodulation distortion products generated when several frequencies interact in some sound system component having a slight nonlinearity (such as a loudspeaker) are another example.

8.8   OTHER NONLINEAR EFFECTS: AURAL HARMONICS AND SUMMATION TONES

A single tone of frequency f, if it is sufficiently loud, should produce additional pitch sensations at $2f$, $3f$, $4f$, etc. These are called *aural harmonics,* and they are due to nonlinear behavior in the auditory system (analogous to nonlinearity in an amplifier or loudspeaker).

Aural harmonics were observed many years ago by Fletcher (1929), who suggested a simple power series for the response of the ear:

$$x = a_0 + a_1 p + a_2 p^2 + a_3 p^3 + \dots, \qquad (8.1)$$

where p is sound pressure and the a's are constants that can be deter-mined by experiment. Such a response predicts that for each 1-dB increase in signal level the second harmonic will increase by 2 dB and the third harmonic by 3 dB. This is consistent with the results of exper-iments by Clack (1977) and others. At a signal level of 70 dB the sec-ond and third aural harmonics have sound levels in the neighborhood of 25 and 15 dB but with wide variation between individuals.

If the response of the ear is nonlinear, summation as well as dif-ference tones ought to be generated, but no one has presented convincing evidence that even simple sum tones ($f_1 + f_2$) can be heard, not to mention other summation tones ($2f_1 + f_2, f_1 + 2f_2$, etc.). According to the theory of Helmholtz (1877), the sum and difference tones, which have frequencies of $f_1 + f_2$ and $f_1 - f_2$ respectively, ought to have amplitudes in the ratio

$$\left(\frac{f_1 - f_2}{f_1 + f_2}\right)^2.$$

Since $f_1 - f_2$ is usually much smaller than $f_2 + f_1$, this ratio is quite small. Lying above f_1 and f_2 in frequency, the weaker summation tones may be masked by the primary tones.

8.9 ORIGIN OF DIFFERENCE TONES

Difference tones have been studied extensively, not only because of their importance in the perception of musical sound, but also because they provide us with a window into the functioning of the auditory system. Intensities of difference tones can be measured by adjusting the amplitude and phase of a "cancellation tone" until it just cancels the difference tone of interest (Zwicker, 1955).

The difference tone with frequency $f_2 - f_1$ (properly called the quadratic difference tone) has the behavior we would expect from a quadratic distortion product (i.e., one that results from the quadratic term in Eq. 8.1). If either one of the primary tones increases 3 dB in level, the difference tone also increases 3 dB; if both primary tones increase 3 dB, the difference tone increases 6 dB. If the cubic difference tone were a result of the cubic term in Eq. 8.1 (i.e., a cubic distortion product), it would be expected to increase 9 dB when each primary tone increased 3 dB. Instead, the cubic difference tone is observed to increase more nearly 3 dB, or about half as much as the quadratic difference tone.

Intensities of the cubic (and higher order) difference tones decrease rapidly with increasing frequency ratio f_2/f_1 (Goldstein, 1967). They are best heard in the region just above $f_2/f_1 = 1.1$ (see Fig. 8.12). The quadratic difference tone $f_2 - f_1$, on the other hand, can be heard over a wider range of frequency, and its intensity varies much less with the frequency ratio f_2/f_1. These results, among others, seem to indicate that the quadratic and cubic difference tones are *not* produced in the same way in the ear. The cubic and higher order difference tones appear to be produced in the frequency selective inner ear, whereas the quadratic difference tone results from a nonlinearity without frequency selectivity, probably in the middle ear.

It would be well to point out that the beats that occur when $f_2 - f_1$ is small (less than about 15 Hz) are distinctly different from difference tones that occur at greater frequency separation. Beats, which are heard as periodic variations in intensity, do *not* require nonlinearity in the ear; audible difference tones *do*. Hall (1981) demonstrates this by presenting tones f_1 and f_2 with and without nonlinear distortion (and also shows the spectra of the audible signals). Without the distortion, masking noise that spans the region of f_1 and f_2 can mask the difference tones, but adding distortion generates a signal at the difference tone frequency that can be heard in the presence of noise.

8.10 CONSONANCE AND DISSONANCE: MUSICAL INTERVALS

Pythagoras of ancient Greece is considered to have discovered that the tones produced by a string vibrating in two parts with simple length ratios such as 2:1, 3:2, or 4:3 sound harmonious. These ratios define the so-called perfect intervals of music, which are considered to have the greatest consonance. Galileo (1638) observed: "Agreeable consonances are pairs of tones which strike the ear with a certain regularity; this regularity consists in the fact that the pulses delivered by the two tones, in the same time, shall be commensurable in number, so as not to keep the eardrum in perpetual torment, bending in two different directions in order to yield to the ever-discordant impulses." The most consonant intervals of music are generally considered to be the following (in descending order of consonance):

2:1	octave	(C/C),
3:2	perfect fifth	(G/C),
4:3	perfect fourth	(F/C),
5:3	major sixth	(A/C),
5:4	major third	(E/C),
8:5	minor sixth	(A♭/C),
6:5	minor third	(E♭/C).

Why are some intervals more consonant than others? Helmholtz (1877) explained *consonance* by referring to Ohm's acoustical law (see Section 7.5), which stated that the ear performs a spectral (Fourier) analysis of sound, separating a complex sound into its various partials. Helmholtz concluded that *dissonance* occurs when partials of the two tones produce 30–40 beats per second. The more the partials of one tone coincide with the partials of the other, the less chance that beats in this range will produce roughness (dissonance). This explains why simple ratios define the most consonant intervals.

More recent studies have shown that Helmholtz was on the right track. A more accurate picture of consonance and dissonance should be based on critical bands in hearing (which, of course, were unknown to Helmholtz). We will briefly consider the consonance of two pure tones, two complex tones, and chords of three or four notes.

Two Pure Tones

Research by Plomp and Levelt in The Netherlands and by Kameoka and Kuriyagowa in Japan led to the same conclusion: The consonance or dissonance of two pure tones sounded together depends upon their frequency difference rather than on their frequency ratio. If the frequency difference between two pure tones is greater than a critical band, they sound consonant; if it is less than a critical band they sound dissonant. Plomp and Levelt (1965) found that maximum dissonance occurs when the frequency difference is approximately $\frac{1}{4}$ of a critical bandwidth, as shown in Fig. 8.15. Kameoka and Kuriyagowa (1969), however, suggested a slightly more complicated dependence on frequency and also found a dependence on sound pressure level. Their empirical expression giving the frequency difference for greatest dissonance Δf_d is:

$$\Delta f_d = 2.27 \left(1 + \frac{L_p - 57}{40}\right) f^{0.447} \tag{8.2}$$

where L_p is sound pressure level and f is the frequency of the primary tone. Note that around 500 Hz, the maximum dissonance predicted in both Eq. 8.2 and Fig. 8.15 corresponds reasonably well with the 30–40 Hz criterion of Helmholtz.

It is instructive to examine a few intervals between pure tones that occur in different octaves. If C_4 (262 Hz) and G_4 (392 Hz) are sounded together, the difference frequency is 130 Hz, which is 40 percent greater than the critical bandwidth (approximately 90 Hz in this octave). Thus they sound consonant. However, an octave lower on the

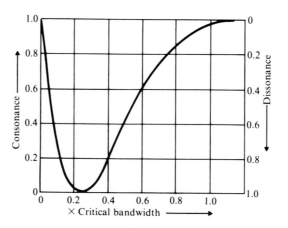

FIG. 8.15
Consonance of two pure tones as a function of frequency separation relative to the critical bandwidth (Plomp and Levelt, 1965).

scale, a perfect fifth is not quite so consonant, since the frequency difference between C_3 (131 Hz) and G_3 (196 Hz) is 65 Hz, which is less than the critical bandwidth. Another octave or so lower, the frequency difference approaches $\frac{1}{4}$ of the critical bandwidth, the criterion for maximum roughness or dissonance. So the degree of dissonance of the interval between two pure tones is strongly dependent on their location on the musical scale. The higher on the scale, the closer two pure tones can be to each other in pitch and still sound consonant when sounded together.

Two Complex Tones

In the case of musical tones, which nearly always have several harmonics, the situation is quite different, however. Roughness can occur between the harmonics of the tones as well as between the fundamentals. Herein lies the reason why certain intervals are inherently more consonant than others. Fewer harmonics of the more consonant intervals have frequency differences within the roughness range.

In Table 8.2, harmonics of four musical intervals are tabulated. We have used frequency ratios from the just scale rather than the scale of equal temperament (see Section 9.1). In the case of the perfect fifth, two of the lower harmonics coincide and two produce frequency differences within the critical bandwidth but well away from the range of maximum roughness. In the case of the minor third, however, there are many interactions that produce roughness, and in the case of the major second, nearly all the harmonics thus interact.

Figure 8.16 shows the dissonance to be expected between two tones, each having six harmonics, making the assumption that dissonances between various harmonics are additive. Maxima in conso-

TABLE 8.2 Interactions between harmonics of two tones separated by different intervals

Perfect fifth C_4:G_4($f_2/f_1 = 3/2$)		Major sixth C_4:A_4 ($f_2/f_1 = 5/3$)		Minor third C_4:E_4 ($f_2/f_1 = 6/5$)		Major whole tone C_4:D_4 ($f_2/f_1 = 9/8$)	
mf_1	nf_2	mf_1	nf_2	mf_1	nf_2	mf_1	nf_2
262	392	262	436	262	314	262	294
523		523		523 R	628	523 R	589
785 =	785	785 R	872	785		785 R	883
1047 R*	1177	1047		1047 R	942	1047 R	1177
1308		1308 =	1308	1308 R	1256	1308	
1570 =	1570	1570		1570 =	1570	1570 R	1472
1831 R	1962	1831 R	1744	1831 R	1884	1831 R	1766
2093 R		2093 R	2180	2093 R	2198	2093 R	2060

*R denotes roughness due to frequency difference within the critical bandwidth.

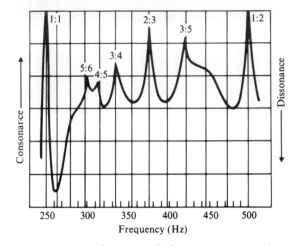

FIG. 8.16
Consonance expected from tone *A* with *f* = 250 Hz and tone *B* with varying frequency, both *A* and *B* having six harmonics. (From Plomp and Levelt, 1965).

nance occur when tone *B* forms consonant intervals with tone *A*. The major sixth (5:3) appears more consonant than the perfect fourth (5:4) under these assumptions.

It certainly should not be implied that consonant intervals are good and dissonant intervals bad. Music written with consonant intervals alone would be exceedingly dull; musicians make a clear distinction between pleasantness and consonance. Sixths (5:3, 8:5) and thirds (5:4, 6:5) are generally found to be pleasant intervals, as are the fourth (5:4) and minor seventh (9:5), even though some of these intervals are not particularly consonant.

Chords

When three or more musical (complex) tones are sounded together in a chord, there are many opportunities for roughness-producing interactions between the various partials that lie within critical bandwidths of one another. Musicians generally consider major and minor chords to be more consonant than diminished and augmented chords, which sound dissonant and need to be resolved (see Fig. 8.17). In a psychoacoustic study of three- and four-note chords, Roberts (1986) found major chords to be the most consonant, followed (in order) by minor, diminished, and augmented chords. Chords in root position were found to be more consonant than in first or second inversion, and chords in equal temperament were found to be more consonant than chords in just or Pythagorean tuning (see Chapter 9). Chords were judged to be more consonant when heard in a traditional musical context.

FIG. 8.17
Examples of: (a) Major triad; (b) Minor triad; (c) Diminished triad; (d) Augmented triad; (e–h) Same chords in first inversion.

An interesting indication of the preferences of musicians resulted from a statistical analysis by Plomp and Levelt (1965) of chords in two musical compositions. They analyzed a movement of J. S. Bach's *Trio Sonata for Organ, No. 3* and a movement of A. Dvořák's *String Quartet, Op. 51,* and determined the interval width that is not exceeded 25%, 50%, and 75% of the time. Figure 8.18 shows the results for each piece, taking into account only the fundamental ($n = 1$) and also the first nine harmonics ($n = 9$). In other words, Fig. 8.18(a) indicates the distribution in the size of intervals in which each of seven notes appears along with another note of higher frequency. Figure 8.18(b) includes many more cases in

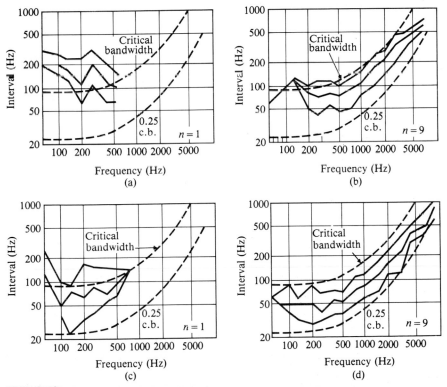

FIG. 8.18

(a) and (b) Statistical analysis of the chords in Bach's *Trio Sonata for Organ, No. 3* (C-minor). (c) and (d) Similar analysis of chords in Dvořák's *String Quartet, Op. 51* (E♭-major). Shown in each graph are the interval widths (in Hz) not exceeded 25%, 50%, and 75% of the time. The dotted lines indicate the critical bandwidth and one-fourth of the critical bandwidth. In (a) and (c) only the fundamental is analyzed; in (b) and (d) the first nine harmonics are included. (From Plomp and Levelt, 1965).

which one of fourteen notes appears as a fundamental or as a harmonic of the fundamental.

Note that in the Bach composition, at least half of the intervals between fundamentals exceed the critical band, and thus would be judged as consonant. Inclusion of the harmonics places many intervals in the musically interesting region of dissonance, but seldom is the maximum degree of dissonance approached. Dvořák's composition, however, shows a much larger percentage of dissonant intervals.

Papers by Houtsma and Goldstein (1972) and by Terhardt (1974) discuss the role of the central processor in the perception of musical intervals. Terhardt feels that tonal meaning plays at least as important a role as roughness in determining consonance and dissonance.

8.11   EFFECT OF PHASE ON TIMBRE

A good chef knows that the quality of the end product depends not only on using the right amount of the various ingredients but also combining them in a prescribed way (likewise the chemistry student is cautioned to pour acid into water, not water into acid!) In the same way, building up complex tones using the same spectrum with different phases between the harmonics can lead to totally different waveforms. An interesting question to consider is: Do these different waveforms, which have the same harmonic spectrum, sound different?

Helmholtz (1877) stated that "the quality of the musical portion of a compound tone depends solely on the number and relative strength of its partial simple tones, and in no respect on their differences of phase." Although Helmholtz felt the influence of phase between various harmonics to be negligibly small in general, he did recognize exceptions in the case of "unmusical" sounds. Experiments by R. König using siren discs, performed about the same time, indicated that timbre has some dependence on phase. Using a generator that controlled both the phase and the amplitude of sixteen harmonics, Licklider (1959) noted that changing the phase of a high-frequency component has more effect on timbre than changing the phase of a low-frequency component.

Plomp (1970) summarizes the most important results of his experiments conducted with H. J. M. Steeneken on phase and timbre:

1. The maximum effect of phase on timbre is the difference between a complex tone in which the harmonics are in phase and one in which alternate harmonics differ in phase by 90°.

2. The effect of lowering each successive harmonic by 2 dB is greater than the maximum phase effect described above.

3. The effect of phase on timbre appears to be independent of the sound level and the spectrum.

Effect of Phase on Waveform

Four of the waveforms used by Plomp and Steeneken in their experiments on phase and timbre are shown in Fig. 8.19. All four of them consist of ten harmonics with amplitudes proportional to $1/n$ but with different phases. In the first two, the harmonics are in phase; in waveforms 3 and 4, alternate harmonics differ in phase by 90°. Mathematical expressions for the waveforms are:

$$\sin \omega t + \tfrac{1}{2} \sin 2\omega t + \tfrac{1}{3} \sin 3\omega t + \ldots \tfrac{1}{10} \sin 10\omega t \qquad (1)$$

$$\cos \omega t + \tfrac{1}{2} \cos 2\omega t + \tfrac{1}{3} \cos 3\omega t + \ldots \tfrac{1}{10} \cos 10\omega t \qquad (2)$$

$$\sin \omega t + \tfrac{1}{2} \cos 2\omega t + \tfrac{1}{3} \sin 3\omega t + \ldots \tfrac{1}{10} \cos 10\omega t \qquad (3)$$

$$\cos \omega t + \tfrac{1}{2} \sin 2\omega t + \tfrac{1}{3} \cos 3\omega t + \ldots \tfrac{1}{10} \sin 10\omega t \qquad (4)$$

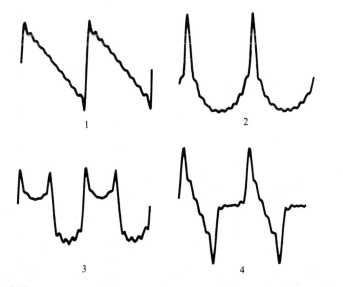

FIG. 8.19
Four waveforms consisting of ten harmonics in different phase patterns. Waveforms 3 and 4 sound different from waveforms 1 and 2 (Plomp and Steeneken, 1969)

Plomp and Steeneken (1969) found that their subjects could easily distinguish waveforms 3 and 4 from 1 and 2, but it was much more difficult to distinguish 1 from 2 and 3 from 4. Additional experiments established that the timbre difference between the two groups is the greatest possible using this spectrum of harmonics; hence the first conclusion above.

The nature of the amplitude envelope can have an effect on timbre. When all the harmonics are in phase, for example, the resulting "spikes" tend to make the tone sound rougher than a complex tone with the same harmonics in random phase (Patterson, 1973). Phase shift that increases linearly with frequency (or with harmonic number) delays the tone but leaves the amplitude envelope unchanged; most observers find the timbre unchanged (an observation of importance to designers of high fidelity sound reproducing equipment, as we will see in Chapter 22).

Adding distortion can produce noticeable changes in the timbre of a complex tone when phase angles are changed (Demonstration 33, Houtsma, Rossing, and Wagenaars, 1987), since the distortion products now interfere constructively or destructively with the harmonics, depending upon their relative phases.

The mechanism for phase perception is not well understood at present. Goldstein (1967) constructed a theory that explains certain experiments on phase perception on the basis of the ear's limited frequency resolution. When two or more spectral components lie within a critical band, the ear is not able to resolve them, so it seeks clues from the time envelope.

It should be mentioned that whereas the effects of phase on timbre are small for steady tones, the ear is in fact quite sensitive to *changes* in phase, especially if they take place at a regular rate. This is illustrated by the phenomenon known as *second-order beats,* to be described in Section 8.12. Thus, in the dynamical case, phase can have an appreciable effect on timbre.

8.12   BEATS OF MISTUNED CONSONANCES

A sensation of beats occurs when the frequencies of two tones f_2 and f_1 are nearly, but not quite, in a simple ratio. If $f_2 = 2f_1 + \delta$, beats are heard at a frequency δ. In general, when $f_2 = {}^n/_m f_1 + \delta$, $m\delta$ beats occur per second. These are sometimes called *second-order beats,* and have been discussed by Helmholtz (1877), Plomp (1967), and many others. They are described by various observers as variations in timbre, changes in loudness of one or both constituent tones, etc., and they are quite prominent even at low sound levels. They are clearly visible on an oscilloscope as periodic changes in pattern of the type shown in Fig. 8.20. These changes in pattern occur in synchronism with the beat sensations heard. Second-order beats between mistuned consonances can be heard up to about 1500 Hz or so, depending on the sound level.

In the case of a mistuned octave, we observe envelope changes as large as 13 percent that correspond to variations in sound level of about 1.3 dB, too small to account for the prominent beats that are observed, however. Apparently the beats of mistuned consonances are related to periodic variations of the waveform. The ear, which is a poor detector of static phase, appears to be quite sensitive to cyclical variations in phase. In the light of modern auditory theory, it would appear that the beats are due to the periodicity of nerve impulses. Nerve impulses are evoked when the displacement of the basilar membrane passes a certain critical value, and thus slow variations in the waveform, corresponding to beats, result in slow variations in the time pattern of the impulses.

Beats of mistuned consonances have long been used by piano tuners, for example, to tune fifths and fourths or even octaves on the piano. Violinists also make use of them in tuning their instruments.

When the two interacting tones include harmonics of the fundamental, second-order beats generally are stronger than when they are pure tones. Most musical tones have harmonics, and in this case ordinary first-order beats can occur between the various harmonics. For example, in a mistuned fifth consisting of tones of 220 and 332 Hz, the third harmonic of f_1 and the second harmonic of f_2 have frequencies of 660 and 664 Hz, and they will produce four beats per second. However, the beats between complex tones do not always appear louder than between pure tones of the same frequency (Rossing and Dols, 1976); this phenomenon needs further study.

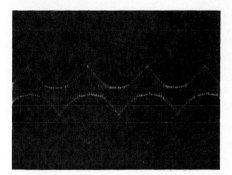

FIG. 8.20
Oscilloscope presentation of beats of a mistuned fifth (pure tones); $f_1 = 332$ Hz, $f_2 = 220$ Hz.

8.13   THE CENTRAL NERVOUS SYSTEM: AUTOCORRELATION AND CROSS-CORRELATION

In Chapters 5–8, we discussed many aspects of the modern theory of hearing. We have not described how auditory data is processed in the central nervous system, however. In Sections 8.13 and 8.14, only a brief summary will be given of this rather complex subject, which has been the focus of much recent research in psychoacoustics.

The basic building block of the nervous system is the *neuron* or nerve cell, which both transmits and processes neural impulses. The neuron has receptors called *dendrites,* which receive information from other neurons, and *axons,* which transmit that information to selected neurons. The "wiring scheme" among our ten billion neurons, which determines which neurons receive pulses from which others, is the key to human intelligence. Although much of the wiring scheme is fixed, in the cerebral cortex the interconnection of neurons is the result of repeated stimulation patterns, which leads to *learning*.

It is possible to attach very small electrodes to nerve fibers and thus observe the rate of neural impulses. Some neurons generate pulses spontaneously at a certain rate that can be either increased or decreased in accordance with the pulses received from other neurons. When several nerve fibers receiving stimuli from one region of the basilar membrane are tied together, the sum of their impulses constitutes a *volley,* which is synchronous with the auditory stimulus.

A given auditory nerve fiber carries two types of information (Roederer, 1975):

1. The fact that it is firing at all indicates that the basilar membrane has been excited at a particular place; this gives information on primary pitch at all frequencies ("place theory").

2. The time distribution of impulses carries information on repetition rate or periodicity ("periodicity pitch") and possibly on the vibration pattern itself; this works only in the lower frequency range.

Two very important processing functions that the nervous system appears able to perform are autocorrelation and cross-correlation. *Autocorrelation* is the comparison of a pulse train with previous pulse trains in order to pick out repetitive features. Autocorrelation could account for the pitch of delayed noise described in Section 7.6, for example. *Cross-correlation,* on the other hand, describes a comparison between signals on two different nerve fibers (from the cochleas of our two ears, for example). Cross-correlation could account for the localization of sound at low frequencies by measuring the time delay between signals from our two ears, for example.

8.14   CEREBRAL DOMINANCE

Information from our left ear passes preferentially to the right side of the brain (right cerebral cortex) and information from our right ear to the left side. (This is consistent with the bilateral symmetry exhibited in nearly all our sensory and motor processing.) There are, of course, pathways from right ear to right cerebral cortex and strong ties between the right and left cortexes, so that both sides of the brain receive processed information from both ears.

Clinical and experimental evidence has shown that the dominant hemisphere of the brain (the left side in 97 percent of the population) is specialized for speech processing and the minor hemisphere for non-linguistic functions such as music. This is apparently related to the way in which most people process speech and music. Speech processing requires analytic and serial processing of incoming information for which the dominant hemisphere is best suited. Most aspects of musical perception, on the other hand, require holistic or synthetic processing, which is better done in the minor hemisphere.

Recognition of melodies apparently requires some of each. It has been found that musically experienced listeners recognize melodies better in their right ear than their left, while the reverse is true for nonmusicians (Bever and Chiarello, 1974). This suggests that musicians learn to process melodies as they do speech in the dominant hemisphere. Patients with severe traumatic speech impediments are sometimes able to sing a song that had been learned before the trauma had arisen, even though they could not speak the same words.

8.15 SUMMARY

Linear superposition of two pure tones with nearly the same frequency produces *beats,* or variations in loudness at the frequency difference. As the frequency difference increases, a sensation of roughness or *dissonance* develops, finally giving way to *consonance* when the interval between the tones exceeds the critical bandwidth.

Combination tones and aural harmonics are indicative of the nonlinear behavior of the auditory system. The most audible combination tones are the difference tone $(f_2 - f_1)$ and the cubic difference tone $(2f_1 - f_2)$. Differences tones are important in music.

Musical intervals appear to be consonant when their frequency ratios are simple numbers or when the differences between the frequencies of most of the various harmonics exceed the critical bandwidth. Maximum dissonance occurs when frequency differences are as small as one-fourth the critical bandwidth. Skilled use of dissonant intervals adds interest to musical composition and performance. Beats between slightly mistuned consonances are evidence of the ear's ability to detect changes in waveform.

Although the ear is a relatively poor detector of phase, the relative phase of various harmonics can affect the timbre of musical sound. Cyclical variations in pattern or phase, which occur in the case of mistuned consonances, can be heard as beats. The central nervous system makes extensive use of both autocorrelation and cross-correlation in processing sound.

References and Suggested Readings

Bever, T. G., and R. J. Chiarello (1974). "Cerebral Dominance in Musicians and Non-Musicians," *Science* **185:** 537.

Clack, T. D. (1977). "Growth of the Second and Third Aural Harmonics of 500 Hz," *J. Acoust. Soc. Am.* **62:** 1060.

Fletcher, H. (1929). *Speech and Hearing.* New York: Van Nostrand.

Galileo Galilei (1638). *Dialogues Concerning Two New Sciences,* Trans., H. Crew and A. de Salvio. New York: Dover, 1952.

Goldstein, J. L. (1967). "Auditory Nonlinearity," *J. Acoust. Soc. Am.* **41:** 676.

Hall, D. E. (1981). "The Difference Between Difference Tones and Rapid Beats," *Am. J. Phys.* **49:** 632.

Helmholtz, H. L. F. von (1877). *On the Sensations of Tone as a Physiological Basis for the Theory of Music,* 4th ed. Trans., A. J. Ellis. New York: Dover, 1954.

Hindemith, P. (1937). *The Craft of Musical Composition.* Trans., A. Mendel. New York: Associated Music Publishers, 1945.

Houtsma, A. J. M., and J. L. Goldstein (1972). "Perception of Musical Intervals: Evidence for the Central Origin of the Pitch of Complex Tones," *J. Acoust. Soc. Am.* **51:** 520.

Houtsma, A. J. M., T. D. Rossing, and W. M. Wagenaars (1987). *Auditory Demonstrations* (Philips Compact Disc #1126–061 and text). Woodbury, N.Y.: Acoustical Society of America.

Kameoka, A., and M. Kuriyagawa (1969). "Consonance Theory I: Consonance of Dyads," *J. Acoust. Soc. Am.* **45:** 1451.

Licklider, J. C. R. (1959). "Three Auditory Theories," in *Psychology: A Study in Science,* Vol. 1. Ed., S. Koch. New York: McGraw-Hill.

Oster, G. (1973). "Auditory Beats in the Brain," *Sci. Am.* **229**(4): 94.

Patterson, R. D. (1973). "The Effects of Relative Phase and the Number of Components on Residue Pitch," *J. Acoust. Soc. Am.* **53:** 1565.

Perrott, D. R., and M. A. Nelson (1969). "Limits for the Detection of Binaural Beats," *J. Acoust. Soc. Am.* **46:** 1477.

Plomp, R. (1965). "Detectibility Threshold for Combination Tones," *J. Acoust. Soc. Am.* **37:** 1110.

Plomp, R. (1967). "Beats of Mistuned Consonances," *J. Acoust. Soc. Am.* **42:** 462.

Plomp, R. (1970). "Timbre as a Multidimensional Attribute of Complex Tones," in *Frequency Analysis and Periodicity Detection in Hearing.* Eds., R. Plomp and G. Smoorenburg. Leiden: Sijthoff.

Plomp, R. (1976). *Aspects of Tone Sensation.* London: Academic Press.

Plomp, R., and W. J. M. Levelt (1965). "Tonal Consonance and Critical Bandwidth," *J. Acoust. Soc. Am.* **38:** 548.

Plomp, R., and H. J. M. Steeneken (1969). "Effect of Phase on the Timbre of Complex Tones," *J. Acoust. Soc. Am.* **46:** 409.

Roberts, L. A. (1986). "Consonance Judgments of Musical Chords by Musicians and Untrained Listeners," *Acustica* **62:** 163.

Roederer, J. G. (1975). *Introduction to the Physics and Psychophysics of Music,* 2nd ed. New York: Springer-Verlag.

Rossing, T. D. (1974). "Subjective Tones from a Toy Whistle," *Am. J. Physics* **42:** 616.

Rossing, T. D. (1979). "Physics and Psychophysics of High-fidelity Sound," *Phys. Teach.* **17:** 563.

Smoorenburg, G. F. (1972). "Audibility Region of Combination Tones," *J. Acoust. Soc. Am.* **52,** 603; "Combination Tones and Their Origin," *J. Acoust. Soc. Am.* **52:** 615.

Stickney, S. E., and T. J. Englert (1975). "The Ghost Flute," *Phys. Teach.* **13:** 518.

Terhardt, E. (1974). "Pitch, Consonance and Harmony," *J. Acoust. Soc. Am.* **55:** 1061.

Wightman, F. L., and D. M. Green (1974). "The Perception of Pitch," *Am. Scientist* **62:** 208.

Zwicker, E. (1955). Der ungewöhnliche Amplitudengang der nichtlinearen Verzerrungen des Ohres." *Acustica* **5:** 67.

Glossary

aural harmonic A harmonic that is generated in the auditory system.

autocorrelation The comparison of a signal with a previous signal in order to pick out repetitive features.

axon That part of a neuron or nerve cell that transmits neural pulses to other neurons.

beats Periodic variations in amplitude that result from the superposition or additon of two tones with nearly the same frequency.

combination tones A secondary tone heard when two primary tones are received. Combination tones are usually difference tones, although summation tones are possible.

consonance Tones presented together with a minimum of roughness.

critical band The range of frequencies over which tones simply add in loudness; the critical bandwidth appears to determine consonance or dissonance.

cross-correlation The comparison of two signals to pick out common features.

dendrite That part of a neuron that receives neural pulses from other neurons.

difference tone When two tones having frequencies f_1 and f_2 are sounded together a difference tone with frequency $f_2 - f_1$ is often heard. (Properly this should be called the quadratic difference tone to distinguish it from the cubic and other difference tones.)

dissonance Roughness that results when tones with appropriate frequency difference are presented simultaneously.

linear superposition Addition of two waves applied simultaneously to a linear system.

major triad A chord of three notes having intervals of a major third and a minor third, respectively (as C : E : G).

minor triad A chord of three notes having intervals of a minor third and a major third, respectively (as C : E^b : G).

musical staff (pl: **staves**) A five-line graph on which musical notes are written. A clef sign shows the exact location of some particular note.

modulate To change some parameter (usually amplitude or frequency) of one signal in proportion to a second signal.

neuron, or nerve cell Building block of the nervous system that both transmits and processes neural pulses.

phase The fractional part of a period through which a waveform has passed. Phase is often expressed as an angle that is an appropriate fraction of 360°.

phase difference The difference in phase angle between two simple harmonic motions or waves. (If the phase difference is zero, they are in phase; if it is 180°, they are in opposite phase.)

second-order beats Beats between two tones whose frequencies are nearly but not quite in a simple ratio; also called **beats between mistuned consonances**.

sidebands Sum and difference tones generated during modulation.

sine wave A waveform that is characteristic of a pure tone (that is, a tone without harmonics or overtones) and also simple harmonic motion.

Questions for Discussion

1. Is there any technical difference between 4/4 time with ♩ = 96 and 2/2 time (also designated as "cut time" ¢) with ♩ = 48? Is there any implied difference in mood, interpretation, etc.?

2. The lowest fifth on a piano is from A$_0$ (27.5 Hz) to E$_1$ (41.2 Hz). By applying the criteria for roughness of Plomp and Levelt, show why it sounds less pleasant than the fifth from A$_4$ (440 Hz) to E$_5$ (660 Hz), for example.

3. Pianos are tuned to a "tempered scale" for which A$_4$ = 440 Hz and E$_5$ = 659 Hz, whereas a "perfect fifth" would require A$_4$ = 440 Hz and E$_5$ = 660 Hz. Describe how a piano tuner could first set A to exactly 440 Hz by using a 440-Hz tuning fork, and then set E$_5$ to 659 Hz by listening for second-order beats. How many beats should the tuner hear? How can he be sure that he has not set E to 661 Hz rather than 659 Hz?

4. The strings of a violin are tuned at intervals of fifths. Is it more likely that they will be tuned to perfect fifths or to the fifths found on a piano? (You may wish to discuss this with a violinist.)

Problems

1. Make a linear superposition of a square-wave motion and a simple harmonic (sine-wave) motion with half the amplitude and twice the frequency of the square wave. This can be done by making a graph of each waveform to scale and then adding the ordinates point by point. (Alternatively, of course, it could be done by computation.)

2. Pure tones with frequencies of 440 and 448 Hz are sounded together. Describe what is heard (pitch of the fused tone, frequency of beats). Do the same for tones with frequencies of 440 and 432 Hz.

3. Verify each of the six amplitudes in Table 8.1 when $A = 3$ and $B = 2$ by drawing to scale vectors A and B (with correct angles between them) to represent the two harmonic motions, and a vector $A + B$ to represent the resultant. Compare the lengths of the resultants to those calculated from Table 8.1.

4. Calculate the first three difference frequencies that result from $f_1 = 900$ Hz and $f_2 = 1000$ Hz.

5. Make a list of possible subjective tones (combination tones, aural harmonics) that might result when tones of 240 and 300 Hz are sounded together. Indicate the most prominent tones by an asterisk (*). Enclose in parentheses those that occur only for loud tones.

6. Square waves of 200 and 301 Hz are sounded together. How many beats are heard? Write out the harmonics of each tone and indicate harmonics that might beat with harmonics of the other tone. Do the same for sawtooth waves at these frequencies. Can second order beats be explained as beats between harmonics of the two tones?

7. Examine the harmonics of two tones for roughness in the manner shown in Table 8.2 when the tones are:

   a) C_3 and F_3 ($f = 130.8$ and 174.4 Hz);

   b) C_2 and E_2 ($f = 65.4$ and 82.4 Hz).

   To what musical intervals do these correspond?

8. Using frequencies of the first three notes in Fig. 8.11 ($B_2^\flat$, $B_5^\flat$, and C_6) as given in Table 9.2, verify that the lowest note nearly corresponds to the difference tone produced by the upper two. What is the melody shown in Fig. 8.11?

9. When an AM radio station amplitude-modulates a carrier wave of 800 kHz with a single pure tone of 1000 Hz, what are the principal sideband frequencies?

10. Using the frequencies in Table 9.2, how many beats per second occur between C_4 and E_4? (Regard them as having an interval of a major third, slightly mistuned from their 5/4 ratio on the just scale.)

Musical instruments are found in nearly all cultures of the world, past and present. Considerable variation is found in the sophistication of design and the level of craftsmanship in construction, as would be expected. Musical instruments have been the main source of musical sound and the most practical means of musical expression through the ages.

There are many ways to classify musical instruments. One of the most common methods places them in three familiar families: string, wind, and percussion. For convenience we describe the wind family in two different chapters: brasses in one and woodwinds in the other. We have also gathered the discussions of various keyboard instruments from different families into one chapter. Our discussion begins with a chapter on musical scales and temperament.

There will probably be a tendency for the reader to read first the chapter that discusses a favorite family of instruments and to spend less time with the others. This is quite understandable, although the reader should be reminded that an understanding of one's own instrument is greatly enhanced by comparing and contrasting it with the acoustical behavior of other families of instruments.

PART THREE

Acoustics of Musical Instruments

CHAPTER 9
Musical Scales and Temperament

The word "scale" is derived from a Latin word (*scala*) meaning "ladder" or "staircase." A musical scale is a succession of notes arranged in ascending or descending order. Most musical composition is based on scales, the most common ones being those with five notes (*pentatonic*), twelve notes (*chromatic*), or seven notes (major and minor *diatonic*, Dorian and Lydian modes, etc.). Western music divides the octave into 12 steps called *semitones*. All the semitones in an octave constitute a *chromatic scale* or 12-tone scale. However, most music makes use of a scale of seven selected notes, designated as either a *major* scale or a *minor* scale and carrying the note name of the lowest note. For example, the C-major scale is played on the piano by beginning with any C and playing white keys until another C is reached.

Other musical cultures use different scales. The pentatonic or five-tone scale, for example, is basic to Chinese music but also appears in Celtic and Native American music. A pentatonic scale can be played on the piano by beginning with C♯ and playing only black keys, or by playing the notes C, D, F, G, A, C. The scales of Indian music are often said to be quite complex because of the abundance of microtonal intervals. However, Benade (1976) points out that Indian music is based on a seven-tone scale quite similar to our own major scale, and that the microtonal decorations are a matter of style, not unlike that of the American jazz player.

There are many ways to construct musical scales. The construction of scales has appealed to mathematicians as well as to musicians since the time of the Greeks. We will discuss three important scales: the just scale, the Pythagorean scale, and the scale of equal temperament.

9.1 THE SCALE OF JUST INTONATION

The *scale of just intonation* (or just diatonic scale) is based on the *major triad,* a group of three notes that sound particularly harmonious (for example, C : E : G). The notes of the major triad are spaced in two intervals: a major third (C : E) and a minor third (E : G). When these intervals are made as consonant as possible, the notes in the major triad are found to have frequencies in the ratios 4:5:6.

The psychophysical basis of consonance and dissonance was discussed in Section 8.9, where the most consonant musical intervals were given as:

2:1	octave,
3:2	perfect fifth,
4:3	perfect fourth,
5:3	major sixth,
5:4	major third,
8:5	minor sixth,
6:5	minor third.

The numbers above are the frequency ratios of the just intervals. These intervals are consonant because the dissonant combinations between harmonics of the two tones are minimal, especially in the case of the so-called perfect intervals (octave, fifth, and fourth), which have the simplest ratios.

The three major triads in a major scale are the tonic, subdominant, and dominant chords (also called the I, IV, and V chords, since they are built on the first, fourth, and fifth notes of the scale). The frequencies of all seven notes in the just diatonic scale can be determined by letting these three triads consist of notes with the frequency ratios 4 : 5 : 6.

Frequency Ratios in the Just Scale

This may be illustrated in the key of C as follows. First we let the notes of the tonic chord (C, E, G) have the ratios 4 : 5 : 6 and set C = 1; we have now determined C = 1, E = 5/4, G = 6/4 = 3/2. Next we let the notes of the dominant chord (G, B, D) be in the ratio 4 : 5 : 6; this determines B = 5/4 × 3/2 = 15/8 and D = 3/2 × 3/2 = 9/4. Dropping D down into the same octave as C makes D = 9/8. Now we require that the subdominant chord (F, A, C) likewise be in the ratio 4 : 5 : 6; in this case we work down from C an octave up (C = 2), and obtain F = 2 ÷ 3/2 = 4/3 and A = 2 ÷ 6/5 = 5/3. Putting this all together, we obtain the frequency ratios for the just scale in C-major:

C	D	E	F	G	A	B	C
1	9/8	5/4	4/3	3/2	5/3	15/8	2

Note that E is a major third above C, F is a perfect fourth above C, and G is a perfect fifth above C. Next consider the intervals between successive notes: D to E is 5/4 ÷ 9/8 = 10/9; E to F is 4/3 ÷ 5/4 = 16/15; G to F is 3/2 ÷ 4/3 = 9/8, and so on. In fact, if we write all the ratios as in Fig. 9.1 we observe that there are only three different intervals and they have ratios 9/8, 10/9, and 16/15. The interval corresponding to 9/8 is called a *major whole tone,* the 10/9 interval a *minor whole tone,* and the 16/15 interval a *semitone.* In the just scale, there are three major whole tones, two minor whole tones, and two semitones.

FIG. 9.1

Frequency ratios of notes in the just diatonic scale. Numbers in the bottom row give intervals between two adjacent notes.

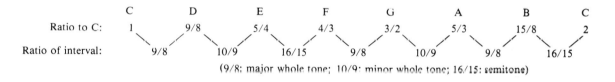

	C	D	E	F	G	A	B	C
Ratio to C:	1	9/8	5/4	4/3	3/2	5/3	15/8	2
Ratio of interval:		9/8	10/9	16/15	9/8	10/9	9/8	16/15

(9/8: major whole tone; 10/9: minor whole tone; 16/15: semitone)

Besides the three major triads, the just scale has two triads (E, G, B and A, C, E) with frequencies in the ratio 10 : 12 : 15 that are called *minor triads.* Minor triads, like major triads, include one major and one minor third, but with the order reversed. That is, the lower interval (12/10 = 6/5) is a minor third and the upper interval (15/12 = 5/4) is a major third. So far so good; there are three major triads and two minor triads, all in "just" intonation. But further examination reveals some shortcomings of the just scale.

Problems with the Just Scale

If the fifths in the just scale are examined, five are seen to be perfect, but the other one, D : A is imperfect (that is, the frequencies are not in the ratio 3 : 2). The same is true of the fourths in the scale; five are perfect but A:D is not.

Further difficulties appear when sharps and flats are added. Requiring that E to G♯ be a major third sets G♯ at 5/4 × 5/4 = 25/16, but requiring A♭ to C to be a major third sets A♭ at 2 ÷ 5/4 = 8/5. Thus A♭ should be a little higher in pitch than G♯ to maintain the justness of the above intervals. (A♭ and G♯ are called *enharmonic* notes; on the piano they are the same note.)

The just scale has never been of much practical use. Because of the two different whole-tone intervals (major and minor), retuning of an instrument would be required at each change of key. The interval C to D, for example, would have the ratio 9/8 in the key of C, but in the key of F, this interval would need to be tuned to the ratio of 10/9. Pipe organs have been built with complicated keyboards that would allow playing in just intonation in several keys, but few are in existence today. An orchestra composed of instruments with just intonation would approach musical chaos.

Example 9.1   Find the frequency ratio for C : F♯ in the scale of just intonation by requiring that the triads D–F♯–A and B–D♯–F♯ have frequency ratios 4 : 5 : 6.

Solution: (a) In order to be a major third above D, F♯ must have the ratio 9/8 × 5/4 = 45/32 with C. (b) In order to be a perfect fifth above B, F♯ must have the ratio 15/16 × 3/2 = 45/32 with C.

9.2   THE PYTHAGOREAN SCALE

Another important musical scale is the *Pythagorean scale,* which creates the largest possible number of perfect fourths and fifths. In order to do this, the other intervals, such as major and minor thirds and sixths, must vary from just tuning.

In constructing the Pythagorean scale based on fourths and fifths, two facts become apparent:

1. An octave is a fourth plus a fifth; therefore, going up a fourth leads to the same letter as going down a fifth and vice versa.

2. All notes on the scale (sharps and flats included) can be reached by going up or down 12 successive fifths or 12 successive fourths.

The second fact above can be represented on the "circle of fifths" shown in Fig. 9.2. Beginning at C, one can follow the outer circle by

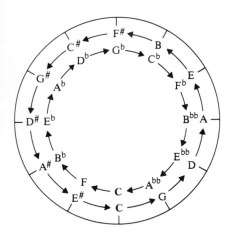

FIG. 9.2
The circle of fifths. The outer circle visits all twelve notes on the chromatic scale by going up by fifths (or down by fourths). The inner circle goes down by fifths (or up by fourths).

going up a fifth (or down a fourth) at each step. Twelve such steps bring one back to C . . . almost. If 3/2 is multiplied by itself 12 times, one obtains 129.7, which means that going up 12 perfect fifths takes one up seven octaves plus one-fourth of a semitone extra. By the same token, going down 12 fifths (or up 12 fourths), one follows the note names along the inner circle, encounters flats instead of sharps, and arrives back at C one-fourth of a semitone flat.

To determine the notes of the C-major scale, we can see clearly from Fig. 9.2 that we should go down a fifth (or up a fourth) to determine F as 4/3, and then go up five successive fifths to determine the other five notes. This gives G = 3/2, D = 3/2 × 3/2 = 9/4, A = $(3/2)^3$ = 27/8, E = 81/16, and B = 243/32. Putting them into the proper octave gives the frequency ratios below:

C	D	E	F	G	A	B	C
1	9/8	81/64	4/3	3/2	27/16	243/128	2

(Note that if we had omitted the last two fifths, E and B would not appear, and we would have the pentatonic scale C–D–F–G–A–C with appropriate intervals.)

Next the intervals between successive notes are calculated in Fig. 9.3. Note that there are only two different intervals; whole tones and semitones. The whole tone is the same as the major whole tone of the just scale (Fig. 9.1), but the semitone is about 20 percent smaller. Also, the semitone is 10 percent less than one-half of a whole tone.

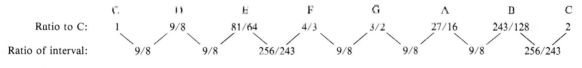

	C	D	E	F	G	A	B	C
Ratio to C:	1	9/8	81/64	4/3	3/2	27/16	243/128	2
Ratio of interval:		9/8	9/8	256/243	9/8	9/8	9/8	256/243

FIG. 9.3
Frequency ratios of notes in the Pythagorean scale. Numbers in the bottom row give intervals between two adjacent notes.

Continuing around the circle of fifths to obtain the frequencies of the flats and sharps leads to another interval. The ratio of F♯ to F turns out to be 2187/2048 = 1.068, which is called a *chromatic semitone*. It is larger than the diatonic semitone (256/243 = 1.053) previously determined. So the Pythagorean scale has one size of whole tone but two different semitones. The ratio of the chromatic to the diatonic semitone is 1.0136, which defines an interval called the *Pythagorean comma*. This is the same interval between 12 fifths and 7 octaves noted on the circle of fifths. It is also the interval between pairs of enharmonic notes (e.g., A♭ and G♯) facing each other on the inner and outer circles. So much for numbers.

The great advantage of the Pythagorean scale is the emphasis on perfect fifths and fourths. A great disadvantage, however, is the poor tuning of thirds. The major thirds exceed the just tuning by a ratio of 1.0125, and the minor thirds are smaller than just tuning by the same ratio. This ratio is called the *syntonic comma.* Pythagorean thirds sound quite out of tune. Nevertheless, studies have shown that many concert violinists tend to favor Pythagorean tuning in their performance, which points out the great importance of fifths and fourths in music.

Example 9.2 Compare the frequency ratios of the major thirds C:E, F:A, and G:B in the Pythagorean scale.

Solution: C:E $\dfrac{81}{64} \div 1 = \dfrac{81}{64}$

F:A $\dfrac{27}{16} \div \dfrac{4}{3} = \dfrac{81}{64}$

G:B $\dfrac{243}{128} \div \dfrac{3}{2} = \dfrac{81}{64}$

They are the same.

9.3 MEANTONE TEMPERAMENT

Because Pythagorean thirds sound out of tune, numerous alterations of the Pythagorean scale have been developed. Nearly all of them flat the third (E in the C-major scale) so that the major third (C to E) and minor third (E to G) are close to the corresponding just intervals. Similar adjustments are made in other notes. Such compromises form the bases of various *meantone temperaments.*

One such meantone temperament, called the *quarter-comma meantone temperament,* raises or lowers various notes by 1/4, 1/2, 3/4, or 5/4 of the syntonic comma δ, as shown in Fig. 9.4. (The syntonic comma, you will recall from Section 9.2, is the amount by which the major third and minor third differ from the corresponding just values.) Note that the fourths and fifths are no longer perfect intervals, although they are not too far from it. The real problems with meantone temperament, however, occur when playing instruments in keys with many sharps or flats. If the instrument is set to meantone temperament in C, it becomes increasingly out of tune as sharps and flats are added to the key signature.

Pythagorean:	C	D	E	F	G	A	B	C
Meantone:	C	D $-1/2\,\delta$	E $-\delta$	F $+\frac{1}{4}\,\delta$	G $-\frac{1}{4}\,\delta$	A $-\frac{3}{4}\,\delta$	B $-\frac{5}{4}\,\delta$	C

FIG. 9.4
Quarter-comma meantone temperament. The diatonic scale is compared to the Pythagorean. δ is the syntonic comma and represents a frequency ratio of 1.0125.

9.4 EQUAL TEMPERAMENT

The most convenient temperament of all is the equal temperament, which makes all semitones the same. The *scale of equal temperament* (often referred to simply as the ''tempered scale'') consists of five equal whole tones and two semitones; the whole tones are twice the size of the semitones. Twelve equal semitones make up an octave.

To determine the frequency ratio of the interval that is exactly 1/12 of an octave is a mathematical rather than a musical exercise. The square root of 2 (written $\sqrt{2}$ or $2^{1/2}$) is a number that can be multiplied by itself to give the product 2. Likewise, the twelfth root of 2 (written $^{12}\sqrt{2}$ or $2^{1/12}$) is a number that multiplied by itself 12 times will give the product 2. Many pocket calculators have a y^x key; if you have access to one, you can verify that $2^{1/12} = 1.05946$. This is the semitone of equal temperament. A whole tone is $(1.05946)^2$ or 1.12246. A fifth is 1.498 and a fourth is 1.335, both very close to the perfect intervals 1.500 and 1.333. A major third is 1.260 and a minor third is 1.189, not very close to the just intervals 1.250 and 1.200, but not as far sharp and flat as the thirds on the Pythagorean scale.

Rather than deal with ratios, it is customary to compare tones by using cents. One *cent* is 1/100 of a semitone in equal temperament. Thus an octave is 1200 cents, a tempered fifth is 700 cents, and so forth. One cent has the ratio $2^{1/1200} = 1.000578$. When we use cents, the comparison between various scales becomes more meaningful. Such a comparison is made in Fig. 9.5.

Note the rather large differences in the third (E) and sixth (A) notes of the scale. A rather surprising feature is the great difference in the sharps and flats in the three scales. Whereas C♯ lies above D♭ in the Pythagorean scale, for example, it lies well below D♭ in the just scale. A performer attempting to play in either scale might very well tend toward some meantone tuning of the sharps and flats.

FIG. 9.5

A comparison of Pythagorean, just, and equally tempered scales on a scale of cents (see Table 9.2).

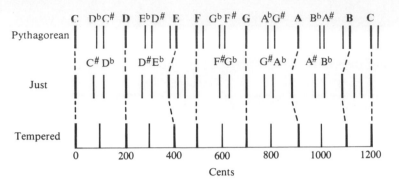

Three tables of intervals and tunings follow. Table 9.1 compares the main intervals, and Table 9.2 has the frequencies of nine octaves of notes in equal temperament with A_4 at 440 Hz. Table 9.3 gives the tunings for all the notes of one octave.

TABLE 9.1   Musical intervals in various tunings

Interval	Tempered		Just		Pythagorean	
	Ratio	Cents	Ratio	Cents	Ratio	Cents
Octave	2.000	1200	2/1 = 2.000	1200	2.000	1200
Fifth	1.498	700	3/2 = 1.500	702	1.500	702
Fourth	1.335	500	4/3 = 1.333	498	1.333	498
Major third	1.260	400	5/4 = 1.250	386	1.265	408
Minor third	1.189	300	6/5 = 1.200	316	1.184	294
Major sixth	1.682	900	5/3 = 1.667	884	1.687	906
Minor sixth	1.587	800	8/5 = 1.600	814	1.580	792

9.5   TUNING TO EQUAL TEMPERAMENT

The tuning of most instruments is based on equal temperament. The exact tuning of keyboard instruments, however, depends on how nearly harmonic are the overtones. Organs, which have harmonic overtones, can be tuned almost exactly to equal temperament; piano tuners take the inharmonicity of the strings into account, however. Pianos will sound better if the intervals are stretched, especially in the upper octaves; small upright pianos require a greater amount of stretch than concert grands do. This subject will be discussed in Chapter 14.

We now present a brief outline of a procedure for "laying the temperament." The first step in tuning an instrument consists in tuning a series of intervals slightly sharp or flat so that each interval beats at a certain rate. This can be done by working around the circle of fifths shown in Fig. 9.2 and setting each one 2¢ (cents) flat (a fre-

TABLE 9.2   Frequencies of notes in tempered scale

C_0	16.352	C_3	130.81	C_6	1046.5
	17.324		138.59		1108.7
D_0	18.354	D_3	146.83	D_6	1174.7
	19.445		155.56		1244.5
E_0	20.602	E_3	164.81	E_6	1318.5
F_0	21.827	F_3	174.61	F_6	1396.9
	23.125		185.00		1480.0
G_0	24.500	G_3	196.00	G_6	1568.0
	25.957		207.65		1661.2
A_0	27.500	A_3	220.00	A_6	1760.0
	29.135		233.08		1864.7
B_0	30.868	B_3	246.94	B_6	1975.5
C_1	32.703	C_4	261.63	C_7	2093.0
	34.648		277.18		2217.5
D_1	36.708	D_4	293.66	D_7	2349.3
	38.891		311.13		2489.0
E_1	41.203	E_4	329.63	E_7	2637.0
F_1	43.654	F_4	349.23	F_7	2793.8
	46.249		369.99		2960.0
G_1	48.999	G_4	392.00	G_7	3136.0
	51.913		415.30		3322.4
A_1	55.000	A_4	440.00	A_7	3520.0
	58.270		466.16		3729.3
B_1	61.735	B_4	493.88	B_7	3951.1
C_2	65.406	C_5	523.25	C_8	4186.0
	69.296		554.37		4434.9
D_2	73.416	D_5	587.33	D_8	4698.6
	77.782		622.25		4978.0
E_2	82.407	E_5	659.26	E_8	5274.0
F_2	87.307	F_5	698.46	F_8	5587.7
	92.499		739.99		5919.9
G_2	97.999	G_5	783.99	G_8	6271.9
	103.83		830.61		6644.9
A_2	110.00	A_5	880.00	A_8	7040.0
	116.54		932.33		7458.6
B_2	123.47	B_5	987.77	B_8	7902.1

quency ratio of 1.4983 rather than 1.5000). Twelve such flatted fifths equal exactly 7 octaves, whereas 12 perfect fifths, you will recall, come to 7 octaves plus 23.5¢ (about one-fourth of a semitone).

Suppose that C_4 (middle C) is first set to exactly 261.63 Hz by means of a tuning fork (one can start with A_4 if a 440-Hz fork is more readily available). Then G_4 is set at 392.00 Hz by counting beats.

Recall from Section 8.11 that in the case of a mistuned fifth, where $f_2 = (3/2)f_1 + \delta$, 2δ beats are heard each second. A little arithmetic shows that if $f_2/f_1 = 1.4983$, this leads to $\delta = 0.0017f_1$ or $0.0034f_1$ beats per second. With $f_1 = 261.63$ Hz, the rate of beats is 0.89 Hz (one beat each 1.12 seconds), so one might begin with a near-perfect fifth and lower it until 8.9 beats are heard each ten seconds.

Next, D_5 would be tuned to give $(0.0034) \times (392) = 1.33$ beats per second (this beat rate can also be calculated as the difference between the second harmonic of D_5 and the third harmonic of G_4). After setting D_5, it is best to tune D_4 exactly an octave below, and then proceed up a fifth to A_4. This procedure eventually tunes the entire octave C_4 to C_5 to equal temperament. The number of beats for each of the 12 intervals would be:

Interval	Beat frequency
C_4 to G_4	0.89
G_4 to D_5	1.33
D_4 to A_4	1.00
A_4 to E_5	1.50
E_4 to B_4	1.12
B_3 to $F_4^{\#}$	0.84
$F_4^{\#}$ to $C_5^{\#}$	1.26
$C_4^{\#}$ to $G_4^{\#}$	0.94
$G_4^{\#}$ to $D_5^{\#}$	1.41
$D_4^{\#}$ to $A_4^{\#}$	1.06
$A_4^{\#}$ to F_5	1.58
F_4 to C_5	1.19

If one is tuning an organ, the other octaves will probably be tuned to match the C_4 to C_5 octave. In the case of a piano, however, the octaves will no doubt be stretched, so the tuning procedure becomes more complicated.

9.6   ELECTRONIC TUNERS

A number of electronic tuners are now available; most of them work on the stroboscopic principle. One of the best known is the Conn Strobotuner, shown in Fig. 9.6. A specially designed disc (see Fig. 33.7) is rotated by a synchronous motor at one of 12 selected speeds, each corresponding to a note on the tempered scale. Neon lamps behind the disc flash on and off at the frequency of amplified sound from the microphone. When the neon lights are synchronized with the spinning disc, an appropriate portion of the pattern appears stationary. If a tone is sharp, the pattern revolves clockwise; if it is flat, the pattern

TABLE 9.3   Notes of scales based on C

Note	Tempered Cents	Just Ratio	Just Cents	Pythagorean Ratio	Pythagorean Cents
C	1200	2.000	1200	2.000	1200
B♯	1200	1.953	1159	2.027	1224
C♭	1100	1.920	1129	1.873	1086
B	1100	1.875	1088	1.898	1110
B♭	1000	1.800	1018	1.778	996
A♯	1000	1.758	977	1.802	1020
A	900	1.667	884	1.688	906
A♭	800	1.600	814	1.580	792
G♯	800	1.563	773	1.602	816
G	700	1.500	702	1.500	702
G♭	600	1.440	631	1.405	588
F♯	600	1.406	590	1.424	612
F	500	1.333	498	1.333	498
E♯	500	1.302	457	1.352	522
F♭	400	1.280	427	1.249	384
E	400	1.250	386	1.266	408
E♭	300	1.200	316	1.185	294
D♯	300	1.172	275	1.201	318
D	200	1.125	204	1.125	204
D♭	100	1.067	112	1.054	90
C♯	100	1.042	71	1.068	114
C	0	1.000	0	1.000	0

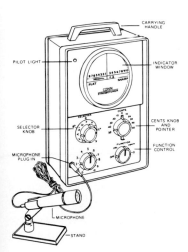

FIG. 9.6
Conn Strobotuner. Shown at the right is the rotating disc with eight concentric patterns of light and dark bars. (Courtesy of C.G. Conn, Ltd.)

revolves counterclockwise. By means of the "cents knob," the Strobotuner can be set at any number of cents sharp or flat. Thus any desired temperament can be set, within the accuracy of the instrument.

Another type of tuner uses computer circuits to produce a standard frequency. The Widner AccuTone tuner, for example, uses punched cards to select frequencies of the desired scale. Cards are available for just, Pythagorean, tempered, and meantone scales. Another set of cards gives various stretched scales for pianos, and it is possible to punch cards to obtain a scale of one's own choosing.

Although electronic tuners are a valuable aid for those who wish to tune their own instruments, few professional piano tuners use them. Along with the systematic shifts in partials due to string stiffness, there are more or less random shifts due to soundboard resonances. The truly skilled piano tuner has learned to take these into account, consciously or unconsciously, by tuning the strings in the various octaves for the best sound from each instrument.

Conversion from Cents to Frequency Ratio

Since most electronic tuners read in "cents," the following conversion formulae will be useful.

To convert from an interval in cents $I (\text{¢})$ to a frequency ratio R:

$$R = 10^{\left(\frac{I \log 2}{1200}\right)} \tag{9.1}$$

To convert from frequency ratio R to interval in cents I ¢:

$$I(\text{¢}) = \frac{1200}{\log 2} \log R \tag{9.2}$$

To convert 45 ¢ to a frequency ratio, for example, you would press the following keys on a typical calculator: 2, log, ÷, 1200, ×, 45, =, inv log (or 10^x); you should obtain 1.026.

9.7   INTONATION

Research on the intervals played or sung by skilled musicians shows substantial variability in the intonation from performance to performance. These variations are frequently larger than the differences between the various scales. Studies of large numbers of performances have shown that deviations from equal temperament are usually in the direction of Pythagorean intervals (Ward, 1970).

Several investigators have found that performers tend to stretch their intervals (including octaves), even when unaccompanied by a piano. Many choral conductors prefer the third in a chord slightly raised, especially in sustained chords or cadences, to avoid any suggestion of "flatting" the chord, a particular nemesis of choirs. How much of the apparent preference for sharpened intervals is due to constant exposure to pianos with stretched tuning is difficult to determine.

9.8 SUMMARY

Most musical composition makes use of scales. Three different ways to construct musical scales have been given. These lead to the just scale, the Pythagorean scale, and the scale of equal temperament; each has its own advantages and disadvantages. In addition, there are various meantone temperaments that are compromises between tunings. Most instruments today are tuned to equal temperament. A procedure for tuning to equal temperament and some electronic instruments for tuning have also been described in this chapter.

References and Suggested Readings

Backus, J. (1969). *The Acoustical Foundations of Music.* New York: Norton (See Chapter 8.)

Barbour, J. M. (1951). *Tuning and Temperament.* East Lansing: Michigan State College Press.

Benade, A. H. (1976). *Fundamentals of Musical Acoustics.* New York: Oxford University Press.

Burns, E. M., and W. D. Ward (1982). "Intervals, Scales, and Tuning," in *The Psychology of Music.* Ed. D. Deutsch. New York: Academic Press.

Martin, D. W. (1962). "Musical Scales Since Pythagoras," *Sound* **1:** 3, 22.

White, B. W. (1943). "Mean-Tone Temperament," *J. Acoust. Soc. Am.* **15:** 12.

Wood, A. (1975). *The Physics of Music,* 7th ed. Rev., J. M. Bowsher. London: Chapman and Hall. (See Chapter 11.)

Glossary

cent 1/100 of a semitone.

chromatic scale An ascending or descending sequence of twelve tones, each separated by a semitone.

diatonic scale A scale of seven whole tones and semitones appropriate to a particular key.

enharmonic notes Two different notes that sound the same on keyboard instruments (e.g., G$^\sharp$ and A$^\flat$).

equal temperament A system of tuning in which all semitones are the same; namely, a frequency ratio of $2^{1/12} = 1.059$.

inharmonic overtones Overtones whose frequencies are not multiples of the fundamental (i.e., harmonics).

just intonation A system of tuning that attempts to make thirds, fourths, and fifths as consonant as possible; it is based on major triads with frequency ratios 4:5:6.

major diatonic scale A scale of seven notes with the following sequence of intervals: two whole tones, one semitone, three whole tones, one semitone.

microtone Any interval smaller than a semitone.

minor scale A scale with one to three notes lowered a semitone from the corresponding major scale. In the key of C minor, the three minor scales are:

> natural: C D E♭ F G A♭ B♭ C;
> harmonic: C D E♭ F G A♭ B C;
> melodic (ascending): C D E♭ F G A B C;
> melodic (descending): C D E♭ F G A♭ B♭ C.

pentatonic scale A scale of five notes used in several musical cultures, such as the Chinese, Native American, and Celtic cultures.

Pythagorean comma The small difference between two kinds of semitones (chromatic and diatonic) in the Pythagorean tuning; a frequency ratio 1.0136 corresponding to 23.5¢.

Pythagorean tuning A system of pitches based on perfect fifths and fourths.

semitone A half step; in equal temperament, a semitone corresponds to 100¢ or to a frequency ratio of 1.059.

stroboscopic tuner A tuning device that makes use of a rotating pattern illuminated by flashing lights.

syntonic comma The small difference between a major or minor third in the Pythagorean and just tunings.

triad A chord of three notes; in the just tuning, a major triad has frequency ratios 4:5:6, while a minor triad has ratios 10:12:15.

Questions for Discussion

1. Why is the fingerboard of a guitar fretted, whereas the fingerboard of a violin is not?

2. Table 8.1 was constructed using frequency ratios appropriate to just intonation. How much would the consonant or dissonant character of intervals change if some other tuning were used?

3. Handel was one of several composers who felt that the key in which a piece of music is written does much to set its mood. (For example, F major sounds pastoral; F minor and F♯ minor are tragic keys; C major expresses vigor or military discipline.) Think of acoustical reasons why this might be true. Do any of these reasons remain valid in this day of tuning to equal temperament?

Problems

1. Verify by direct multiplication that a major third in equal temperament has the ratio of 1.26 and a minor third the ratio 1.19.

2. From your knowledge of equal temperament, show that if you invest money at an interest rate of 5.9 percent compounded annually, your investment doubles in twelve years!

3. An octave-band sound analyzer measures the sound level in ten octave bands with center frequencies 31.5, 63, 125, 250, 500, 1000, 2000, 4000, 8000, and 16,000 Hz. What are the closest notes on the musical scale?

4. The sounds used in a touch-tone telephone have the following frequencies: 697, 770, 850, 941, 1209, 1337, and 1477 Hz. What are the closest notes on the musical scale?

5. Verify by multiplication that a fifth plus a fourth equals an octave in any tuning, as does a major sixth plus a minor third.

6. According to the data in Fig. 7.6, the jnd in frequency up to about 700 Hz (roughly the lower half of the piano keyboard) is about 3 Hz. Convert $\Delta f/f = 3/400$ to cents. Then refer to Table 9.2 (or Fig. 9.5) and answer the following.

a) Can one normally hear a difference between A_4 on the just, Pythagorean, and tempered scales based on C?

b) How about E_4, F_4, and A_4?

7. Using the frequency ratios given in Fig. 9.1, verify that the intervals C:G, E:B, F:C, G:D, and A:E are perfect fifths in the just diatonic scale. Determine the frequency ratios for the imperfect fifth D:A.

8. Find the frequency ratio that corresponds to 25¢. What are the frequencies of the notes: $A_4 + 25$¢? $A_4 - 25$¢?

9. Some tuning forks are designed to a scale in which the C's have frequencies that are powers of 2 (128, 256, 512 Hz, etc.). How many cents flat are they compared to the International standard frequencies given in Table 9.2?

CHAPTER 10
String Instruments

The string instruments have held a special place in music for many years. Bowed string instruments are the backbone of the symphony orchestra, and plucked string instruments fulfill a similar role in folk music as well as rock music. The violin has been the most popular instrument for solo performances as well as chamber music for several generations.

Bowed string instruments date back to medieval times, and perhaps even earlier. The *viola da gamba* or *viol* developed in the late Middle Ages and reached its zenith in the seventeenth century. Of the once large viol family, the four sizes in general use today have six strings and fretted fingerboards. The six strings are tuned in fourths except for the middle two, which are separated by a major third (e.g., the soprano viol tunes to D_3, G_3, C_4, E_4, A_4, D_5).

The modern violin was developed in Italy in the sixteenth century, largely by Gasparo da Salo and the Amati family. In the eighteenth century, Antonio Stradivari, a pupil of Nicolo Amati, and Guiseppi Guarneri created instruments with great brilliance that have set the standard for violin-makers since that time.

Violins have probably had more attention from historians, musicians, and scientists than any other musical instrument. Outstanding contributions to our understanding of violin acoustics have been made by such distinguished scientists as Felix Savart (France, 1791–1841),

(a)

(b)

FIG. 10.1
Bowed string instruments:
(a) violin and viola;
(b) string bass and cello.

Hermann von Helmholtz (Germany, 1821–1894), Lord Rayleigh (England, 1842–1919), C. V. Raman (India, 1888–1970), and Frederick Saunders (United States, 1875–1963). More recently, the work of Professor Saunders has been continued by members of the Catgut Acoustical Society, an organization that publishes an excellent journal describing current research on the acoustics of string instruments.

Interesting questions still remain unanswered: Do the best instruments of today compare to the great instruments of the Italian masters: Did Stradivari, Guarneri, and the Amatis possess secrets still undiscovered today? Most experts think not, but the question can arouse some lively discussion.

It is impossible to discuss in detail all the members of the string family, even the principal bowed string instruments shown in Fig. 10.1. We will first describe how plucked and bowed strings vibrate, applying principles learned in Chapters 3 and 4, and then focus on some of the acoustical features of the violin and the guitar. Brief mention will be made of other string instruments.

10.1   CONSTRUCTION OF THE VIOLIN

A violin has four strings, of steel or gut, tuned to G_3, D_4, A_4, and E_5 (196, 294, 440, and 660 Hz). Strings of small diameter displace very little air as they vibrate; therefore, they radiate very little sound. The vibrating strings, however, set the body of the violin into vibration, which results in radiated sound of considerable strength. The body of the violin is designed to vibrate over a very wide range of frequency, and in this playing range it will have a great many resonances, a few of which will be discussed in Section 10.4.

An exploded view of a violin is shown in Fig. 10.2. The top plate or *belly* of the body is usually made of spruce, the ribs and back plate of curly maple. Two openings, called *f-holes,* are cut in the top plate, and a *bass bar* is glued to the top plate directly under one foot of the *bridge.* Near the other foot of the bridge is a short wooden stick called the *sound post,* which extends from the top to the back plate. The strings are attached to the tail piece, pass over the bridge, along the fingerboard, and over the *nut,* and finally are attached to wooden pegs inserted in the peg box.

In order to vibrate properly, the top plate must be made very thin (its thickness is typically in the range of 2 to 4 mm). When tuned to normal pitch, the combined tension of the strings is about 220 newtons (50 lb). The downward force exerted on the top plate by the bridge is nearly 100 newtons. Without the sound post and the bass bar, the fragile top plate would not be able to support this load for long. These two important structural members also serve important acoustic functions, as will be discussed in Section 10.4.

The bow consists of many strands of horsehair held under tension by a wood stick (pernambuco, a very dense, stiff wood, is the preferred material). Violin bows typically have a length of about 73 cm and a mass of about 60 grams (Reder, 1970). The tension of the hair is adjusted by moving the frog with a screw. Although violin players attach a great deal of importance to the bow, its properties have not been studied nearly as carefully as those of the violin body.

10.2 VIBRATIONS OF A PLUCKED STRING

When the string of a musical instrument is excited by bowing, plucking, or striking, the resulting vibration can be considered to be a combination of the normal modes of vibration or resonances. For example, if the string is plucked at its center, the resulting vibration will consist of the fundamental plus the odd-numbered harmonics.

Figure 10.3 illustrates how the modes associated with the odd-numbered harmonics, when each is present in the right proportion, can add up at one instant in time to give the initial shape of the string. Modes 3, 7, 11, etc., must be opposite in phase from modes 1, 5, and 9 in order to give a maximum in the center.

Since all the modes shown in Fig. 10.3 have different frequencies of vibration, they will quickly get out of phase, and the shape of the string changes rapidly after plucking. What happens is that two identical pulses propagate in opposite directions away from the center, as shown in Fig. 10.4. The shape of the string at each moment in time can still be obtained by adding up the modes in the proportions shown in Fig. 10.4, but it is more difficult to do so, because each of the modes will be at a different point in its cycle. The resolution of the string motion into two pulses, shown by the dotted lines in Fig. 10.4, results in a simpler analysis.

If the string is plucked at a point other than its center, the recipe of the constituent modes is different, of course. For example, if the string is plucked one-fifth the distance from one end, the recipe of mode amplitudes is as shown in Fig. 10.5. Note that the fifth harmonic is missing. Plucking it one-fourth the distance from the end

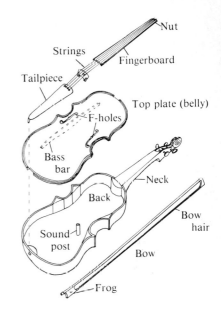

FIG. 10.2
Parts of a violin.

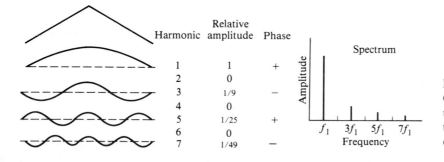

Harmonic	Relative amplitude	Phase
1	1	+
2	0	
3	1/9	−
4	0	
5	1/25	+
6	0	
7	1/49	−

FIG. 10.3
Odd-numbered modes of vibration add up in appropriate amplitude and phase to the shape of a string plucked at its center.

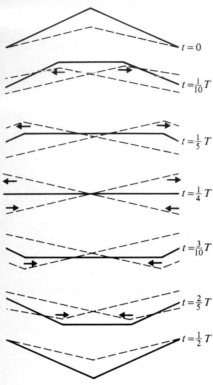

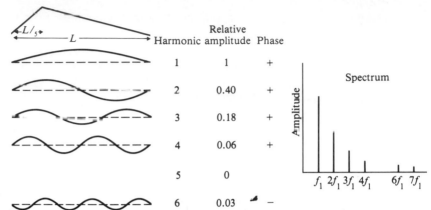

FIG. 10.5

The addition of modes to obtain the shape of a string plucked at one-fifth its length. It should be noted that the spectra in Fig. 10.3 and the figure above show the relative amplitudes of the different modes of vibration. The spectra of the radiated sound will have the same frequencies but their relative amplitudes will be quite different due to the acoustical properties of the instrument.

FIG. 10.4

The motion of a string plucked at its midpoint through one half cycle. Motion can be thought of as due to two pulses traveling in opposite directions.

suppresses the fourth harmonic, and so on. (In Fig. 10.3 it can be noted that plucking it at one-half the distance eliminated the second harmonic, as well as other even-numbered ones.)

10.3   VIBRATIONS OF A BOWED STRING

As the bow is drawn across the string of a violin, the string appears to vibrate back and forth smoothly between two curved boundaries, like a free string vibrating in its fundamental mode (see Fig. 4.4 or the second curve in Fig. 10.3). But this appearance of simplicity is deceiving. If we took a high-speed photograph of the bowed violin string, we would find that the string is nearly straight with a sharp bend at one point; at certain times it resembles the initial shapes of the plucked strings shown in Figs. 10.3 and 10.5. Over a hundred years ago, Hermann von Helmholtz (who contributed so much to our understanding of physics, anatomy, physiology, and the arts) discovered what really happens. The sharp bend racing along the bowed string follows the curved path that we see; because of its speed, our eye sees only the curved envelope.

During the greater part of each vibration, the string is carried along by the bow. Then it suddenly detaches itself and moves rapidly back until it is caught by the moving bow again. The motion of the string at the point of bowing is shown in Fig. 10.6.

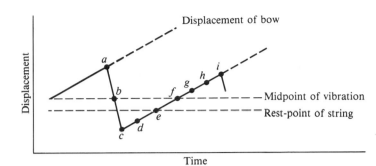

FIG. 10.6
Displacement of bow and string at the point of contact with the bow. Note that the midpont of the vibration is displaced slightly from the rest position of the string.

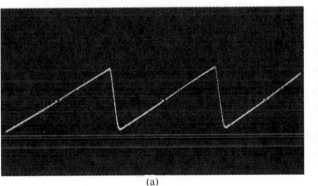

(a)

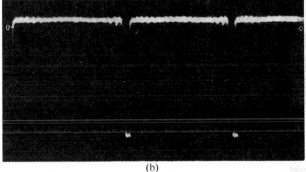

(b)

FIG. 10.7
(a) Displacement and (b) velocity of a bowed string. The velocity (b) at every point in time equals the slope of the displacement curve. (From Schelleng, 1974).

Up to the point of release *a* and again from *c* to *i*, the string moves at the constant speed of the bow (recall that speed is represented by the slope of the displacement curve). From *a* to *c*, the string makes a rapid return until caught up by a different point on the bow. The displacement and velocity (speed) of a very flexible string during bowing are shown in the oscillograms in Fig. 10.7. The oscillogram on the left, which shows the displacement versus time, is almost identical to Fig. 10.6, whereas the curve of velocity versus time (on the right) shows rather narrow spikes, which represent the large velocity of the string when it slides rapidly along the bow (*a-b-c* in Fig. 10.6).

The connection between the sawtoothlike motion shown in Figs. 10.6 and 10.7 and the motion of the bends racing back and forth on the string can be understood by referring to Fig. 10.8, which shows the entire string at successive times in the vibration cycle. At the moment of release, shown in Fig. 10.8(a), the bend has just passed the bow. In (b), the bend has reached the bridge, from which it will reflect back down the string in (c), (d), and (e), until it reaches the nut in (f) and is again reflected. At point *c* in Fig. 10.6 and also in frame (c), in Fig. 10.8, the string is captured by the bow, and once again moves upward at the speed of the bow. In the set of diagrams in the right-

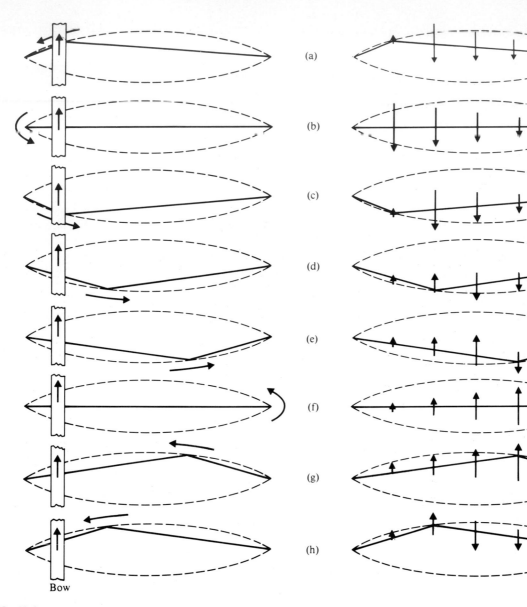

(a)

(b)

(c)

(d)

(e)

(f)

(g)

(h)

Bow

FIG. 10.8
The motion of a bowed string at successive times during the vibration cycle. The configurations shown in (a)–(h) correspond to the points *a–h* in Fig. 10.6. The bend races around a curved boundary, which appears to be the profile of the string viewed during bowing.

hand column of Fig. 10.8, the bow has been deleted, and the velocity of the string at several points is indicated by the arrows.

The "stick" and "slip" of the string against the moving bow are determined partly by the friction between the horsehair of the bow and the string. It is well known that the force of friction between two objects is less when they are sliding past each other than when they move together without slippage. Once the string begins to slip, it moves rather freely until it is once again captured by the bow. It is important to note, however, that the beginning and the end of the slipping are triggered by the arrival of the bend (slipping begins when the bend arrives from the nut and ends when it arrives again from the bridge). Because the time required for one round trip depends only on the string length and wave velocity (which, in turn, depends on tension and mass), the vibration frequency of the string remains the same under widely varying bowing conditions. If friction alone determined the beginning and the end of slipping, the vibrations would be irregular rather than regular.

The limits on the bowing conditions are the limits on the conditions at which the bend can trigger the beginning and the end of slippage between bow and string. For each position of the bow, there is a maximum and minimum bowing force, as shown in Fig. 10.9. The closer to the bridge the instrument is bowed, the less leeway the violinist has between minimum and maximum bowing force. Bowing close to the bridge (*sul ponticello*) gives a loud, bright tone, but requires considerable bowing force and the steady hand of an experienced player. Bowing further from the bridge (*sul tasto*) produces a gentle tone with less brilliance.

Further study of Fig. 10.8 will help us understand how amplitude of the vibration is determined by the speed and position of the bow. Since the speed of the bend around its curved path is essentially independent of the speed and position of the bow, the amplitude of vibration can increase either by increasing the bow speed or by bowing closer to the bridge.

Figures 10.7 and 10.8 illustrate the motion of the string at the point of contact with the bow. The displacement and velocity at two other points on the string are shown as functions of time in Fig. 10.10.

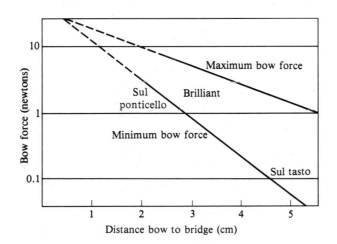

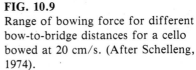

FIG. 10.9
Range of bowing force for different bow-to-bridge distances for a cello bowed at 20 cm/s. (After Schelleng, 1974).

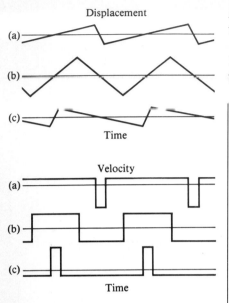

FIG. 10.10
Displacement and velocity of a bowed string as a function of time at three positions: (a) near the bridge; (b) at the center; (c) near the nut. For bowing in the opposite direction, exchange (a) and (c).

At the center of the string, the velocity (speed) of the string (slope of the displacement curve) is the same in both directions.

When the bow moves in the opposite direction, the curves in Figs. 10.10(a) and (c) are exchanged. Also the bend moves clockwise, rather than counterclockwise as shown in Fig. 10.8.

The difference in motion at the three points can be demonstrated by a "following bow" experiment (Schelleng, 1974). (See Fig. 10.11.) With the violin supported horizontally, a second bow is suspended by a string at its heavy end, the other end resting lightly on the bowed string. As the string is bowed loudly, the hanging bow moves in the direction of string motion during the longer part of each cycle. Placed near the bridge, it moves in the same direction as the driving bow; near the nut, it moves in the opposite direction. At the center, there is little motion in either direction.

The behavior of the following bow can be understood by examining the velocity of the string near the bow and the nut as shown in Fig. 10.10. The suspended bow follows the slow motion of the string in one direction but is unable to follow the rapid return of the string in the other direction; thus it undergoes a net motion during each cycle of vibration.

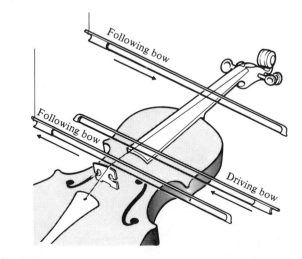

FIG. 10.11
The following-bow experiment. A second bow hanging by a string indicates the direction of string motion during the longer part of each vibration cycle.

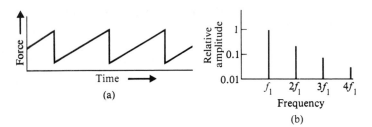

(a)

(b)

FIG. 10.12
(a) Idealized waveform of the force exerted on the bridge by a bowed string. (b) Spectrum of the force showing harmonics decreasing as $1/n$.

The vibrating string exerts a sideways force on the bridge, which in turn transmits this force to the top plate. In the ideal case of a completely flexible string vibrating between two fixed end supports, this force has a sawtooth waveform with a spectrum of harmonics varying in amplitude as $1/n$, as shown in Fig. 10.12. In actual practice, the waveform of the force is modified by the string stiffness, mechanical properties of the bridge (see Hacklinger, 1978), and other factors.

The motion of the top plate, which is the source of most of the violin sound, is the result of a complex interaction between the driving force from the bridge and the various resonances of the violin body. These will now be considered.

10.4 VIBRATIONS OF THE VIOLIN BODY

An important factor in the sound quality and playability of a violin is the vibrational behavior of its body. We often describe this in terms of normal modes of vibration, which consist of coupled motions of the top plate, the back plate, and the enclosed air. Smaller contributions are made by the ribs, neck, and fingerboard.

To determine the normal modes of vibration, it is customary to apply an oscillating force to the bridge and observe the motion of the various parts of the violin. This may be done electrically, by means of accelerometers or optically, by means of holographic interferometry.

Six low-frequency modes of vibration of a violin are shown in Fig. 10.13. The C_1 and C_2 body ("corpus") modes radiate little sound, although they contribute to the "feel" of the violin. The other four modes are indicated on the frequency response curve in Fig. 10.14, where the vertical coordinate is the velocity divided by the force (called *mechanical admittance* or *mobility*).

The A_0 ("air") mode and the T_1 ("top") mode both involve considerable motion of air in and out of the f-holes. Both modes radiate sound efficiently, and they dominate the low-frequency sound spectra of most violins.

Above 1 kHz, the vibrational modes or resonances are bunched together, as can be seen in Fig. 10.14, and it is difficult to identify the individual modes. Of considerable interest is the concentration of

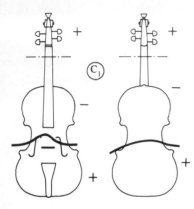
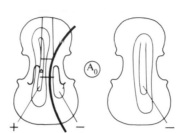
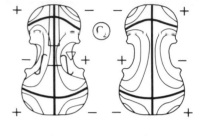
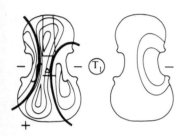
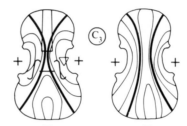
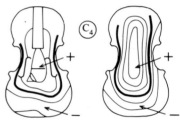

FIG. 10.13

Modal shapes for six modes in a violin.

C₁ (185 Hz): one-dimensional bending;

A₀ (275 Hz): air flows in and out of the *f*-holes;

C₂ (405 Hz): two-dimensional flexure;

T₁ (460 Hz): mainly motion of the top plate;

C₃ (530 Hz), C₄ (700 Hz): two-dimensional flexure.

Top plate and back plate are shown for each mode. The heavy lines are nodal lines; the direction of motion is indicated by + or −. The drive point is indicated by a small triangle. (after Alonso Moral and Jansson, 1982a).

FIG. 10.14

Input admittance (driving point mobility) of a Guarneri violin driven on the bass bar side (Alonso Moral and Jansson, 1982b).

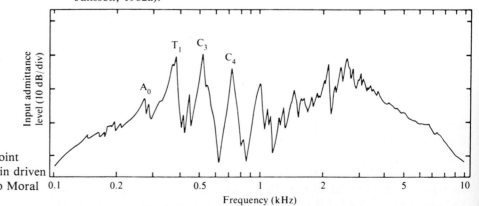

resonances around 2 to 3 kHz, which resembles the "singer's formant" found in the spectra of most opera singers (see Chapter 17). Dunnwald (1983) found this concentration to be particularly characteristic of violins made by the old Italian masters (Stradivari, Guarneri, etc.).

10.5   TUNING THE TOP AND BACK PLATES

The carving of the top and back plates is one of the most critical steps in constructing a violin. Since both plates have an arch, they are carved from blocks of wood thicker at the center, as shown at O–O in Fig. 10.15.

As the plates near their desired shapes, the violin-maker tests them by listening to "tap tones." To hear these, the plate is held in a particular way and tapped at certain spots. The trained ear of the violin-maker can extract useful information by noting the pitch and the decay time of the tap tones.

In recent years, the analysis of the vibrations of the plates has been refined considerably by the application of electronics and optics. One rewarding technique has been the observation of individual modes of vibration by *Chladni patterns* of the type shown in Fig. 2.16. The plate is usually mounted horizontally and excited by a loudspeaker. The two most important modes of vibration are those shown in Fig. 10.16. (The plates shown are for a viola, but violin and cello plates show similar patterns.) If the frequency of the lower mode (a) in the top plate matches the back plate, then tuning the upper modes (b) to the same frequency or a semitone different will usually produce plates of high quality (Hutchins, 1977). Of course, the shape of the nodal lines is equally as important as the frequency of the modes.

Other methods of vibration analysis give rather detailed information on the modes but are less convenient to set up in the violin shop. One such method involves setting a plate into vibration and scanning

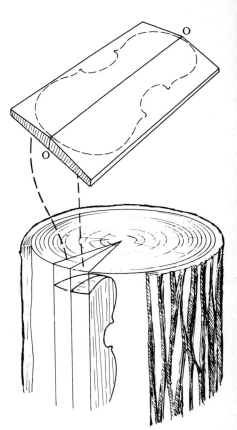

FIG. 10.15
Wood for a top or back plate. The top plate is spruce; the back is generally maple.

Back	Top		Back	Top

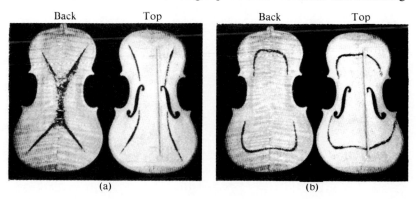

(a)   (b)

FIG. 10.16
Chladni patterns showing modes of vibration in the viola top and back plates (From Hutchins, 1977. Reprinted by permission).

it with a beam of light. The light reflected from the plate falls on a photocell, where it creates an electrical signal that indicates the character of vibration at that point. An elegant method, called *hologram interferometry,* employs laser light. Figure 10.17 is a photograph of the vibrational modes of a violin top plate and back plate using this technique (Hutchins, Stetson, and Taylor, 1971). The dark lines indicate areas that vibrate with the same amplitude; along the center of each large white area is a nodal line. Mode II in Fig. 10.17 is the mode shown in Fig. 10.16(a), whereas mode V corresponds to the mode in Fig. 10.16(b).

The vibrational characteristics of wood plates depend not only on the size, shape, thickness, and arching of the wood but also on the density, stiffness, and internal damping of the wood. These properties vary from sample to sample and even from day to day, as the temperature and humidity change. Varnish fortifies the wood to some extent against the effects of changing humidity, but the varnish also affects the wood vibrations. Varnish adds some mass and stiffness to the wood plates (resonances in a violin top plate may shift by as much as a quarter tone), but its greatest effect is to increase the damping. The best advice on varnish may be "not too soft, not too hard, not too much" (Schelleng, 1968).

One particular property of wood that deserves mention is its *anisotropy;* its stiffness along the grain is much greater than that across the grain. Spruce has a particularly high ratio (more than 10:1), which is one of the reasons it is used in a number of musical instruments

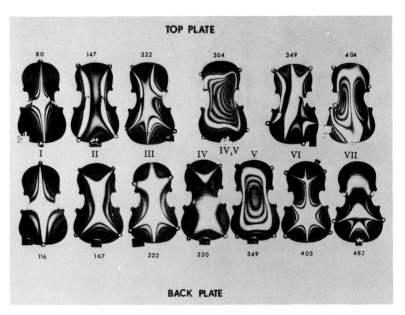

FIG. 10.17
Modes of vibration of a violin top and back plates visualized by hologram interferometry. (From Hutchins, Stetson and Taylor, 1971. Reprinted by permission.)

(for top plates in violins and guitars, for soundboards in pianos and harpsichords, etc). This anisotropy obviously has a large influence on its modes of vibration.

Attempts have been made to develop a synthetic material that would substitute for spruce. One of the most successful materials has been a sandwich with an inner core of cardboard overlaid with graphite fibers set in epoxy. The fibers are carefully aligned in one direction to achieve the desired anisotropy (like wood grain). A violin and a guitar with top plates of this material have been constructed and demonstrated at meetings of the Acoustical Society of America (Haines, Chang, and Hutchins, 1975).

10.6   THE BRIDGE

The primary role of the bridge is to transform the motion of the vibrating strings into periodic driving forces applied by its two feet to the top plate of the instrument. However, generations of luthiers have discovered that shaping the bridge is a convenient way to alter the frequency response of a violin as well.

Violin bridges typically have strong resonances around 3000 and 6000 Hz, as shown in Fig. 10.18. The lower resonance is due to a rocking-bending motion, whereas the upper one is due to a symmetrical up-and-down motion. Clearly, these resonance frequencies can easily be changed by cutting away wood at the appropriate places.

In order to "darken" the sound of a violin, the player may attach a mute to the bridge. The additional mass of the mute shifts the bridge resonances to lower frequencies, changing the sound spectrum of the instrument. Typical violin mutes have masses of about 1.5 grams.

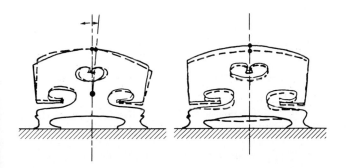

FIG. 10.18
First two vibrational modes (resonances) of a violin bridge.
(After Reinecke, 1973).

10.7   OTHER BOWED STRING INSTRUMENTS

The other string instruments of the orchestra are the viola, cello (violoncello), and string bass (also known as contrabass, double bass, or bass viol). The general shape and construction of the viola and cello are similar to those of the violin. The strings, like those of the violin, are tuned in intervals of fifths: C_3, G_3, D_4, A_4 in the viola and C_2, G_2, D_3, A_3 in the cello. Although the viola is tuned a fifth below the violin, its main air resonance is typically less than a third below that of the violin, and the main top plate resonance is a little over a third lower. Thus the lowest notes on a viola lack carrying power. Tunings and resonances of the violin, viola, and cello are shown in Fig. 10.19.

The lower resonances of a cello are related to its strings in about the same way as those of the viola. The tall bridge of the cello, however, results in a large driving force and a strong response near the main wood resonance. Also, the second air resonance is nearly an octave above the main air resonance and thus reinforces this resonance by strengthening the second harmonic.

The string bass is tuned in fourths rather than fifths, which reduces the distances that the hand must travel along the long fingerboard. It has relatively narrow sloping shoulders, and its flat back saves a substantial amount of wood, as compared to what would be needed for an arched back like that of the violin or cello.

FIG. 10.19
Tunings and resonances of violin, viola, cello, and string bass. The main air resonance is denoted by A, the main top plate resonance by T. The strings are given by ○.

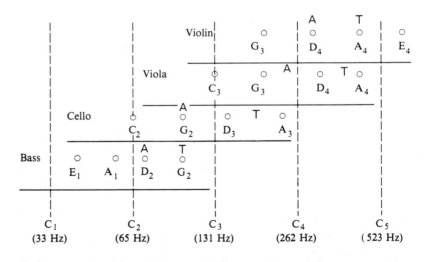

10.8   MUSIC AND PHYSICS: A NEW FAMILY OF FIDDLES

In 1958, Carleen Hutchins and other members of the Catgut Acoustical Society set about to develop a new family of fiddles, with resonances scaled to those of a fine violin. Their research led to a family of eight instruments shown in Figs. 10.20 and 10.21. The family has two instruments pitched above the violin: the treble and the soprano. The alto, which is tuned like the viola, is held either under the chin (if one has long arms) or played vertically on a peg in the fashion of a cello. The tenor tunes between the viola and cello, and the baritone has cello tuning. Finally there are two basses, with their strings tuned in fourths.

These new instruments have appeared in several concerts and have aroused considerable interest, both musically and scientifically. The alto violin, especially, is considered to have increased power and tone quality over the viola. "That is the sound I have always wanted from the violas in my orchestra," declared conductor Leopold Stokowski upon hearing the alto violin (Hutchins, 1967).

FIG. 10.20
The eight instruments in the new fiddle family. (Photograph by John Castronovo. Reprinted by permission.)

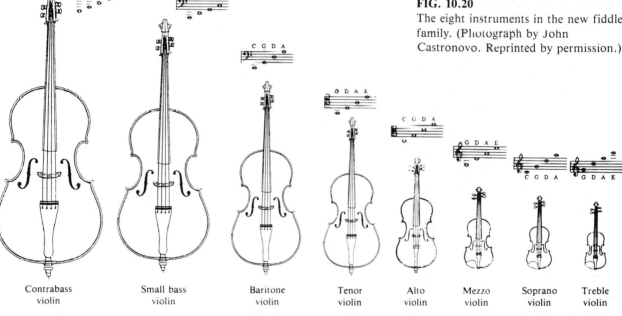

| Contrabass violin | Small bass violin | Baritone violin | Tenor violin | Alto violin | Mezzo violin | Soprano violin | Treble violin |

10.9   CONSTRUCTION OF THE GUITAR

The guitar has its origins in antiquity, probably more than 3000 years ago in Egypt. Its ancestors include the lute and the vihuela. Although the guitar appeared in several countries of Europe (Antonio Stradivari constructed about a dozen guitars), it was in Spain that it developed great popularity. By the end of the thirteenth century, guitar making was a developing art in Andalusia in southern Spain. Perhaps the most

FIG. 10.21
Comparative sizes and tuning of the new violin family. (Reprinted from *The Instrumentalist,* June 1967, p. 41. © The Instrumentalist Company 1989. Used by permission of The Instrumentalist Co.)

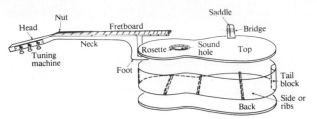

FIG. 10.22
An exploded view of a guitar, showing details of construction.

celebrated of the Spanish luthiers was Antonio de Torres Jurado, whose innovations established a Spanish school of guitar making in the late nineteenth century.

It is in the twentieth century, however, that the guitar has become one of the most popular of all musical instruments. In the United States alone, there are an estimated 15 to 20 million guitars. Guitar music has developed along at least five different lines: classical, flamenco, folk, jazz, and rock.

Throughout the years, the guitar has undergone many changes in design, including the size and shape of the body and the number of strings. Torres developed the larger body, fan-braced sound board, and 65-cm string length that are popular today. The modern guitar, shown in Fig. 10.22, has six strings tuned to E_2, A_2, D_3, G_3, B_3, and E_4. The strings, which lie in a single plane, are fastened directly to the bridge. The long fingerboard is fitted with frets, which greatly simplify the playing of chords.

The top is usually cut from spruce, planed to a thickness of about 2.5 mm (3/32″). The back is usually of a hardwood, such as rosewood, mahogany, or maple, also about 2.5 mm thick. Both top and back are braced, the bracing of the top being one of the critical parameters in guitar design. Braces strengthen the fragile plate and also transmit vibrations of the bridge to various parts of the sound board. Figure 10.23 shows several different designs of bracing used in guitars.

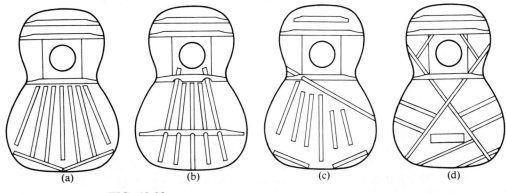

FIG. 10.23
Various designs for bracing guitar sound board: (a) traditional fan bracing; (b) Bouchet (France); (c) Ramirez (Spain); (d) crossed bracing.

(a)   (b)   (C)

FIG. 10.24
Styles of acoustic guitars: (a) classical;
(b) flattop; (c) archtop. (Courtesy of
Gibson Guitar Corp., Nashville, TN.
Reprinted by permission.)

Acoustic guitars usually fall into one of four families of design: classical, flamenco, flattop (or folk), and archtop. Classical and flamenco guitars have nylon strings; flattop and archtop guitars have steel strings. Since steel strings are under greater tension, flattop guitars usually have a steel rod imbedded inside the neck, and their sound boards are provided with crossed bracing. The fingerboards of flattop and archtop guitars are narrower than those of classical and flamenco guitars. Steel strings tend to give a louder sound than nylon strings. Classical, flattop, and archtop guitars are shown in Fig. 10.24.

Some flattop guitars carry twelve strings, which are positioned and played in pairs. In the two pairs of highest pitch (B, E), the strings are tuned in unison, but in the other four pairs, one string is tuned an octave higher than normal for greater brilliance (occasionally one of the low E strings is tuned two octaves higher than the other).

10.10   THE GUITAR AS A VIBRATING SYSTEM

The guitar can be considered to be a system of coupled oscillators. The plucked strings, which were discussed in Section 10.2, radiate only a small amount of sound directly, but they excite the bridge and top plate, which in turn transmit vibrational energy to the air cavity and back plate. Sound is radiated efficiently by the vibrating plates and through the sound hole.

Figure 10.25 is a simple schematic of a guitar. At low frequency the top plate transmits energy to the back plate and the sound hole

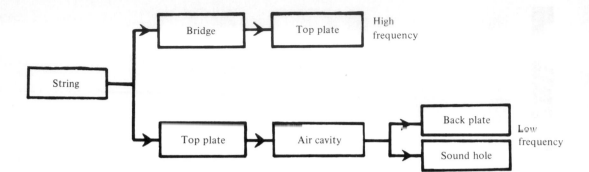

FIG. 10.25
Simple schematic of a guitar. At low frequency sound is radiated by the top and back plates and the sound hole. At high frequency most of the sound is radiated by the top plate.

via the air cavity; the bridge essentially acts as part of the top plate. At high frequency, however, most of the sound is radiated by the top plate, and the mechanical properties of the bridge becomes significant. We will expand this simple model in the following sections.

The waveform of the actual bridge force is strongly influenced by the stiffness and damping of the string and the manner in which it is plucked. If the string is plucked with the finger or a soft plectrum, the spectrum of the force will have less prominent high harmonics.

10.11 VIBRATIONS OF THE TOP PLATE, BACK PLATE, AND AIR CAVITY

Before considering the way in which the guitar body vibrates, it is well to consider the vibrational modes of the individual parts. In describing these modes, it is important to specify the exact conditions under which they are measured. A top plate with a free edge, for example, has completely different vibrational modes than a top plate attached to guitar ribs.

The first four modes of a guitar top plate are illustrated in Fig. 10.26, using the technique of hologram interferometry. (These interferograms were made with the strings and back plate removed and with the sides clamped in place.) The second mode (Fig. 10.26(b)) has a nodal line running in the direction of the grain, and the next mode (Fig. 10.26(c)) has a node just above the bridge. For the top plate shown in Fig. 10.26 the modal frequencies were 185, 287, 460, and 508 Hz.

In our laboratory, we usually observe the top plate modes with the back plate in place but heavily damped in a sandbox and with the sound hole closed. This changes the mode frequencies slightly, but the mode shapes remain nearly the same as in the backless guitar of Fig. 10.26.

Vibrational modes of the top plate, the back plate, and the air cavity of a folk guitar are shown in Fig. 10.27. Note that in this folk guitar with crossed bracing (as in Fig. 10.23(d)), the cross-grain bend-

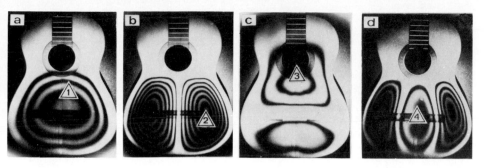

FIG. 10.26
Holographic interferograms of the first
four modes of vibration of the top
plate of a guitar without back and
strings. Along each line, the amplitude
of the vibrational motion is a constant.
(From Jansson, © 1971 by Acustica.
Reprinted by permission.)

ing mode (designated as the (1,0) mode in Fig. 10.27) is observed at a
higher frequency than the long-grain (0,1) bending mode. This behavior, opposite to that observed in Fig. 10.26, occurs because the crossed
bracing in the folk guitar adds considerable stiffness across the grain.

The lowest mode of the air cavity is the familiar Helmholtz resonance (see Fig. 4.10(b)), whose frequency is determined by the cavity
volume and sound hole diameter. Higher modes resemble the standing
waves in a rectangular box.

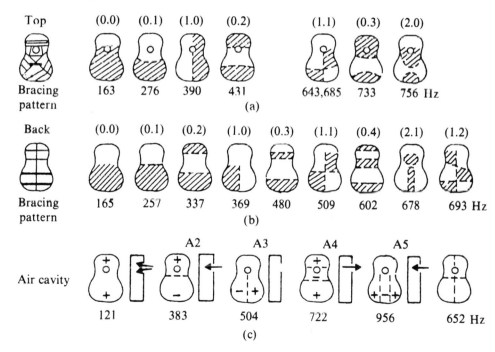

FIG. 10.27
(a) Modes of a folk
guitar top (Martin
D-28) with the back and
ribs in sand; (b) Modes
of the back with the top
and ribs in sand; (c)
Modes of the air cavity
with the guitar body in
sand. Modal
designations are given
above the figures and
modal frequencies
below (Rossing, Popp,
and Polstein, 1985).

10.12   RESONANCES OF THE GUITAR BODY

Most of the prominent low-frequency resonances of a guitar can be
attributed to coupled motion of the top plate, the back plate, and the
enclosed air. Coupling between the lowest modes of each of these in

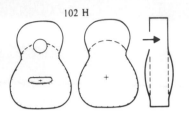
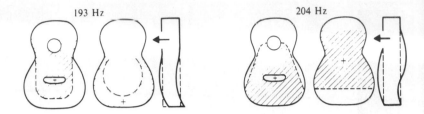

FIG. 10.28
Vibrational motion of a freely-supported Martin D-28 folk guitar at three strong resonances in the low frequency range.

the guitar shown in Fig. 10.27 leads to resonances at 102, 193, and 204 Hz, as shown in Fig. 10.28.

At the lowest of these three resonances, the top and back plates move in opposite directions, causing the guitar to breathe in and out of the sound hole. At the second resonance, the plates move in the same direction. At the third resonance, they again move in opposite directions, but the air in the sound hole moves in a direction opposite to its motion in the first mode. This is somewhat like the highest mode of the three-mass vibrator in Fig. 2.10(c).

Note that the resonance frequencies in Fig. 10.28 are given for a guitar freely supported on rubber bands. Fixing the ribs lowers the second resonance from 193 to 169 Hz, but the first and third resonances remain essentially unchanged because they involve but little motion of the ribs. This illustrates the dependence of the vibrational modes on the method of support, and suggests that the timbre of the instrument depends upon the way it is held by the player.

The (1,0) modes in the top plate, back plate, and air cavity combine to give at least one strong resonance between 250 and 300 Hz in a classical guitar but closer to 400 Hz in a cross-braced folk guitar. Motion of the plates at two such resonances in a Martin D-28 folk guitar are shown in Fig. 10.29.

Above 400 Hz, the coupling between top and back plate modes appears to be weaker, so the observed resonances are due mainly to resonances in one or the other of the two plates. A fairly prominent

FIG. 10.29
Vibrational configurations of a Martin D-28 guitar at two resonances resulting from ''see-saw'' motion of the (1,0) type.

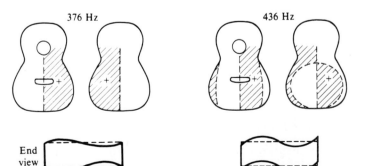

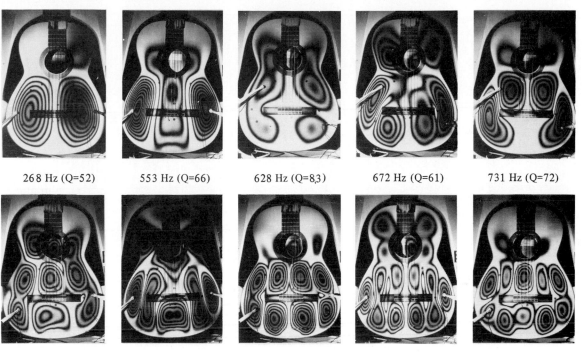

268 Hz (Q=52)   553 Hz (Q=66)   628 Hz (Q=83)   672 Hz (Q=61)   731 Hz (Q=72)

873 Hz (Q=75)   980 Hz (Q=48)   1010 Hz (Q=80)   1174 Hz (Q=58)   1194 Hz (Q=39)

(2,0) top plate resonance is generally observed in classical guitars around 550 Hz, but this mode is generally less prominent in folk guitars. Vibrational configurations of a classical guitar top plate at several resonances are illustrated by the holographic interferograms in Fig. 10.30. Q-values are a measure of the sharpness of each resonance.

10.13   SOUND RADIATION

At low frequencies, considerable sound is radiated by the top plate, the back plate, and the sound hole. The sound spectrum depends upon the direction in which it is observed. Figure 10.31 shows how the radiated sound varies with frequency in one particular direction (straight ahead of the sound hole) in an anechoic (echo-free) room. Figure 10.32 shows how the sound level varies with direction at the frequencies of four of the resonances shown in Figs. 10.28 and 10.29. At 102 and 204 Hz, the guitar radiates well in all directions, but at 376 and 436 Hz it does not. (At 376 Hz, the radiation pattern takes on a dipole character, not unlike the field of a bar magnet with two poles; at 436 Hz a quadrupole character is apparent.)

FIG. 10.30
Holographic interferograms of a classical guitar top plate at several resonances. Resonance frequencies and Q-values (a measure of the sharpness of the resonance) are given (From Richardson and Roberts, (1985). "The Adjustment of Mode Frequencies in Guitars: A Study by Means of Holographic Interferometry and Fiite Element Analysis, "PROC. SMAC 1983. Reprinted by permission.)

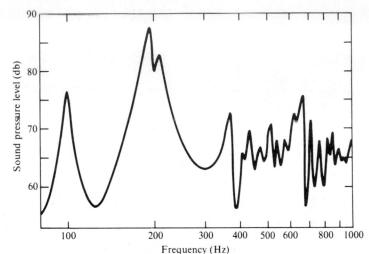

FIG. 10.31

Sound pressure level one meter in front of a folk guitar (Martin D-28) driven by a force of constant amplitude applied to the treble side of the bridge.

In an ordinary room, the directionality of the sound radiation pattern is well obscured by reflections from the walls, ceiling, and other surfaces. Nevertheless, a different sound spectrum occurs at every location in the room. It is clear why many guitarists take great care to hold their instruments in such a way that the back is free to vibrate.

For the sound spectra and radiation patterns in Figs. 10.31 and 10.32, a sinusoidal force was applied to the treble side of the bridge perpendicular to the top plate. When a guitar string is plucked, the force on the bridge has spectral components at many frequencies (see Figs. 10.3 and 10.5), and furthermore the force is not exactly perpendicular to the top plate. Thus the sound spectra become more complicated.

One way to study the playing characteristics of a guitar is to record the sound pressure for each note of the scale plucked with as consistent an initial displacement as possible. Such a curve for a classical guitar is shown in Fig. 10.33. Playing curves of this type will be

FIG. 10.32

Sound radiation patterns at four resonance frequencies in a Martin D-28 folk guitar (compare with Figs. 10.28 and 10.29 which show the corresponding vibrational configurations).

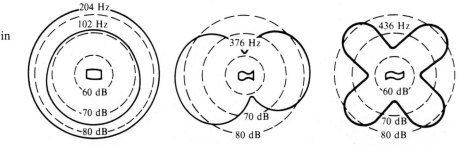

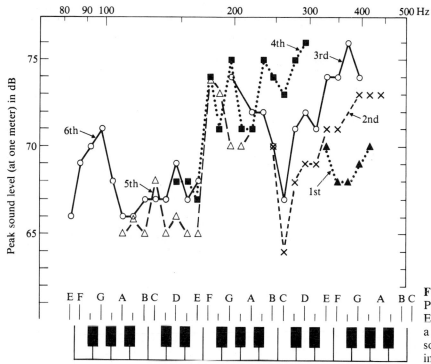

FIG. 10.33
Playing curve for a classical guitar. Each note is plucked with as consistent a displacement as possible, and the sound level is recorded one meter away in a fairly reverberant room.

quite different from frequency response curves, such as Fig. 10.31, because plucking the string excites a large number of harmonics. Driving the guitar sinusoidally at 87 Hz, for example, gives a weak response, but the note F_2 comes through fairly well because its harmonics are near enough to guitar resonances. There is some difference in the sound levels due to playing the same note on different strings. The heavier strings tend to drive the guitar a little more strongly, as might be expected.

A folk guitar with steel strings will produce sound levels that are typically 5 to 10 dB greater than those of a nylon-strung classical guitar. Steel strings normally have approximately twice the tension and mass of the corresponding nylon strings.

10.14   THE ELECTRIC GUITAR

Although it is possible to attach a contact microphone to the body of an acoustic guitar, an electric guitar nearly always uses magnetic pickups in which the vibrating strings induce electric signals directly.

Electric guitars may have a solid wood body or a hollow body, the solid design being the more common. Vibrations of the body have much less influence on tone in the electric guitar than in its acoustic

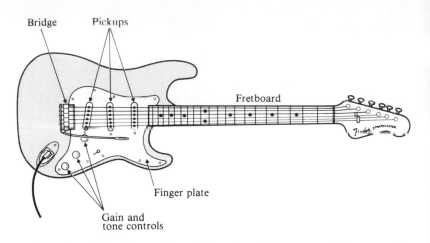

Bridge   Pickups

Fretboard

Finger plate

Gain and
tone controls

FIG. 10.34
An electric guitar.

cousin. The solid guitar, although heavier, is less susceptible to acoustic feedback (from the loudspeaker to the guitar) and it also allows the strings to continue vibrating for a slightly longer time. Figure 10.34 shows the main features of an electric guitar.

The electromagnetic pickup consists of a coil with a magnetic core. The vibrating steel string causes changes in the magnetic flux through the core, thus inducing an electrical signal in the coil. The principle of the guitar pickup is similar to the magnetic phonograph pickup (see Section 21.4). Most electric guitars have at least two pickups for each string, some have three. These pickups, located at different points along the string, sample different strengths of the various harmonics, as shown in Fig. 10.35(b). The front pickup (nearest the fretboard) generates the strongest signal at the fundamental frequency, whereas the rear pickup (nearest the bridge) is most sensitive to the higher harmonics (the resulting tones are sometimes characterized as "mellow" and "gutsy," respectively). Switches or individual gain controls allow the guitarist to mix together the signals from the pickups as desired.

Most pickups have a threaded pole piece that can be adjusted in height by screwing it in or not. Adjusting the pole piece closer to the string will usually increase the volume; if it is too close to the string, however, distortion will result due to the force exerted on the string by the magnet. The distortion becomes especially noticeable when fingering beyond the twelfth fret, which brings the string down close to the pickup.

Electrical circuits for guitar pickups may incorporate a wide variety of different features. Many guitars include tone controls that adjust the high-frequency response. Others have a switch to reverse the phase of the signal from one pickup with respect to the others. *"Humbucking" pickups* have two coils wound in such a way that stray mag-

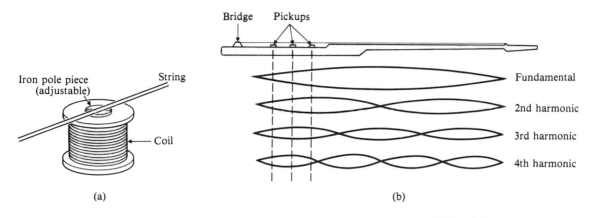

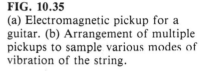

(a)   (b)

netic fields (from power cords, lights, etc.) will induce opposing electrical signals in the two coils; thus the hum they produce will be minimized.

A special type of electric guitar is the *bass guitar* or *electric bass* widely used in rock and jazz bands. Besides being tuned lower (E_1, A_1, D_2, G_2), it differs from the ordinary electric guitar in that it has only four strings and a longer fretboard (90 cm instead of 65 cm). An electric bass is shown in Fig. 10.36.

FIG. 10.35
(a) Electromagnetic pickup for a guitar. (b) Arrangement of multiple pickups to sample various modes of vibration of the string.

10.15   STRINGS, FRETS, AND COMPENSATION

Strings for the modern classical and flamenco guitars are made of nylon, replacing the gut strings used in the past. The three highest strings are usually monofilament nylon, while the three lowest strings have nylon cores wrapped with a metal winding. Some flamenco players substitute second and third strings with plastic windings around nylon cores for a slightly more brilliant sound.

Flattop or folk guitars use steel wire for the highest two strings and sometimes the third, whereas the remaining strings have steel cores wrapped with steel, nickel-steel, or bronze. Usually the wrapping is composed of a fine wire of circular cross section ("roundwound" strings), but sometimes a flat ribbon of stainless steel is used for the wrapping ("flatwound" strings). Other variations are the "flatground" string (wound with round wire that is then ground flat), and compound strings with a winding of silk between the steel core and metal outer windings.

Spacing the frets on the fretboard presents some interesting design problems. Semitone intervals on the scale of equal temperament correspond to frequency ratios of 1.05946 (see Section 9.4). This is very near the ratio 18:17, which has led to the well-known *rule of eighteen*. This rule states that each fret should be placed 1/18 of the remaining

FIG. 10.36
An electric bass. (Photo courtesy of Sky Computers.)

FIG. 10.37
Fret placement. According to the rule of eighteen, each fret is placed 1/18 of the remaining distance d to the bridge saddle, or $x = d/18$. (Greater accuracy is obtained by using 17.817 rather than 18.)

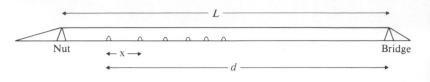

distance to the bridge, as shown in Fig. 10.37. Obviously the fret spacing x decreases as one moves down the fingerboard.

Since the ratio 18/17 equals 1.05882 rather than 1.05946 (an error of about 0.06 percent), each semitone interval will be slightly flat if the rule of eighteen is used to locate the frets. By the time the twelfth fret is reached, the octave will be 12 cents (12/100 semitone) flat, which is noticeable to the ear. Thus for best tuning the exact figure 17.817 should be used in place of 18; in other words, each fret should be placed 0.05613 of the remaining distance to the bridge.

Another problem in guitar design is the fact that pressing down a string against a fret increases the tension slightly. This effect is much greater in steel strings than nylon, since a much greater force is required to produce the same elongation. Fretted notes will tend to be sharp compared to open ones. The greater the clearance between strings and frets, the greater this sharpening effect will be.

To compensate for this change in tension in fingering fretted notes, the actual distance from the nut to the saddle is made slightly greater than the scale length used to determine the fret spacings. This small extra length is called the string *compensation,* and it usually ranges from between 1–5 mm on acoustic guitars to several centimeters on an electric bass. Bass strings require more compensation than treble strings, and steel strings require considerably more than nylon strings. A guitar with high action (larger clearance between strings and frets) requires more compensation than one with a lower action. Some electric guitars have bridges that allow easy adjustment of compensation for each individual string.

10.16   SUMMARY

The vibrating strings of a violin radiate very little sound directly, but they transmit their vibrations to the violin body. Plucking a string at the center tends to excite modes that are odd-numbered harmonics of the fundamental. Bowing a string causes a bend to propagate along the string, and the displacement of each point on the string describes a sawtooth curve as a function of time. As the bend passes the point of contact, it initiates "sticking" and "slipping" of the string and bow.

The violin body has two strong resonances, called the main air and main wood resonances. Tuning the top and back plates during construction is made much easier by using Chladni patterns to observe the modes of vibration. The resonances of violas and cellos lie somewhat higher in relationship to their open strings than those of a violin. Recently a new family of eight fiddles has been designed to have resonances scaled to those of the violin.

The guitar also has two resonances called "air" and "main wood," which largely determine the low-frequency characteristics. The frequencies of these two resonances are determined by the frequencies of fundamental vibration mode of the top plate and the Helmholtz air resonance plus the amount of coupling between them. The vibration modes of the top plate depend on the plate stiffness and bracing. Electric guitars have pickups, which sense the vibrations of the strings directly. Guitars may have either steel or nylon strings, plain or wrapped. Strings are made slightly longer than the calculated scale length in order to compensate for the change in tension when a string is pressed against a fret.

References and Suggested Readings

Alonso Moral, J., and E. Jansson (1982a). "Eigenmodes, Input Admittance, and the Function of the Violin," *Acustica* **50**: 329.

Alonso Moral, J., and E. Jansson (1982b). "Input Admittance, Eigenmodes, and Quality of Violins," Report STL QPSR 2-3/1982, Speech Transmission Laboratory, Royal Inst. of Tech. (KTH), Stockholm, p. 60.

Caldersmith, G. (1977). "Low Range Guitar Function and Design," *Catgut Acoust. Soc. Newsletter* **23**: 19. A more complete but more mathematical paper appears in *J. Acoust. Soc. Am.* **63**: 1566 (1978).

Dünnwald, H. (1983). "Auswertung von Geigenfrequenzgangen," *Proc. 11th ICA, Paris,* **4**: 373.

Fletcher, N. H., and T. D. Rossing (1989). *The Physics of Musical Instruments.* New York: Springer Verlag. Chapters 9 and 10.

Hacklinger, M. (1978). "Violin Timbre and Bridge Frequency Response," *Acustica* **39**: 323.

Haines, D. W., N. Chang, and C. M. Hutchins (1975). "Violin with Graphite-Epoxy Top Plate" (abstract), *J. Acoust. Soc. Am.* **57**: S21.

Hutchins, C. M. (1967). "Founding a Family of Fiddles," *Physics Today* **20** (2): 23.

Hutchins, C. M. (1973). "Instrumentation and Methods for Violin Testing," *J. Audio Eng. Soc.* **21**: 563.

Hutchins, C. M. (1977). "Another Piece of the Free Plate Tap Tone Puzzle," *Catgut Acoust. Soc. Newsletter* **28**: 22.

Hutchins, C. M., K. A. Stetson, and P. A. Taylor (1971). "Clarification of 'Free Plate Tap Tones' by Hologram Interferometry," *Catgut Acoust. Soc. Newsletter* **16**: 15. Also reprinted in *Musical Acoustics, Part 2.* Ed., C. M. Hutchins. Stroudsburg, Pennsylvania: Dowd, Hutchinson and Ross, 1976.

Jansson, E. V. (1971). "A Study of Acoustical and Hologram Interferometric Measurements of the Top Plate Vibrations of a Guitar," *Acustica* **25**: 95.

Reder, O. (1970). "The Search for the Perfect Bow," *Catgut Acoust. Soc. Newsletter* **13**: 21.

Reinecke, W. (1973). "Ubertragungseigenschaften des Streichinstrumentenstegs," *Catgut Acoust. Soc. Newsletter* **19**: 26.

Richardson, B. E., and G. W. Roberts (1985). "The Adjustment of Mode Frequencies in Guitars: A Study by Means of Holographic Interferometry and Finite Element Analysis," *Proc. SMAC 83* (Royal Swedish Academy of Music, Stockholm).

Rossing, T. D., J. Popp, and D. Polstein (1985). "Acoustical Response of Guitars," *Proc. SMAC 83* (Royal Swedish Academy of Music, Stockholm).

Schelleng, J. C. (1968). "Acoustical Effects of Violin Varnish," *J. Acoust. Soc. Am.* **44**: 1175.

Schelleng, J. C. (1974). "The Physics of the Bowed String," *Sci. Am.* **230** (1): 87.

Sloane, I. (1966). *Classic Guitar Construction.* New York: Dutton.

Glossary

anisotropy The difference in some property when measured in different directions (such as the stiffness of wood along and across the grain).

bass bar The wood strip that stiffens the top plate of a violin or other string instrument and distributes the vibrations of the bridge up and down the plate.

belly The top plate of a violin.

bridge The wood piece that transmits string vibrations to the sound board or top plate.

Chladni pattern A means for studying vibrational modes of a plate by making nodal lines visible with powder.

compensation (string) An extra length of string added because tension changes when a string is pressed against a fret.

f-holes The openings in the top plate of a string instrument shaped like the letter "f."

Helmholtz resonator A vibrator consisting of a volume of enclosed air with an open neck or port (see Sections 2.3 and 4.7).

humbucking pickup A magnetic pickup with two coils designed to minimized hum caused by stray magnetic fields.

mobility (mechanical admittance) The ratio of velocity to force (called input admittance or driving point mobility if velocity and force are measured at the same point).

nut The strip of hard material that supports the strings at the head end.

purfling The thin wood strip near the edge of the top or back plate of a string instrument.

saddle The strip of hard material (ivory or bone) that supports the string at the bridge of a guitar.

sinusoidal force A smoothly varying force with a single frequency; the waveform is described as a sine wave.

sound hole (rose hole) The round hole in the top plate of a guitar that plays an important role in determining the lower resonances of the body.

sound post The short round stick (of spruce) connecting the top and back plates of a violin or other string instrument.

sul ponticello Bowing near the bridge.

sul tasto Bowing near the fingerboard.

viol An early bowed string instrument usually having six strings and a fretted fingerboard.

viola da gamba A viol played in an upright position.

wolf tone An undesirable tone due to the interaction of a violin or cello resonance with the vibrations of the string.

Questions for Discussion

1. Does increasing the force on a violin bow increase the loudness of the tone? Explain why.

2. Why does a folk guitar with steel strings play more loudly than the same guitar with nylon strings?

3. How might the resonances of a hollow-body electric guitar affect the tonal output? (Remember that the pickups sense string motion only.)

4. Electric guitars (especially those with a hollow body) are susceptible to acoustic feedback (see Section 19.6), even though they have no microphone. Explain why this occurs and how it can be prevented.

Problems

1. Continue the sketches in Fig. 10.4 to show the shape of the plucked string during the next half cycle from $t = \frac{1}{2}T$ to $t = T$.

2. In Figs. 10.3 and 10.5, note that plucking a string one-fifth the distance from one end suppresses the fifth harmonic, and plucking it at the midpoint (one-half the distance) suppresses the second harmonic. Also note that the phase of the harmonics (indicated by + or −) changes in going through a zero. Using this information, draw a similar diagram to show the addition of modes to obtain the shape of a string plucked at one-third its length.

3. Assuming a frequency of 440 Hz and a bow speed of 0.2 m/s in Fig. 10.7, what is the displacement of the string at its midpoint? What is the speed of the string when it leaves the bow and "snaps" back? (Hint: first determine the time during which the string moves at the speed of the bow.)

4. Note the similarity between the Chladni patterns in Fig. 10.16 and the holograms in Fig. 10.17 for modes II and V. Sketch a Chladni pattern that one might expect for mode I in Fig. 10.17.

5. If the main resonances (A and T in Fig. 10.19) of the alto violin in the new family of fiddles are scaled to those of the conventional violin, near what notes will they lie? (The alto is tuned a fifth below the violin.) Compare these resonances to those of the viola.

6. Determine the musical intervals between the strings of the guitar (see Secton 10.9) and those of the electric bass (see Section 10.14).

7. Calculate the two lowest resonance frequencies of a pipe 46 cm long closed at both ends (they are the same as those of a pipe open at both ends; see Section 4.5). Do the same for a pipe 33 cm long. Now compare these frequencies to those given for the resonances A_2 and A_4 (longer pipe) and A_3 and A_5 (shorter pipe) in Fig. 10.27(c). Discuss the significance of the similarity.

8. Carefully measure the distance of each fret of a guitar from the saddle of the bridge, and determine how closely the rule of eighteen has been followed. If done carefully, you can determine how much compensation is included for each string.

CHAPTER 11
Brass Instruments

In Chapters 11 and 12, we will discuss a wide variety of wind instruments, or aerophones, which differ widely in their construction and in their acoustical properties. It is customary to classify them into two families, the brasses and the woodwinds. When a brass instrument is played, the player's lips act as a valve, introducing puffs of air at just the right time to maintain oscillations of the air column. When a woodwind is played, an oscillating air stream or an oscillating reed excites the air column. The brass instruments radiate sound from a flared end of the tube, called the *bell;* woodwinds usually radiate sound from several holes in the sides of the air column. Other differences will become clear in these next two chapters.

In ancient times, men apparently blew animal horns, which served as sound sources for religous ceremonies and as warnings of danger. Attempts to use these horns to play music, however, apparently began during the Middle Ages. Later, wood and metal tubing were substituted for the horns of animals, and mouthpieces were added to simplify playing. Among the brass instruments that have survived from the Renaissance and Baroque eras are the valveless posthorn; the cornett, played with side holes; and the sackbut, forerunner of the trombone.

11.1 INSTRUMENTS OF THE BRASS FAMILY

The principal members of the brass family are the trumpet, French horn, trombone, and tuba (see Fig. 11.1(a)). Their playing ranges are indicated in Fig. 11.1(b). Other important brass instruments are the cornet, fluegelhorn, bugle, and baritone horn.

Brass instruments have four sections: a *mouthpiece,* a tapered *mouthpipe,* a *cylindrical* section, and a *bell,* as shown in Fig. 11.2. The trumpet, French horn, and trombone have cylindrical sections of considerable length, as indicated in Table 11.1, while the fluegelhorn,

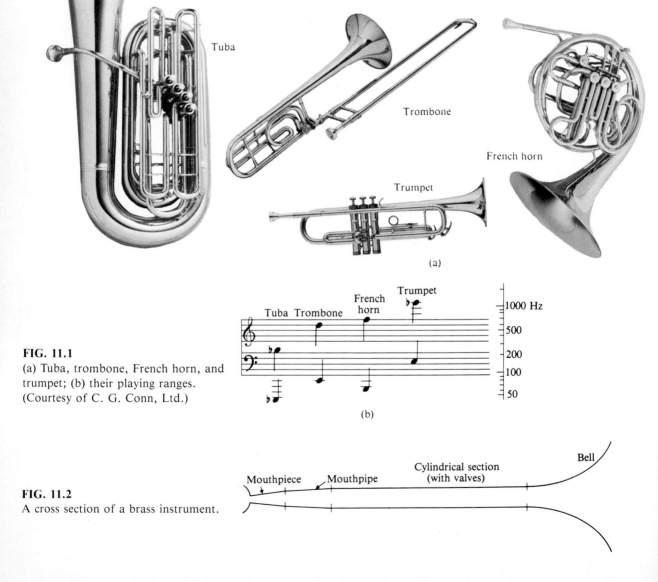

FIG. 11.1
(a) Tuba, trombone, French horn, and trumpet; (b) their playing ranges. (Courtesy of C. G. Conn, Ltd.)

FIG. 11.2
A cross section of a brass instrument.

TABLE 11.1 Comparison of brass instruments

	Trumpet	French horn	Trombone	Tuba	Baritone
Fundamental	$B^\flat_2$	F_1	$B^\flat_1$	$B^\flat_0$	$B^\flat_1$
Lowest "open" note	$B^\flat_3$	F_2^*	$B^\flat_2$	$B^\flat_1$	$B^\flat_2$
Length	140 cm	375 cm	275 cm	536 cm	264 cm
Bore (diameter) of main tube	1.1 cm	1.1 cm	1.2 cm	1.8 cm	1.3 cm
Cylindrical portion	53 cm	193 cm	170 cm		
Bell diameter	11 cm	32 cm	18 cm	35–60 cm	25 cm

*Many French horns are "double horns" composed of two horns tuned to F_2 and $B^\flat_2$.

baritone, and tuba, often referred to as instruments of conical bore, are tapered throughout much of their length.

11.2 OSCILLATIONS IN A PIPE

The air column of a $B^\flat$ trumpet has a length of about 140 cm. As we discussed in Section 4.5, the resonances of a closed cylindrical pipe of this length should be

$$f_n = n\frac{v}{4L} = n\frac{343}{4(1.4)} = 61.3n \qquad (n = 1, 3, 5, 7...)$$

$$= 61, 184, 306, 429, ... \text{ Hz.}$$

If we blow softly on the end of the pipe in the manner of a brass player (this may be easier if a smooth ring is attached to the end), the frequencies obtainable are very near those calculated above. (The fundamental, which is difficult to blow with the lips, can be sounded by attaching a clarinet mouthpiece.)

One way to study the resonances of a pipe (or a wind instrument) is to make a graph of its *acoustic impedance* as a function of frequency. (Acoustic impedance, defined in Section 4.6, is the sound pressure divided by the volume velocity.) An apparatus for doing so is shown in Fig. 11.3. A loudspeaker driver forces sound through a capillary tube designed to produce a volume velocity of constant amplitude. A pressure microphone next to the mouthpiece detects pressure variations; since the velocity amplitude is kept essentially constant, the pressure peaks and valleys will be a replica of the impedance peaks and valleys.

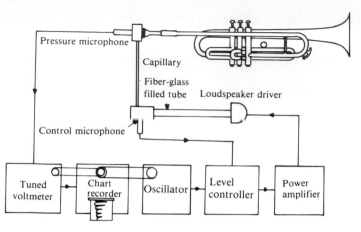

FIG. 11.3
Apparatus for graphing acoustic impedance of a wind instrument, as used by Benade (1973), Backus (1976), and other investigators. (Copyright © 1976 by Virginia Benade. Reprinted by permission.)

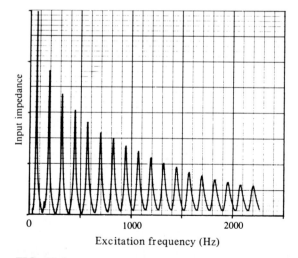

FIG. 11.4
Impedance curve for a cylindrical pipe 140 cm long (about the length of a trumpet). (From *Fundamentals of Musical Acoustics* by Arthur H. Benade. Copyright © 1976 by Oxford University Press, Inc. Reprinted by permission.)

Figure 11.4 shows the impedance peaks of a trumpet-length cylindrical tube 140 cm long.* Note that frequencies are the odd-numbered harmonics of the fundamental, as shown in Fig. 4.8. At the frequency of one of these pressure (impedance) peaks, the tube "cooperates" with the lips, because the excess pressure during a portion of the cycle helps to open the lips to admit air at the right time. Energy is thus supplied to the vibrating air column, which makes up for various losses (including the radiation of sound) and sustains the oscillations. The blowing pressure in the mouth must be maintained higher than the highest pressure peak, but this is easy to do, because the pressure peaks are not very large.

11.3 PRESSURE-CONTROLLED VALVES

Consider what happens when a brass player's lips open for a moment to admit a puff of air into the mouthpiece. A standing wave builds up in the horn, but it immediately begins to die out, like a plucked guitar string, as the energy is dissipated at the walls of the tube or radiated as sound from the bell. In order to sustain the oscillation, the player must continue to supply puffs of air at appropriate times, not unlike the regular pushes applied to a child in a swing.

A steady flow of air into the instrument will not sustain the oscillation any more than a steady force applied to a swing, because energy would be added during half the cycle and removed during the other half. To maintain oscillation, air must be added at the appropriate part of each cycle: when the mouthpiece pressure is high. It would be difficult, if not impossible, for the player to synchronize his/her lip opening by muscular action alone; fortunately, this is not necessary. Pressure pulses reflected back from the horn tend to force the player's lips open at the right time during the cycle of oscillation. This *regenerative* or *positive feedback* is a characteristic of all oscillators, mechanical, acoustical, or electrical.

The cooperation between lips and air column to sustain oscillations is similar to the electrical feedback that sustains oscillations in an audio generator (see Chapter 19). It is equivalent to having a pressure-controlled valve that admits air whenever the pressure is high, as shown in Fig. 11.5(a). An analysis of the cooperation between a pressure-controlled valve and a vibrating air column is given by Helm-

---

* Graphs showing impedance vs frequency, called "impedance curves," should not be confused with spectra, which show sound output or vibration amplitude vs frequency (as in Fig. 7.11 or Fig. 10.24, for example).

FIG. 11.5

(a) A pressure-controlled valve (player's lips) admits air at the time when air pressure is a maximum, thus sustaining the oscillation. (b) A "water trumpet," which illustrates the same principle with sloshing water in a trough. (From *Fundamentals of Musical Acoustics* by Arthur H. Benade. (Copyright © 1976 by Virginia Benade. Reprinted by permission.)

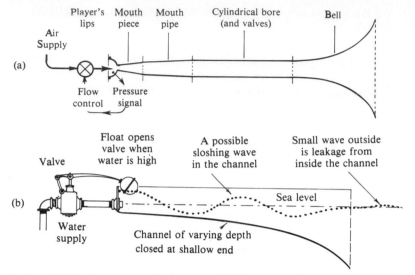

holtz (1877), but a much more lucid discussion is given by Benade (1973, 1976).

Benade's "water trumpet," shown in Fig. 11.5(b), illustrates the principle of pressure-controlled feedback with sloshing water substituted for the longitudinally vibrating column of air. When the water level rises at the "mouthpiece" end, a valve opens to admit more water, and the water piles up into a larger wavecrest. A steady water pressure in the supply pipe has been used to add energy to the water waves in the trough and thus compensate for the losses incurred due to friction, leakage at the "bell" end, and so on.

We will not go into the mathematical details of the pressure feedback that controls the player's lips. It will suffice to point out that stable oscillation can be maintained when the horn oscillation frequency is *above* the natural resonance frequency of the lips. The player's lips, which were described by Helmholtz (1877) as an "outward-striking reed," have a natural resonance frequency determined by their mass and tension (which can be adjusted by the player).

A very important principle in describing the cooperation of a pressure-controlled valve with a vibrating air column is the following.

> If the flow rate is not proportional to the pressure (that is, if the valve is nonlinear), oscillation is favored when the air column has one or more resonances that correspond to the partials (overtones) of the tone being produced.

Except at the softest level of playing, the vibrating air column of a brass instrument will have many harmonics. If the pressure peaks of each of these partials reach the lips at different times, cooperation between the lips and the air column will be difficult. However, if these pressure peaks arrive at the same time, they add together, and the oscillation is stabilized. This is equivalent to the statement in the box above: that oscillations are favored when these partials correspond to the resonances of the air column itself. The oscillation of an air column with nonlinear excitation has been carefully studied by several investigators and is summarized in Chapter 20 of Benade (1976), by Fletcher (1979), and by Fletcher and Rossing (1989).

11.4 THE BELL AND MOUTHPIECE

All brass instruments have flared bells. The bell has several important effects on the acoustics of the brass instruments:

1. It changes both the frequencies and the heights of the impedance peaks (resonances).

2. It changes the radiation pattern of the horn, making it more directional at high frequencies.

3. It changes the spectrum of the radiated sound.

4. It allows more efficient radiation of sound by matching the high pressure inside the horn to the lower pressure outside.

Figure 11.6 shows the impedance curve of a length of trumpet tubing with a trumpet bell attached. Comparison with Fig. 11.4 indi-

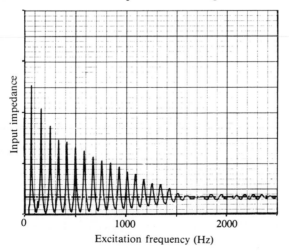

FIG. 11.6
Impedance curve for a 140-cm length of trumpet tubing with a trumpet bell attached. Compare this with Fig. 11.4 (From *Fundamentals of Musical Acoustics* by Arthur H. Benade. Copyright © 1976 by Oxford University Press, Inc. Reprinted by permission.)

cates that the peaks are shifted toward lower frequencies, but in a slightly different way than would occur from a mere lengthening of the tube. The higher peaks have been lowered proportionately more than the lower peaks. This is due to a basic property of flared horns: The effective length increases with frequency. The manner in which the effective length changes with frequency depends on the flare of the bell (Pyle, 1975). The *turning point** of the wave can be observed by noting the sound level at a small microphone as it is slowly inserted into the bell; at sufficiently loud playing levels, one can feel the turning point by inserting a finger into the bell.

Another experiment is to disconnect the bell by removing the tuning slide of a trumpet or trombone. The instrument now plays out of tune, and would do so even if enough tubing were added to match the effective length of the instrument including the bell.

Also apparent in Fig. 11.6 is the virtual disappearance of peaks above 1500 Hz when a bell is added to the pipe (compare with Fig. 11.4). At the same time, the higher frequencies become more prominent in the radiated sound. These two related effects both result from the more efficient radiation of higher frequencies by the bell. At the frequencies of the lower peaks, most of the sound is reflected back from the bell, and the amplitude builds up; at higher frequencies, however, a substantial part is radiated into the room.

The mouthpiece also shifts the frequencies of the impedance peaks (resonances), although it has a more dramatic effect on the peak heights. Peaks (as well as valleys) in the vicinity of the lowest resonance of the mouthpiece are greatly enhanced. The frequency of the lowest resonance of the mouthpiece, which usually occurs in the range of 750–850 Hz, is often called its *popping frequency,* because it can be determined by slapping the rim with the palm of the hand and noting the frequency of the resulting pop.

The impedance curve of a B♭ trumpet is shown in Fig. 11.7. Comparisons with Figs. 11.4 and 11.6 point out the enhancement of impedance (pressure) peaks in the vicinity of the mouthpiece resonance. These large pressure peaks greatly enhance the ability of the air column to control the rather massive lips of the player. Skillful design of the bell and the mouthpiece have now brought the peak frequencies, except for the lowest one, very nearly into a harmonic relationship, so that the air column resonances will correspond to the partials of the desired tone.

Also indicated in Fig. 11.7 are the peaks (2, 4, 6, and 8) that correspond to the partials of B♭$_3$ (f = 233 Hz; this is the written C$_4$ for a

---

*The point of furthest penetration of the standing sound wave into the bell.

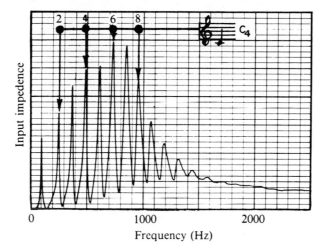

FIG. 11.7
Impedance curve of a B♭ trumpet, showing peaks that form a "regime of oscillation" for C₄. (From *Fundamentals of Musical Acoustics* by Arthur H. Benade. Copyright © 1976 by Oxford University Press, Inc. Reprinted by permission.)

B♭ instrument). When the note is blown very softly, the player's lips vibrate nearly sinusoidally, and only peak 2 is of importance. As the level of loudness increases, more harmonics are produced, which excite peaks 4, 6, and 8. The tone takes on increasing stability as the lips become subject to control by pressure impulses corresponding to all four peaks.

Note that the lowest peak in Fig. 11.7 lies below the fundamental frequency (about 117 Hz, corresponding to B♭₂). It is possible to play the "missing" fundamental note, sometimes called the pedal tone, by relying on feedback from other peaks having harmonically related frequencies. It is even possible to play at the frequency of the lowest peak, but that is not a useful note at all. Figure 11.8 shows the approximate pressure distribution for each of the first four resonances in Fig. 11.7. Note the way in which the *turning point* in the bell changes with frequency.

It is clear from Fig. 11.7 why the notes above F₅ (written G₅) or so become increasingly difficult to play. The peaks that correspond to the harmonics are small, which indicates that the corresponding resonances are too weak to be of much help in controlling the lips and stabilizing the tone. F₅ is quite easy to play softly, because peak 6 is very tall, indicating a strong resonance; little increase in stability is noted during crescendo, however, because peak 12, which lies near the second harmonic, is small. Above B♭₆ (written C₆), even the fundamental is weakly supported by the air column, and the lips must provide their own stability, which makes great demands on the player's ability to control his or her lips. This control makes use, to some extent, of a velocity-dependent force called the *Bernoulli* force.

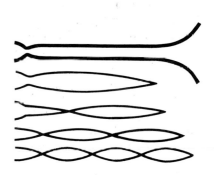

FIG. 11.8
Approximate pressure distribution for the first four modes in a trumpet. Note that the "turning point" moves outward in the bell as the frequency increases. Mode frequencies are nearly in the ratios 0.8 : 2 : 3 : 4.

The Bernoulli Force

When a fluid flowing through a pipe enters a region in which
the area of the pipe decreases, the speed of the fluid increases.
To thus accelerate the fluid requires a net force, so the pressure in the larger pipe behind must be greater than in the
smaller pipe ahead, as shown in Fig. 11.9(a). The reduced
pressure in the center section of the tube causes the liquid in
the U-tube to stand at a higher level. This effect (i.e., reduction in pressure when flow velocity increases) was described
by Daniel Bernoulli in 1738 and now carries his name.

   Simple experiments that illustrate the Bernoulli effect are
shown in Figs. 11.9(b) and (c). In each case, the reduced pressure in the moving air stream gives rise to a net upward force
F_B, which is often referred to as the *Bernoulli* force.

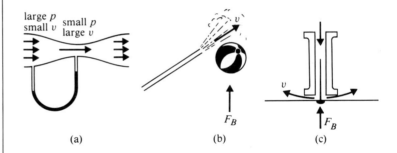

(a)   (b)   (c)

FIG. 11.9
The Bernoulli effect. (a) The reduced pressure in the center section
of the tube causes the liquid in the U-tube to stand at a higher
level. (b) The reduced pressure in the air jet from a vacuum
cleaner hose gives rise to a net upward force F_B, which can
support a beach ball. (c) Blowing downward through the spool
causes air to flow outward, supporting a card and pin by means
of the Bernoulli force.

11.5   VALVES AND SLIDES: FILLING IN BETWEEN THE MODES

The instrument we have been discussing thus far is a valveless trumpet
like the familiar military bugle. It is quite possible to play eight or
more notes, corresponding to the peaks in Fig. 11.7, and a few more
may be added by skillful playing. To play the remaining notes of the
scale, however, the acoustical length of the instrument must be varied.
This is done by moving a slide (trombone) or by inserting lengths of
additional tubing by means of valves (trumpet, French horn, baritone,

tuba, etc.). A French horn normally uses rotary valves, but most other brass instruments in the United States use piston valves of the type shown in Fig. 11.10. In the raised position, a hole in the piston, called the *windway,* extends the air column straight through the valve. When the valve is depressed, however, two other windways connect the air column to an additional length of tubing.

The playing positions of a trombone slide are shown in Fig. 11.11. Seven positions of the slide are needed to play all desired notes of the scale; note that the seven positions are not spaced equally. Going down a semitone decreases the frequency by about six percent, so the instrument should be lengthened by about this same amount. But as the instrument gets longer, a six-percent increase in length also becomes progressively longer.

Another complication arises in determining the lengths of tubing to be added. The additional length of tubing added by moving a trombone slide or by depressing a trumpet piston valve must be cylindrical. Thus the average flare of the entire horn is reduced. The result is that the addition of a piece of tubing makes a larger percentage change in the lower modes than in the higher ones.

Another fundamental problem arises in designing the additional lengths of tubing for each valve. If the lengths of tubing are designed so that the first and second valves can lower the pitch by a whole tone and a semitone, respectively, then the combination will be inadequate to lower the pitch by three semitones. This can be illustrated by the following simple arithmetic.

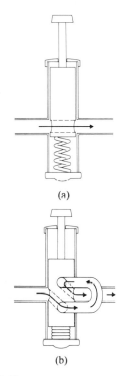

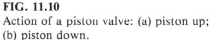

FIG. 11.10
Action of a piston valve: (a) piston up; (b) piston down.

To lower the pitch a whole tone, the first valve must add a length of tubing 12.2 percent of the total length L, or $0.122L$. To lower the pitch a semitone requires the addition of 5.9 percent of the total length, or $0.059L$. Thus the two valves together increase the length by 18.1 percent, or $0.181L$. But three semitones requires an increase of 18.9 percent (see Table 9.1).

The same conclusion can be reached by examining Fig. 11.10. The length of tubing needed to go one semitone down from the second to the third position is greater than that needed to go from the first to the second position.

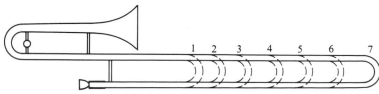

FIG. 11.11
Playing positions of a trombone slide.

Various compromises have been proposed to deal with this tuning problem (Young, 1967). The valve slides for the first and second valves may be made slightly longer than the optimum length, thus accepting a slight flattening of the first two semitones in order to reduce the sharpening of the third one. Similarly, the third valve slide may be designed to give a shift slightly more than three semitones. Since the discrepancy is usually greatest when the third valve is used in combination with the others, most trumpets have a provision for moving the third valve slide with the left hand while the instrument is being played. Beyond that, the brass player must "lip" the various notes into tune, that is, make small corrections in frequency by changing the tension of the lips.

The valve problem becomes increasingly difficult in the larger brass instruments. A fourth valve is frequently added to the tuba; it lowers the pitch by a fourth, and thus substitutes for the troublesome combination of valves 1 and 3. Even a fifth or sixth valve is sometimes added, their function varying in the different tubas.

11.6   THE FRENCH HORN

The French horn in common use is a "double horn," composed of two horns tuned to F_2 and $B^\flat_2$, which share the same mouthpiece and bell. The three main valves each have two sets of windways and valve slides, and a fourth valve is used to switch from one horn to the other. The $B^\flat$ horn is preferred for playing high notes, because it has stronger resonances at the higher frequencies, and they are spaced further apart.

The length of the F-horn is about 375 cm (more than 12 feet), 30 percent longer than the trombone. It has many resonances, and much of the time the horn is played at a pitch corresponding to one of the higher resonances. This means that many notes on a French horn lack the stability of notes normally played on a trumpet or trombone, where there are several prominent resonances corresponding to the overtones of the note played.

Placing the hand in the bell of the horn makes it easier to play the higher notes. By inhibiting the radiation and increasing the reflection of the higher frequencies, it allows standing waves to build up and leads to more usable resonances at the higher frequencies. Figure 11.12 shows the impedance curves of a French horn with and without a hand in the bell. The additional resonances in Fig. 11.12(a) stabilize the higher notes by providing feedback at the overtone frequencies.

Placing the hand in the bell also lowers the pitch. If the hand is inserted as far as possible, a "stopped" tone is produced, which appears to be about a semitone higher than normal. What actually hap-

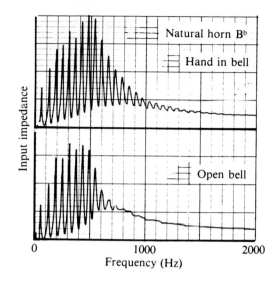

FIG. 11.12
Impedance curves of a B♭ horn with and without a hand in the bell. (From *Fundamentals of Musical Acoustics* by Arthur H. Benade. Copyright © 1976 by Oxford University Press, Inc. Reprinted by permission.)

pens is that all the resonances are lowered to the point where the next higher resonance is only a semitone above the desired note. This can be confirmed by holding a note as the hand is slowly inserted to the stopped position. If no attempt is made to hold the original pitch with the lips, the pitch will fall slowly. If the player tries to hold the pitch, however, the note jumps up to the next higher resonance at some point as the hand is inserted (Backus, 1977).

11.7   THE TROMBONE

The trombone, like the trumpet, has a bore that is predominently cylindrical. Its tubing has twice the length of a trumpet, so its main resonances occur at frequencies that are approximately harmonics of a missing fundamental near B♭₁ (58 Hz). The lowest note that is played in "first" position (i.e., with the slide fully retracted, see Fig. 11.11) is normally B♭₂, although the B♭₁ pedal note in a trombone is a more useful note than in a trumpet, and occasionally is called for in musical scores.

The playing range of the standard tenor trombone extends up to D₅ (the tenth harmonic of the fundamental B♭₁). The bass trombone has a slightly larger bore and a mouthpiece of larger volume than the tenor, making it easier to play the lower notes. Most bass trombones, and some tenor trombones as well, include a rotary valve and additional tubing to lower the fundamental to F₁.

Valve trombones use valves to change the effective length rather than a slide, but they are much less common.

11.8   TUBA, BARITONE, AND FLUEGELHORN

The tuba, baritone horn, and fluegelhorn are examples of instruments that are essentially conical throughout their entire length. The baritone horn has roughly the same playing range as a trombone, but with a conical bore, it has a different timbre.

The small tuba in $B\flat_2$ has much the same playing range as the bass trombone. Larger tubas tune to $E\flat_2$ and $B\flat_1$. One version of the large tuba, popular in marching bands, is called a *sousaphone* in honor of bandsman John Philip Sousa. It coils around the player's body, which makes it easier to carry.

The fluegelhorn is the soprano of the conical brass instrument family. It has a playing range similar to that of the trumpet, and its mellow tone has made it a popular solo instrument in jazz ensembles.

11.9   THE SPECTRA OF BRASS INSTRUMENTS

"Typical" sound spectra of the various instruments frequently appear in books and articles about musical instruments. These can be a little misleading, however, because the spectrum of an instrument changes substantially with changes in pitch and loudness. This statement is especially true of the brass instruments.

We have already discussed (in Section 11.3), how pianissimo playing on a trumpet uses mainly the resonance that matches the fundamental frequency of the played note, but increasing the sound level brings additional resonances into prominence. Thus the spectrum of the sound inside the instrument takes on more and more harmonics as the playing level increases. However, the internal spectrum is not what we hear.

In any wind instrument, the spectrum of the radiated sound depends both on the spectrum of the standing waves within the instrument and the portion of the sound energy that leaks to the outside. In a brass instrument, this portion is determined by the radiation efficiency of the bell, and it changes markedly with frequency.

For a cylindrical pipe with an open end, the radiation efficiency increases with f^2 up to a certain frequency f_c and then remains more or less constant. The value f_c is called the *cutoff frequency* and is approximately $f_c = c/\pi a$, where c is the speed of sound and a is the radius of the pipe. The radiation efficiency of a small pipe thus quadruples each time the frequency goes up one octave, giving a "treble boost" of 6 dB per octave.

For a pipe equivalent to the trumpet bore the cutoff frequency will be nearly 20 kHz. Adding a bell lowers the cutoff frequency several octaves, however.

Figure 11.13 illustrates how the internal spectrum of a trumpet may be combined with the radiation curve (sometimes called the spectrum transformation function) to obtain the spectrum of the radiated sound. Note that the internal and external spectra move up and down the frequency axis as the pitch changes; the radiation curve does not, because it is determined by the shape of the instrument. Thus the various harmonics receive different amounts of treble boost for different notes on the scale. The curves in Figs. 11.13(a) and (c) are for $B^b{}_4$ (466 Hz); they would be quite different for notes in other parts of the playing range. The curve in Fig. 11.13(b) would not change, however.

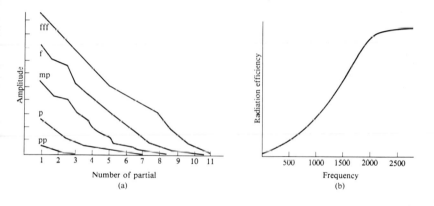

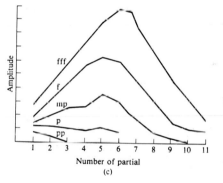

FIG. 11.13
(a) Spectrum of sound inside a trumpet for $B^b{}_3$ (233 Hz).(b) Radiation curve for a trumpet. (c) Spectrum of radiated sound obtained by combining (a) and (b). (From *Fundamentals of Musical Acoustics* by Arthur H. Benade. Copyright © 1976 by Oxford University Press, Inc. Reprinted by permission.)

Recording spectra of the brass instruments is further complicated by the fact that sounds of high frequency are radiated straight ahead in a relatively narrow beam, whereas sounds of low frequency are radiated in all directions, due to the phenomenon of diffraction (see Section 3.10). In addition, the acoustics of the room have a strong influence on the sound spectrum if the microphone is any appreciable distance away from the bell. In spite of these uncertainties, the increasing prominence of the higher harmonics in the louder sounds is clear.

11.10   MUTES

Mutes of varying designs are used to alter the timbre of brass instruments. Three types of trumpet mutes are shown in Fig. 11.14. The straight mute is a truncated cone closed at the large end with three

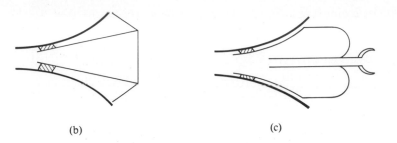

(a)   (b)   (c)

FIG. 11.14
Three types of trumpet mute:
(a) straight mute; (b) cup mute; (c)
Harmon, or "wah-wah," mute.

narrow cork pads spaced around the cone so that an air space of about
3 mm in thickness is maintained between the bell and the mute. The
cup mute adds a second cone at the large end of the original cone.
The Harmon, or "wah-wah," mute is a metal enclosure of a special
shape with a small adjustable bell at one end.

Backus (1976) has studied the acoustical effects of trumpet mutes.
His results can be summarized as follows:

1. The mute changes the radiation characteristics of the bell, reducing the radiation at low frequencies much more than the radiation
   at high frequencies.

2. An extra peak is added to the impedance curve around 100 Hz,
   associated with the resonance of the mute itself. Since this peak
   is below the playing range of the instrument, it has little or no
   effect on its playing characteristics.

FIG. 11.15
Spectra of trumpet tones with and
without mutes. Adding a mute reduces
the strengths of the low harmonics
slightly, but gives a boost to harmonics
in the range of 1500–3000 Hz. Note the
filtering by the resonances (indicated by
arrows).

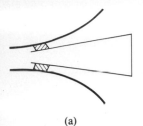

F_3, no mute

F_4, no mute

F_3, straight mute

F_4, straight mute

F_3, cup mute

F_4, cup mute

1   2   3   4
kHz

1   2   3   4
kHz

3. The straight mute acts as a high-pass filter, letting through frequencies above 1800 Hz; the cup mute passes a band of frequencies in the range of 800–1200 Hz; the Harmon mute passes a band of frequencies from 1500–2000 Hz.

4. Several tiny peaks in the impedance curve above 1300 Hz are enhanced; by reducing the radiation from the bell, which is usually large at these frequencies, build-up of standing waves is encouraged.

The spectra of trumpet tones with and without mutes in Fig. 11.15 were recorded by Richard Ross in our laboratory. They show substantial, but not complete, agreement with Backus's results given above.

11.11 WALL MATERIAL

The question of how the playing qualities of a brass instrument depend on the material used in its construction has been debated for many years. Brass instruments are nearly always made from thin-walled brass tubing, although bells have been crafted from sterling silver and other materials. Large tubas (sousaphones) of synthetic material are used in marching bands. Experimental brass instruments made of other metals and even wood are reported to be indistinguishable from instruments of brass. It is probably safe to say that the quality of the craftsmanship in a brass instrument is more important than the type of material used.

The walls of the brass tubing vibrate during playing, and whereas the walls radiate a negligible amount of sound, wall vibrations can contribute to damping. The friction of air against the walls is also a factor. However, if the walls of the tubing were made thick enough to minimize the effect of wall vibrations, and the inside of the tubing were smooth enough to minimize friction, it is doubtful whether the nature of the material would be of importance.

Careful studies by Smith (1978), by Watkinson and Bowsher (1982), and by other investigators have shown how the bells of brass instruments vibrate at high playing amplitudes. Smith (1978) found that the vibration amplitude increases rapidly when the wall thickness is reduced below 0.4 mm. These vibrations have little effect on the radiated sound along the axis of the bell, but the sound radiated laterally (which is heard mainly by the player) is augmented. Thus the player probably would notice a difference even though the audience would not. Vibrations of a 0.3-mm thick trombone bell at 240 and 630 Hz are illustrated by the holographic interferograms in Fig. 11.16.

FIG. 11.16
Holographic interferograms showing
the vibrations of a 0.3 mm-thick
trombone bell driven acoustically at
240 and 630 Hz.

(Courtesy of Richard Smith/Crown
copyright, NPL. Reprinted by
permission.)

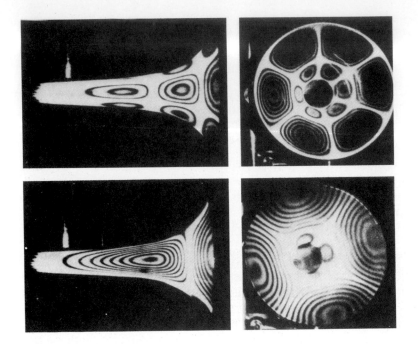

11.12   SUMMARY

Brass instruments have a mouthpiece, various lengths of straight and
tapered tubing, and a flared bell. The player's lips act as a valve, controlled to some extent by pressure (impedance) peaks due to the various resonances of the air column. Cooperation between the lips and
the resonances of the air column sustains oscillations at a stable pitch,
and the stability is greatest when the air column has strong resonances
at several partials of the tone being produced.

The bell changes the frequencies of the pressure peaks (resonances) of the air column and also determines the radiation characteristics of the instrument. The mouthpiece also shifts the frequencies of
the pressure peaks, and it increases the heights of peaks in the vicinity
of its own resonance frequency. Valves or slides are used to change
the lengths of brass instruments to fill in notes of the chromatic scale
that lie between modes in the "open" instrument. The radiated spectrum of a brass instrument is determined by the internal spectrum of
standing waves and the radiation characteristic of its bell. The resonances of a French horn can be adjusted by placing a hand in the bell.
Mutes act as high-pass or band-pass filters and thus alter the timbre
and also the loudness of brass instruments.

References and Suggested Readings

Backus, J. (1976). "Impedance Curves for the Brass Instruments," *J. Acoust. Soc. Am.* **60**: 470.

Backus, J. (1977). *The Acoustical Foundations of Music.* 2nd ed. New York: Norton. (See Chapter 12.)

Benade, A. H. (1973). "The Physics of Brasses," *Sci. Am.* **229**(1): 24.

Benade, A. H. (1976). *Fundamentals of Musical Acoustics.* New York: Oxford. (See Chapter 20.)

Campbell, M., and C. Greated (1987). *The Musician's Guide to Acoustics.* New York: Schirmer Books. Chapter 9.

Fletcher, N. H. (1979). "Air Flow and Sound Generation in Musical Wind Instruments," *Ann. Rev. Fluid Mech.* **11**: 123. (See also *Acustica* **43**: 63.)

Fletcher, N. H., and T. D. Rossing (1989). *The Physics of Musical Instruments.* New York: Springer Verlag. Chapters 13 and 14.

Helmholtz, H. von (1877). *On the Sensations of Tone.* Trans., A. Ellis. New York: Dover, 1954. (See Appendix VII.)

Martin, D. W. (1942). "Lip Vibrations in a Cornet Mouthpiece," *J. Acoust. Soc. Am.* **13**: 305.

Pratt, R. L., S. J. Elliott, and J. M. Bowsher (1977). "The Measurement of the Acoustic Impedance of Brass Instruments," *Acustica* **38**: 242.

Pyle, R. W., Jr. (1975). "Effective Length of Horns," *J. Acoust. Soc. Am.* **57**: 1309.

Smith, R. A. (1978). "Recent Developments in Brass Design," *International Trumpet Guild Journal* **3**: 27.

Smithers, D., K. Wogram, and J. Bowsher (1986). "Playing the Baroque Trumpet," *Sci. Am.* **254**(4): 108.

Watkinson, P. S., and J. M. Bowsher (1982). "Vibration Characteristics of Brass Instrument Bells," *J. Sound and Vibration* **85**: 1.

Young, R. W. (1967). "Optimum Lengths of Valve Tubes for Brass Wind Instrument," *J. Acoust. Soc. Am.* **42**: 224.

Glossary

acoustic impedance The ratio of sound pressure to volume velocity (see Section 4.6). A graph of acoustic impedance of a musical instrument as a function of frequency shows peaks that correspond to the resonances of the air column.

bell The flared section that terminates all brass instruments and determines their radiation characteristics.

Bernoulli effect The pressure in a fluid is decreased when the flow velocity is increased.

cutoff frequency The frequency above which an instrument radiates so efficiently that standing waves inside the instrument are weak. In brass instruments the cutoff frequency is determined mainly by the shape of the bell.

feedback The addition of a part of the output of a system to the input; positive feedback is used to sustain oscillations in wind instruments, audio generators, etc. (See Chapter 19.)

filters (high-pass and band-pass) Acoustic elements that allow certain frequencies to be transmitted while attenuating others. A high-pass filter allows all components above a cutoff frequency to be transmitted; a band-pass filter allows frequencies within a certain band to pass. Electrical filters are discussed in Chapter 18.

mouthpiece The part of a brass instrument that couples the vibrating lips to the air column.

mouthpipe Tapered tubing that connects the mouthpiece to the main section of a brass instrument.

mute An acoustic device that alters the timbre and loudness of a musical instrument.

"popping" frequency The lowest resonance of a brass instrument mouthpiece.

radiation curve, or characteristic A graph showing what portion of the internal sound is radiated by the bell or other part of the instrument.

turning point The point at which reflection of a wave occurs at the open end of the bell or tubing.

volume velocity The rate of air flow in a tube, expressed in units of volume per unit of time (such as m^3/s).

Questions for Discussion

1. A slide trombone can be played in perfect intonation, whereas a valve-operated instrument requires adjustments in the pitch by "lipping." Explain why slide trumpets, having this advantage of a slide trombone, are not used in an orchestra.

2. Make a tracing of the trumpet impedance curve shown in Fig. 11.7. Indicate which peaks form "regimes of oscillation" for F_4 ($f = 349$ Hz) and F_5 ($f = 698$ Hz). Comment on the stability of these two notes played *pp* and *ff*.

3. The trumpet spectrum in Fig. 11.11(c) is typical of what would be heard immediately in front of the trumpet bell. What change would occur in the spectrum heard at some distance (say, 50 feet) away:
   a) in the direction in which the bell is pointed;
   b) at right angles to this direction (assume no room reflections).

4. Dents in the tubing of a brass instrument will cause the reflection of sound waves. How will these show up the impedance curves? How will they affect the playing characteristics of the instrument?

Problems

1. Assume that the length of a trombone is 275 cm in first ("open") position. How far should the slide have to be moved to lower the pitch one semitone? (Remember the length is increased by twice this amount.) If possible, compare this with the slide motion of an actual trombone.

2. Calculate the frequencies of the first three resonances of closed tubes having lengths of 275 cm and 375 cm. Compare these to the resonances shown in the impedance curves of the French horn and trombone.

3. Determine the frequencies of the trumpet resonances as accurately as you can from Fig. 11.7. How closely do they correspond to the "bugle" notes: B^b_3, F_4, B^b_4, D_5, F_5, A^b_5, and B^b_5 (written C_4, G_4, C_5, E_5, G_5, B^b_5, and C_6)?

4. Pressing the first valve of a trumpet or tuba increases the acoustic length by 12.2 percent and lowers the pitch a whole tone. How long must the first valve slide be in each of these instruments? (In other words, how many cm of tubing should be added to produce the pitch change?) If possible, compare your answers to the measured lengths in actual instruments.

5. Find the frequency ratios of the first five peaks in Fig. 11.4 to the corresponding peaks in Fig. 11.6.

CHAPTER 12
Woodwind Instruments

The woodwinds are a large family of instruments with great diversity in size, shape, and design. The principal woodwinds in the symphony orchestra are the flute, clarinet, oboe, and bassoon; auxiliary instruments include the piccolo, English horn, bass clarinet, and contrabassoon. The saxophone is the principal woodwind in jazz bands and also plays an important role in marching and concert bands and, occasionally, in symphony orchestras. Old instruments, such as recorders and crumhorns, have recently experienced a revival in interest.

Woodwinds were originally constructed from wood, with few exceptions, and hence their family name. Modern woodwinds may be either wood, metal, or plastic, although wood is still the preferred material for woodwinds other than flutes, piccolos, and saxophones. Like the brass instruments, woodwinds make use of feedback from an oscillating air column to control the flow of air input and maintain the oscillations. The flow-control valve may be either a vibrating reed or an oscillating stream of air. The feedback mechanisms in these two cases are quite different, as will be discussed in this chapter.

In woodwinds, the resonances of the air column are tuned by opening and closing tone holes with the fingers and with mechanical keys. Sound is radiated from the open tone holes, so that the radiation

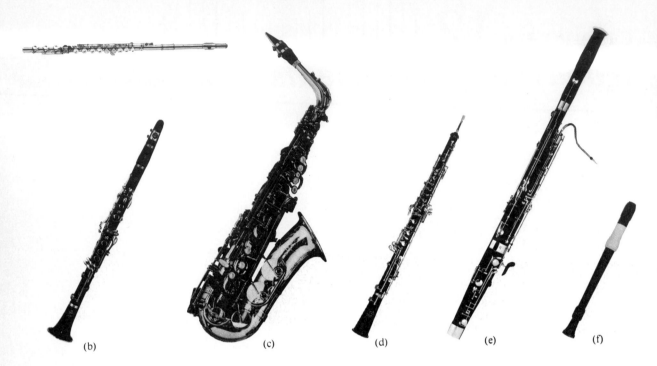

(b)   (c)   (d)   (e)   (f)

FIG. 12.1
Woodwind instruments: (a) flute;
(b) clarinet; (c) saxophone; (d) oboe;
(e) bassoon; (f) recorder. (Courtesy of
Selmer, Elkhart, Indiana.)

pattern becomes more complex than that of the brass instruments, which radiate virtually all their sound from the bell.

Woodwinds are often classified into three groups: the single reeds, the double reeds, and the air reeds, or flute-type instruments. Some modern woodwind instruments are shown in Fig. 12.1.

12.1 HOW A PIPE-REED SYSTEM VIBRATES

Imagine a clarinet mouthpiece attached to a cylindrical pipe of a length that the total acoustical length of the pipe plus mouthpiece is L. Now as blowing pressure is applied, the reed valve allows a puff of air to flow into the instrument and at the same time starts the reed swinging "shut" toward the mouthpiece, as shown in Fig. 12.2(a).

The puff of air (or pulse of positive pressure) travels down the pipe until it comes to the open end, where the pressure rather abruptly drops to zero (see Fig. 3.11). This causes a negative pressure pulse to propagate back up the pipe toward the mouthpiece, as shown in Figs. 12.2(b) and (c). When it arrives, the reed is just completing its swing toward the mouthpiece, and the negative pressure pulse "pulls" the reed valve a little farther shut (Fig. 12.2d). Since the reed valve is now closed or nearly closed, very little air enters, so a negative pressure pulse starts back down the tube toward the open end (Fig. 12.2e).

Now we reverse the chain of events described in the preceding paragraph. The negative pressure pulse arrives at the open end, the pressure suddenly rises to zero (actually, to normal atmospheric pressure), and a positive pressure pulse begins its journey back toward the mouthpiece (Figs. 12.2f and g). When it arrives, the reed is swinging open (h) and the pressure pulse pushes it farther open, so that a new puff of air can be introduced from the player's mouth (i).

Note that the reed has received one pull and one push as the pulse makes two trips up and down the pipe (once as a positive pulse, once as a negative pulse). This is analogous to giving brief pushes to a child in a swing to make up for energy lost during a cycle of oscillation. Often called *regenerative feedback* or *positive feedback*, it is analogous to the positive feedback in an electrical oscillator. Actually, the pulses are spread out in time, and the air pressure at any point in the pipe varies continuously from positive to negative, measured with respect to atmospheric pressure.

The air in the pipe has appreciable mass, and is able to pretty much force the vibrating reed to lock in on the natural frequency of the air column. Thus the reed in a woodwind has little to say about the selection of a frequency at which the system will vibrate, at least when compared to brass instruments, where the vibrating lips of the player have considerable mass, allowing the player a substantial amount of lip control of frequency. One has only to observe the sound of a reed and mouthpiece blown by themselves to realize that the reed is forced by the air column to vibrate at an unfamiliar frequency.

There is another way in which the reed could have "locked in" with the air column by vibrating three times as fast as in the mode just described. In this case, the reed would close, open, and close again during one round trip of the pressure pulse, so that once again it would be ready to receive a "pull" when the negative pulse returned. If the embouchure pressure is great, there is a tendency to jump into this mode, which has a frequency three times that of the fundamental mode (provided the pipe is cylindrical). Normally, reed instruments shift into this mode of higher frequency only when a register key is opened, however (see Section 12.5).

The frequencies of the modes of oscillation can be calculated easily. In the fundamental mode, the pulse or wave travels down and back the length of the pipe during each half cycle of the reed vibration, so the period of the lowest mode is

$$T_1 = \frac{4L}{v} \quad \text{and} \quad f_1 = \frac{1}{T_1} = \frac{v}{4L},$$

where T and f represent the period and frequency of vibration, L is the acoustical length of the tube plus mouthpiece, and v is the speed (or

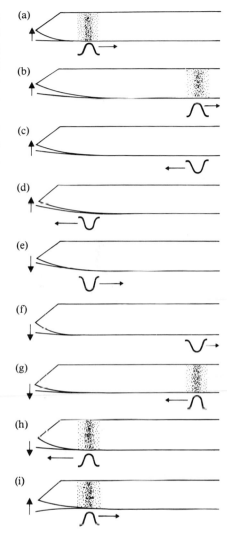

FIG. 12.2
Vibration of a pipe-reed system. A pulse of excess pressure propagates down the pipe (a, b) and is reflected as a pulse of underpressure (c, d), which returns and helps to draw the reed valve shut. In (e)–(h) the process repeats with this negative pulse, which reflects as a positive pressure pulse.

velocity) of sound. The second mode has a frequency $f_3 = 3f_1$, and the modes of higher frequency are the odd-numbered harmonics of the fundamental frequency (see Section 4.5):

$$f_n = nf_1 = n\frac{v}{4L}, \quad n = 1, 3, 5, \ldots \quad (12.1)$$

A Note on Reeds

Musical instrument reeds are often described as either free or striking. Woodwind reed instruments and most organ reed pipes use *striking* reeds, so named because they "strike" against some surface as they vibrate. *Free* reeds, which do not strike a surface, are used in the harmonium (reed organ), the harmonica (mouth organ), and some organ pipes. The lips of the brass player are sometimes referred to as a *lip reed*, and the oscillating air jet of a flute as an *air reed*. (The brass player's lips would be classified as a striking reed.)

The reeds of the clarinet, oboe, bassoon, and organ pipe *strike inward*; that is, they close by moving with the flow of air. The lips of the brass player, on the other hand, *strike outward*; that is, they move against the flow of air in order to close. In the clarinet, the maximum flow of air outward must coincide with the maximum displacement of the reed inward (Helmholtz, 1877). In this respect, woodwind reed instruments differ from brass instruments, where the maximum flow of air outward coincides with the maximum displacement of the lip reed outward.

A detailed discussion of the flow-control characteristics and elastic properties of reeds is beyond the scope of this book. Figure 12.3 shows the way in which the flow velocity varies with the pressure across the reed. Since the clarinet reed strikes inward, the desired flow control occurs on the descending slope of the curve. The entire curve may be lowered by increasing embouchure pressure, since the reed opening is thus decreased.

The compliance of the reed in this graph is about $8 \times 10^{-8} \text{m}^3/\text{N}$, which is typical of clarinet reeds. (Compliance is the inverse of stiffness.) The maximum Bernoulli force (see Section 11.3) is estimated to be $1.2 \times 10^{-2} \text{N}$, which is less than two percent of the force necessary to close the reed, and in fact moves the reed only $7 \times 10^{-6} \text{m}$ (Worman, 1971). Thus the Bernoulli force can be ignored in the clarinet reed, although it plays a fairly important role in brass instruments and apparently also in double reeds.

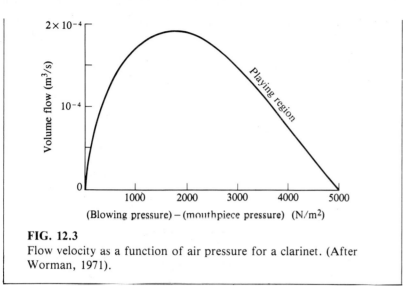

FIG. 12.3
Flow velocity as a function of air pressure for a clarinet. (After Worman, 1971).

12.2   TONE HOLES

A pipe fitted with a clarinet mouthpiece and having a length and a bore comparable to those of a clarinet produces a sound resembling that of a clarinet. However, as the pipe is shortened, it sounds less clarinetlike. The difference lies in the tone holes of the clarinet. Both the open and closed tone holes of the clarinet affect its acoustical behavior, as we shall see.

One function of the tone holes is to change the effective or acoustical length of the clarinet. In the case of a single tone hole, the larger the hole, the more the effective length is shortened, as shown in Fig. 12.4. When the size of the tone hole matches the bore, the pipe effectively ends at the open tone hole. Opening and closing tone holes permits the clarinetist to change the pitch of his or her instrument.

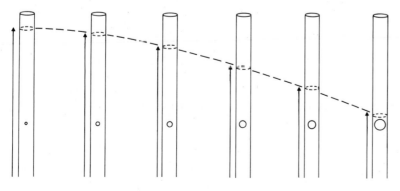

FIG. 12.4
Effective length of a pipe with open tone holes of different diameters.

When a pipe has more than one open hole, its acoustical behavior exhibits several interesting features. If the open holes are regularly spaced, they constitute a *tone-hole lattice*, not unlike a lattice of atoms in a crystal or a line of beads spaced equally on a string. The open tone-hole lattice acts as a filter that transmits waves of high frequency but reflects those of low frequency. The critical frequency above which sound waves can propagate through a lattice of tone holes is called the *cutoff frequency* of the lattice, which has been found to be an important factor in determining the timbre of a woodwind instrument.

The cutoff frequency of a lattice of tone holes depends on their size, shape, and spacing. The formula for calculating the cutoff frequency (Benade, 1976) is

$$f_c = 0.11 \frac{b}{a} \frac{c}{\sqrt{s(t + 1.5b)}},$$

where c is the speed of sound (344 m/s), and a, b, s, and t (expressed in meters) are physical parameters, shown in Fig. 12.5.

FIG. 12.5
An open tone hole lattice indicating the parameters to be used in calculating

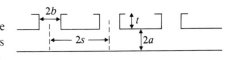

the cutoff frequency: a = the radius of the bore, b = the radius of the tone hole, $2s$ = the tone hole spacing, and t = the tone hole height.

The effective length of a pipe with open tone holes is found to vary with frequency in much the same manner as a tube terminated with a trumpet bell (see Section 11.3). As the frequency increases, the turning point moves further down the pipe; the pipe acts longer at high frequency than at low frequency. Thus the upper resonances are lowered slightly with respect to the lowest one, as can be seen in the impedance curves shown in Fig. 12.6.

More important, however, is the fact that the resonances above the cutoff frequency become very weak. Above the cutoff frequency, the open tone holes radiate so well that very little sound is reflected back toward the mouthpiece. The spectrum of the sound recorded by a small microphone inside the mouthpiece will have peaks that resemble those found on the impedance curve. The spectrum measured outside the instrument does not show the same weakening of upper partials, however, because of the greater efficiency of radiation from the open tone holes above the cutoff frequency.

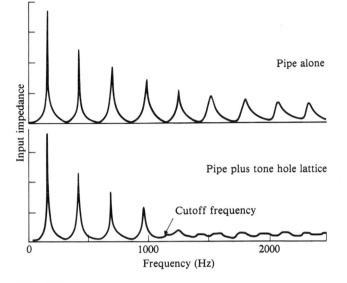

FIG. 12.6
Impedance curves for a piece of cylindrical clarinet tubing with and without tone holes. (From *Fundamentals of Musical Aoustics* by Arthur H. Benade. Copyright © 1976 by Oxford University Press, Inc. Reprinted by permission.)

Closed tone holes also affect the acoustics of a woodwind air column. The increased volume of air at each closed tone hole reduces the velocity of the sound wave down the tube, and lowers the resonance frequencies slightly. Another way of expressing this is to say that a pipe with regularly spaced bumps appears longer acoustically than a smooth pipe of the same length.

12.3   BORE TYPES

Woodwinds are designed with cylindrical or conical bores, because these shapes have resonance frequencies that are harmonically related to the fundamental. The bores of the flute and clarinet are essentially cylindrical; most other woodwinds (oboe, English horn, bassoon, saxophones, etc.) are essentially conical. The cone angles of the oboe and bassoon are small ($1.4°$ and $0.8°$), whereas those of the saxophone are quite large ($3°$ to $4°$) (Nederveen, 1969).

The resonances of a cone have essentially the *same frequencies as an open pipe of the same length*. This statement is true even if the cone is truncated, which may seem paradoxical at first glance. As a sound wave travels toward the small end of a cone, its pressure must increase. Pressure distributions for the first three modes of a cone are shown in Fig. 12.7, along with the corresponding modes of a pipe.

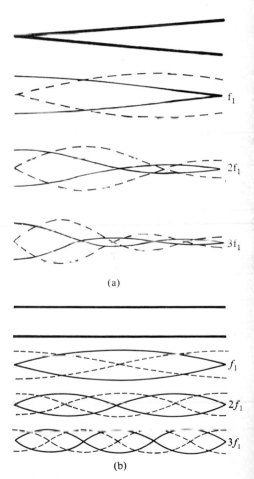

FIG. 12.7
Pressure (solid line) and velocity (dashed line) distributions for the first three modes of a cone (a) and an open pipe (b). (For modes of a closed pipe see Fig. 4.8.)

TABLE 12.1 Mode frequencies in cylindrical, conical, and flared tubes (Strong and Plitnik, 1983)

Mode	Cylindrical	Conical	Flared
1	136.4 Hz	255.6 Hz	281 Hz
2	409.2	512.0	520
3	682.0	771.0	767
4	954.9	1032.0	1021

The four lowest mode frequencies of a cylindrical tube (closed at one end), a cone, and a flared horn are compared in Table 12.1. The mode frequencies for a cylindrical tube have the ratios 1:3:5:7, while those of a cone are nearly in the ratios 1:2:3:4. Wavefronts in the cone are spherical sections, but they travel at the same speed as the plane waves in the cylindrical tube. When the cone takes on a flare, however, the wave speed depends on frequency, and thus the mode frequencies are no longer harmonic.

The input impedance for a conical pipe is different from that of a cylindrical pipe (as shown in Fig. 12.6). The impedance peaks occur at frequencies having the ratios 1:3:5:7: . . . , as in Table 12.1, but the impedance minima do not lie midway between the peaks, as in a cylindrical pipe. Impedance curves for cylindrical and conical pipes are compared by Strong and Plitnik (1983) and by Ayers, et al. (1985).

12.4   THE CLARINET

A cross-sectional view of the clarinet is shown in Fig. 12.8. The cylindrical bore has a diameter of about 15 mm. A single cane reed is clamped by the ligature against a specially designed surface of the mouthpiece called the *table*. The vibrating portion of the reed is shaved down to a wedge shape with a thickness of about 0.1 mm.

The opening between the tip of the mouthpiece and the reed is typically about 1 mm. When the instrument is played, the lower lip pushes the reed in to about half this distance, and it vibrates around this position. For soft tones, the tip of the reed does not touch the

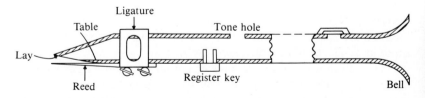

FIG. 12.8
A cross-sectional view of a clarinet.

mouthpiece; therefore, the flow of air is not interrupted. As the blowing pressure is increased, the amplitude of the reed increases, until for loud tones the tip of the reed touches the mouthpiece during about half of the cycle (Backus, 1961). Thus the flow of air, which is more or less sinusoidal (single frequency), takes on more harmonics as the blowing pressure increases, as shown in Fig. 12.9.

A clarinet reed has a natural frequency of its own at around 2000–3000 Hz; oscillations at this frequency are prevented largely by the damping action of the lower lip pressed against the reed. If the clarinet is blown with the teeth against the reed rather than the lip, unpleasant squeaks and squeals appear.

The clarinet is an important instrument in the orchestra, but plays an even more important role in the band, where B♭ clarinets form a principal choir much as violins do in the orchestra. The B♭ clarinet has a wide playing range of 3½ octaves divided into three registers, as shown in Fig. 12.10. The low or *chalumeau* register extends from D_3 to E_4, and a second register, the *clarion*, plays a twelfth above this, since the first overtone of a cylindrical closed tube is the third harmonic of the fundamental. Notes above $B♭_5$ can be played in the *altissimo* register, which uses the third mode (fifth harmonic) of the air column. Often the "throat tones" (G_4 to A''_4), which open keys in the throat of the clarinet, are included in the chalumeau register.

The bass clarinet, which plays an octave lower than the B♭ clarinet, is also used in orchestras and bands. Orchestral clarinetists alternate between clarinets tuned in A and B♭, the choice usually depending on the key signature of the music. Other clarinets are the E♭ soprano and E♭ alto. The C clarinet, used in the Classical period, is rarely seen today.

The open tone-hole cutoff frequency in most good clarinets is around 1500 Hz. A clarinet with a higher cutoff frequency tends to have a "bright" tone; an instrument with a lower cutoff frequency has a "dark" tone. In an experiment in which two matching clarinets were reworked to raise the cutoff frequency slightly in one and to lower it in the other, it was found that players of classical music preferred the one with the lower cutoff frequency, while jazz clarinetists chose the one with the higher cutoff and "bright" tone (Benade, 1976).

Impedance curves are shown for several clarinet fingerings in Fig. 12.11. In fingering E_3, all tone holes are closed, and the resonances are

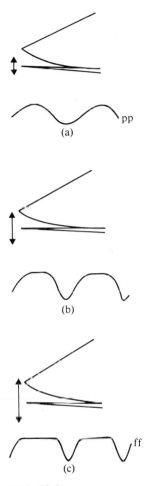

FIG. 12.9
Vibration of a clarinet reed at different dynamic levels: (a) pp; (b) mf; (c) ff.

FIG. 12.10
Three registers of the clarinet. Shown are the notes fingered; notes sounded on a B♭ clarinet are one whole tone lower.

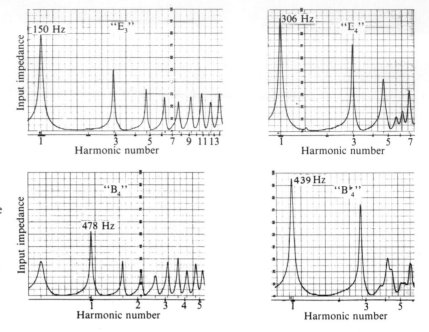

FIG. 12.11

Impedance curves for several clarinet fingerings: E_3 and B_4 have the same fingering except for the open register hole. Actual notes on a B♭ clarinet are a whole note lower than the fingered note. (From J. Backus, (1961). "Vibrations of the Reed and the Air Column in the Clarinet," *J. Acoust. Soc. Am.* **33**: 860. Reprinted by permission of the Amer. Inst. of Physics.)

not greatly different from those of the pipe with a clarinet mouthpiece shown in Fig. 12.6. The differences are mainly due to the bell and the effect of the closed tone holes. In E_4 and B♭$_4$ cases, a cutoff due to the open tone-hole lattice is apparent. B_4 has the same fingering as E_3, except for an open register hole, which "spoils" the lowest resonance.

12.5   REGISTERS AND REGISTER HOLES

All the woodwinds play in at least two different registers. In a clarinet, corresponding notes in the lower two registers differ by a musical twelfth (see Fig. 12.10), whereas in the other woodwinds there is an octave difference. In the lower register, the note sounded corresponds to the first resonance. On the impedance curve of a clarinet, this peak is the tallest one, although in the case of the conical instruments, it is usually not. Nevertheless to shift to a high register, it is necessary to reduce the strength of this lowest resonance or to shift its frequency in a way that discourages cooperation with the other modes.

The easiest way to reduce the strength of a resonance is to leak a little air at a point of maximum pressure for that mode, which can be done with a register hole. This function of a register hole can be demonstrated by fitting a clarinet mouthpiece to a length of pipe; at about one-third the distance from the reed end of the pipe (considering the mouthpiece as part of the pipe), a tiny hole is drilled. Uncovering this hole will cause the oscillation to shift up a musical twelfth from the first to the second register (i.e., to the third harmonic). A similar hole at about one-fifth the distance down the pipe will cause it to shift up an additional sixth to the fifth harmonic, because this mode of oscillation has a pressure node at that point, whereas the two lower modes do not, as shown in Fig. 12.12.

A practical difficulty arises when an entire register is considered. In order to be able to open a register hole near a pressure node for every note in the register, a large number of register holes would be needed. An oboe effectively has three register holes; a saxophone has two; but a clarinet has a single register hole for transitions between the chalumeau and clarion registers. It becomes somewhat of a challenge to design a single register hole that functions well throughout the entire register. An open tone hole near the top end of the clarinet (ordinarily closed by the first finger) serves as a register hole for the altissimo register.

12.6   THE DOUBLE REEDS

The family of orchestral double reeds includes the *oboe*, *English horn*, *bassoon*, and *contrabassoon*. Basically, they are conical tubes with the tip of the cone cut off and a reed attached. All these instruments use a double reed consisting of two halves of cane beating against each other. Small mouthpieces with a single reed have been used, but have not become very popular because they change the quality of the instrument. Cross sections of the oboe and bassoon are shown in Fig. 12.13.

The oboe bore is a nearly straight cone about 60 cm in length. Since the resonances of a cone are an octave apart, the registers of the oboe are an octave apart, extending from D_4 to C_5 and D_5 to C_6. Additional keys and cross fingerings extend the playing range from $B^\flat_3$ to G_6. Oboes are usually constructed from three pieces of wood: a top section, a bottom section, and a bell, all three sections having tone holes.

Impedance curves for four different oboe notes by two different investigators are shown in Fig. 12.14. Curve (b) for E_4 shows the effect of an open tone-hole cutoff at about 1200 Hz; in curve (c) a cutoff is not apparent. Comparison of (c) and (d) shows how opening the register hole weakens the first resonance.

Quality oboes have cutoff frequencies that are nearly constant throughout their playing range, and for different instruments vary from about 1100 to 1500 Hz (Benade, 1976). A higher cutoff frequency results in a bright tone, and a lower cutoff frequency in a dark tone, just as in the case of clarinets. The spectra of oboe sounds show substantial amounts of the higher harmonics (see Fig. 12.15) along

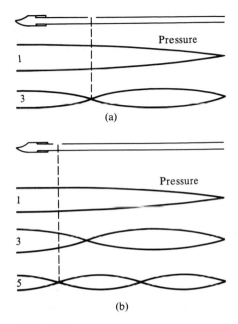

FIG. 12.12
The effect of register holes on the modes of a pipe attached to a clarinet mouthpiece. (a) The hole one-third of the way down the pipe damps the first mode, and encourages the third harmonic. (b) The hole one-fifth of the way down the pipe damps the first and third harmonics, and encourages oscillation at the fifth harmonic.

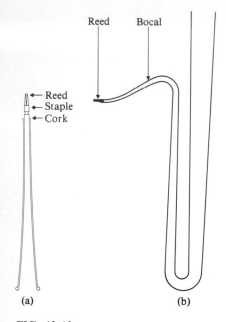

FIG. 12.13

Cross section of (a) an oboe; (b) a bassoon.

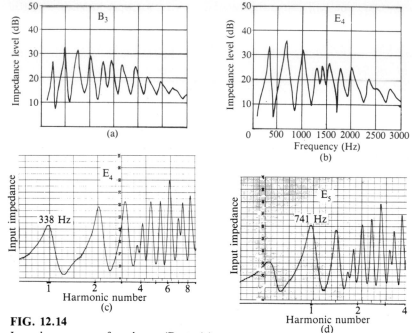

FIG. 12.14

Impedance curves for oboes. (Parts (a) and (b) from Plitnik and Strong, 1979; parts (c) and (d) from Backus, 1974. Reprinted by permission of the American Institute of Physics.)

with broad resonances or *formants* in the neighborhood of 1000 and 3000 Hz. Fransson (1967) attributes these formants to the mechanical properties of the reed.

The *English horn* or cor Anglais, is an alto version of the oboe with a pear-shaped bell that gives a distinctive tone quality to certain notes near its resonance. Composers have taken note of this distinctive tone quality in writing solos for this instrument. The English horn is tuned in F, a fifth below the oboe.

The *bassoon* has a nearly conical bore with a total length of about

FIG. 12.15

Spectra of a 250-Hz oboe tone played at dynamic markings of p, mf, and ff. These may be compared to Fig. 12.14 (a), which is the impedance curve for B_3. Note the "formants" around 1000 and 3000. (From Strong and Plitnik, 1977).

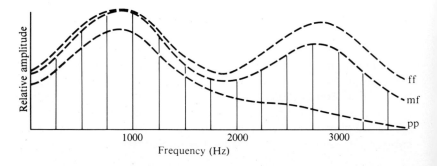

254 cm. The tube makes several bends, as shown in Fig. 12.13(b) in order to be of a manageable size. The playing range of the bassoon extends from B♭₁ to about C₅ (58–523 Hz), while the *contrabassoon* plays an octave lower. Open tone-hole cutoff frequencies for quality bassoons range from about 350 to 500 Hz (Benade, 1976). Like the oboe, the bassoon has many sharp resonances, and the tone is rich in harmonic overtones.

An extensive study of the bassoon played by eleven professionals revealed a rather strong formant extending from about 440 to 494 Hz (Lehman, 1962). Presumably this is attributable to the reed, as in the oboe. Spectra of the bassoon for pianissimo and fortissimo playing are shown in Fig. 12.16. (These spectra were recorded in an anechoic (echo-free) room with a single microphone and are therefore influenced by variation in the radiation pattern of the various harmonics; nevertheless, the comparison between spectra at the two dynamic levels is probably significant.)

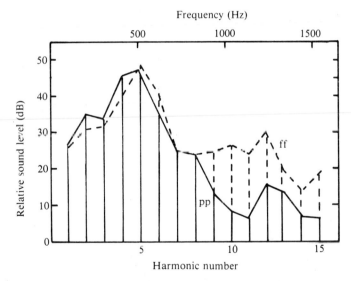

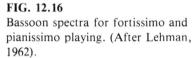
FIG. 12.16
Bassoon spectra for fortissimo and pianissimo playing. (After Lehman, 1962).

12.7 THE SAXOPHONE

The saxophone is a conical-bore single-reed instrument invented in 1846 by Adolphe Sax. Although used occasionally in symphony orchestras, its main use has been in all types of bands and in ensembles that play jazz and popular music. The four main members of the saxophone family are the soprano in B♭, the alto in E♭, the tenor in B♭, and the baritone in E♭. Saxophones are characterized by a large bore diameter, a low input impedance, and a louder sound than the other

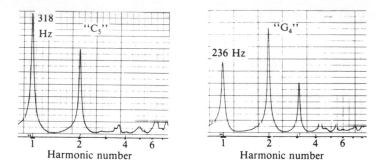

FIG. 12.17
Impedance curves for an E♭ alto saxophone. (From Backus, 1974).

woodwinds. The impedance curves for two fingerings on an alto saxophone (see Fig. 12.17) show few modes of high frequency.

Since the saxophone has a conical bore, its spectrum has even-numbered harmonics as well as odd-numbered ones. Its cone angle is greater than that of the oboe, however, so saxophone tone has fewer prominent harmonics. Its registers are an octave apart, and so it uses oboelike fingerings.

12.8   OSCILLATING AIR STREAMS AND WHISTLES

In nearly all musical instruments, sound production depends in some way on the vibration of some mechanical element. In a string or percussion instrument, the mechanical properties of the vibrating element determine its vibration frequency. The vibrating reed of the clarinet or oboe and the brass player's lips, on the other hand, are strongly influenced by the resonances of the vibrating air column of the instrument, as we have seen.

In flutes, recorders, flue organ pipes, and toy whistles, the vibrating element is a jet of air. The oscillating air stream in these aerodynamic whistles is sometimes called an "air reed," since its behavior bears some resemblance to the cane reed of the clarinet or the "lip reed" of the trumpet. There is an important difference, however: The input flow or air reed is controlled not by the feedback of pressure pulses from the air column but by the direction of air flow due to standing waves in the air column. We say that the input is *flow*-controlled rather than *pressure*-controlled.

A familiar example of a flow-controlled valve is the excitation of oscillations in a bottle by blowing across its mouth. If the blowing is done at the proper angle and pressure, air in the neck of the bottle will be set in motion and sound is radiated (see Section 4.7). Flow control by blowing across a bottle is illustrated in Fig. 12.18. When the air in the neck of the bottle is flowing inward, a part of the air jet is directed inward also. When the air in the neck is flowing outward, the jet of

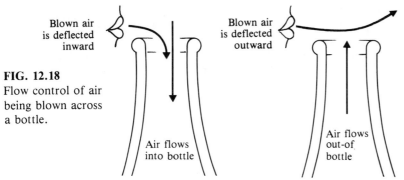

FIG. 12.18
Flow control of air being blown across a bottle.

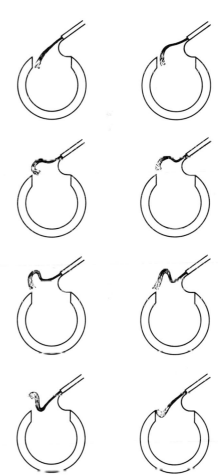

blown air is directed away from the bottle. Thus the energy supplied from the lungs via the steady jet of air is converted to energy of oscillation through the mechanism of flow control.

In most aerodynamic whistles, the frequency of oscillation varies almost linearly with flow speed (Chanaud, 1970). Examples are the "aeolian tones" generated by a wind blowing past telephone wires or through trees, the "edge tones" generated when a jet of air encounters a wedge-shaped obstacle, and the hole tone of a whistling tea kettle.

The manner in which a stream of air interacts with the embouchure hole in a flute was studied several years ago by Coltman (1968). Using an artificial lip, he injected smoke into the center of the jet stream and photographed it by using a strobe light. The sketches in Fig. 12.19 are based on his photographs.

12.9   THE FLUTE

A flow-controlled air stream, such as that which occurs in a blown bottle or a flute, collaborates with an air column to oscillate at frequencies at which the air column has dips in its impedance curve (that is, where the pressure is at a minimum and, therefore, the flow-controlling jet deflection is at a maximum). This is opposite to the woodwind reed or brass player's lips, which are controlled by pressure peaks.

The flute is basically a cylindrical pipe, about 1.9 cm in diameter, open at both ends. (See Fig. 12.20.) Its vibration modes thus form a series of frequencies that includes all the harmonics of the fundamental. The resonance frequencies of the pipe are influenced by the presence of the stream, however. In particular, the frequency of the lowest resonance is raised when the blowing pressure increases, whereas the second resonance remains more or less unchanged. The range of the flute is normally B_3 to D_7.

FIG. 12.19
Oscillation of air stream in flute embouchure hole. Compare the oscillation of the jet stream in and out of the hole with the flow of air in and out of the bottle in Fig. 12.17. (After Coltman, 1968).

FIG. 12.20
Construction of the flute.

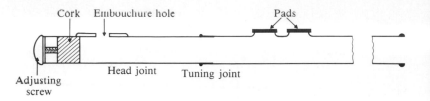

The narrowing of the distance between these peaks with increasing pressure must be counteracted by shaping the air column, if the flute is to play in tune at all dynamic levels. In most modern flutes, this is accomplished by tapering the first section of the air column, called the *head joint*. Baroque flutes, on the other hand, have a cylindrical head and a slightly tapered bore.

The flute plays in three registers, and the player must select the desired register, without the aid of register keys, by adjusting the technique of blowing. The three parameters at the player's control are the blowing pressure, the length of the air jet, and the area of lip opening. The technique used by most flute players includes adjustments in all three of these (Fletcher, 1974).

The most efficient excitation of the fundamental takes place when the time for the air jet to travel across the embouchure hole is about half the period of an oscillation. If the jet travel time is shortened much below this, the fundamental will not sound. Thus to sound the second register, the player moves the lips forward to decrease the jet length and/or increases the blowing pressure. At the same time the blowing pressure is increased, the size of the lip opening is decreased to maintain the loudness and tuning.

Studies of a number of flute players by Fletcher (1974) show that to shift to a register an octave higher, a flute player will typically double the blowing pressure, reduce the jet length by 20 percent, and reduce the lip opening by about 30 percent. (Doubling the blowing pressure increases the jet speed by 40 percent, since jet speed is approximately proportional to the square root of pressure.)

The angle of blowing depends on the shape of the player's lips, but typically ranges from 25° to 40° below the horizontal. Most players use a shallower angle for low notes and for loud playing.

The narrow end of the head joint is closed by a cork plug whose position can be adjusted by a screw. Changing the length of the small cavity between the cork and the embouchure hole can have a substantial effect on the tuning and "playability" of certain notes.

The *piccolo* is about one-half the length of the flute, and it sounds one octave higher. An alto flute, which plays a fourth lower than the standard flute, is occasionally seen.

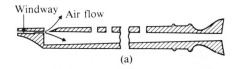
(a)

12.10 THE RECORDER

The *recorder*, also known as the Blockflöte, flauto dolce, and English flute, is an early member of the flute family, now experiencing a revival. It has a reverse conical bore, tapering inward toward the foot, and a whistle-type mouthpiece with a fixed windway, as shown in Fig. 12.21. The three most commonly played recorders are the descant (soprano) in C, the treble (alto) in F, and the tenor in C. A bass in F, a sopranino in F, and a great bass in C complete the family. Each instrument has a two-octave normal playing range.

The fixed windway makes the recorder easy to sound but somewhat difficult to play in tune at different dynamic levels. Without the flexibility to adjust the embouchure in the manner of the flautist, the recorder player changes to the upper register by half-closing the thumb hole and increasing the blowing pressure slightly. Wind pressure in an alto recorder varies from about 100 N/m² for the lowest note to about 500 N/m² for the highest (Herman, 1959). Figure 12.22 compares the blowing pressures of the alto recorder with those of the flute. In both instruments, the blowing pressure is approximately proportional to frequency.

The recorder has only eight tone holes. Therefore, in order to play a full chromatic scale, liberal use is made of cross-fingerings. Cross-fingerings leave one or more tone holes open and close other tone holes below these. In general, cross-fingered notes are not as stable as other notes, and the recorder player often changes the fingering slightly from instrument to instrument. Since the pitch produced by a given fingering may rise as much as 100 cents from low to high blowing pressure (Bak, 1969), a skilled recorder player often changes fingerings with dynamic level.

(b)

FIG. 12.21
(a) Cross section of a recorder. (b) An alto recorder.

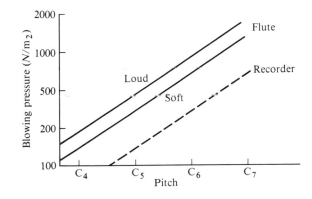

FIG. 12.22
Variation in blowing pressure with frequency for the flute and recorder (1000 N/m² is 1/100 of atmospheric pressure).

12.11 ORGAN PIPES

Flue organ pipes are similar to flute-type woodwind instruments. The tone-producing mechanism is similar to that of a recorder. A narrow stream of air impinges on the upper lip and oscillates back and forth, sometimes directed into the pipe, sometimes into the room. A variety of pipe resonators are used, including closed and open cylinders and cones. The pipe organ will be discussed in Chapter 14.

12.12 SUMMARY

Woodwind instruments use feedback from an oscillating air column to control the flow of air into the instrument. The feedback control may be applied to a vibrating reed (clarinet, oboe) or to an air jet (flute). In woodwinds with reeds, the feedback is pressure-controlled, and the instruments play at the frequencies of one or more impedance (pressure) peaks. In flute-type woodwinds, the input is flow-controlled, and the instruments play at impedance (pressure) minima. The open tone holes radiate sound, and their size and spacing determine the cutoff frequency of the instrument.

The air column of a clarinet is essentially a closed cylinder, that of a flute an open cylinder, whereas the oboe, bassoon, and saxophone are conical (closed). The registers of a clarinet are thus spaced a twelfth apart, whereas those of the other instruments are an octave apart. In all instruments except the flute, one or more register keys help the player change registers. The flute player changes registers by adjusting blowing pressure and the position and shape of the lips.

The recorder is a flute-type instrument with a reverse conical bore and a fixed windway. The blowing pressure is substantially less than that of the flute; blowing pressure increases with frequency in both instruments. Cross-fingerings are used to play a chromatic scale on the recorder.

References and Suggested Readings

Ayers, R. D., L. J. Eliason, and D. Mahgerefteh (1985). "The Conical Bore in Musical Acoustics," *J. Acoust. Soc. Am.* **53**: 528.

Backus, J. (1961). "Vibrations of the Reed and the Air Column in the Clarinet," *J. Acoust. Soc. Am.* **33**: 806.

Backus, J. (1974). "Input Impedance Curves for the Reed Woodwind Instruments," *J. Acoust. Soc. Am.* **56**: 1266.

Bak, N. (1969/70). "Pitch, Temperature and Blowing Pressure in Recorder Playing. Study of Treble Recorders," *Acustica* **22**: 295.

Benade, A. H. (1976). *Fundamentals of Musical Acoustics.*

New York: Oxford. (See Chapters 21–22.)

Chanaud, R. C. (1970). "Aerodynamic Whistles," *Sci. Am.* **222**(1): 40.

Coltman, J. W. (1968). "Acoustics of the Flute," *Physics Today* **21**(11): 25.

Fletcher, N. H. (1974). "Some Acoustical Principles of Flute Technique," *The Instrumentalist* **28**(7): 57. A more mathematical version appears in *J. Acoust. Soc. Am.* **57**: 233 (1975).

Fletcher, N. H. (1979). "Air Flow and Sound Generation in Musical Wind Instruments," *Ann. Rev. Fluid Mech.* **11**: 123.

Fletcher, N. H., and T. D. Rossing (1989). *The Physics of Musical Instruments*. New York: Springer Verlag. Chapters 15 and 16.

Fransson, F. (1967). "The Source Spectrum of Double-Reed Woodwind Instruments," Report STL-QPSR 1, 25, Royal Institute of Technology, Stockholm.

Helmholtz, H. L. F. von (1877). *On the Sensations of Tone as a Physiological Basis for the Theory of Music,* 4th ed. Trans., A. J. Ellis. New York: Dover, 1954.

Herman, R. (1959). "Observations on the Acoustical Characteristics of the English Flute," *Am. J. Physics* **27**: 22.

Lehman, P. R. (1962). "The Harmonic Structure of the Tone of the Bassoon." Ph.D. dissertation, University of Michigan, Ann Arbor.

Nederveen, C. J. (1969). *Acoustical Aspects of Woodwind Instruments,* Amsterdam: F. Knuf.

Plitnik, G. R., and W. J. Strong (1979). "Numerical Method for Calculating Input Impedances of the Oboe," *J. Acoust. Soc. Am.* **65**: 816.

Strong, W. J., and G. R. Plitnik (1983). *Music, Speech, and High Fidelity,* 2nd ed. Provo, Utah: Soundprint. Chapter 6.

Worman, W. E. (1971). "Self-Sustained Nonlinear Oscillations of Medium Amplitude in Clarinet-like Systems." Ph.D. dissertation, Case Western Reserve University, Cleveland.

Glossary

aerodynamic Having to do with the flow of air and its interaction with other bodies.

Bessel horn A family of horns of different shapes, including cylindrical and conical.

cutoff frequency The frequency above which the sound loss due to radiation through a lattice of tone holes is large, so that the resonances of an air column are weak.

embouchure The lip position used in playing a wind instrument. The embouchure hole in a flute is the hole through which the lips blow air.

feedback Use of an output signal to control or influence the input. Positive feedback, if great enough, can cause a system to oscillate.

formant A range of frequency to which a system responds preferentially or which is emphasized in its output.

impedance The ratio of the pressure to the velocity in a sound wave.

lay The slightly curved portion of a clarinet or saxophone mouthpiece that faces the reed.

register A group of related notes on a musical instrument; one register, for example, may include all notes whose pitch corresponds to the lowest resonance of an air column.

register hole A hole that can be opened in order to cause an instrument to play in a higher register.

sinusoidal Pertaining to a sine wave; thus a pure tone or single frequency of vibration.

tone hole A hole that can be opened to raise the pitch of an instrument.

turning point The point in a musical instrument at which most of the sound wave is reflected back toward the mouthpiece.

Questions for Discussion

1. Why must a clarinet have more keys than a flute?

2. Discuss the acoustical implications of playing the clarinet with a stiff reed rather than a soft one.

3. If a clarinet were played in a pure helium atmosphere, what would its lowest note be? (See Table 3.1.) Would you expect this same pitch if a clarinet player completely filled his or her lungs with helium before blowing?

4. Can you "overblow" a bottle to obtain a higher note in a different register? Try it.

Problems

1. A B♭ clarinet is about 67 cm long.

   a) What is the lowest resonance frequency of a closed pipe of that length?

   b) The lowest note on a B♭ clarinet is D_3. Compare its frequency to your answer in (a).

   c) Can you explain the difference between these two frequencies?

2. It is possible to fit a flute-type head joint to a clarinet, so that it plays in the manner of a flute. What would you expect the lowest note to be?

3. a) Making use of the fact that the speed of an air jet is approximately proportional to the square root of the blowing pressure, show that doubling the blowing pressure increases the jet speed by 40 percent.

   b) How much does the jet speed increase when the blowing pressure is tripled?

4. Show that the frequencies of the notes in the different clarinet registers that are fingered nearly the same are in the ratio 1:3:5 (see Fig. 12.10; the frequencies of the notes can be found in Table 9.2).

5. The length of an alto recorder (from the windway to the bell) is about 42 cm.

   a) What is the lowest resonance frequency of an open pipe of that length?

   b) The lowest note on the recorder is F_4. Compare its frequency to your answer in (a).

   c) Explain the difference between these two frequencies.

CHAPTER 13
Percussion Instruments

Percussion instruments may be our oldest musical instruments (with the exception of the human voice), but recently they have experienced a new surge in interest and popularity. Many novel percussion instruments have been developed recently and more are in the experimental stage. What is often termed "contemporary sound" makes extensive use of percussion instruments. Yet, relatively little material has been published on the acoustics of percussion instruments.

There are many instruments in the percussion family and a number of ways in which to classify them. Sometimes they are classified into four groups: idiophones (xylophone, marimba, chimes, cymbals, gongs, etc.); membranophones (drums); aerophones (whistles, sirens); and chordophones (piano, harpsichord). Another system divides them into two groups: those that have definite pitch and those that do not. In this chapter we will discuss various membranophones and idiophones; chordophones will be discussed in Chapter 14.

Percussion instruments generally use one or more of the following basic types of vibrators: strings, bars, membranes, plates, air columns, or air chambers. The first four are mechanical, and the latter two are pneumatic. Two of them (the string and the air column) tend to produce harmonic overtones; the others, in general, do not. Bars, membranes, and plates are three classes of vibrators whose modes of vibration are not related harmonically. Thus the overtones they sound will not be harmonics of the fundamental tone. The inharmonic overtones of these complex vibrators give percussion instruments their distinctive timbres.

13.1 VIBRATIONS OF BARS

Vibrations of bars were discussed briefly in Section 2.6. A bar (or rod or tube) can vibrate either longitudinally (by expanding and contracting in length) or transversely (by bending at right angles to its length). In percussion instruments, the transverse modes of vibration are nearly always used (one exception being the aluminum stroke rods, which are excited longitudinally by stroking with a rosined cloth or gloves). Longitudinal vibrations are much higher in frequency than transverse vibrations, and the various longitudinal modes are related harmonically. The frequency of vibration depends on the length of the rod and on the elasticity of the material from which it is fabricated but is independent of the thickness, surprisingly enough. In fact, the frequency of vibration of a longitudinal rod is expressed by the simple formula of $f_n = nv_L/2L$, where v_L is the speed of sound in the rod, L is its length, and $n = 1, 2, 3, \ldots$, denotes the number of the harmonic (beginning with $n = 1$ for the fundamental frequency). The velocity $v_L = \sqrt{E/\rho}$, where E is Young's elastic modulus and ρ is the density.

When a bar vibrates longitudinally, its motion is almost identical to the movement of the air in a pipe that is open at both ends (a flute, for example). Maximum movement occurs at the ends of the bar or air column (called antinodes), and one or more points in between have a minimum of movement (called nodes). For the fundamental mode, there is one node, and that occurs at the center. For the first overtone (harmonic number 2), there are two nodes with another antinode at the center. One can selectively excite any desired mode of vibration in a bar by clamping it at the location of one of the nodes for that particular mode. If the fundamental mode is desired, the bar should be supported at the center only.

Transverse vibrations in a bar are a little more complicated. First of all, there are three possible end conditions for the bar: clamped, simply supported, and free. However, nearly all percussion instruments use bars with free ends, so only the free bar will be considered. (One exception is the electronic carillon, which sometimes uses bars or rods that are free at one end and clamped at the other.)

The frequency of a bar in either longitudinal or transverse vibration depends on its length and the density and elasticity of the material, but in transverse vibration, the frequency depends on the thickness of the bar as well. The frequency of transverse vibrations in a bar with free ends is given by the formula $f_n = (\pi v_L K/8L^2)m^2$, where v_L is the speed of sound, L is the length of the bar, m is a sequence number to be defined in a moment, and K is the *radius of gyration,* which is related to the size and shape of the bar. (For a flat bar, K is the thickness divided by 3.46; values for other shapes are given in Table 13.1.) Note the following differences from the longitudinally vibrating bar.

TABLE 13.1   Formulas for vibration frequencies of strings and bars

---

String

  Transverse vibration: $f_n = n\dfrac{v}{2L}$

  v = speed of waves in the string = $\sqrt{T/\mu}$
  L = length of the string
  n = 1, 2, 3, . . .
  T = tension
  μ = mass per unit length

Longitudinal vibration: $f_n = n\dfrac{v_L}{2L}$

  v_L = speed of sound in the string = $\sqrt{E/\rho}$
      (does not change with tension)
  E = Young's modulus of elasticity
  ρ = density

Bar with free ends

  Transverse vibration: $f_n = \dfrac{\pi v_L K}{8L^2}m^2$

  v_L = speed of sound = $\sqrt{E/\rho}$
  E = Young's modulus of elasticity
  ρ = density
  L = length
  m = 3.0112, 5, 7, . . . , $(2n+1)$
  K = radius of gyration
        $K = t/\sqrt{12} = t/3.46$ for a rectangular bar
        t = thickness
        $K = \frac{1}{2}\sqrt{a^2 + b^2}$ for a tube
        a = inner radius
        b = outer radius

Longitudinal vibration: $f_n = n\dfrac{v_L}{2L}$
  v_L = speed of sound = $\sqrt{E/\rho}$
  L = length
  n = 1, 2, 3 . . .

---

1. The frequency depends on L^2 rather than L.

2. The mode frequencies are not harmonic, but increase as m^2.

3. The frequency depends on the shape of the bar through the factor K.

The frequencies of the modes are in proportion to the squares of the odd integers—almost. The number m begins with 3.0112 and then continues with the simple values 5, 7, 9, . . . , $(2n + 1)$. Transverse vibrations of a bar are illustrated in Fig. 2.17.

The frequencies of the transverse modes of vibration of a uniform bar have the ratios 1.0 : 2.76 : 5.40 : 8.90, and so on, which are anything but harmonic and, in fact, match no intervals on the musical scale. They give a distinctive timbre to instruments such as chimes, orchestra bells, and triangles, which are nearly uniform bars. (The bars of marimbas, xylophones, and related instruments are not uniform; they have been cut to have quite a different set of mode frequency ratios and thus a different timbre.)

Longitudinal and transverse vibrations may be strikingly demonstrated with aluminum rods about 1 cm in diameter. Longitudinal vibrations can be excited by stroking the rod with a rosined cloth or tapping lightly on the end with a hammer; transverse vibrations, which in a long rod occur at a much lower frequency, can be excited by hitting the rod with one's hand near the center. Since the frequency of transverse vibrations varies with $1/L^2$ but the frequency of longitudinal vibrations only varies with $1/L$ (see Table 13.1), it is possible to cut a rod to a length such that the two fundamental frequencies coincide. Note that the nodes for longitudinal and transverse vibration occur at different places, so one's grip must be changed. Aluminum stroke rods are now being used as percussion instruments in the performance of contemporary music.

13.2   RECTANGULAR BARS: THE GLOCKENSPIEL

The glockenspiel, or orchestra bells, use rectangular steel bars 2.5 to 3.2 cm (1 to 1¼ inches) wide and 0.61 to 1 cm (¼ to ⅜ inch) thick. Its range is customarily from G_5 ($f = 784$ Hz) to C_8 ($f = 4186$ Hz), although it is scored two octaves lower than it sounds. The glockenspiel is usually played with brass or hard plastic mallets. The bell lyra, a portable version that uses aluminum bars and usually covers the range A_5 ($f = 880$ Hz) to A_7 ($f = 3520$ Hz), is sometimes used in marching bands. Glockenspiel and bell lyra are shown in Fig. 13.1.

When struck with a hard mallet, a glockenspiel bar produces a crisp, metallic sound, which quickly gives way to a clear ring at the

(a)

(b)

FIG. 13.1
(a) Orchestra bells or glockenspiel;
(b) bell lyra.

designated pitch. Because the overtones have very high frequencies and die out quickly, they are of relatively less importance in determining the timbre of the glockenspiel than are the overtones of the marimba or xylophone, for example. For this reason, no effort is made to bring the inharmonic overtones of a glockenspiel into a harmonic relationship through overtone tuning.

The frequencies for transverse vibrations in a bar with free ends, given in Section 13.1, were shown to be inharmonic; thus the glockenspiel has no harmonic overtones. However, for even the lowest bar the first overtone has a frequency of 2160 Hz, which is getting into the range in which the pitch discrimination of human listeners is diminished. (This will not be the case for other bar percussion instruments, however, as we will see.)

The vibrational modes of a glockenspiel bar are shown in Fig. 13.2. These modes can be excited individually by passing an electric current from an audio amplifier through the bar while it is in the field of a magnet. Besides the transverse modes, which are labeled 1, 2, 3, 4, and 5, there are torsional or "twisting" modes labeled *a, b, c, d;* a longitudinal mode *l;* and transverse modes in the plane of the bar, *1x* and *2x*. Recording a spectrum of the strike sound requires a real-time spectrum analyzer.

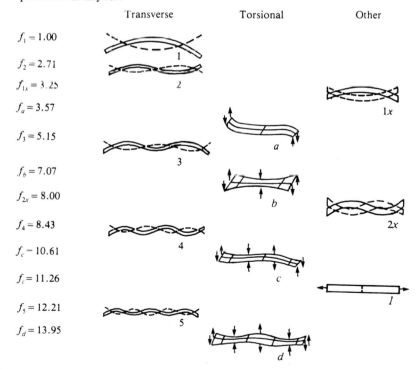

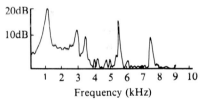

FIG. 13.2

Modes of C_6 glockenspiel bar, which lie within the audible range of frequency. The transverse modes are labeled 1, 2, 3, 4, and 5; *a, b, c,* and *d* are torsional modes; *l* is a longitudinal mode; and *1x* and *2x* are transverse modes excited edgewise. Mode frequencies are given as ratios to the fundamental. The sound spectrum above shows the relative amplitudes of the partials associated with these modes. (From Rossing, 1976. Reprinted by permission from the Amer. Assoc. of Physics Teachers.)

The transverse modes of vibration of the glockenspiel bar are spaced a little closer together than predicted by the simple theory of thin bars. The more sophisticated theory for thick bars (called Timoshenko bars in engineering textbooks) takes into account other factors such as the moment of inertia of the bar and the shear stresses in the bar.

13.3   THE MARIMBA, XYLOPHONE, AND VIBES

The three most common bar percussion instruments are the marimba, the xylophone, and the vibraphone or vibraharp (commonly called *vibes*). All three instruments consist of tuned bars with tubular resonators. They are played with mallets of varying hardness.

The *marimba* typically includes 3 to $4\frac{1}{3}$ octaves of tuned bars of rosewood or fiberglass synthetic (typical tradenames: Kelon, Klyperon), graduated in width from about 4.5 to 6.4 cm ($1\frac{3}{4}$ to $2\frac{1}{2}$ inches). Beneath each bar is a tubular resonator tuned to the fundamental frequency of that bar (see Fig. 13.3(a)). When the marimba is played with soft mallets, it produces a rich mellow tone. The playing range of a large concert marimba is A_2 to C_7 ($f = 110$ to 2093 Hz), although bass marimbas extend to C_2 ($f = 65$ Hz).

A deep arch is cut in the underside of marimba bars, particularly in the low register. This arch serves two useful purposes: It reduces the length of bar required to reach the low pitches, and it allows tuning of the overtones (the first overtone is nominally tuned two octaves above the fundamental). Figure 13.4(a) shows a scale drawing of a marimba bar, and also indicates the positions of the nodes for each of the first seven modes of vibration. (This may be compared to Fig. 13.2, which shows the nodes of a glockenspiel bar without a cut arch.) The ratios of the frequencies of these modes are also indicated. Note that the second partial (first overtone) of this bar has a frequency 3.9 times that of the fundamental, which is close to a two-octave interval (a ratio of 4.0).

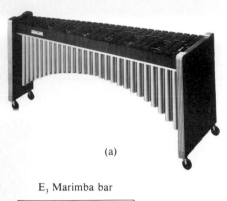

(a)

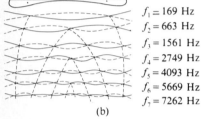

E_3 Marimba bar

$f_1 = 169$ Hz
$f_2 = 663$ Hz
$f_3 = 1561$ Hz
$f_4 = 2749$ Hz
$f_5 = 4093$ Hz
$f_6 = 5669$ Hz
$f_7 = 7262$ Hz

(b)

FIG. 13.3

(a) A marimba. (b) Scale drawing of a marimba bar tuned to E_3 ($f = 265$ Hz). The dashed lines locate the nodes of the first seven modes. (Photograph courtesy of J.C. Deagan Co.)

FIG. 13.4

Sound spectrum for E_3 marimba bar. The partials are also indicated on music staffs, which include a "supertreble" clef, two octaves above the treble clef. (From Rossing, 1976). Percussion Instruments, Part I," *Phys. Teach.* 14: 546. Reprinted by permission from the Amer. Assoc. of Physics Teachers.)

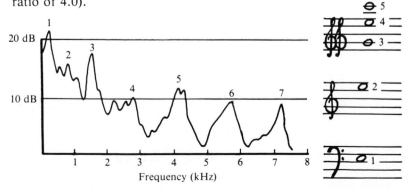

The sound spectrum of the E_3 marimba bar, shown in Fig. 13.4 along with the partials on musical staffs, indicates the presence of a strong third partial, which has a frequency of 9.2 times the fundamental (about three octaves plus a minor third above). The relative strengths of the partials, of course, depend on where the bar is struck and what type of mallet is used. To emphasize a particular partial, the bar should be struck at a point of maximum amplitude for that mode, as indicated in Fig. 13.3.

Marimba resonators are cylindrical pipes tuned to the fundamental mode of the corresponding bars. A pipe with one closed end and one open end resonates when its acoustical length is one-fourth of a wavelength of the sound. The purpose of the tubular resonators is to emphasize the fundamental and also to increase the loudness, which is done at the expense of shortening the decay time of the sound. We have measured the decay time (60 dB) of a rosewood bar in the low register (E_3) to be 1.5 s with the resonator and 3.2 s without it. Decay times in the upper register are generally shorter; we measured 0.4 s and 0.5 s for an E_6 bar with and without the resonator, respectively. The corresponding decay times for synthetic bars are somewhat longer.

It is not difficult to test the overtone tuning of a marimba by noting the position of the nodes as shown in Fig. 13.3(b). To suppress the fundamental mode and emphasize the first overtone, one should touch the bar firmly at the center and strike the bar either at one end or at a point about one-third of the way from the center to either end.

As an experiment, the author designed a marimba with variable timbre which has a second set of resonators tuned to the first overtone of the bars. Each resonator is equipped with a vane that can partially or completely close the mouth of the tube; thus the timbre can be varied by adjusting the amount of closure. [This instrument was first played at a meeting of the Acoustical Society of America.]

The *xylophone* is a close cousin to the marimba (or perhaps "uncle" is a better term, since the xylophone apparently has a longer history). Xylophones typically cover a range of 3 to $3\frac{1}{2}$ octaves extending from F_4 or C_5 to C_8 ($f = 349$–4186 Hz) and may have bars of synthetic material or rosewood. Modern xylophones are nearly always equipped with tubular resonators to increase the loudness of the tone (see Fig. 2.18).

Xylophone bars are also cut with an arch on the underside, but the arch is not as deep as that of the marimba, since the first overtone is tuned to a musical twelfth above the fundamental (that is, three times the frequency of the fundamental). Since a pipe closed at one end can also resonate at three times its fundamental resonant frequency, the twelfth will also be reinforced by the resonator. This overtone boost, plus the hard mallets used to play it, give the xylophone a much crisper, brighter sound than the marimba. We have found that careful overtone tuning is usually ignored in the upper register, as in the case of the marimba.

The xylomarimba is a large xylophone with a $4\frac{1}{2}$ to 5 octave range (C_3 or F_4 to C_8), and is occasionally used in solo work or in modern scores. Bass xylophones and keyboard xylophones have also been constructed (Brindle, 1970).

A very popular bar percussion instrument is the *vibraphone* or *vibraharp,* as they are designated by different manufacturers. *Vibes,* as they are popularly called, usually consist of aluminum bars tuned over a three-octave range from F_3 to F_6 ($f = 175 - 1397$ Hz). The bars are deeply arched so that the first overtone has four times the frequency of the fundamental, as in the marimba. The aluminum bars tend to have a much longer decay time than the wood or synthetic bars of the marimba or xylophone, and so vibes are equipped with pedal-operated dampers.

The most distinctive feature of vibes, however, is the vibrato introduced by motor-driven discs at the top of the resonators, which alternately open and close the tubes. The vibrato produced by these rotating discs or pulsators consists of rather substantial fluctuation in amplitude ("intensity vibrato") and a barely detectable change in frequency ("pitch vibrato"). The speed of rotation of the discs may be adjusted to produce a slow vibe or a fast vibe. Often vibes are played without vibrato by switching off the motor. They are usually played with soft mallets or beaters, which produce a mellow tone, although some passages call for harder beaters.

Because vibraphone bars have a much longer decay time than do marimba and xylophone bars, the effect of the tubular resonators on decay time is more dramatic. At 220 Hz (A_3), for example, we measure a decay time (60 dB) of 40 s without the resonator and 9 s with the tube full open. For A_5, we measure 24 s with the resonator closed and 8 s with it open. In the recording of sound level shown in Fig. 13.5, the intensity modulation, as well as the slow decay of the sound, can be clearly seen.

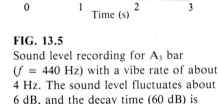

FIG. 13.5
Sound level recording for A_3 bar ($f = 440$ Hz) with a vibe rate of about 4 Hz. The sound level fluctuates about 6 dB, and the decay time (60 dB) is seven seconds.

13.4 CHIMES

Chimes, or tubular bells, are usually fabricated from lengths of brass tubing $1\frac{1}{4}$ to $1\frac{1}{2}$ inches in diameter. The upper end of each tube is partially or completely closed by a brass plug with a protruding rim. This rim forms a convenient and durable strike point.

One of the interesting characteristics of chimes is that there is no mode of vibration with a frequency at, or even near, the pitch of the strike tone one hears. The frequencies excited when a chime is struck are very nearly those of a free bar described earlier. Modes 4, 5, and 6 appear to determine the strike tone. This can be understood by noting that these modes for a free bar have frequencies nearly in the ratio $9^2:11^2:13^2$, or $81:121:169$, which are close enough to the ratio $2:3:4$ for the ear to consider them nearly harmonic and to use them as a basis for establishing a pitch (see Section 7.4). The largest near-common factor in the numbers 81, 121, and 169 is 41.

Figure 13.6 is a graph of the frequencies of the G-chime as a function of m, along with those predicted by the thin-bar theory of Lord Rayleigh (1894) and also the more detailed thick-bar theory (Flugge, 1962). Also shown are the frequencies of vibration with the end plug removed. Note that the end plug lowers the frequencies of the first few modes but has little effect on the higher modes. The strike tone, which lies one octave below the fourth mode, is also indicated.

The ratios of the modal frequencies of a chime tube with and without a load at one end are given in Table 13.2. Also given are the ratios considered desirable for a tuned carillon bell. Note the similarity between the partials of a chime and those of a carillon bell. Adding a load to one end of a chime lowers the frequencies of the lower modes more than the higher ones (see also Fig. 13.6) and thus "stretches" the modes into a more favorable ratio. The end plug also adds to the durability of the chime and helps to damp out the very high modes.

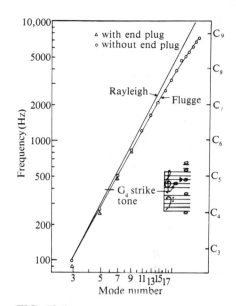

FIG. 13.6
Frequencies of a G-chime as a function of mode number m, along with those predicted by thin-bar theory (Rayleigh) and thick-bar theory (Flugge). The first five modes are shown on musical staffs, along with the subjective strike tone. (From "Acoustics of Percussion Instruments, Part I," *Phys. Teach.* 14: 546. Reprinted by permission from the Amer. Assoc. of Physics Teachers.)

TABLE 13.2 Ratios of mode frequencies for loaded and unloaded chime tube (compared with the strike tone)

n	Thin rod	Tube	Loaded with 193 g	435 g	666 g	Tuned bell
1	0.22	0.24	0.23	0.22		0.5
2	0.61	0.64	0.63	0.62	0.61	1
3	1.21	1.23	1.22	1.22	1.22	1.2 or 1.25 (1.5)
4	2	2	2	2	2	2 (2.5)
5	2.99	2.91	2.93	2.95	2.94	3
6	4.17	3.96	4.01	4.04	4.03	4
7	5.56	5.12	5.21	5.21	5.18	5.33
8	7.14	6.37	6.50	6.43	6.37	6.67
Strike tone		416 Hz	393 Hz (G_4)	383 Hz	381 Hz	

(All modes are compared with the strike tone.)

Source: Rossing (1976)

The bell-like quality of chimes is well known, of course, and has been used in many compositions for band and orchestra (Tchaikovsky's *1812 Overture,* for example). This bell-like timbre can be maximized by selecting the optimum size of end plug for each chime. For the G-chime in Table 13.2, the 193-g plug, with which the chime was originally fitted, is quite near the optimum. Most chime-makers use the same size plug throughout the entire set of chimes, however, so the timbre changes up and down the scale, typically being optimum only near the center. A set of chimes "scaled" to have the same timbre throughout would have some advantages.

A well-tuned chime not only has its overtones tuned to resemble those of a carillon bell, but also is free of "beats" between modes with nearly, but not quite, the same frequencies of vibration. These beating modes occur when the chime tube is not perfectly round or its wall thickness is not perfectly uniform. As a result, the transverse vibrations will have slightly different frequencies in two different transverse directions, resulting in beats when both modes are excited. These undesired beats can be eliminated by squeezing the chime in a vise or thinning the wall slightly on one side to bring the modes into tune.

13.5   TRIANGLES

Because of their many modes of vibration, triangles are characterized as having an indefinite pitch. They are normally steel rods bent into a triangle (usually, but not always, equilateral) with one open corner. Triangles are suspended by a cord from one of the closed corners, and are struck with a steel rod or hard beater.

Triangles are typically available in 15 cm, 20 cm, and 25 cm (6-, 8-, and 10-inch) sizes, although other sizes are also used. Sometimes one end of the rod is bent into a hook, or the ends may be turned down to smaller diameters than the rest of the triangle to alter the modes of vibration. The sound of the triangle depends on the strike

FIG. 13.7
Sound spectra for a 25-cm steel triangle (a) struck in the plane; (b) struck perpendicular to the plane. Two frequency and amplitude ranges are shown in each case. (Reprinted by permission of the Percussive Arts Society).

(a)

(b)

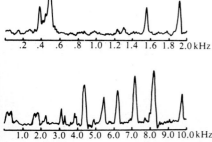

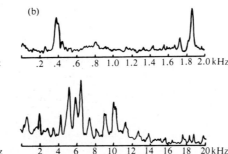

point as well as the hardness of the beater. Single strokes are usually played on the base of the triangle, whereas the rapid strokes of a tremolo are made on the inside of the triangle near the upper angle.

The modes of the triangle are many, and they are not harmonically related. Generally, they can be characterized by vibrations in the plane of the triangle and perpendicular to the plane. Normal strokes will tend to emphasize vibrations in the plane of the triangle. Figure 13.7 shows sound spectra for a 10-inch triangle at two different strike points with strokes parallel and perpendicular to the plane of the triangle.

Since triangles are made of steel, it is easy to excite single modes of vibration. A choke coil with a ferrite core or a ferrite antenna from a radio can be driven by an audio amplifier to produce an oscillating magnetic field. At each resonance frequency, a small microphone can be used to locate the nodes and antinodes and thus identify the mode. Some of the modes of vibration of a 10-inch triangle are shown in Fig. 13.8, along with those calculated for a steel rod of the same length and diameter. The triangle modes show a surprisingly close correspondence to those of a straight rod.

13.6 VIBRATIONS OF MEMBRANES

The vibrations of an ideal membrane were discussed briefly in Section 2.6. We pointed out that a membrane may be thought of as a two-dimensional string, in that the restoring force necessary for it to vibrate is supplied by tension applied from the edge. A membrane, like a string, can be tuned by changing its tension. One major difference between vibrations in the membrane and in the string, however, is that the mode frequencies in the ideal string are harmonics of the fundamental, but in the membrane they are not. Another difference is that in the membrane, nodal lines replace the nodes that occur along the string. These nodal lines are circles and diameters, as shown in Fig. 13.9.

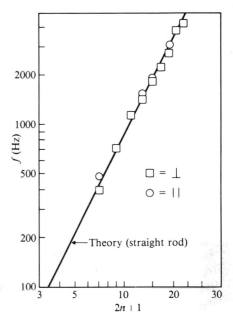

FIG. 13.8
Mode frequencies for a 25-cm steel triangle driven in the plane and perpendicular to the plane. The line gives the predicted frequencies for a steel rod of the same diameter and length. (Reprinted by permission of the Percussive Arts Society).

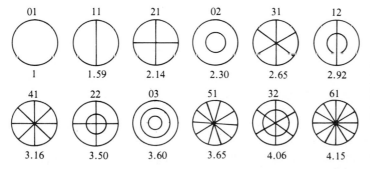

FIG. 13.9
Modes of an ideal membrane, showing radial and circular nodes and the customary mode designation (the first number gives the number of radial modes, and the second number the circular nodes, including the one at the edge). The number below each mode diagram gives the frequency of that mode compared to the fundamental (01) mode.

> The formula for the frequencies of the various vibrational modes of an ideal membrane is
>
> $$f_{mn} = \frac{1}{2a}\sqrt{\frac{T}{\sigma}}\beta_{mn} \, ,$$
>
> a = the radius of the membrane (meters), T = the tension applied to the membrane (newtons), σ = the area density (kg/m²), and β_{mn} = values for which a certain mathematical function (the Bessel function) becomes zero.
>
> The formula itself is not particularly important to the musician, except that it shows how the pitch of a vibrating membrane varies with radius, tension, and thickness. The values m and n for each mode of vibration indicate the number of nodal diameters and the number of nodal circles, respectively.

The first twelve modes of vibration of a circular membrane are shown in Fig. 13.9. Above each sketch are given the values of m and n, and below it is the frequency of vibration for that mode divided by the frequency of the lowest (01) mode. For example, the (31) mode has 3 nodal diameters and one nodal circle (around the edge) and vibrates at a frequency 2.65 times that of the lowest mode. The four modes shown in Fig. 2.14 are the (01), (02), (11), and (21) modes.

13.7   TIMPANI

Drums consist of membranes of animal skin or synthetic material stretched over some type of air enclosure. Some drums (e.g., timpani, tabla, boobams) sound a definite pitch; others convey almost no sense of pitch at all. Drums are important in nearly all musical cultures and constitute one of the most universally used types of musical instrument throughout history.

The timpani or kettledrums are the most important drums in the orchestra, one member of the percussion section usually devoting attention exclusively to them. During the last century, various mechanisms were developed for changing the tension to tune the drumheads rapidly. Most modern timpani have a pedal-operated tensioning mechanism in addition to six or eight tensioning screws around the rim of the kettle. The pedal typically allows the player to vary the tension over a range of 2:1, which corresponds to a tuning range of about a musical fourth. A modern pedal-equipped kettledrum is shown in Fig. 13.10.

FIG. 13.10
A kettledrum with suspended bowl and pedal-operated mechanism for tuning.

At one time all timpani heads were calfskin, but this material has gradually given way to Mylar (polyethylene terephthalate). Calfskin heads require a great deal of hand labor to prepare and great skill to tune properly. Some orchestral timpanists prefer them for concert work under conditions of controlled humidity, but use Mylar when touring. Mylar is insensitive to humidity and easier to tune, due to its homogeneity. A thickness of 0.19 mm (0.0075 inch) is considered standard for Mylar timpani heads. Timpani kettles are roughly hemispherical; copper is the preferred material, although fiberglass and other materials are frequently used.

Although the modes of vibration of an ideal membrane are not harmonic, a carefully tuned kettledrum will sound a strong principal note plus two or more harmonic overtones. Rayleigh (1894) recognized the principal note as coming from the (11) mode and identified overtones about a perfect fifth ($f:f_1 = 3:2$), a major seventh (15:8), and an octave (2:1) above the principal tone. Taylor (1964) identified a tenth (octave plus a third of $f:f_1 = 5:2$) by humming near the drumhead, a technique some timpanists use to fine-tune their instruments.

How are the inharmonic modes of the ideal membrane coaxed into a harmonic relationship? Three effects contribute: (1) the membrane vibrates in a "sea of air," and the mass of this air "sloshing" back and forth lowers the frequency of the principal modes of vibration; (2) the air enclosed by the kettle has resonances of its own that will interact with the modes of the membrane that have similar shapes; (3) the stiffness of the membrane, like the stiffness of piano strings, raises the frequencies of the higher overtones. Our studies show that the first effect (air loading) is mainly responsible for establishing the harmonic relationship of kettledrum modes; the other two effects only "fine tune" the frequencies but may have considerable effect on the rate of decay of the sound (Rossing, 1982a).

Stiffness in a two-dimensional membrane has a little different meaning than in a one-dimensional string. A plastic membrane, like a sheet of paper, offers little resistance to bending along a line. However, it strongly resists the type of distortion needed to wrap it around a ball without wrinkling, for example. This resistance is characterized as *stiffness to shear*. Stiffness to shear affects the frequencies and the shapes of vibrational modes in the same way that stiffness to bending affects the modes of a vibrating string (see Section 14.2).

We should point out that the frequencies of the fundamental (01) and other symmetrical modes (02, 03, etc.) will be raised by the "stiff-

TABLE 13.3 Vibration frequencies of a kettledrum, a drumhead without the kettle, and an ideal membrane

Mode	Kettledrum		Drumhead alone		Ideal membrane
	f	f/f_{11}	f	f/f_{11}	f/f_{11}
01	127 Hz	0.85	82 Hz	0.53	0.63
11	150	1.00	155	1.00	1.00
21	127	1.51	229	1.48	1.34
02	252	1.68	241	1.55	1.44
31	298	1.99	297	1.92	1.66
12	314	2.09	323	2.08	1.83
41	366	2.44	366	2.36	1.98
22	401	2.67	402	2.59	2.20
03	418	2.79	407	2.63	2.26
51	434	2.89	431	2.78	2.29
32	448	2.99	479	3.09	2.55
61	462	3.08	484	3.12	2.61
13	478	3.19	497	3.21	2.66
42			515	3.32	2.89

ness" of the enclosed air in the kettle. (This is not unlike the increase in the resonance frequency of a loudspeaker when it is mounted in an airtight enclosure; see Chapter 20.) The other modes are not thus affected, however, because their net displacement of air is zero.

Table 13.3 shows the vibration frequencies of a 65-cm (26-in.) kettledrum and an identical drum without the kettle. In both cases, the ratios of the frequencies to the principal (11) mode are given. Note that the (21) and (31) modes will radiate overtones a fifth and an octave above the fundamental, as noted by Rayleigh.

The sound spectra obtained by striking the drum in its "normal" place (about one-fourth of the way from edge to center) and at the center are shown in Fig. 13.11. Note that the fundamental mode (01) appears much stronger when the drum is struck at the center, as do the other symmetrical modes (02, 03). These modes damp out rather quickly, however, so they do not produce much of a drum sound. In fact, striking the drum at the center produces quite a dull, thumping sound.

Normal striking technique produces prominent partials with frequencies in the ratios 0.85 : 1 : 1.5 : 1.99 : 2.44 : 2.89. If we ignore the heavily damped fundamental, the others are nearly in the ratios 1 : 1.5 : 2 : 2.5, a harmonic series built on a nonexistent fundamental an octave below the principal tone. Measurements on timpani of other sizes give similar results (Rossing and Kvistad, 1976). It is a little surprising that the pitch of timpani corresponds to the pitch of the princi-

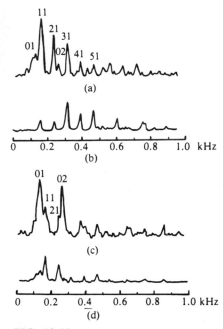

FIG. 13.11
Sound spectra from a 65-cm timpani tuned to E₃ (a) approximately 0.03 s after striking at the normal point; (b) approximately 1 s later; (c) approximately 0.03 s after striking at the center; (d) approximately 1 s later.

pal tone rather than the missing fundamental of the harmonic series, which would be an octave lower. Apparently the strengths and durations of the overtones are insufficient, compared to the principal tone, to establish the harmonic series of the missing fundamental.

13.8   BASS DRUM AND SNARE DRUM

Among drums that convey an indefinite pitch or none at all are the bass drum, snare drum, tenor drum, tomtom, conga, and bongo. Many drums of this type have two heads, and each head is given a slightly different tension. Thus the many inharmonic partials of the two heads produce the indefinite pitch that will blend with music in any key.

Bass drums are typically 50 to 100 cm in diameter. Single-headed or "gong" drums produce a mellow sound, although two-headed drums with their more indefinite pitch are usually preferred in bands and orchestras. The bass drum is capable of making the loudest sound of all the instruments in the orchestra.

Most drummers tune the *batter* or beating head to a great tension than the *carry* or vibrating head; some percussionists suggest that the difference in tension be as great as 75% (giving an interval of about a musical fourth). Quite a different timbre results from setting both heads to the same tension.

In Fig. 13.12 are sound spectra of an 82-cm (32-inch) diameter drum with the carry head set at a lower tension than the batter head (a) and with both heads at the same tension (b). In both cases, the lowest partial, radiated by the (01) mode in which the two heads vibrate in phase, is the strongest one. In Fig. 10.13(b), the partial at

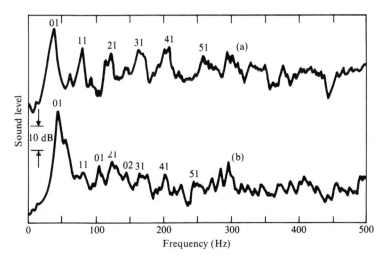

FIG. 13.12
Sound spectra of an 82-cm diameter bass drum (a) with the carry head at a lower tension than the batter head; (b) with both heads at the same tension.

104 Hz is identified with the higher component in a (01) doublet in which the batter and carry heads move in opposite directions. (Additional restoring force due to compression of the enclosed air raises the frequency of this component, as in the third mode of the guitar shown in Fig. 10.28).

Frequencies of the (11), (21), (31), (41), and (51) modes have nearly harmonic ratios, as in the timpani, and if their partials were the only ones heard, the bass drum would be expected to have a rather definite pitch. In the frequency range above 200 Hz, however, there are many inharmonic partials that sound louder since the ear is more sensitive to them than to low-frequency sounds. Fletcher and Bassett (1978) found some 160 partials in the frequency range 200–1100 Hz.

Another drum with indefinite pitch is the *snare* or *side drum*. The orchestral snare drum is a double-headed instrument about 35 cm (14 in) in diameter and 13 to 20 cm (5 to 8 in) deep. Strands of wire or gut are stretched across the lower or *snare* head. When the upper or *batter* head is struck, the lower head vibrates against the snares. Alternatively, the snares can be moved away from the lower head.

Side drum heads are nearly always made of polyester. Batter heads are 0.13 to 0.36 mm thick, while snare heads are much thinner, typically 0.08 mm in thickness. Frequencies of a few modes of vibration in a 36-cm drum are given in Table 13.4. In order to observe the modes of the separate heads, the opposite head was damped (with sandbags); thus the frequencies are for the drumhead backed by the enclosed air. Note the (01) doublet, similar to that observed in a bass drum. The modal frequencies are not harmonically related, which accounts for the indefinite pitch.

Although Mylar has replaced calfskin in most drumheads, many musicians prefer the sound of calfskin and are willing to put up with its disadvantages. Even when prepared by an expert craftsman, calfskin heads are not uniform in density and strength. However, in studying Mylar and calfskin heads on bass and field drums, Hardy and Ancell (1963) confirmed in the laboratory what many experienced drummers know: Calfskin is capable of a larger range of tension and, under some conditions, has more damping. The peak sound spectra

TABLE 13.4   Modal frequencies in a 36–cm side drum

Mode	Batter head	Snare head	Drum
01	245 Hz	350 Hz	195,307 Hz
11	280	520	280
21	405	617	410
31	520	870	520
02	602	770	565
41	607		635

were found to be quite similar, although Mylar had a slightly louder sound during a drum roll due to lower damping. The principal advantage of Mylar, however, is its insensitivity to humidity and moisture.

13.9 OTHER DRUMS

Tomtoms are unsnared drums that are made in a number of sizes, ranging from 20 to 45 cm in diameter. Tomtoms may have either one or two heads; the more indefinite pitch of the two-headed type is usually preferred for orchestral work. A drum set, such as the one shown in Fig. 13.13, usually includes two or three tomtoms of different sizes.

Bongos and *congas* are two popular types of thick-skin drums played with the hands and used extensively in Latin American music. Bongos are typically from 15 to 20 cm in diameter and about 12 cm deep. Conga drums are larger than bongos, having diameters of 25 to 30 cm and thick, tapered shells 60 to 75 cm long (see Fig. 13.14).

Many types of drums from Africa and Asia and other parts of the world are becoming popular in American music. One drum that is particularly interesting acoustically is the *tabla* from north India, a single-headed drum with a closed resonating chamber. The head is tensioned by straps that stretch vertically from top to bottom as shown in Fig. 13.15. Cylindrical pieces of wood are inserted between the straps, and by moving these up and down the drum can be finely tuned.

FIG. 13.14
Conga drums.

FIG. 13.15
Indian drums of the tabla family: *tabla* on the left and *banya* on the right. Note the tensioning straps and the drumhead loaded with black paste at its center.

FIG. 13.13
Jazz or trap drum set including snare drum, pedal-operated bass drum, three tomtoms, two cymbals, and a pedal-operated "hi-hat" (pair of cymbals). (Courtesy of Ludwig Industries.)

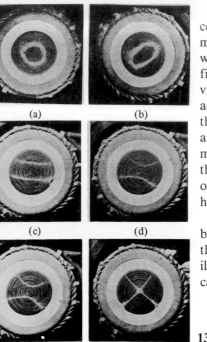

(a)   (b)

(c)   (d)

(e)   (f)

FIG. 13.16
Chladni patterns of six different modes of vibration in a tabla all of which have frequencies near the third harmonic. The (02) and (21) normal modes are shown in (a) and (f), respectively; the other modes are combinations of these two. (Raman, 1934).

The tabla is loaded at the center by a paste of iron-oxide, charcoal, starch, and gum, which hardens but remains flexible. C. V. Raman (who won a Nobel prize for his work in spectroscopy but also wrote several articles on the acoustics of violins) observed that the first four overtones of the tabla are harmonics of the fundamental by virtue of the extra load. Later, he discovered that these five harmonics actually result from nine modes of vibration, several of which have the same frequencies; the (11) mode is the second harmonic, the (21) and (02) modes both vibrate at the third harmonic, the (12) and (31) modes at the fourth harmonic, and the (03), (22), and (41) modes at the fifth harmonic (Raman, 1934). Fig. 13.16 shows Chladni patterns of the (02) and (21) modes and several combinations of them which have frequencies near the third harmonic.

The two-headed *mridanga* from south India is similar in acoustic behavior to the tabla. We recorded sound spectra at 35 stages during the process of preparing a mridanga head by loading it in order to illustrate how 9 modes in the unloaded head develop into 5 harmonically-related modes in the finished head (Rossing and Sykes, 1982).

13.10 VIBRATIONS OF PLATES

Vibrating plates bear much the same relationship to membranes that vibrating bars do to strings. Whereas in strings and membranes the restoring force results from the tension, in bars and plates it results from the stiffness of the solid material. In plates and bars, the overtones tend to be substantially higher than the fundamental. They can vibrate with a variety of boundary conditions, including clamped and free edges.

The vibrations of plates have fascinated physicists for many years. Nearly 200 years ago E. F. F. Chladni published his first book describing his well-known method of sprinkling sand on vibrating plates to make the nodal lines visible. Chladni's lectures throughout Europe attracted many famous persons, including Napoleon. Chladni patterns are used even today to observe the nodal lines of vibrating plates, although some type of electromechanical driver is usually preferred to Chladni's violin bow.

Plates can vibrate in a large number of modes. The modes of a circular plate are often given the labels m and n, like those of a circular membrane, to designate the number of nodal diameters and nodal circles, respectively. Chladni patterns of a circular plate are shown in Fig. 2.19, and those of the top and back plates of a violin in Fig. 10.17. Chladni observed that the frequencies of the various modes in a plate are nearly proportional to $(m + 2n)^2$, a relationship that has been called "Chladni's law" (Rossing, 1982b).

13.11   CYMBALS

Cymbals are among the oldest of musical instruments and have had both religious and military use in a number of cultures. The Turkish cymbals generally used in orchestras and bands are saucer-shaped with a small dome in the center, in contrast to Chinese cymbals, which have a turned-up edge more like a tamtam. Orchestral cymbals are usually between 40 and 55 cm in diameter and are made of bronze. The leading manufacturer of cymbals, the Avedis Zildjian Company, claims that its secret process for treating cymbal alloys was discovered in 1623.

Many different types of cymbals are used in orchestras, marching bands, concert bands, and jazz bands. Orchestral cymbals (see Fig. 2.20 (a)) are often designated as "French," "Viennese," and "Germanic" in order of increasing thickness. Jazz drummers use cymbals designated by such onomatopoeic terms as "crash," "ride," "swish," "splash," "ping," and "pang." Cymbals range from 20 cm to 75 cm (8 to 30 in.) in diameter.

Given the wide ranges in diameter and thickness, considerable variety of cymbal tone is available to the percussionist. Furthermore, a good cymbal can be made to produce many different tones by using a variety of sticks and striking it at several different places. A large cymbal struck gently near the rim produces a low sound not unlike

FIG. 13.17
Modes of vibration of a 38-cm (15-inch) cymbal. The first six modes resemble those of a flat plate, but after that the resonances tend to be combinations of two or more modes. (Rossing and Peterson, 1982).

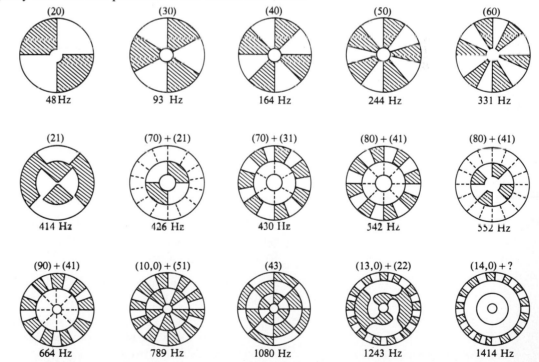

(20)	(30)	(40)	(50)	(60)
48 Hz	93 Hz	164 Hz	244 Hz	331 Hz

(21)	(70) + (21)	(70) + (31)	(80) + (41)	(80) + (41)
414 Hz	426 Hz	430 Hz	542 Hz	552 Hz

(90) + (41)	(10,0) + (51)	(43)	(13,0) + (22)	(14,0) + ?
664 Hz	789 Hz	1080 Hz	1243 Hz	1414 Hz

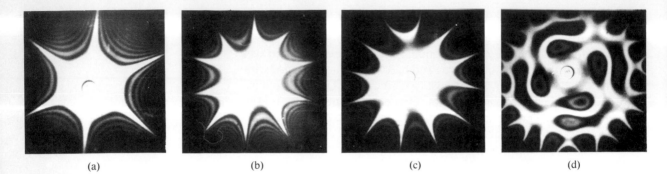

(a)   (b)   (c)   (d)

FIG. 13.18
Hologram interferograms of four of the modes of vibration shown in Fig. 13.17:(a) 30; (b) 50; (c) 60;
(d) 13, 0 + 22 (a combination). (Rossing and Peterson, 1982).

that of a small tamtam. The fullest sound is obtained by a glancing blow about one-third of the way in from the rim.

The individual modes of vibration of cymbals, shown in Fig. 13.17, are basically those of a circular plate (see Fig. 2.19), but altered by the saucerlike shape of the cymbal. Note that after the (60) mode, the cymbal resonances are combinations of two or more single modes. The mode at 1243 Hz, for example, is a combination of the (13,0) and (22) modes (see Fig. 13.18(d)). Hologram interferograms of four of these modes are shown in Fig. 13.18.

A cymbal may be excited in many different ways. It may be struck at various points with a wooden stick, a soft beater, or another cymbal. The onset of sound is quite dependent on the manner of excitation. The coupling between vibrational modes in a cymbal is strong, however, so that a large number of partials quickly appear in the spectrum, however it is excited.

Sound spectra of a 40-cm cymbal immediately after striking and after intervals of 0.05, 1.0 and 2.0 seconds are shown in Fig. 13.19. From this and other similar spectra we can make the following observations:

1. The sound level below about 700 Hz shows a rather rapid decrease during the first 200 ms, after which it decays slowly. This is apparently due to conversion of energy into modes of higher frequency.

2. Several strong peaks in the 700–1000 Hz range build up between 10 and 20 ms, then decay.

3. Sound energy in the important 3 to 5 kHz range peaks about 50 to 100 ms after striking.

4. Sound in the range of 3 to 5 kHz, which gives the cymbal its "shimmer," is often the most prominent feature from about one to four seconds after striking.

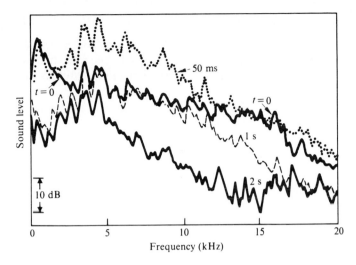

FIG. 13.19
Sound spectra of a 40-cm cymbal immediately after striking and after intervals of 0.05, 1.0, and 2.0 seconds. (Rossing and Shepherd, 1983).

5. The low frequencies again dominate the lingering sound, but at a much lower level, so that they are rather inconspicuous.

13.12   GONGS AND TAMTAMS

Gongs play a very important role in Oriental music, but they enjoy considerable popularity in Western music as well. They are usually cast of bronze with a deep rim and a protruding dome. Gongs used in symphony orchestras usually range from 0.5 to 1 m (20 to 38 in.) in diameter, and are tuned to a definite pitch. When struck near the center with a massive soft mallet, the sound builds up relatively slowly and continues for a long time if the gong is not damped.

Massive gongs are central to every gamelan (ensemble) in Indonesia. Of considerable interest to the acoustician are the gongs used in Chinese opera orchestras, which glide upward or downward in pitch after being struck, due to their special shapes (Rossing and Fletcher, 1983).

Tamtams are similar to gongs in appearance, and are often confused with them. The main differences between the two are that tamtams do not have the dome of the gong, their rim is not as deep, and their metal is thinner. Tamtams sound a much less definite pitch than do gongs. In fact, the sound of a tamtam may be described as somewhere between the sounds of a gong and a cymbal. The sound of a large tamtam develops slowly, changing from a sound of low pitch at strike to a collection of high-frequency vibrations, which are described as "shimmer." These high-frequency modes fail to develop if the tamtam is not hit hard enough, indicating that the conversion of energy takes place through a nonlinear process. Figure 13.20 shows the shapes of a number of different gongs and tamtams.

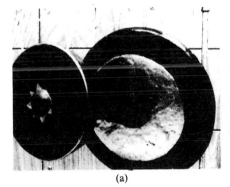

(a)

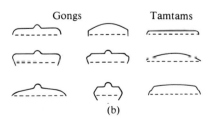

(b)

FIG. 13.20
(a) Gong and tamtam. (b) Typical shapes of gongs and tamtams. Note the deeper rims and tuning domes of the gongs.

13.13   STEEL DRUMS

An instrument of rather recent origin that has achieved great popularity is the Caribbean steel drum. It has become the foremost musical instrument in its home country (Trinidad) and in other Caribbean countries; steel bands are becoming increasingly common in Europe and the United States as well. Steel drums are generally played in ensembles ranging in size from six or eight players to one hundred or more. The instrument is occasionally used as a solo instrument accompanied by an orchestral ensemble.

Steel drums originated in Trinidad around 1940, partly because of a ban on "bamboo tamboo" stick bands, whose rhythm had become a feature of the annual Carnival festivities. (Unfortunately, the large bamboo sticks were being used not only as musical instruments but as weapons in fights between rival bands and with police.) But the enterprising musicians were not to be denied, and they turned to garbage cans, buckets, brake drums, or whatever was available. Thus, the first "steel drums" were definitely rhythmic rather than melodic. The development of tuned steel drums took place in the years following the end of World War II, when the annual celebration of Carnival resumed with great enthusiasm.

Steel drums or pans are usually fabricated from 55-gallon oil drums. The drums in a steel band are known by various names such as soprano, ping-pong, double tenor, guitar, cello and bass. The soprano or ping-pong has from 26 to 32 different notes, but each bass drum has only 3 or 4; hence the bass drummer typically plays on 6 drums in the manner of a timpanist.

The first step in making a steel drum is to hammer the end of the oil barrel to the shape of a shallow basin. Then a pattern of grooves is cut with a nail punch in order to define sections of the various notes. Next, each section is "ponged up" with a hammer, and after heating and hardening the drum, each section is tuned by the skillful use of the hammer.

Different drum makers use different patterns for their drums, and these patterns change from time to time. A steel band may typically employ the following instruments.

1. Soprano pan (also called lead, single tenor, or ping-pong): one drum of $2\frac{1}{2}$ octaves (28 notes) ranging from D_4 to F_6.

2. Alto (also called double tenor): two drums of 15 notes each from $A^{\flat}_3$ to $C^{\sharp}_6$.

3. Tenor (also called double second): two drums of 14 notes each from $F^{\sharp}_3$ to A_5.

4. Cello: three drums of seven notes each from B_2 to G_4.

5. Bass: six drums of three notes each from C_2 to F_3.

The design of the individual drums is shown in Fig. 13.21.

FIG. 13.21

Design of a typical steel drum set, showing locations of various notes. (a) Soprano (single tenor); (b) alto (double tenor); (c) double second; (d) cello; (d) bass. (Courtesy of Clifford Alexis.)

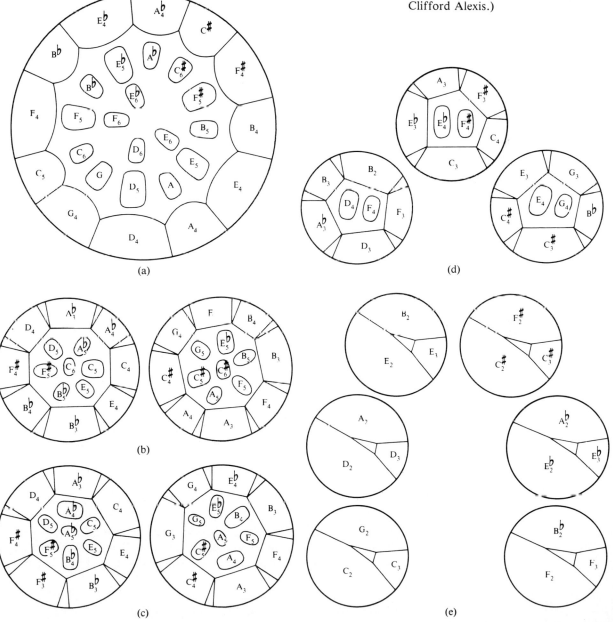

Other designs may include a guitar consisting of two drums of 10 notes each, a tenor bass consisting of four drums with five notes each, and a bass set of six drums with four notes each.

A skilled steel drum maker tunes at least one overtone of each note to a harmonic of the fundamental (usually the octave). Sometimes it is possible to tune another mode to the third or fourth harmonic by adjusting internal "tension" along a boundary.

The sound spectra of steel drums are surprisingly rich in harmonic overtones. These harmonic overtones appear to have three different physical origins:

1. Radiation from higher modes of vibration tuned harmonically by the tuner;

2. Radiation from nearby notes whose frequencies are harmonically related to the struck note;

3. Nonlinear motion of the note area vibrating at its fundamental frequencies.

13.14 BELLS AND CARILLONS

Bells have been a part of nearly every culture in history. Bells existed in the Near East before 1000 B.C., and a number of Chinese bells from the time of the Shang dynasty (1600–1100 B.C.) are found in museums throughout the world. A set of tuned bells from the 5th century B.C. was recently discovered in the Chinese province of Hubei.

Bells developed as Western musical instruments in the 17th century when bell founders discovered how to tune their partials harmonically. The founders in the Low Countries, especially the Hemony brothers (François and Pieter) and Jacob van Eyck, took the lead in tuning bells, and many of their fine bells are found in carillons today.

The carillon also developed in the Low Countries. Chiming bells by pulling ropes attached to the clappers had been practiced for some time before the idea of attaching these ropes to a keyboard or handclavier occurred to bell ringers in the 16th century. Many mechanical improvements during the 17th and 18th centuries, including the breached wire system and the addition of foot pedals for playing the larger bells, led to development of the modern carillon. Today, the term carillon is reserved for an instrument of 23 (two octaves) or more tuned bells played from a clavier (smaller sets are called "chimes"). The largest carillon in existence is the 74-bell (6 octave) carillon at Riverside Church in New York whose largest bell is more than 18,000 kg (20 tons).

Perhaps it is stretching the imagination a bit to think of a bell as being a plate, but the general principles of its vibrational behavior are similar. Although the mathematical description of the vibrations of a bell are understandably complex, the principal modes, at least, can be described by specifying the number of circular nodes and meridian nodes. The lowest mode of vibration, (called the *hum* tone), for example, has four meridian nodes, so that alternate quarters of the bell essentially move inward and outward.

Figure 13.22 shows the principal vibrational modes for a carillon bell. The mode called the third is tuned a minor third above the strike tone, whereas the upper third is usually a major third above the octave. The *strike tone* is determined by the octave, the twelfth, and the upper octave, whose frequencies have the ratios 2:3:4, just as in chimes (see Section 13.4). Unlike chimes, however, carillon bells have a mode called the prime or fundamental with a frequency at or near the strike tone. Careful studies have shown, however, that the pitch of the strike tone is determined by the three modes mentioned above rather than by the prime.

FIG. 13.22
The first eight vibrational modes of a tuned bell. Dotted lines indicate the approximate locations of nodes. Frequencies relative to the strike tone are given. The notes that correspond to these in a C_4 bell are shown on a musical staff. In some bells only five modes are tuned harmonically.

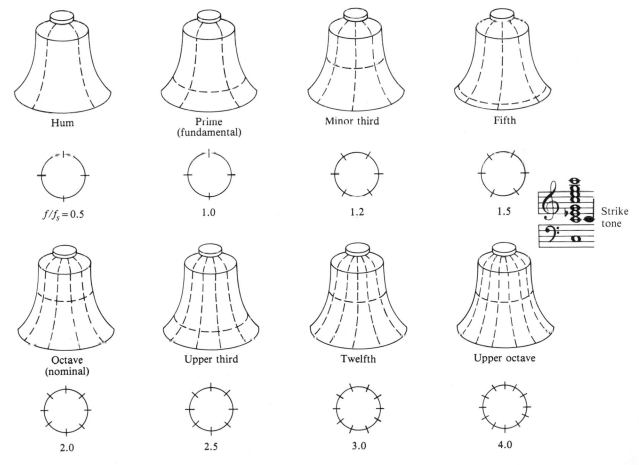

Hum   Prime (fundamental)   Minor third   Fifth

$f/f_s = 0.5$ 1.0   1.2   1.5   Strike tone

Octave (nominal)   Upper third   Twelfth   Upper octave

2.0   2.5   3.0   4.0

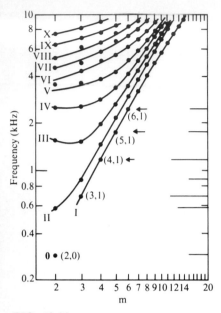

FIG. 13.23
Vibrational frequencies of groups 0–IX in a D_5 church bell. Also shown on the right are the relative strengths at impact of several partials in the bell sound. Arrows denote the three partials in group I that determine the strike note. (From Rossing and Perrin, 1986).

Vibrational frequencies of groups 0–IX in a church bell with a D_5 strike note are shown in Fig. 13.23. Horizontal lines on the axis at the right indicate relative strengths of several partials in the bell sound (these vary considerably with location, of course). Arrows denote the three partials (octave or nominal, fifth (twelfth), and upper octave) that determine the strike note.

A new type of carillon bell has been developed at the Royal Eijsbouts Bellfoundry in the The Netherlands. The new bell replaces the strong minor-third partial with a major-third partial, thus changing the tonal character of the bell sound from minor to major. This requires an entirely new bell profile. The new bell design evolved partly from the use of a technique for structural optimization using finite element methods on a digital computer. This technique allows a designer to make changes in the profile of an existing structure, and then to compute the resulting changes in the vibrational modes (Lehr, 1987).

13.15   HANDBELLS

Handbells also date back at least several centuries B.C., although tuned handbells of the present-day type were developed in England in the 18th century. One early use of handbells was to provide tower bellringers with a convenient means to practice change ringing. In more recent years, handbell choirs have become popular in schools and churches–some 2000 choirs are reported in the United States.

Although they are cast from the same bronze material and cover roughly the same range of pitch, the sounds of church bells, carillon bells, and handbells have distinctly different timbres. In a handbell, only two modes of vibration are tuned (although there are three harmonic partials in the sound), whereas in a church bell or carillon bell at least five modes are tuned harmonically. A church bell or carillon bell is struck by a heavy metal clapper in order to radiate a sound that can be heard at a great distance, whereas the gentle sound of a handbell requires a relatively soft clapper.

In the so-called English tuning of handbells, followed by most handbell makers in England and the United States, the (3,0) mode is tuned to three times the frequency of the (2,0) mode. The fundamental (2,0) mode radiates a rather strong second harmonic partial, however, so that the sound spectrum has prominent partials at the first three harmonics (Rossing and Sathoff, 1980). Some Dutch founders aim at tuning the (3,0) mode in handbells to 2.4 times the frequency of the fundamental, giving their handbell sound a minor-third character somewhat like a church bell. Such bells are usually thicker and heavier than bells with the English-type tuning.

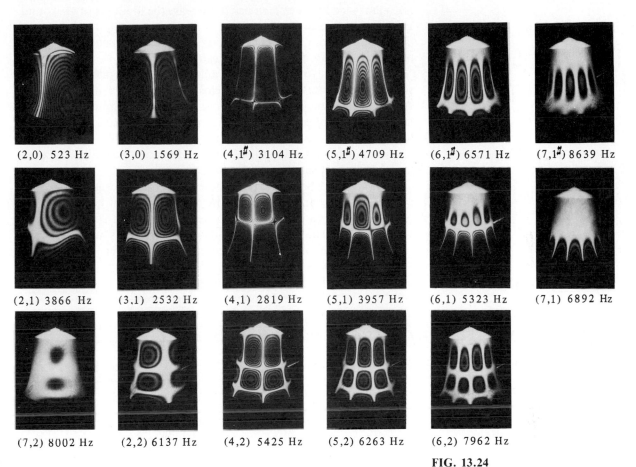

(2,0) 523 Hz   (3,0) 1569 Hz   (4,1$^\sharp$) 3104 Hz   (5,1$^\sharp$) 4709 Hz   (6,1$^\sharp$) 6571 Hz   (7,1$^\sharp$) 8639 Hz

(2,1) 3866 Hz   (3,1) 2532 Hz   (4,1) 2819 Hz   (5,1) 3957 Hz   (6,1) 5323 Hz   (7,1) 6892 Hz

(7,2) 8002 Hz   (2,2) 6137 Hz   (4,2) 5425 Hz   (5,2) 6263 Hz   (6,2) 7962 Hz

Hologram interferograms of a number of the modes are shown in Fig. 13.24. The "bull's eyes" locate the antinodes. Note that the upper half of the bell moves very little in the (7,1) mode; the same is true in (m,1) modes when m > 7.

FIG. 13.24
Time-average hologram interferograms of inextensional modes in a C_5 handbell. (From Rossing et al., 1984).

13.16 SUMMARY

Percussion instruments have experienced a surge of interest, especially in contemporary music. Bars, membranes, and plates vibrate in modes that are not harmonically related to the fundamental, and this gives percussion instruments a distinctive timbre. The lower bars of marimbas, xylophones, and vibes have arches cut on the underside in order to tune the first overtone to a harmonic of the fundamental (fourth harmonic, in the case of the marimba and vibes; third harmonic, in the case of the xylophone). These three instruments have a tubular resonator placed under each bar to increase the loudness of the sound. In vibes, these resonators can be opened and closed rapidly to generate a type of vibrato. Chimes have no mode of vibration at the frequency of the strike tone, which is a subjective tone.

Drums have membranes of animal skin or synthetic material stretched over some type of air enclosure. The modes of vibration are shifted in frequency away from those of an ideal membrane by the air loading. Timpani convey a strong sense of pitch; the first overtone is tuned to a fifth above the principal tone. Other drums, such as the bass drum, have an indefinite pitch. The Indian tabla has several overtones that are harmonics of the fundamental.

Examples of percussion instruments that are vibrating plates are cymbals, gongs, and tamtams. The sound spectrum of cymbals are characterized by a buildup and subsequent decay of sound in the 3–10 kHz range. The spectra of Caribbean steel drums are surprisingly rich in harmonic overtones. Tuned carillon bells and handbells may also be described as platelike. Carillon bells have five to eight modes whose frequencies are harmonics of a fundamental. In a small handbell, only two modes are usually tuned harmonically.

References and Suggested Readings

Bork, I., and J. Meyer (1982). "Zur klanglichen Bewertung von Xylophonen," *Das Musikinstrument* **31**(8): 1076. English translation in *Percussive Notes* **23**(6): 48 (1985).

Brindle, R. S. (1970). *Contemporary Percussion*. London: Oxford University Press.

Fletcher, H., and I. G. Bassett (1978). "Some Experiments with the Bass Drum," *J. Acoust. Soc. Am.* **64**: 1570.

Fletcher, N. H., and T. D. Rossing (1989). *The Physics of Musical Instruments*. New York: Springer-Verlag. Chapters 18–20.

Flugge, W. (1962). *Statik und Dynamik der Schalen*. New York: Springer-Verlag.

Hardy, H. C., and J. F. Ancell (1963). "Comparison of the Acoustical Performance of Calfskin and Plastic Drumheads," *J. Acoust. Soc. Am.* **33**: 1391.

Lehr, A. (1987). "From Theory to Practice," *Music Perception* **4**: 267.

Raman, C. V. (1934). "The Indian Musical Drums," *Proc. Indian Acad. Sci.* **A1**: 179.

Lord Rayleigh (J. W. Strutt) (1894). *The Theory of Sound,* Vol. 1, 2nd ed. London: Macmillan. Reprinted by Dover, New York, 1945.

Rossing, T. D. (1976). "Acoustics of Percussion Instruments, Part I," *Phys. Teach.* **14**: 546.

Rossing, T. D. (1977). "Acoustics of Percussion Instruments, Part II," *Phys. Teach.* **15**: 278.

Rossing, T. D. (1982a). "The Physics of Kettledrums," *Sci. Am.* **247**(5): 172.

Rossing, T. D. (1982b). "Chladni's Law for Vibrating Plates," *Am. J. Phys.* **50**: 271.

Rossing, T. D. (1984a). *The Acoustics of Bells*. Stroudsburg, PA: Van Nostrand Reinhold.

Rossing, T. D. (1984b). "The Acoustics of Bells," *Am. Scientist* **72**: 440.

Rossing, T. D. (1987). "Accoustical Behavior of a Bass Drum," *J. Acoust. Soc. Am.* **82**: S69.

Rossing, T. D., and N. H. Fletcher (1982). "Acoustics of a Tamtam," *Bull. Australian Acoust. Soc.* **10** (1): 21.

Rossing, T. D., and N. H. Fletcher (1983). "Nonlinear Vibrations in Plates and Gongs," *J. Acoust. Soc. Am.* **73**: 345.

Rossing T. D., and G. Kvistad (1976). "Acoustics of Timpani: Preliminary Studies," *The Percussionist* **13**: 90.

Rossing, T. D., and R. Perrin (1986). "Vibrations of Bells," *Applied Acoustics* **20**: 41.

Rossing, T. D., R. Perrin, H. J. Sathoff and R. W. Peterson (1984). "Vibrational Modes of a Tuned Handbell," *J. Acoust. Soc. Am.* **76**: 1263.

Rossing, T. D., and R. W. Peterson (1982). "Vibrations of

Plates, Gongs, and Cymbals," *Percussive Notes* **19**(3): 31.

Rossing, T. D., and H. J. Sathoff (1980). "Modes of Vibration and Sound Radiation from Tuned Handbells, *J. Acoust. Soc. Am.* **68**: 1600.

Rossing, T. D., and R. B. Shepherd (1983). "Acoustics of

Cymbals," *Proc. 11th Intl. Congress on Acoustics* (Paris): 329.

Rossing, T. D., and W. A. Sykes (1982). "Acoustics of Indian Drums," *Percussive Notes* **19**(3): 58.

Taylor, H. W. (1964). *The Art and Science of the Timpani.* London: Baker.

Glossary

harmonics A series of partials with frequencies that are simple multiples of a fundamental frequency. (In a harmonic series, the first harmonic would be the fundamental, the second harmonic the first overtone.)

inharmonic partials Overtones or partials that are not harmonics of the fundamental.

nodes Points or lines that do not move when a body vibrates in one of its modes.

overtones Upper partials or all components of a tone except the fundamental.

radius of gyration A measure of the difficulty of rotating a body of a given mass.

strike tone The subjective tone that determines the pitch of a bell or chime; in most tuned bells it corresponds closely to one of the partials, but in chimes it does not.

tension The force applied to the two ends of a string, or around the periphery of a membrane, that provides a restoring force during vibration.

torsional mode An oscillatory motion that involves twisting of the vibrating member.

vibrato Frequency modulation (FM) that may or may not have amplitude modulation (AM) associated with it. Some musicians speak of "intensity vibrato," "pitch vibrato," and "timbre vibrato" as separate entities; others understand vibrato to include all three. Sometimes the term **tremolo** is used to describe AM, but this is not recommended because tremolo is used to describe other things, such as a rapid reiteration of a note or even a trill.

Young's modulus The ratio of stress to strain; also called **modulus of elasticity.**

Questions for Discussion

1. What is wrong with the statement "The resonators of a marimba prolong the sound"?

2. The lowest mode of vibration of the triangle shown in Fig. 13.8 has four nodes. Make a sketch of the type in Figs. 13.2 or 13.3 to illustrate how it vibrates.

3. Explain why a microphone placed some distance above the center of a kettledrum picks up very little of

the principal tone (except, of course, by reflection from the walls, etc.) What modes would be picked up best by a microphone in such a location?

4. Compare the frequencies of the modes of the kettledrum given in Table 13.3 with those given by Rayleigh (Section 13.7).

Problems

1. Compare the ratios of the frequencies of transverse vibrations in the glockenspiel bar in Fig. 13.2 with the theoretical ratios for a thin rectangular bar given in Section 13.1. Can you account for the difference? (*Hint:* Compare Fig. 13.6.)

2. Write an expression for the frequency (f_1) of the lowest mode of vibration of a rectangular bar in terms

of its length L and the speed of sound v.

3. Write an expression for the frequency ratio of the lowest transverse mode in a bar to the lowest longitudinal mode. Find this ratio for a glockenspiel bar 21.4 cm long and 0.90 cm thick ($K = t/3.46$). Compare this ratio with the ratio of the corresponding values given in Fig. 13.2.

4. What are the actual ratios of the numbers 81, 121, and 169 (discussed in Section 13.4)? How close are they to the ratios 2:3:4?

5. Using the formulas for longitudinal and transverse vibrations in a bar (Table 13.1), show that the lowest longitudinal and transverse modes will have the same frequency when $L = (9/4) \pi K$. Find the ratio of length to diameter for a bar (rod) of circular cross section having the frequency for its lowest transverse and longitudinal vibrations.

6. Determine the frequencies of the (41), (51), and (61) modes of the bell in Fig. 13.23. Show that they are nearly in the ratios 2:3:4, and that they will produce a strike note (virtual pitch) corresponding to D_5 (see Section 7.4). Is there a vibrational mode with this frequency?

7. In large bells, the (61), (71), (81), and (91) modes create a secondary strike note. Determine the frequencies of these modes in Fig. 13.23, and estimate their virtual pitch (see Section 7.4). What note on the scale is this nearest (see Table 9.2)?

8. If a bell does not have perfect circular symmetry (perfection is seldom achieved in practice), one or more modes of vibration will show "doublet" behavior: that is, there will be two modes of vibration with slightly different frequencies. Suppose that such a doublet exists, having components with frequencies of 440 and 442 Hz. At what rate will the bell "warble"? How is it possible to minimize this warble?

CHAPTER **14**
Keyboard Instruments

This chapter features three different types of keyboard instruments. In the piano, strings are set into vibration by striking them with hammers; in the harpsichord, the strings are plucked; and in the pipe organ, sound is produced by blowing air into tuned pipes.

14.1 CONSTRUCTION OF THE PIANO

Invented in 1709 by Bartolomeo Cristofori in Florence, the piano has become the most popular and versatile of all instruments. It has a range of more than seven octaves (A_0 to C_8) and a wide dynamic range. Pianos vary in size from small home uprights or spinets to large concert grands.

The main parts of the piano are the keyboard, the action, the strings, the soundboard, and the frame. A simplified diagram of a piano is shown in Fig. 14.1. Over 200 strings extend from the pin

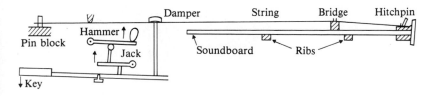

FIG. 14.1
A simplified diagram of the piano. When a key is depressed, the damper is raised, and the hammer is "thrown" against the string. Vibrations of the string are transmitted to the soundboard by the bridge.

287

block or wrest plank across the bridge to the hitch-pin rail at the far
end. When a key is depressed, the damper is raised, and the hammer
is "thrown" against the string, setting it into vibration. Vibrations of
the string are transmitted to the soundboard by the bridge.

A typical concert grand piano has 243 strings, varying in length
from about 2 m at the bass end to about 5 cm at the treble end. Included are eight single strings wrapped with one or two layers of wire,
five pairs of strings also wrapped, seven sets of three wrapped strings,
and 68 sets of three unwrapped strings. Smaller pianos may have
fewer strings but they play the same number of notes: 88. A small
grand piano with 226 strings is shown in Fig. 14.2. Note that the bass
strings overlap the middle strings, which allows them to act nearer the
center of the soundboard. The acoustical advantages of wrapped

FIG. 14.2
The top view of a studio grand piano
showing the cast-iron frame, the
overlapping strings, hammers, and
dampers. The cutaway portion at the
treble end shows the tuning pin block.
(Courtesy of Baldwin Piano and Organ
Co.)

strings and multiple strings for most notes will be discussed in Section 14.2.

The *soundboard* is nearly always made of spruce, about 1 cm thick, with its grain running the length of the piano. Ribs on the underside of the soundboard stiffen it in the cross-grain direction. The soundboard is the main source of radiated sound, just as is the top plate of a violin or cello.

To obtain the desired loudness, piano strings are held at high tensions which may exceed 1000 newtons (220 lb). The total force of all the strings in a concert grand is over 20 tons! In order to withstand this force and maintain stability of tuning, grand pianos have sturdy frames of cast iron.

The rather complicated action of a grand piano is shown in Fig. 14.3. When a key is pressed down, the damper is raised. At the same time, the capstan sets the whippen into rotation around its pivot. The rotating whippen causes the jack to push against the hammer, knuckle or roller, starting the hammer on its journey toward the string. Just before the hammer strikes the string, the lower end of the jack strikes the jack regulator and rotates away from the knuckle. The freely rotating hammer now strikes the string, rebounds immediately (in order not to damp the string), and falls back to the repetition level. The back check prevents the hammer from bouncing back to strike the string a second time. When the key is lifted slightly, a spring pulls the jack back under the knuckle, so that pressing the key a second time repeats the note. Most upright pianos do not have this feature; thus a note cannot be repeated unless the key returns to its starting position.

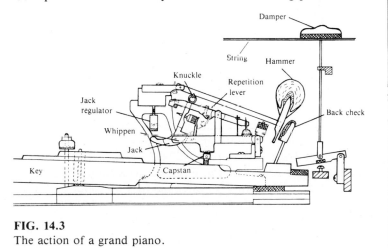

FIG. 14.3
The action of a grand piano.

The upright piano developed about the middle of the nineteenth century. The nearly rectangular soundboard and the strings in upright pianos are vertical; the hammers travel horizontally. Thus the action is different from that of the grand piano in Fig. 14.3. In full-size upright pianos, which stand 130 to 150 cm (4½ to 5 feet) in height, the striking mechanism or action is located some distance above the keys and connected to them mechanically by stickers. In studio uprights or console pianos, which stand about 100 to 130 cm (3¼ to 4¼ feet) in height, the action is mounted directly over the keys without stickers. In small spinet pianos (less than 100 cm in height), the action is below the keys and drop stickers transmit key motion to the action.

Pianos may have two or three pedals. The right pedal is called the *sustaining* pedal. It raises all the dampers, which allows the struck strings to continue vibrating after the keys are released. The left pedal is some type of expression pedal. In most grand pianos, it shifts the entire action sideways, causing the treble hammers to strike only two of their three strings. This shifting type of pedal is called the *una corda* pedal. In vertical pianos, and a few grands, the left pedal is a *soft* pedal, which moves the hammers closer to the strings, decreasing their travel and thus their striking force.

Many pianos have a third pedal. On most grands and a few uprights, the center pedal is a *sostenuto* pedal, which sustains only those notes which are depressed prior to depressing the pedal, and does not sustain subsequent notes. On a few pianos, the center pedal is a bass sustaining pedal, which lifts only the bass dampers. On a few uprights, the center pedal is a *practice* pedal, which lowers a piece of felt between the hammers and the strings, muffling the tone.

14.2 PIANO STRINGS

The strings are the heart of the piano. They convert some of the kinetic energy of the moving hammers into vibrational energy and pass it on to the bridges and soundboard in a manner that determines the sound quality of the piano.

Piano strings make use of high-strength steel wire. Efficiency of sound production calls for the highest string tension possible, while at the same time minimizing inharmonicity calls for using the smallest string diameter (core diameter in a wrapped string) possible. This results in tensile stresses of around 1000 N/mm^2, which is about half the yield strength of steel wire. For steel with an elastic modulus of 2×10^{11} N/m^2, this results in an elongation of about ½% when the string is under tension. Fortunately, when strings break, it is usually near the keyboard end, so that the broken strings recoil away from the pianist.

An ideal string vibrates in a series of modes that are harmonics of a fundamental (see Section 4.4). Actual strings have some stiffness, which provides a restoring force (in addition to the tension), slightly raising the frequency of all the modes. The additional restoring force is greater in the case of the higher modes because the string makes more bends. Thus the modes are spread apart in frequency and are no longer exact harmonics of a fundamental. In other words, a real string with stiffness is partly stringlike and partly barlike.

The inharmonicity of strings (that is, the amount by which the actual mode frequencies differ from a harmonic series) is found to

---

The formula may be written

$$f_n = nf_1[1 + (n^2 - 1)A],$$

where f_n = the frequency of the nth harmonic and f_1 = the frequency of the fundamental. For a solid wire without wrapping,

$$A = \frac{\pi^3 r^4 E}{8TL^2},$$

where r = the radius of the string, E = Young's modulus, T = the tension, and L = the length of the string. Thus the inharmonicity is smallest for long, thin wires under great tension (large L and T, small r).

---

vary with the square of the partial number (Fletcher, 1964). Thus the second harmonic is shifted four times as much as the fundamental.

Since the stiffness of a string increases sharply with its radius (the factor A in the box above increases as r^4), inharmonicity is especially noticcable in the case of the large bass strings. Because wrapped strings are more flexible than are solid strings of the same diameter, the inharmonicity of the bass strings is reduced substantially by the use of wrapped rather than solid strings of the same weight. (The lower strings on a guitar and violin are wrapped rather than solid, for the same reason.)

A small amount of inharmonicity of string partials is considered desirable in pianos. One study of synthesized piano sounds demonstrated the preference of listeners, both musicians and nonmusicians, for tones with inharmonic partials. Tones synthesized with harmonic partials were described as "lacking warmth" and sounding much less like piano tones than those synthesized from slightly inharmonic partials (Fletcher, Blackham, and Stratton, 1962). Inharmonicity may also help disguise small tuning errors in the same way that vibrato serves the players of other instruments.

The inharmonicity in strings is the main reason why pianos are "stretch-tuned" (see Section 9.5). If they were not, the upper partials of a note would be slightly sharp with respect to the notes in the upper octaves to which they correspond, and undesirable beats would result when chords are played.

Since pianos are usually tuned by minimizing beats between notes, a stretched scale will automatically be the result. Figure 14.4 shows the deviations from equal temperament that resulted when a spinet piano was aurally tuned by a fine-tuner at a piano factory. Also shown is the *Railsback stretch,* which is an average from 16 different pianos measured by O. L. Railsback. The aural tuning results follow the Railsback curve generally, but with a few deviations that are probably attributable to soundboard resonances. A jury of listeners showed little preference between tuning done electronically according to the Railsback curve and tuning done aurally by the skilled tuner (Martin and Ward, 1961).

A large grand piano with long bass strings will show less stretch in the bass than does the small spinet piano in Fig. 14.4. However, the stretch in the upper octaves will not be much different, since the upper strings are not appreciably different in a spinet and a concert grand.

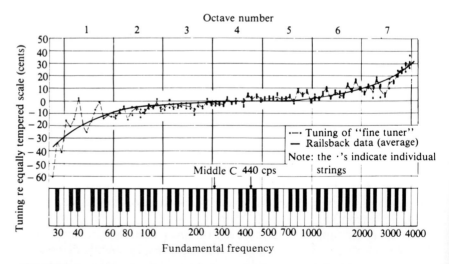

FIG. 14.4
Deviations from equal temperament in a small piano. (From Martin and Ward, 1961. Reprinted by permission of the Amer. Inst. of Physics.)

14.3   THE TUNING OF UNISONS

Over most of its playing range, the piano has three strings for each note. Studies have shown that the best piano sound results from tuning these strings one to two cents different from each other (Kirk, 1959). If the strings are tuned to exactly the same frequency, the transfer of energy from the strings to the soundboard takes place rapidly, and the decay time of the sound is too small. If the unison strings are tuned too far apart, prominent beats are heard, and what we commonly call a "barroom piano" sound is the result.

When the unison strings are tuned to be a few cents different, the decay curve takes on two different slopes. The hammer sets all three strings into vibration with the same phase, and energy is rapidly transferred to the soundboard initially. Since there are small differences in their frequencies, however, the strings soon get out of phase, and the rate of sound decay slows down, leading to a second slope in the decay curve ("aftersound"). If the frequencies differ by two cents, for example, about 400 vibrations would be required for the strings to fall out of phase. The actual coupling between the strings depends, in a somewhat complicated way, on the mechanical properties of the bridge (Weinreich, 1977), and the uneven decay of coupled strings is an important characteristic of piano sound. Fine tuning of the unisons is a means of regulating the amount of aftersound. Hammer irregularities can also affect the aftersound, and a skilled piano tuner can probably compensate for these to some extent by adjusting the unisons.

14.4   PIANO SOUND

Production of sound by a piano is a rather complicated process. The strings are set into vibration by the hammer, and they in turn act on the bridge to set the soundboard into vibration. Vibrational waves travel in many directions on the soundboard, and this leads to a complicated pattern of sound radiation. Eventually, the various components of the sound, which come from different parts of the soundboard, reach our ears; we process them and identify them as coming from a piano.

The various partials in a piano sound build up quite rapidly (typically in three milliseconds) and decay at widely different rates. Thus the sound spectrum of the piano is constantly changing with time, as in the case of the percussion instruments, which we discussed in Chapter 13. Many of the clues we use to identify a sound as coming from a piano are contained in the initial interval. This can be demonstrated by playing backward a tape recording of a piano. Now the sounds,

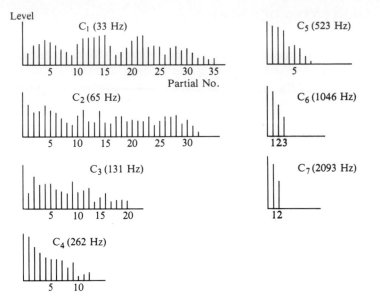

FIG. 14.5
The spectrum of high and low piano notes. (After Fletcher et al., 1962).

which still have the same overall spectrum as before, build up slowly and decay rapidly, and they are more suggestive of a reed organ than a piano (Demo. 29, Houtsma et al., 1987).

Figure 14.5 shows how the spectrum changes over the wide range of the piano (Fletcher, Blackham, and Stratton, 1962). In the lowest octave, as many as 45 partials can be detected; at the highest notes, only two or three partials can be detected.

14.5   HAMMER-STRING INTERACTION

The dynamics of the hammer-string interaction has been the subject of considerable research, beginning with Helmholtz (1877). The problem drew the attention of a number of Indian researchers, including Nobel laureate C. V. Raman, in the 1920s and 1930s, and it has recently been taken up by Hall (1986, 1987).

When the hammer has less mass than the string, it will most likely be thrown clear of the string by the first reflected pulse. The theoretical spectrum is missing harmonics numbered n/β (where β is the fraction of the string length at which the hammer strikes). If the hammer mass is not too small, the spectrum envelope falls off as $1/n$ (6 dB/octave) above a certain mode number. A heavier hammer is less easily stopped and thrown back by the string. It may remain in contact with the string during the arrival of several reflected pulses. Analytical models of hammer behavior are virtually impossible to construct, but computer simulations can be of value.

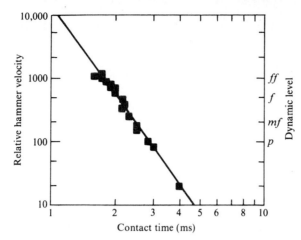

FIG. 14.6
Hammer velocity and hammer-string contact time at various dynamic levels for the C$_4$ note on a grand piano. (Askenfelt and Jansson, 1988).

Hall (1987) has considered the cases of a hard, narrow hammer and a soft, narrow hammer. In the case of the hard hammer, the mode spectrum envelope takes on a slope of −6 dB/octave at high frequencies. For the treble strings, where the hammer mass exceeds the string mass, the spectrum envelope may have a slope as steep as −12 dB/octave at high frequency.

Figure 14.6 shows a rather clear relationship between hammer-string contact time and hammer velocity at different dynamic levels. Striking the key with greater force increases the hammer velocity and decreases the contact time.

14.6 THE CLAVICHORD

The clavichord, like the piano, depends on struck strings for its sound; but there the similarity ends. A clavichord is a portable keyboard instrument with a soft, delicate sound, well suited for small living rooms but not for the concert hall. Its great virtue is its sensitivity; the tone can be varied in loudness and even given a vibrato by varying the force on the key.

The action of a clavichord is shown in Fig. 14.7. A small square piece of brass, called a tangent, is attached to the end of each key. When a key is depressed, the tangent strikes a string (or pair of strings) and causes the portion between the tangent and the bridge to vibrate. A damper prevents the other portion from vibrating and also damps the vibration of the entire string when the tangent is released from the string. The tangent will normally oscillate up and down at a frequency of a few hertz (determined by the mass of the string, the key, and the player's finger; the tension of the string; and the force applied to the key). This up-and-down motion varies the string tension and generates

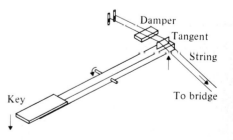

FIG. 14.7
Clavichord action (simplified).

a vibrato. The playing range of the clavichord is typically four octaves, from C_2 to C_6.

The soundboard of a typical clavichord is a nearly rectangular slab of spruce 2 to 3 mm thick. The first two resonances, at about 140 and 330 Hz in one clavichord, are due to coupling between the fundamental modes of the soundboard and the enclosed air, as in a guitar or violin (Thwaites and Fletcher, 1981). The next two resonances, around 470 and 570 Hz, are due to coupling between the second modes of the soundboard and the enclosed air. The clavichord is found to have a fairly uniform sound output except in the lowest octave where it is understandably weak.

14.7 THE HARPSICHORD

Another keyboard instrument that was very popular in the Baroque period is the harpsichord. Like a number of instruments from that period, it has experienced a revival of interest in recent years. Many modern harpsichords are patterned after the best of the early instruments, such as those built by the Ruckers family early in the seventeenth century.

The action of a harpsichord is shown in Fig. 14.8. Pressing the key raises the jack so that the plectrum plucks the string. Plectra were originally carved from birds' quills, but plastic plectra are being used in many modern instruments. The plectrum is attached to a hinged tongue, so that when the jack is lowered, the plectrum contacts the string only briefly before swinging back out of the way. Lowering the jack (by releasing the key) also causes a damper felt to touch the string and damp its vibrations.

The construction of a harpsichord is not unlike that of a grand piano, except that everything is smaller and lighter. The soundboard of a harpsichord is typically 2.5 to 3 mm thick, compared to 10 mm in a typical piano. Strings have about one-third the diameter and about one-tenth the tension of the corresponding piano strings. The thinner strings of the harpsichord have much less inharmonicity (recall from Section 14.2 that inharmonicity is proportional to the fourth power of the radius), even though its bass strings are solid wire.

The string dimensions of harpsichords are usually scaled in some regular fashion from bass to treble. In one modern harpsichord design, for example, the string lengths are proportional to $1/f$ over the top half of the range, with the rate of increase of length decreasing toward the bass end. The bass strings are brass, and their diameters are proportional to $1/f$; the treble strings are steel, and their diameter varies as $(1/f)^{0.3}$. The strings are plucked at a distance ℓ from the nut, where ℓ is a fraction of the total length L given by $f^{0.6}$ at the treble

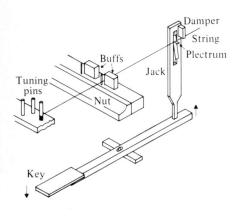

FIG. 14.8
Harpsichord action (simplified).

end. At the lower end $\ell/L = 0.13$. In general, the harmonic content of string vibration increases as the plucking point moves away from the center (Section 10.2), so that the harmonic development of the harpsichord increases in the lower notes (Fletcher, 1977a).

In order to vary the loudness and timbre, most harpsichords add one to three additional sets of strings to the original set. The selection of the strings to be played is made by means of *stops*. Borrowing from organ terminology, we call a set of strings that play an octave higher a *four-foot* set, and those that play an octave lower *sixteen-foot;* one or more sets of *eight-foot* strings play at the pitch of the key pressed. Some harpsichords add a second keyboard, and some of the largest instruments have a pedalboard as well. Couplers between the keyboards provide the harpsichordist with even more tonal options. Many harpsichords have a *buff stop* in which a piece of felt or soft leather is gently pressed against the string near its end to damp it and provide a short, distinctive sound. Sometimes a special set of jacks is provided to pluck the strings very close to the end in order to produce a nasal timbre (*lute stop*).

The *virginal* or *spinet* is a small instrument like a harpsichord with strings running lengthwise in a rectangular case. The keyboard is built into one side of the case, making it a portable tabletop instrument. Virginals were particularly popular in Elizabethan England. An elaborately decorated virginal is shown in Fig. 14.9.

FIG. 14.9
A Flemish virginal built by Hans Ruckers, 1581. (Reprinted by permission. All rights reserved, The Metropolitan Museum of Art.)

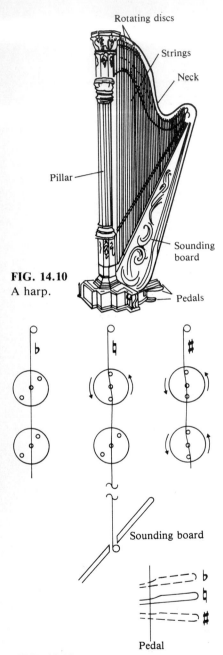

FIG. 14.10
A harp.

FIG. 14.11
Mechanism for tuning the harp.
Depressing one of the tuning pedals
increases the tension on all the strings
with that note name by one (for
naturals) or two (for sharps) units.

14.8 THE HARP

Although it does not have a keyboard, the harp is related to the harp-sichord and other string instruments described in this chapter, so a brief description of it is appropriate here. The modern harp, shown in Fig. 14.10, has 47 strings tuned to the notes of the diatonic scale, plus seven pedals by which each string can be raised or lowered a semitone in order to play the chromatics. With all the pedals in their middle position, the harp plays in the key of C major. Depressing the C-pedal raises all the C's on the harp to C♯. Figure 14.11 illustrates how the pedals raise and lower the pitch of the harp by means of two rotating discs attached to the neck of the instrument, which change the length of the vibrating part of the string.

The strings are attached to a slanted sounding-board at the bottom of the instrument. The nominal range, from C_1 to G_7, is almost as great as that of the piano. The most familiar sound of a harp is a sweeping glissando. By using the pedals to raise some notes and lower others, it is possible to make this glissando sound like a chord. The C-strings and F-strings are usually colored differently from the others in other to help the player locate the desired notes.

14.9 THE PIPE ORGAN: ITS CONSTRUCTION

The pipe organ has been called the "king of musical instruments." No other instrument can match it in size, in range of tone, in loudness, or in complexity. The world's largest organ, in the Convention Hall in Atlantic City, has over 32,000 pipes of various sizes and shapes. No two pipe organs in the world are exactly alike.

The modern pipe organ consists of a large variety of pipes arranged into divisions or organs. Each division is controlled by a separate keyboard or manual, including a pedalboard that is operated by the organist's feet. The pipes in the *swell* division are usually enclosed behind a set of shutters that can be opened or closed (by means of the swell pedal) to change the loudness. The other divisions are not usually enclosed, and their loudness can only be changed by adding or subtracting pipes through the use of *stops*. The principal division of an organ is called the *great organ,* and it usually contains the most stops. Most organs have couplers that allow certain pipes in one division to be controlled from the manual of another.

The windchests contain valves that can be opened to admit air into the pipes. The three main types of windchest actions used today are shown in Fig. 14.12. The oldest type is the *tracker* or mechanical action, in which the keyboards are connected to the chests by rather complicated mechanical linkages. With the second type, the *direct electric,* pressing the key energizes an electromagnet, which in turn opens the valve to allow air to flow into the pipe. The third type is called *electropneumatic,* because electromagnets are used to exhaust air from the bellows, which open the valve into the pipe. Full pneumatic actions, in which the key controls an air valve, are becoming rare.

Organ pipes are organized into *ranks* of similar pipes. One rank of pipes will include one pipe for each note (61 in the divisions operated from a keyboard, 32 in the pedal division). Each stop on the organ usually corresponds to one rank of pipes, except in the case of mixture stops, which involve several ranks. On smaller organs, a rank of pipes may be included in more than one stop.

14.10   ORGAN PIPES

There are two basic types of organ pipes: *flue* (labial) pipes and *reed* (lingual) pipes. Flue pipes produce sound by means of a vibrating air jet, in a manner similar to the flute and the recorder (see Section 12.11). Reed pipes use a vibrating brass reed to modulate the air stream.

The essential parts of a metal flue pipe are shown in Fig. 14.13(a). The air jet passes through a flue or windway, a narrow opening between the languid and the lower lip. Sound is produced when the air jet encounters the upper lip and oscillates back and forth, sometimes blowing into the pipe, sometimes blowing out through the mouth of the pipe. The large flue pipes usually have ears on either side of the mouth to guide the air jet.

The stopped wooden flue pipe, shown in Fig. 14.13(b), uses a similar mechanism to produce sound. Since the resonator is now a closed pipe, it need be only half as long as an open pipe in order to sound

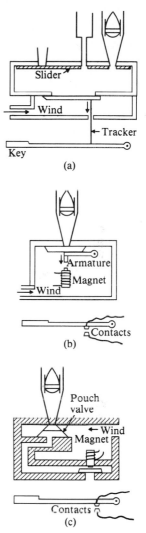

FIG. 14.12
Windchest actions: (a) tracker; (b) direct electric; (c) electropneumatic.

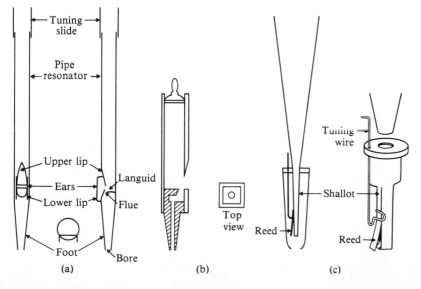

FIG. 14.13
Organ pipes: (a) open flue pipe of metal; (b) stopped wood flue pipe; (c) reed pipe.

the same pitch. Wooden pipes usually have a square cross section, and produce a sound with a flutelike quality.

There are three families of flue pipes: (1) diapasons, (2) flutes, and (3) strings. Pipes of the flute family usually have the least overtone content, while the bright-sounding strings have the most. String pipes are generally slender cylinders, whereas diapasons are open cylinders of somewhat greater diameter. Flute pipes come in several sizes and shapes, are constructed of either wood or metal, and may be open or closed. Closed pipes sound mainly the odd-numbered harmonics of the fundamental.

The reed pipe, shown in Fig. 14.13(c), has a vibrating reed or tongue, which modulates the flow of air passing through the shallot into the resonator. The reed is pressed against the open side of the shallot by a wire that can be adjusted up and down to tune the vibrating reed. The resonator is usually tuned to the same frequency as the reed. Reed pipes are rich in harmonics. If the resonator is cylindrical, the odd-numbered harmonics will be favored, as in a clarinet. Conical resonators, however, will reinforce all harmonics, both odd and even.

Figure 14.14 shows several different pipes and the names given to the corresponding organ stops. Resonators of several different shapes are shown, including open and closed cylinders, cones, inverted cones, rectangular cylinders, and closed pipes with chimneys. (The chimney is tuned to boost one of the upper harmonics, around the fifth or sixth.)

The *scale* of a rank of pipes refers to the ratio of diameter to length for the pipe of lowest pitch. Large-scale (large-diameter) pipes

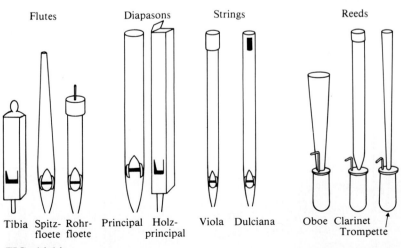

FIG. 14.14
Organ pipes of various families with the names of the corresponding stops.

tend to have a dominant fundamental and fewer harmonics, whereas small-scale pipes have more harmonics. The reason for this is related to a frequency dependence of the end correction of a pipe. For an open cylindrical pipe, the end correction at the open end is approximately 0.6 times the radius (see Section 4.5), whereas at the mouth of a flue pipe, one adds approximately 2.7 times the radius to calculate the effective length (Strong and Plitnik, 1977). The end correction decreases with frequency, so the pipe effectively shortens for the higher partials; thus the pipe resonances are slightly less than an octave apart. However, the spectrum of the sound source (the oscillating jet) has exact harmonics. Thus in a large-scale pipe, only the first few harmonics in the source are reinforced by the natural frequencies of the pipe. For pipes of small scale, such as the strings, the end correction is much smaller to start with; thus the pipe resonances more nearly match the harmonics of the oscillating jet.

Each rank of pipes is usually graduated in diameter according to a fixed relationship for that rank. For large-scale pipes, the diameter may reduce to one-half at the seventeenth note of the scale; for small-scale pipes, the diameter will reduce at a slower rate than this.

Normally, the mouth width, lip cut-up, and width of the flue opening follow the same scale as the pipe diameter. The mouth width, for example, may vary from 0.6 times the pipe diameter (large-scale flute) to 0.8 times the diameter (string or diapason). Similarly, the cut-up may vary from 0.25 times the mouth width (diapason) to 0.4 times the width (flute); the flue width is typically about 0.03 times the mouth width (Fletcher, 1977b).

The pitch of a rank of pipes is expressed as the approximate length of an open pipe having the same pitch as the lowest pipe in that rank. Thus the lowest pipe of an eight-foot rank sounds C_2 ($f = 65.4$ Hz), and this is sounded by the lowest key on the manual. The octave from C_4 (middle C) to C_5 then occupies the same central position that it does on the piano keyboard. A four-foot stop plays an octave higher, and a sixteen-foot stop an octave lower. Table 14.1 lists the lowest note and the harmonic number of various ranks.

TABLE 14.1   Pitch and harmonic number of pipe ranks

Equivalent length (in feet)	32	16	8	4	$2\frac{2}{3}$	2	$1\frac{3}{5}$	$1\frac{1}{3}$	$1\frac{1}{7}$	1
Harmonic number			1	2	3	4	5	6	7	8
Lowest note	C_0	C_1	C_2	C_3	G_3	C_4	E_4	G_4	$B\flat_4$	C5
"Organ" notation		CCC	CC	C	G	c^1	e^1	g^1	$\flat\flat^1$	c^2
Lowest pitch (in Hz)	16	33	65	131	196	261	330	392	466	523

14.11   SOUND GENERATION IN FLUE PIPES

When air is admitted at the foot of a flue organ pipe, it flows upward and forms a sheetlike jet as it emerges from the flue (see Fig. 14.13). The jet flows across the mouth of the pipe and strikes the upper lip, where it interacts both with the lip itself and with the air in the pipe resonator. The physics of this interaction, which leads to steady oscillation in the pipe, is rather complicated, but it has been elegantly simplified in an article by Fletcher and Thwaites (1983), which we follow in this brief description. A more complete discussion can be found in Chapter 16 of Fletcher and Rossing (1989).

Jets can be described as either *laminar* (arranged in layers or streamlines) or *turbulent* (characterized by eddies or vortices). Jets in organ pipes are almost completely turbulent, which makes it quite difficult to describe them mathematically. Somewhat paradoxically, however, a fully turbulent jet in an organ pipe is more stable than a laminar one, and organ builders go to some lengths to ensure fully turbulent jets by cutting fine nicks along the edge of the languid, for example.

A fully turbulent jet mixes gradually with the surrounding air, thus slowing down and broadening out in a rather simple way. The jet velocity, which is maximum at the center, falls off in a bell-shaped curve whose width increases linearly with distance from the flue. Conservation of flow momentum requires that the velocity decrease with the square root of the distance from the flue slit.

Assuming that standing waves already exist in the pipe resonator, the air motion at the mouth associated with these standing waves will be essentially at right angles to the jet. This sets up a wavelike disturbance in the jet (compare Fig. 12.19) which propagates along the jet at a little less than half the speed of the air at the center of the jet. The amplitude of this wavelike disturbance grows nearly at an exponential rate along the jet, typically doubling during each millimeter of travel. The growth rate is found to be greatest when the wavelength along the jet is about six times the width of the jet at that location. On the other hand, when the length of the wave is less than the width of the jet, it does not grow at all, so only long wavelengths can propagate with large amplitude along long jets.

The large acoustic standing waves in the pipe drive the tip of the jet alternately inside and outside the upper lip. If the puffs of air enter the pipe in step with the standing waves, energy will be added. The important relationship between the acoustical air motion in the pipe mouth and the time the air pulse arrives inside the upper lip is determined by the time required for a wave on the jet to travel from the flue slit to the upper lip. This is, in turn, determined by the mouth height ("cut-up") and the blowing pressure.

Two great 19th century physicists, Hermann von Helmholtz and Lord Rayleigh, reached opposite conclusions about how the air pulses interact with the air column and thus about the required phase relationship between sound waves in the pipe resonator and wavelike disturbances in the jet. Helmholtz thought that the most important factor was the flow volume and thus the puffs should enter the pipe at times when the acoustic pressure inside the upper lip is maximum. Rayleigh, on the other hand, thought that the jet should enter the pipe when the acoustic flow, rather than the pressure, is maximum. This is analogous to giving a child in a swing a push when the swing is at the top of its arc (Helmholtz) or at the bottom of its arc (Rayleigh).

Both Helmholtz and Rayleigh were partly correct. The Helmholtz mechanism dominates for low blowing pressures and high frequencies, while the Rayleigh mechanism dominates at high blowing pressures and low frequencies. In most organ pipes, the Helmholtz mechanism is the more important.

14.12   TUNING AND VOICING ORGAN PIPES

There are several ways of tuning different types of organ pipes. Reed pipes are tuned by moving the tuning wire up and down, thus allowing a longer or shorter length of the reed to vibrate. Flue pipes, on the other hand, are tuned by changing the effective length of the pipe in some way. In the case of a closed pipe, this is accomplished by moving the stopper up or down to change the pipe length. Many open pipes have a tuning sleeve that slides up and down; others have an adjustable slot near the open end. Open pipes without such a tuning device can be tuned by the use of tuning cones. To raise the pitch, the apex of the cone is inserted and tapped to widen the end of the pipe slightly. To lower the pitch, the cup end of the cone is placed over the pipe and tapped to close the end slightly.

One of the most critical of all the steps in organ building is *voicing* the pipes, which means making coarse and fine adjustments in the various parts of the pipe so that it "speaks" properly. Much of the voicing can be done in the organ-builder's shop, but the final voicing or finishing is done after the organ is installed, and it takes into account the acoustics of the room, as well as the acoustics of the organ. One of the objectives of voicing is to achieve a uniformity of loudness and timbre within each rank of pipes. Another is to adjust the initial transient or attack of each pipe.

The attack time varies substantially from one rank of pipes to another and also from large to small pipes within a rank. In general, the number of cycles necessary to build the tone to its steady state remains about the same within a given rank, so the attack time doubles in going one octave down the scale. Reed pipes have shorter attack times than do flue pipes of the same pitch. In many pipes, the attack time is shorter for the upper harmonics than for the fundamental, so the initial sound is richer in harmonic content than the steady state. (Figure 7.15 shows a diapason pipe in which the second harmonic builds up rapidly.) Octave or mixture ranks are often coupled to slow-speaking pedal flue pipes to decrease the attack time. Also important during the attacks are certain characteristic sounds, such as *chiff.* Figs. 7.15, 7.16, and 7.17 show how the timbre of various organ pipes changes during the attack.

The main parameters that are adjusted during voicing of flue pipes are (Mercer, 1951):

1. *The size of the foot bore.* An increase in bore diameter increases the air flow and produces greater power and greater harmonic content.

2. *The condition of the bore.* Rounding the edges of the bore can change the tone.

3. *Nicking.* Nicks on the languid or the lower lip tend to remove chiff and other extraneous sounds during the attack.

4. *The width of the flue.* This affects both the attack and steady state in a way that is not well understood.

5. *Obstruction near the mouth.* A bar or "roller beard" placed between the ears is sometimes necessary to prevent a pipe of high pressure or small scale from jumping up an octave in pitch.

6. *The height of the languid.* This controls mainly the articulation or attack; lowering the languid speeds up the attack.

7. *The height of the mouth or cut-up.* Increasing the cut-up reduces the harmomic content of the tone.

8. *Setting of the upper lip.* Moving the upper lip farther in increases the attack time.

9. *The condition of the upper lip.* Bevelling the upper lip increases the harmonic content of the tone, whereas rounding it decreases harmonic content.

These voicing adjustments tend to be interrelated. For example, increasing the cut-up reduces the harmonic content, whereas increasing the bore diameter produces greater harmonic development and more power. Thus, by combining the two, it is possible to obtain approximately the same tone at increased power.

Laboratory studies (Nolle, 1979; Fletcher and Douglas, 1980) of the effects of mouth height (cut-up) and mouth position (which relates to adjustments 6 and 8) show that:

1. A high cut-up causes the harmonic content to decrease, the ratio of high to low harmonics to increase, and the overall level to decrease.

2. A low cut-up causes the harmonic content to increase, the ratio of high to low harmonics to decrease, and parasitic oscillations due to the edgetone mechanism (see Section 30.4).

3. Within the optimum operating range, slightly increasing the mouth height favors a "ping" transient, whereas decreasing the height favors a "chiff" transient.

4. A frequency jump to a higher mode is encouraged by a high cut-up or an inward mouth position.

5. Adding ears to a pipe of small scale (diameter to length ratio) increases the range of mouth height and position over which stable oscillation occurs.

Voicing of reed pipes involves adjustments to the reed as well as to the resonator. One of the most important adjustments is in the curvature of the reed. A curved reed does not cut off the flow of air as abruptly as a straight reed does. Also, since tuning is done at the reed, changes in the resonator of a reed pipe will mainly adjust the loudness and timbre of the pipe independently of the pitch.

Many organs have one rank of pipes that is deliberately tuned sharp so that it will produce beats when played with another similar rank. The stop associated with this is called the *voix celeste*.

One or more windchests on an organ are usually equipped with a mechanical device called a *tremolo* or *tremulant,* which causes the air pressure to fluctuate at a regular rate and thus to produce a vibrato. Electronic organs are discussed in Chapters 26 and 28.

14.13 SUMMARY

The *piano,* with a playing range of over seven octaves and a wide dynamic range as well, has become the most versatile and popular of all musical instruments. The struck strings transfer vibrational energy through the bridge to the soundboard, which radiates most of the sound. Piano strings have "stretched" partials because of their stiffness; this makes stretch tuning desirable. Most notes on the piano have three strings, and their tuning in relation to each other affects the decay rate as well as the timbre of the note.

The *clavichord,* which also uses struck strings, is a portable instrument with a soft, delicate tone. The strings are struck by tangents, which stay in contact with the strings after striking. The *harpsichord,* which plucks the strings, is experiencing a revival in popularity. It is strung to a much lower tension than the piano, and has a much lighter soundboard. Large instruments have several sets of strings or stops, which may play in different octaves. The strings of a *harp* are plucked by the player rather than by mechanical plectra (as in the harpsichord). Harp strings are tuned to the diatonic scale; to play sharps or flats, the pitch of a string is raised or lowered a semitone by pedals.

The *pipe organ* is the largest of all instruments and has an extremely wide range of tone and dynamics. Its many pipes are organized into ranks and divisions. The main families of pipes (in order of increasing harmonic content) are the flutes, diapasons, strings, and reeds. Flutes, diapasons, and strings use flue pipes with a sounding mechanism similar to that of the flute and the recorder. Reed pipes have a metal reed that modulates the air flow to produce sound. Organ pipe resonators may be cylindrical, conical, or rectangular, with open or closed ends, and made of metal or wood. The pitch of a rank of pipes is denoted by the length of an open pipe sounding the same pitch as the lowest pipe in the rank. Pitch designations range from 1 foot to 32 feet. Voicing of the individual pipes is a very important step in organ building; voicing determines both the initial attack and the steady-state timbre.

References and Suggested Readings

Askenfelt, A., and E. Jansson (1988). "From Touch to String Vibrations—the Initial Course of Piano Tone," Report STL/QPSR 1/88, 31-109, Royal Inst. Technology (KTH), Stockholm.

Blackham, E. D. (1965). "The Physics of the Piano," *Sci. Am.* **99**(6): 88.

Fletcher, H. (1964). "Normal Vibration Frequencies of a Stiff Piano String," *J. Acoust. Soc. Am.* **36**: 203.

Fletcher, H., E. D. Blackham, and R. Stratton (1962). "Quality of Piano Tones," *J. Acoust. Soc. Am.* **34**: 749.

Fletcher, N. H. (1974). "Nonlinear Interactions in Organ Flue Pipes," *J. Acoust. Soc. Am.* **56**: 645.

Fletcher, N. H. (1977a). "Analysis of the Design and Performance of Harpsichords," *Acustica* **37**: 139.

Fletcher, N. H. (1977b). "Scaling Rules for Organ Flue Pipe Ranks," *Acustica* **37**: 133.

Fletcher, N. H., and L. M. Douglas (1980). "Harmonic Generation in Organ Pipes, Recorders, and Flutes," *J. Acoust. Soc. Am.* **68**: 767.

Fletcher, N. H., and T. D. Rossing (1989). *The Physics of Musical Instruments.* New York: Springer-Verlag. Chapters 11, 12, 16, and 17.

Fletcher, N. H., and S. Thwaites (1983). "The Physics of Organ Pipes," *Sci. Am.* **248** (1): 94.

Hall, D. E. (1986). "Piano String Excitation in the Case of Small Hammer Mass," *J. Acoust. Soc. Am.* **79**: 141.

Hall, D. E. (1987). "Piano String Excitation II: General Solution for a Hard Narrow Hammer, and III: General Solutions for a Soft Narrow Hammer,'' *J. Acoust. Soc. Am.* **81:** 535 and 547.

Helmholtz, H. L. F. (1877). *On the Sensations of Tone,* 4th ed. Trans., A. J. Ellis. New York: Dover, 1954.

Houtsma, A. J. M., T. D. Rossing, and W. M. Wagenaars (1987). *Auditory Demonstrations* (Phillips Compact Disc #1126–061 and text). Woodbury, N.Y.: Acoustical Society of America.

Kirk, R. E. (1959). ''Tuning Preferences for Piano Unison Groups,'' *J. Acoust. Soc. Am.* **31:** 1644.

Klotz, H. (1961). *The Organ Handbook.* St. Louis: Concordia.

Martin, D. W., and W. D. Ward (1961). ''Subjective Evaluation of Musical Scale Temperament in Pianos,'' *J. Acoust. Soc. Am.* **33:** 582.

Mercer, D. M. A. (1951). ''The Voicing of Organ Flue Pipes,'' *J. Acoust. Soc. Am.* **23:** 45.

Nolle, A. W. (1979). ''Some Voicing Adjustments of Flue Organ Pipes,'' *J. Acoust. Soc. Am.* **66:** 1612.

Pollard, H. F. (1978). ''Loudness of Pipe Organ Sounds,'' I and II, *Acustica* **41:** 65 and 75.

Strong, W. J., and G. R. Plitnik (1983). *Music, Speech and High Fidelity,* 2nd ed. Provo, Utah: Soundprint. Chapters 35, 37, and 38.

Thwaites, S., and N. H. Fletcher (1981). ''Some Notes on the Clavichord,'' *J. Acoust. Soc. Am.* **69:** 1476.

Weinreich, G. (1977). ''Coupled Piano Strings,'' *J. Acoust. Soc. Am.* **62:** 1474.

Weinreich, G. (1979). ''The Coupled Motions of Piano Strings,'' *Sci. Am.* **240**(1): 118.

Glossary

aftersound Second portion of a sound decay having a longer decay time.

chiff A chirplike sound that occurs during attack, especially in flute pipes on an organ.

clavichord A small portable instrument that produces a soft, delicate sound by means of struck strings.

cut-up Height of the mouth opening in an organ pipe; distance from the lower lip to the upper lip.

edgetone The sound produced when an air jet encounters a sharp edge or wedge and oscillates back and forth, first passing on one side, then the other.

flue The narrow windway between the languid and lower lip in a flue pipe.

flue pipe An organ pipe that produces sound by means of a jet of air passing through the flue and striking the upper lip.

harpsichord A keyboard instrument in which strings are plucked by mechanical plectra.

inharmonicity The departure of the frequencies of partials from those of a harmonic series.

jack A device in piano and harpsichord actions that moves up and down when a key is pressed. In the piano, the jack sets the hammer in motion; in the harpsichord, it carries the plectrum.

labial A flue pipe.

laminar flow Fluid flow in which entire layers have the same velocity.

languid A plate that partially blocks an organ pipe and forms one side of the flue.

lingual A reed pipe.

plectrum The small tongue of quill, leather, or plastic that plucks the string of a harpsichord.

soundboard The wooden plate that radiates much of the sound in string instruments.

stretch tuning Tuning octaves slightly larger than a 2:1 ratio.

sustaining pedal Right hand pedal of a piano which raises all the dampers, allowing the strings to continue vibrating after the keys are released.

tangent The small metal square that strikes the string of a clavichord.

tremulo, tremulant A device on an organ that produces a vibrato, usually by varying the air pressure.

turbulent flow Fluid flow characterized by eddies and vortices; the flow velocity tends to vary randomly.

una corda pedal Pedal on grand pianos which shifts the entire action sideways, causing the treble hammers to strike only two of the three unison strings.

virginal A small plucked string instrument in which the strings run parallel to the keyboard.

voicing Adjusting organ pipes to have the desired sound.

voix celeste An organ stop that uses two ranks of pipes with slightly different tunings so that they produce beats.

windchest The important part of the organ that distributes air to selected pipes to make them sound.

Questions for Discussion

1. When a chord is played on a piano while the sustaining pedal is depressed, the tone sounds richer than the same chord played without the sustaining pedal. Can you explain why?

2. If a piano key is depressed slowly enough, the hammer fails to contact the string at all. Explain why.

3. Will the two notes shown in Fig. 14.15 sound exactly the same when played on a piano?

4. Can you think of any advantages a tracker action might have over a direct electric action in an organ? any disadvantages?

5. In small pipe organs, the sixteen-foot pipes are almost always stopped wooden pipes. Explain why.

FIG. 14.15

Problems

1. Suppose that all the strings of a piano were of the same material and also had the same diameter and tension. If the longest string (A_0) were 2 m in length, how long would the highest A-string (A_7) have to be? Is this practical?

2. Show that two unison strings, tuned two cents (0.12 percent) different and initially in phase, will fall out of phase after about 400 vibrations.

3. If a particular string on a harpsichord has one-third the diameter and one-tenth the tension of the corresponding string on a piano, which string will be longer? What will the ratio of lengths be? (See Sections 3.2 and 4.4; assume that both strings are steel.)

4. If the diameter of pipes in a particular rank is reduced by a factor of two every seventeen notes, show that the diameter reduces by four over a range of about three octaves.

5. A piano tuner finds that two of the strings tuned to C_4 give about one beat per second when sounded together. What is the ratio of their frequencies? Show that their pitches differ by about seven cents. (One cent is 1/100 of a semitone and corresponds to a frequency ratio of approximately 1.0006.)

6. By noting the weak partials in the C_1 spectrum in Fig. 14.5, estimate the fraction of the string length β at which the hammer strikes.

7. Calculate the acoustical lengths of open organ pipes tuned to C_0, C_1, and C_2 (frequencies are given in Table 9.2, also Table 14.1). Compare these to the "equivalent lengths" in Table 14.1.

It is difficult to overstate the importance of the human voice. Of all the members of the animal kingdom, we alone have the power of articulate speech. Speech is our chief means of communication. In addition, the human voice is our oldest musical instrument.

The human voice and the human ear are very well matched to each other. The ear has its maximum sensitivity in the frequency range from 1000 to 4000 Hz, and that is the range of frequency in which the resonances of the vocal tract occur. These resonances (called formants), we will learn, are the acoustical bases for all the vowel sounds and many of the consonants in speech and singing.

Since speech and singing are such closely related functions of the human voice, it is recommended that all three chapters (15-17) be considered together. It is possible for a singer to begin with Chapter 17, but frequent reference will probably need to be made to Chapters 15 and 16.

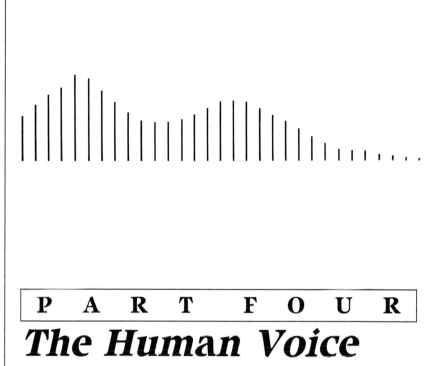

P A R T   F O U R
The Human Voice

CHAPTER **15**
Speech Production

Throughout human history, the principal mode of communication has been the spoken word. The systems in the human body that send and receive oral messages are sophisticated in design and complex in function. Our understanding of both hearing and speech has progressed dramatically in recent years, largely because of new techniques for making acoustical as well as physiological measurements. The auditory system and its function were discussed in Chapter 5, and now it is appropriate to devote the same attention to the vocal organs.

15.1 THE VOCAL ORGANS

The human vocal organs, as well as a representation of the main acoustical features, are shown in Fig. 15.1. The lungs serve as both a reservoir of air and an energy source. In speaking, as in exhaling, air is forced from the lungs through the larynx into the three main cavities of the vocal tract: the pharynx and the nasal and oral cavities. From the nasal and oral cavities, the air exits through the nose and mouth, respectively.

Air can be inhaled or exhaled with little generation of sound if desired. In order to produce speech sounds, the flow of air is interrupted by the vocal cords or by constrictions in the vocal tract (made with the tongue or lips, for example). The sounds from the interrupted

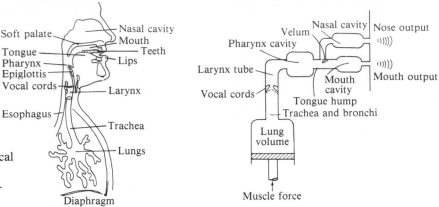

FIG. 15.1
Human vocal organs and a representation of their main acoustical features. (After Flanagan, 1965; reprinted by permission of Springer-Verlag.)

flow are appropriately modified by various cavities in the vocal tract and are eventually radiated as speech from the mouth and, in some cases, the nose.

15.2   THE LARYNX AND THE VOCAL CORDS

The most important sound source in the vocal system is the *larynx,* which contains the *vocal cords* or *vocal folds*. The larynx is constructed mainly of cartilages, several of which are shown in Fig. 15.2. One of the cartilages, the thyroid, forms the projection on the front of the neck known as the Adam's apple.

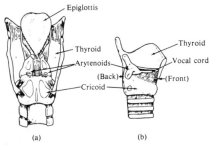

FIG. 15.2
Various views of the larynx: (a) back; (b) side.

The vocal cords are not at all like cords or strings, but consist rather of folds of ligament extending from the thyroid cartilage in the front to the arytenoid cartilages at the back. The arytenoid cartilages are movable and control the size of the V-shaped opening between the vocal cords, which is called the *glottis*. Figure 15.3 shows how the arytenoids control the size of the glottis. Normally, the arytenoids are positioned well apart from each other to permit breathing; however, they come together when sound is produced by the vocal cords.

The vocal cords may act on the air stream in several different ways during speech. From a completely closed position in which they cut off the flow of air, they may open suddenly as in a light cough or a glottal stop (such as the glottal "h" that occurs in Cockney English). A glottal stop may also give a hard beginning to a vowel sound, such as the word "Idiot!" expressed vehemently. On the other hand, the vocal cords may be completely open for unvoiced consonants such as "s," "sh," "f," etc. An intermediate position occurs in the "h" sound, where the air stream interacts lightly as it passes between the vocal cords.

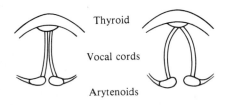

FIG. 15.3
Control of the glottal opening by the arytenoids.

The most useful function of the vocal cords, however, is to modulate the air flow by rapidly opening and closing. This rapid vibration produces a buzzing sound from which vowels and voiced consonants are created. These functions of the vocal cords are somewhat analogous to the functions of the lips. The glottal stop corresponds to the action of the lips in a plosive consonant such as p; the light friction of the cords in producing the ''h'' sound corresponds to the action of the lips in pronouncing ''f.'' The rapid vibration of the vocal cords is similar to the rolling noise made by a child's lips to imitate a motor, or the sound used to indicate coldness (''b'rrr''), or a trumpet player buzzing his or her lips in a practice exercise. You can feel the vibrations set up by the vocal cords by placing a finger lightly against your Adam's apple. Make sounds ''zzzzzz'' and ''sssss'' alternately to turn the vibrations on and off (these are examples of voiced and unvoiced consonants, respectively).

The rate of vibration of the vocal cords is determined primarily by their mass and tension, although the pressure and velocity of the air do contribute in a smaller way. The vocal cords are typically longer and heavier in the adult male than in the female and, therefore, vibrate at a lower frequency (pitch). During normal speech, the vibration rate may vary over a 2:1 ratio (one octave), although the range of a singer's voice is more than two octaves. Typical frequencies used in speech are 110 Hz in the male, 220 Hz in the female, and 300 Hz in the child, with wide variations from one individual to another.

Speech scientists describe three different modes in which the vocal cords can vibrate. In the normal mode, the vocal cords open and close completely during the cycle and generate puffs of air roughly triangular in shape when air flow is plotted against time. In the open phase mode, the cords do not close completely over their entire length, so the air flow does not go to zero. This produces a ''breathy'' voice, sometimes used to express shock (''No!'') or passion (''I love you''). A third mode, in which a minimum of air passes in short puffs, gives rise to a ''creaky'' voice, such as might result if you attempt to talk while lifting a heavy weight. A fourth mode, called *head voice* or *falsetto*, is normally not used in speech, and will be discussed in Section 17.5.

The vocal cords are caused to open by air pressure in the trachea, which tends to blow them upward and outward. As the air velocity increases, the pressure decreases between them, and they are pulled back together by the Bernoulli force (see Section 11.3). Ordinarily, however, the restoring force supplied by the muscles exceeds the Bernoulli force. Feedback from the vocal tract has relatively little influence on the vocal cord vibrations (as compared with the close

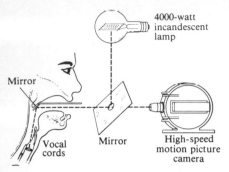

FIG. 15.4
Technique for high-speed motion
picture photography of the vocal
cords. (From Flanagan, 1965; reprinted
by permission of Springer-Verlag).

cooperation between the air column in a brass instrument and the player's lips, for example; see Section 11.2).

Another use of the vocal cords is in the production of a whisper. For a quiet whisper, the vocal cords are in much the same position as they are for an "h" sound. For louder whispers, the cords are brought closer so as to interfere more strongly with the flow of air. Say the word "hat" in a loud whisper and note the different rate of air flow during the "h" and the "a" by holding your hand in front of your mouth.

Just as the frequency of vibration determines pitch, the amplitude of vibration of the vocal cords relates to the loudness. Loud sounds require extra air pressure from the lungs and a large amplitude of vibration of the vocal cords. Lung pressures vary from about 400 to 2000 N/m² (4 to 20 cm) of water during speech, which represents about 0.4 to 2 percent increase above atmospheric pressure.

The vocal cords can be observed by placing a small dental mirror far back in the mouth. Using this technique, several investigators have taken high-speed motion pictures (4000 frames/second) of the vocal cords in vibration. Figure 15.4 illustrates this technique, and Fig. 15.5 shows one cycle of vocal cord vibration at a frequency of about 125Hz. Less obtrusive fiber optic probes inserted into the throat allow continuous observation of the vocal cords.

The flow of air through the glottis is roughly (though not exactly) proportional to the area of the glottal opening. For normal vocal effort, the waveform of the air flow is roughly triangular in shape with a duty factor (that is, the ratio of time open to the total period of a vibration) of 30 to 70 percent, as shown in Fig. 15.6. The sound resulting from this interrupted air flow is characterized as a "buzz" and is rich in overtones. A triangular waveform is composed of har-

FIG. 15.5
Successive phases in one cycle of vocal
cord vibration. The total elapsed time
is approximately 8 ms. (From
Flanagan, 1965; reprinted by
permission of Springer-Verlag).

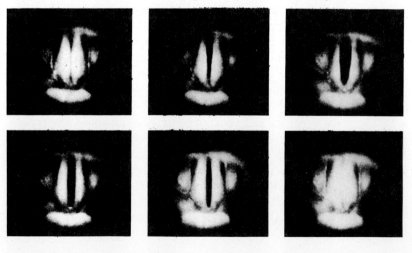

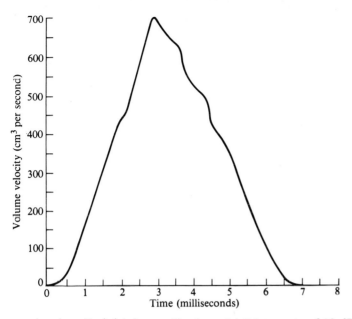

FIG. 15.6
The variation of air flow in a glottal puff. The curve repeats once every 8 ms (a frequency of 125 Hz).

monics that diminish in amplitude as $1/n^2$ (at a rate of 12 dB/octave), and the sound spectrum of the output of the larynx shows approximately this character for the higher harmonics, as seen in Fig. 15.7.

Although the vocal cords serve as the principal source of sound in speech, other sources are used, especially in the production of unvoiced consonant sounds. Sounds such as "f," "th," "s," "sh," (fricative consonants) and "l" are produced by a turbulent flow of air through a constriction somewhere in the vocal tract. The spectrum of such turbulence is quite broad, with many overtones that are not harmonic. Another source of sound is generated by a sudden release of pressure, such as that used in the plosive consonants p, t, and k. The sounds of consonants will be discussed in Section 15.4.

15.3 THE VOCAL TRACT

The function of the vocal tract is a most remarkable one; it transforms the "buzzes" and "whooshes" from the vocal cords and other sources into the intricate, subtle sounds of speech. This demanding function is accomplished by changes in shape to produce various acoustic resonances. Intensive studies in recent years have produced a great deal of information concerning the details of how this is accomplished. This information is the heart of a branch of science called *acoustical phonetics.*

The vocal tract, as shown in Fig. 15.1, can be considered a single tube extending from the vocal cords to the lips, with a side branch

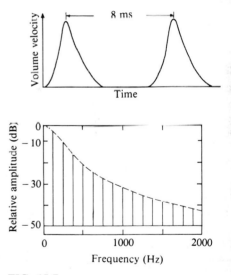

FIG. 15.7
A typical waveform of the volume velocity of the glottal output for a fundamental frequency of 125 Hz, and a Fourier spectrum corresponding to this type of waveform. (From Stevens and House, 1961, *Journal of Speech and Hearing Research,* 4, 303–320. Reprinted by permission of ASHA.)

leading to the nasal cavity. The length of the tube is typically about 17 cm, which can be varied slightly by raising or lowering the larynx and by shaping the lips. For the most part, however, the resonances in the vocal tract are tuned by changing its cross-sectional area at one or more points.

The *pharynx* connects the larynx with the oral cavity. It is not easily varied in shape, although its length can be changed slightly by raising or lowering the larynx at one end and the soft palate at the other end. The soft palate also acts as a valve to isolate or connect the nasal cavity to the pharynx. Since food also passes through the pharynx on its way to the esophagus, valves are necessary at the lower end to prevent food from reaching the larynx and to isolate the esophagus acoustically from the vocal tract. The *epiglottis* serves as such a valve, with the "false vocal cords" at the top of the larynx serving as a backup in case some food gets past the epiglottis. The epiglottis, false vocal cords, and vocal cords are open during normal breathing but closed during swallowing, thus forming a triple barrier to protect the windpipe. The epiglottis and false vocal cords do not appear to play any significant role in the production of speech.

The *nasal cavity* has fixed dimensions and shape, so that it is virtually untunable. In the adult male, the cavity has a length of about 12 cm and a volume on the order of 60 cm^3. The soft palate serves as a valve to control the flow of air from the pharynx into the nasal cavity. If the soft palate is lowered, air and sound waves flow into the nasal cavity and a "nasal" effect results from resonance within the nasal cavity. If, at the same time, flow through the mouth is blocked, air and sound exit through the nose, and humming results. Nasalized vowel sounds, which are common in French, are made by allowing sound to exit through both the mouth and the nose.

You can observe the soft palate in a mirror as it moves up and down (closing the nasal cavity in the up position). Say "ah" and the soft palate will rise; relax, and it will lower to normal breathing position. A cleft palate is a defect that allows air into the nasal cavity for all sounds, even those that should be entirely oral.

The *oral cavity*, or mouth, is probably the most important single part of the vocal tract because its size and shape can be varied by adjusting the relative positions of the palate, the tongue, the lips, and the teeth. The *tongue* is very flexible; its tip and edges can be moved independently or the entire tongue can move forward, backward, up and down. Movement of the lips, cheeks, and teeth also changes the size, shape, and acoustics of the oral cavity.

The lips control the size and shape of the mouth opening through which sound is radiated. Since the mouth opening is small compared to the wavelength of most components of the radiated sound, the size and shape of the opening are not of particular significance, except as they affect the all-important resonance frequencies of the oral cavity (this will be discussed further in Chapter 17). The mouth radiates more efficiently at higher frequencies where the wavelength approaches the size of the opening. In fact, a rise of 6 dB per octave in radiation efficiency is a good approximation to this effect.

The spectrum envelope of speech sound can be thought of as the product of three components:

Speech sound = Source × Filter function × Radiation efficiency.

If each of the above quantities is expressed in decibels, then the contributions are added, rather than multiplied. When the source consists of the vocal chords vibrating in their usual manner, the source function decreases in strength approximately 12 dB per octave (see Fig. 15.7). To this should be added the radiation efficiency of the mouth (which rises approximately 6 dB per octave) giving a net decrease of 6 dB per octave due to the first and last terms in the equation above. It remains to consider the more complicated way in which the filter function of the vocal tract varies with frequency, and that is the subject of Section 15.5.

15.4 ARTICULATION OF SPEECH

Before discussing the resonances of the vocal tract, it is appropriate to briefly describe the articulation of English speech sounds, or *phonemes*. In speech structure, one or more phonemes combine to form a syllable, and one or more syllables to form a word. Phonemes can be divided into two groups: vowels and consonants. Vowel sounds are always voiced; that is, they are produced with the vocal cords in vibration. Consonant sounds may be either voiced or unvoiced.

Various speech scientists list from 12 to 21 different vowel sounds used in the English language. This discrepancy in number comes about partly because of a difference of opinion as to what constitutes a pure vowel sound rather than a diphthong (a combination of two or more vowels into one phoneme). Table 15.1 lists the vowel sounds of Great American, the dialect of English spoken throughout most of western and midwestern United States. Also given are the corresponding symbols from the International Phonetic Alphabet (Denes and Pinson, 1973). Figure 15.8 shows the approximate tongue positions for articulating these vowels.

Whereas the vowel sounds are more or less steady for the duration of the phoneme, consonants involve very rapid, sometimes subtle, changes in sound. Thus consonants tend to be more difficult to analyze and to describe acoustically.

Consonants may be classified according to their *manner of articulation* as plosive, fricative, nasal, liquid, and semivowel. The

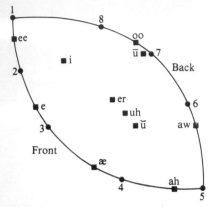

FIG. 15.8
Approximate tongue positions for articulating the vowels listed in Table 15.1. Number 1–8 are the eight cardinal vowels, which serve as a standard of comparison between languages.

TABLE 15.1   The vowels of Great American English

Pure vowels								Diphthongs		
ee	heat	/i/		aw	call	/ ɔ /		ou	tone	/ou/
i	hit	/ɪ/		u̇	put	/u/		ei	take	/eɪ/
e	head	/ɛ/		oo	cool	/u/		aɪ	might	/aɪ/
ae	had	/æ/		ŭ	ton	/ ʌ /		au	shout	/au/
uh	the	/ ə /		er	bird	/ ɜ /		oi	toil	/ɔɪ/
ah	father	/ ɑ /						ju	fuse	/ju/

plosive or stop consonants (p, b, t, etc.) are produced by blocking the flow of air somewhere in the vocal tract (usually in the mouth) and releasing the pressure rather suddenly. The *fricatives* (f, s, sh, etc.) are made by constricting the air flow to produce turbulence. The *nasals* (m, n, ng) are made by lowering the soft palate to connect the nasal cavity to the pharynx and then blocking the mouth cavity at some point along its length. The *semivowels* or glide consonants (w, y) are produced by keeping the vocal tract briefly in a vowel position and then changing it rapidly to the vowel sound that follows; thus, semivowels are always followed by a vowel. In sounding the *liquids* (r, l), the tip of the tongue is raised and the oral cavity is somewhat constricted.

Consonants are further classified according to their *place of articulation*, primarily the lips, the teeth, the gums, the palate, and the glottis. Terms used by speech scientists to denote place of articulation include *labial* (lips), *dental* (teeth), *alveolar* (gums), *palatal* (palate), *velar* (soft palate), *glottal* (glottis), and *labiodental* (lips and teeth). Finally, consonants are classified as to whether they are *voiced* or *unvoiced*.

TABLE 15.2   The classification of English consonants

Place of articulation	Manner of articulation						
	Plosive		Fricative		Nasal	Semivowel	Liquids
	Unvoiced	Voiced	Unvoiced	Voiced			
Labial (lips)	p	b			m	w	
Labiodental (lips and teeth)			f	v			
Dental (teeth)			th /θ/ (thin)	th / ð / (then)			
Alveolar (gums)	t	d	s	z	n	y/j/	l, r
Palatal (palate)			sh /ʃ/	zh/ ʒ /			
Velar (soft palate)	k	g			ng/ ŋ /		
Glottal (glottis)			h				

Phonetic symbols are given where they differ from the English letter.

Twenty-four consonants of English are thus classified in Table 15.2. Note the seven pairs of voiced/unvoiced consonants. In addition, the pair /tʃ, dʒ/, which refer to the "ch" (church) and "j" (judge) sounds, are sometimes included as separate consonants, although each of them consists of a plosive followed by a fricative (ch ≃ t + sh; j ≃ d + zh). Consonants are more independent of language and dialect than vowels are.

15.5   RESONANCES OF THE VOCAL TRACT: FORMANTS

The *vocal tract* consists of three main sections: the pharynx, the mouth, and the nasal cavity. These can be shaped by movements of other vocal organs, such as the tongue, the lips, and the soft palate (see Fig. 15.1).

Although the pitch and intensity of speech sounds are determined mainly by the vibrations of the vocal cords, the spectrum of these sounds is strongly shaped by the resonances of the vocal tract. It is the character of these resonances that distinguishes one phoneme from another.

The peaks that occur in the sound spectra of the vowels, independent of the pitch, are called *formants*. They appear as envelopes that modify the amplitudes of the various harmonics of the source sound. Each formant corresponds to one or more resonances in the vocal tract. Formant frequencies are virtually independent of the source spectrum.

Figure 15.9 illustrates the effect of formants on the source sound from the larynx. Both the waveform and the spectrum of the source sound are shown along with the waveform and the spectrum of the transmitted speech sound. (Note that in the waveform graphs, the horizontal axis is time; in the spectra, the horizontal axis is frequency.)

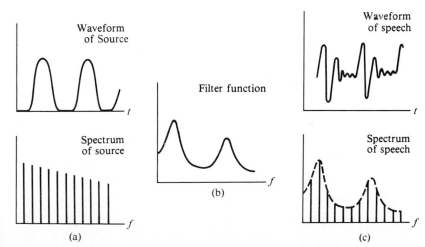

FIG. 15.9

The effect of formants on sound: (a) waveform and spectrum of source sound; (b) filter function showing two formants (resonances); (c) waveform and spectrum of transmitted sound.

t = time

f = frequency

TABLE 15.3 Formant frequencies and amplitude of vowels averaged for 76 speakers

Formant frequencies (Hz)		/i/ (ee)	/I/ (i)	/ε/ (e)	/æ/ (ae)	/ɑ/ (ah)	/ɔ/ (aw)	/ʊ/ (ú)	/u/ (oo)	/ʌ/ (u)	/ɜ/ (er)
F_1	M	270	390	530	660	730	570	440	300	640	490
	W	310	430	610	860	850	590	470	370	760	500
	Ch	370	530	690	1010	1030	680	560	430	850	560
F_2	M	2290	1990	1840	1720	1090	840	1020	870	1190	1350
	W	2790	2480	2330	2050	1220	920	1160	950	1400	1640
	Ch	3200	2730	2610	2320	1370	1060	1410	1170	1590	1820
F_3	M	3010	2550	2480	2410	2440	2410	2240	2240	2390	1690
	W	3310	3070	2990	2850	2810	2710	2680	2670	2780	1960
	Ch	3730	3600	3570	3320	3170	3180	3310	3260	3360	2160
Formant amplitudes (dB)		−4	−3	−2	−1	−1	0	−1	−3	−1	−5
		−24	−23	−17	−12	−5	−7	−12	−19	−10	−15
		−28	−27	−24	−22	−28	−34	−34	−43	−27	−20

Source: Peterson and Barney (1952).

Table 15.3 gives the average formant frequencies for the vowel sounds of men, women, and children. The first nine rows give the average frequencies of the first three formants. The last three rows indicate the relative strengths of the three formants for each vowel. For example, /ɑ/ (ah) has the strongest second formant, only 4 dB weaker than the first formant. For /i/ (ee), on the other hand, the second formant is 20 dB below the first.

15.6 MODELS OF THE VOCAL TRACT

Although the vocal tract, with its many curves and bends, is a rather complex acoustical system, simple models help us to understand the origin of the various formants or resonances.

The simplest acoustic model of the vocal tract is a pipe closed at one end (by the glottis) and open at the other end (the lips). Such a pipe has resonances (see Fig. 4.8) given by $f_1 = v/4L$, $f_3 = 3v/4/L$, $\ldots, f_n = nv/4L$ ($n = 1, 3, 5, \ldots$). For a pipe 17 cm long (the typical length of a vocal tract), the resonances occur at approximately 500, 1500, and 2500 Hz, which are surprisingly close to the peaks in the spectrum of the vowel sound /ε/ (typically at 500, 1800, and 2500 Hz).

Suppose we fasten a small loudspeaker to one end of a 17-cm pipe and place a microphone near the open end, as shown in Fig. 15.10. If the loudspeaker were driven by a pure tone (sine wave) of varying frequency, we would note strong resonances at about 500, 1500, and 2500 Hz, as we just discussed. If the loudspeaker were then driven with a sawtooth waveform (or some other waveform with many harmonics) having a frequency of 100 Hz, and the output of the

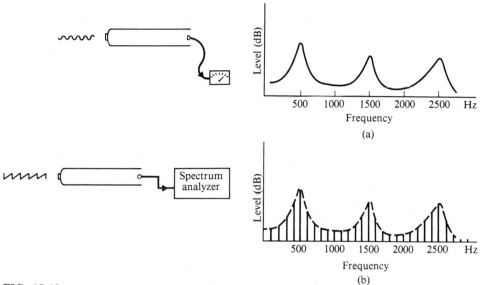

FIG. 15.10

Response of a closed-pipe model of the vocal tract: (a) resonances of a
17-cm pipe excited with a pure tone of varying frequency; (b) spectrum of a
100-Hz sawtooth wave shaped by the resonances (formants) of the pipe.

microphone were displayed on a spectrum analyzer, we would see
something similar to the spectrum shown in Fig. 15.10(b). The heights
of the various harmonics in the source spectrum have now been
shaped by the resonances of the pipe. Formants are present at 500,
1500, and 2500 Hz.

In addition to frequency, the parameters *amplitude* and *bandwidth* are used to describe a formant. The amplitude describes the
height of a resonance (see the last three rows in Table 15.3), and the
bandwidth describes its breadth.

Simple models for the vocal tract add to our understanding of
how various phonemes might be articulated. Models for the vowel
sounds /ɑ/, /i/, and /u/ ("ah," "ee," and "oo") using tubes of two
diameters are shown in Figs. 15.11, 15.12, and 15.13 (Stevens, 1972).

It is more difficult to construct simple models adequate to
describe the articulation of consonant sounds. Stevens (1972) shows,
however, that the simple constriction illustrated in Fig. 15.14, if
moved to different positions in the vocal tract, can approximate the
formants associated with several consonant sounds. The corresponding resonances are shown in Fig. 15.15. For the back portion of the
tube, which is essentially closed at both ends, the resonance frequencies are $f_b = nv/2l_b$. For the front portion, the frequencies are
$f_f = mv/4l_f$, with $m = 1, 3, 5, \ldots$.

(a)   (a)   (a)

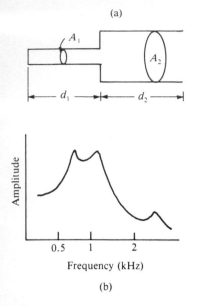

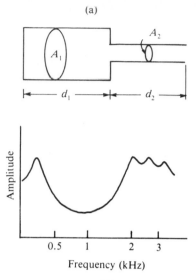

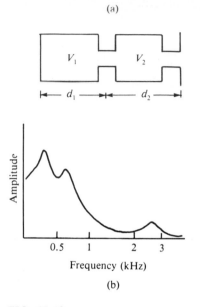

(b)   (b)   (b)

FIG. 15.11
(a) Two-tube approximation of vocal-tract configuration for the vowel /ɑ/ (ah). (b) Approximate form of spectrum envelope of vowel generated by the configuration in (a). (From Stevens, 1972. Reprinted by permission.)

FIG. 15.12
(a) Two-tube approximation of vocal-tract configuration for the vowel /i/ (ee). (b) Approximate form of spectrum envelope of vowel generated by the configuration in (a). (From Stevens, 1972. Reprinted by permission.)

FIG. 15.13
(a) Approximation of vocal-tract configuration for the vowel /u/ (oo). V_1 and V_2 represent the volumes of the two cavities. (b) Approximate form of spectrum envelope of vowel generated by the configuration in (a). (From Stevens, 1972. Reprinted by permission.)

FIG. 15.15
Relations between natural frequencies and the position of the constriction for the configuration shown in Fig. 15.14. The overall length of the tube is 16 cm and the length of the constriction is 3 cm. The dashed lines represent the lowest two resonances of the front cavity (anterior to the constriction); the solid lines

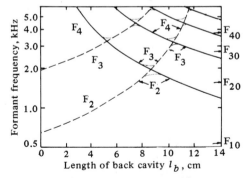

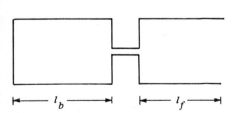

FIG. 15.14
Idealized model of constricted vocal-tract configuration corresponding to a consonant. The constriction is adjusted to different positions to represent different places of articulation. (From Stevens, 1972. Reprinted by permission.)

represent the lowest four resonances of the back cavity. The dotted lines near the points of coincidence of two resonances represent the resonant frequencies for the case in which there is a small amount of coupling between front and back cavities. The resonances of a 16-cm tube with no constriction are shown by the arrows at the right. The curves are labeled with the appropriate formant numbers. (From Stevens, 1972. Reprinted by permission.)

For a given value of the distance l_b from the glottis to the constriction, a vertical line can be drawn and the formant frequencies determined from the intersections with the various curves. Note that for $l_b < 8$ cm, the front cavity (dashed curve in Fig. 15.15) provides the lowest resonance, whereas for $l_b > 10$ cm, the back cavity (solid curve) does so. Near the crossover points, the resonances interact, as represented by the dotted curves. The first formant is not indicated on the graph, but the lowest resonance of the model would be a Helmholtz resonance whose frequency depends on the volume of the back cavity along with the length and cross section of the constriction.

Constriction positions corresponding to $l_b < 8$ cm represent configurations for uvular and pharyngeal consonants that do not ordinarily occur in English. Rather abruptly moving the constriction from left to right through the $F_2 - F_3$ crossover point ($l_b \approx 8.5$ cm) and at the same time enlarging the constriction toward the vowel configuration resembles the articulation of the velar consonants /g/ or /k/. Similar abrupt shifts near the $F_3 - F_4$ crossover point correspond to the fricatives /s/ and / ʃ / ("sh"). The labial and labiodental consonants would have $l_b > 13$ cm.

15.7   STUDIES OF THE VOCAL TRACT

A number of speech scientists have made X-ray photographs of the vocal tract during the production of speech sounds. From these photographs, profiles of the vocal tract can be constructed. It is interesting to compare the profiles for the vowel sounds, a, i, and u ("ah," "ee," "oo"), shown in Fig. 15.16, with the simple models shown in Figs. 15.11–15.13.

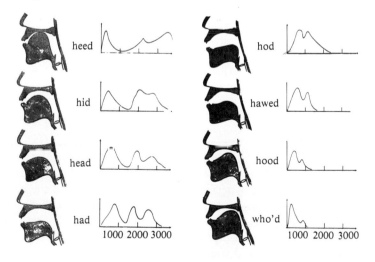

FIG. 15.16
The positions of the vocal organs (based on data from X-ray photographs of the author) and the spectra of the vowel sounds in the middle of the words *heed, hid, head, had, hod, hawed, hood, who'd.* Compare the sounds *hod, heed,* and *who'd* with the corresponding two-tube models of the vocal tract in Figs. 15.11, 15.12, and 15.13. (From P. Ladefoged (1962). *Elements of Acoustic Phonetics.* Published by the University of Chicago Press. Reprinted by permission.)

FIG. 15.17
Profiles of the vocal tract showing place of articulation of the stop or plosive consonants.

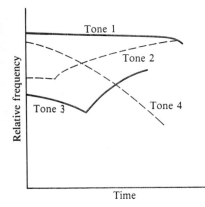

FIG. 15.18
Frequency changes with time for tones in Mandarin Chinese. (After Luchsinger and Arnold, 1965).

In the case of consonants, the profile of the vocal tract depends to some extent on the vowel sounds that precede and follow the consonant. The location of the constriction (the place of articulation) changes very little, however. Profiles for the stop or plosive consonants are shown in Fig. 15.17. Note that the voiced consonants are nearly identical to the corresponding unvoiced consonants. The only difference between the words "to" and "do," for example, is that the vocal cords vibrate during the /d/ sound, but do not begin vibrating until after the /t/ sound has been articulated.

15.8   PROSODIC FEATURES OF SPEECH

Prosodic features are characteristics of speech that convey meaning, emphasis, and emotion without actually changing the phonemes. They include pitch, rhythm, and accent. In English, prosodic features play a secondary role to that of phonemes in the communication of information.

In certain languages, such as Chinese, however, a phoneme can take on several different meanings depending on its "tone." The manner in which the frequency changes with time for the four tones of Mandarin Chinese is shown in Fig. 15.18.

One of the common uses of prosodic features is to change a declarative sentence ("You are going home.") into a question ("You are going home?"). This is done mainly by raising the pitch of the final word. The same sentence could be made imperative by adding stress (increase in both loudness and pitch) to the second word ("You *are* going home!").

Prosodic features tend to indicate the emotional state of the speaker. "Raising one's voice" in anger, for example, increases both loudness and pitch. A state of excitement frequently causes an increase in the rate of speaking. Several attempts have been made to accomplish acoustic "lie detection" by analyzing the prosodic features of recorded speech for evidence of stress.

15.9   DEMONSTRATION EXPERIMENTS

Formant Tube

The experiment illustrated in Fig. 15.10 can be expanded to simulate other vocal tract resonances, as shown in Fig 15.19. A pure tone is varied over the frequency range of 200–4000 Hz, and the resonances of the tube are observed. Listening to a sawtooth waveform with a fundamental frequency of 100–150 Hz should suggest vowellike sounds as the constriction is moved (b) or tubes of different diameters and lengths are joined together (c).

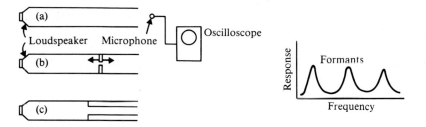

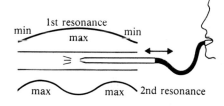

FIG. 15.19
Formant tube, which can be used to demonstrate resonances of the vocal tract.

Movable Constriction

The effect of constricting the air flow at different places in the vocal tract can be simulated by inserting a small nozzle into a short length of pipe, as shown in Fig. 15.20. As the constriction moves up and down the pipe, the sound changes in character, growing louder when the source reaches the position of a pressure maximum for one of the pipe resonances.

FIG. 15.20
Demonstration to show the effect of moving a constriction up and down the vocal tract.

Whispered Vowels

During a whisper, the vocal cords produce broadband ("white") noise, which contains a wide range of frequencies and virtually no sensation of pitch. Whispering vowel sounds, however, shapes the vocal tract so that bands of noise near the formant frequencies are emphasized. A rather faint sense of pitch develops, which usually corresponds to the second formant frequency (Thomas, 1969).

Professor J. F. Schouten of the Netherlands is well known for his demonstrations of acoustic phenomena as well as his research in hearing in speech. He demonstrates the whispered-vowel phenomenon by whispering the following four lines of vowels to produce the well-known Westminster chime:

$$\phi - I - \epsilon - a$$
$$\phi - \epsilon - I - \phi$$
$$I - \phi - \epsilon - a$$
$$a - \epsilon - I - \phi$$

The second formant of ϕ (a vowel sound common in Scandinavian and Germanic languages) is around 1760 Hz, which is two octaves above the musical standard $A_4 = 440$ Hz (Schouten, 1962).

Single Formants

The character of vowel sounds is mainly determined by the first and second formants. The waveform and spectrum /i/ are shown in Fig.

15.21(a). Filtering out one of the formants changes the vowel sound appreciably. The first formant, produced by itself, sounds much like /u/. The second formant by itself has a sharp timbre that produces no phonetic association, since the human voice is unable to produce it singly (Schouten, 1962).

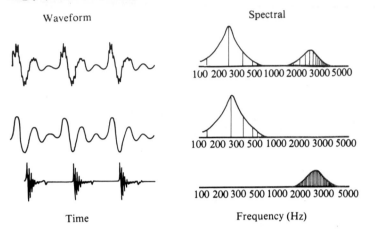

Waveform   Spectral

Time   Frequency (Hz)

FIG. 15.21
(a) Waveform of vowel /i/ and its spectral pattern showing two formants (b) First formant only gives an /u/-like sound. (c) Second formant only. (After Schouten, 1962).

Whistling

When one whistles, the vocal tract is excited from the end opposite the vocal cords. The normal whistling range is from about F_5 to F_7 (700 to 2800 Hz), although with practice many people can whistle down to 500 Hz and below. This suggests that the vocal tract acts somewhat like a closed cylindrical pipe (see Fig. 15.10). Whistled sound is very nearly a pure tone, with exceptionally weak overtones. Incidentally, the lowest whistled note for most people is near the pitch of their highest sung note (in falsetto for a male voice); thus, the human vocal system can emit sounds over a total range of about five octaves.

15.10   SUMMARY

The principal parts of the vocal tract are the larynx and vocal cords, pharynx, nasal cavity, oral cavity, tongue, lips, and teeth. Speech sounds originate with the vibrations of the vocal cords or with a constriction of the air flow, are filtered in the vocal tract, and finally are radiated through the lips or nose. Resonances of the vocal tract, called *formants*, determine the vowel sounds, the first and second formant being the most important. Consonants involve rapid changes in sound generated by changing a constriction somewhere in the vocal tract. Consonants can be classified according to their place and manner of articulation.

Simple models of the vocal tract, constructed from tubes of different lengths and diameters, help us understand the acoustical behavior of the vocal tract. Profiles of the vocal tract can be drawn from X-ray photographs. Prosodic features, such as pitch, rhythm, and accent, convey meaning, emphasis, and emotion.

References and Suggested Readings

David, E. E., Jr., and P. B. Denes, eds. (1972). *Human Communication: A Unified View.* New York: McGraw-Hill.

Denes, P. B., and E. N. Pinson (1973). *The Speech Chain.* New York: Anchor-Doubleday.

Fant, G. (1960). *Acoustic Theory of Speech Production.* The Hague: Mouton.

Flanagan, J. L. (1965). *Speech Analysis, Synthesis and Perception.* New York: Academic.

Ladefoged, P. (1962). *Elements of Acoustic Phonetics.* Chicago: University of Chicago Press.

Lehiste, I., ed. (1967). *Readings in Acoustic Phonetics.* Cambridge, Mass.: MIT Press.

Luchsinger, R., and G. E. Arnold (1965). *Voice-Speech-Language.* Belmont, Calif.: Wadsworth.

Peterson, G. E., and H. L. Barney (1952). "Control Methods Used in a Study of Vowels," *J. Acoust. Soc. Am.* **24**: 175.

Schouten, J. F. (1962). "On the Perception of Sound and Speech," Congress Report, 4th ICA, Copenhagen, 196.

Stevens, K. N. (1972). "The Quantal Nature of Speech: Evidence from Articulatory-Acoustic Data," in *Human Communication: A Unified View*, p. 51. Eds., E. E. David, Jr., and P. B. Denes. New York: McGraw-Hill.

Stevens, K. N., and A. S. House (1971). "An Acoustical Theory of Vowel Production and Some of its Implications," *J. Speech & Hearing Research* **4**(4): 75.

Strong, W. J, and G. R. Plitnik (1983). *Music, Speech and High Fidelity,* 2nd ed. Provo, Utah: Soundprint. Chaps. 23–27.

Sundberg, J. (1977). "The Acoustics of the Singing Voice," *Sci. Am.* **236**(3): 82.

Thomas, I. B. (1969). "Perceived Pitch of Whispered Vowels," *J. Acoust. Soc. Am.* **46**: 468.

Glossary

cardinal vowels Eight vowel sounds that serve as a standard of comparison for the vowels of various languages.

epiglottis A thin piece of cartilage that protects the glottis during swallowing.

filter A device that allows signals in a certain frequency range to pass and attenuates others.

formants Vocal tract resonances that determine speech sounds.

fricatives Consonants that are formed by constricting air flow in the vocal tract (such as f, v, s, z, th, sh, etc.).

glottis The V-shaped opening between the vocal cords.

larynx The section of the vocal tract, composed mainly of cartilage, that contains the vocal cords.

nasals Consonants that make use of resonance of the nasal cavity (m, n, ng).

palate The roof of the mouth.

pharynx Lower part of the vocal tract which connects the mouth to the trachea.

phonemes Individual units of sound that make up speech.

phonetics The study of speech sounds.

plosives Consonants that are produced by suddenly removing a constriction in the vocal tract (p, b, t, d, k, g).

prosodic feature A characteristic of speech, such as pitch, rhythm, and accent, that is used to convey meaning, emphasis, and emotion.

semivowels Consonants for which the vocal tract is formed in a configuration generally used for vowels (w, y).

vocal cords Folds of ligament extending across the larynx that interrupt the flow of air to produce sound.

Questions for Discussion

1. Discuss the function of each of the principal parts of the vocal tract.

2. If a person partially fills his or her lungs with helium and then speaks, the speech sounds distorted (it is sometimes described as sounding like Donald Duck). Explain this on the basis of formants (the information in Table 3.1 may be helpful). (This type of distortion will be discussed in Chapter 16.)

3. Discuss the acoustics of
   a) a "hoarse" throat;

b) a stuffed nose;
c) swollen tonsils.

4. Although the vibrations of the vocal cords are similar to the vibrations of a trumpeter's lips, the control of frequency by the air column through feedback is all but missing in the case of the vocal cords. Can you explain why? (Consider the mass of the vibrating members, the sharpness of the air resonances, and damping in each case.)

Problems

1. Calculate the first three resonances of a tube 11 cm long (the approximate length of a child's vocal tract) open at one end and closed at the other. Compare these to the formant frequencies for /ɛ/ given in Table 15.3.

2. Make a graph of the second formant frequency (vertical axis) vs. the first formant frequency (horizontal axis) for the ten vowel sounds given in Table 15.3. Do this for either the average male or female voice. Select a scale for each axis that is appropriate for the data you intend to graph.

3. Take a simple sentence (e.g., "You always give the right answers") and attempt to give it several meanings by changing prosodic features. For each different way of speaking the sentence, indicate the pattern of pitch and loudness used.

4. Express in newtons/meter² the maximum and minimum lung pressures used in speech (4 cm and 20 cm of water). Atmospheric pressure (10^5 N/m²) corresponds to a manometer pressure of about 34 feet of water.

5. Calculate the frequencies of resonance for a tube 16 cm long closed at one end and open at the other, and show that they correspond to F_{10}, F_{20}, F_{30}, in Fig. 15.15.

6. Determine whether there is a "scaling factor" relating male and female vowel formants by the following calculations. Determine the ratios of the female-to-male formant frequencies for the vowels given in Table 15.3. Find the average ratio. Could recording male speech and playing it back 16 percent faster make it resemble female speech?

7. Suppose a vocal tract 17 cm long were filled with helium ($v = 970$ m/s). What formant frequencies would occur in a "neutral" tract?

8. Estimate relative male and female vocal tract lengths by:

   a) Averaging the ratios of a male and female formant frequencies for several vowels;

   b) Assuming they have about the same ratio as male and female heights.

9. The resonance frequency of a Helmholtz resonator (Section 2.3) is

$$f = \frac{v}{2\pi} \sqrt{\frac{A}{lV}},$$

where v is the velocity of sound, A and l are the cross-sectional area and length of the neck, and V is the volume of the resonator. For the model in Fig. 15.14, assume a neck area of 0.6 cm², a length of 3 cm, a volume V of 20 cm³, and a sound velocity of 344 m/s, and calculate the resonance frequency. Compare this to F_{10}, calculated in Problem 5.

CHAPTER 16

Speech Recognition, Analysis, and Synthesis

Our ability to recognize the sounds of language is truly phenomenal. Speech can be followed at rates as high as 400 words per minute. If we assume an average of five phonemes or individual sounds per word, this means recognizing over 30 phonemes per second; even normal conversation requires the recognition of 10 to 15 phonemes per second. In this chapter, we will consider the way in which speech recognition takes place, particularly through certain types of *cues* in the complex speech sounds we hear. Before we consider speech recognition, however, it is appropriate to discuss the acoustical analysis of speech sounds.

16.1 THE ANALYSIS OF SPEECH

Some speech sounds change rapidly and, therefore, require special techniques for analysis. Graphs of sound level vs. time and graphs of sound level vs. frequency (sound spectra) are useful but inadequate. It is more useful to display the sound level as a function of both frequency and time. Various techniques have been used for creating such displays.

One way to display three variables is on a three-dimensional graph. When comparing sound level, frequency, and time, this can be

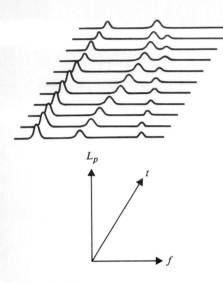

FIG. 16.1
Three-dimensional display of sound level vs. frequency and time.

accomplished by making multiple graphs of sound level vs. frequency, each one displaced slightly in time to create perspective. Such a three-dimensional display is illustrated in Fig. 16.1.

An instrument that rapidly analyzes the spectrum of sound is known as a *real-time spectrum analyzer*. One type, for example, traces out a new spectrum (that is, a graph of sound level vs. frequency) every 40 ms, so that 25 such spectra can be recorded per second. The display in Fig. 16.1 was created with a real-time spectrum analyzer of this type.

Digital computers are frequently used to create three-dimensional displays. The complex waveforms of speech sounds are converted into long sequences of numbers that the computer uses to calculate sound spectra. The spectra can be displayed on an oscilloscope screen or used to draw a three-dimensional graph by means of an $X-Y$ plotter.

An instrument that is particularly useful for speech analysis is a type of sound spectrograph originally developed at the Bell Laboratories around 1945, and which has appeared in several commercial versions. This instrument records a sound-level–frequency–time plot for a brief sample of speech on which the third dimension, sound level, is represented by the degree of blackness in a two-dimensional time-frequency graph.

A functional diagram of a sound spectrograph is shown in Fig. 16.2. First a sample of speech (generally from 2 to 2.5 s in length) is recorded on the magnetic disc. On the same shaft as the disc is a drum holding a piece of electrically sensitive paper. A stylus is in contact with the paper as it rotates rapidly. Electric current passing through the stylus blackens the paper in proportion to the sound level. A band-pass filter is tuned slowly, as the stylus moves up the paper, so frequency is recorded on the vertical axis (in actual practice, the filter frequency remains fixed and the signal spectrum "slides" past the filter with the same end result).

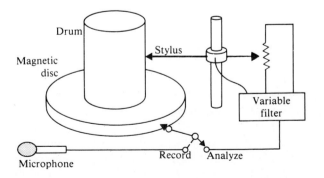

FIG. 16.2
A schematic diagram of a sound spectrograph.

A speech spectrogram is shown in Fig. 16.3. The horizontal axis is time, and the vertical axis frequency. The vertical striations show the fundamental period of the vocal cord vibrations. Two filter bandwidths are customarily used with the instrument, 45 and 300 Hz. The broader band gives better time resolution at the expense of frequency resolution. Because the speech sample must be played many times to record a spectrogram, playback is at a much higher speed than that used for recording on the magnetic disc.

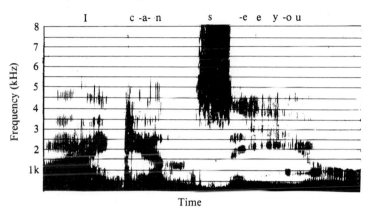

FIG. 16.3
A speech spectrogram. The vertical axis is frequency, and the horizontal axis is time. Sound level is indicated by darkness.

16.2 THE RECOGNITION OF VOWELS

Although spectrograms of vowel sounds may show four or five formants, the first two or three formants are generally sufficient to identify vowel sounds. On the other hand, experiments have shown that, under some conditions, vowels can be recognized from only the higher formants when the lowest two formants are missing. Thus, in normal speech, there are multiple acoustic cues to aid in the recognition of vowel sounds. Some of these extra cues make it possible to determine vowel sounds even when distortion and interference are present, as the following examples illustrate.

One familiar type of distortion occurs when speech is recorded at one speed and played back at a faster speed, producing what has been called "duck talk." Even when the pitch and all formants are raised by an octave or more (by changing phonograph speed from 33⅓ to 78 rpm, for example), it is possible to understand most of what is being said. Apparently our speech processing system is able to "scale" the entire structure of speech. That is, when we hear speech at a higher pitch, we also look for formants in a higher frequency range. How this is accomplished is not very well understood at the present time.

Another example of distortion caused by formant transposition is *helium speech*. The velocity of sound in pure helium is nearly three times greater than in air (see Table 3.1). If one takes a deep breath of helium, the resonances of the vocal tract (formants) will increase in frequency by some factor, which is typically 1.5 rather than 3, since our exhalation will contain a mixture of helium with nitrogen, carbon dioxide, etc. Speech produced under these conditions sounds quite similar to "duck talk" although the nature of the distortion is quite different. Analysis of helium speech indicates that the fundamental pitch is virtually unchanged, since the mixture of gas in the vocal tract has little effect on the vibration frequency of the vocal cords. To understand helium speech, then, we must recognize the raised formants even though the pitch of the vowel sounds corresponds to the normal formants we are accustomed to hearing.

In order to prevent nitrogen narcosis, deep-sea divers breathe mixtures of helium, nitrogen, and oxygen at high pressure, and speech becomes unnatural or even unintelligible. (In Sealab II, for example, the inside pressure was maintained at 6.8 times atmospheric and the gas mixture at 80 percent helium, 15 percent nitrogen, and 5 percent oxygen.) Thus the problem of helium speech has attracted considerable attention, and several experimental speech processors ("formant restorers") have been developed.

The vocal tracts of young children are considerably smaller than those of adults; hence, the formant structure is considerably different. The pharynx tends to be proportionately shorter than the oral cavity, so the formant configuration is not scaled to that of an adult. Yet our speech decoder enables us to recognize a vowel spoken by a young child as being the same vowel spoken by an adult. Some of the advantages of multiple speech cues become more apparent in light of these considerations.

16.3   THE RECOGNITION OF CONSONANTS

Unlike vowel sounds, which change slowly, consonant sounds change very rapidly. As we described in Section 15.4, consonants are articulated by constricting or blocking the flow of air somewhere in the vocal tract. The sound cues by which the consonant is recognized often occur in the first few milliseconds after the block is released and air is allowed to flow through the vocal tract.

Figure 16.4 shows a sound spectrogram of a simple phrase, "this is a sound spectrogram." The vowel formants appear as dark horizontal bars, whereas some of the up-and-down movements of these formants signal the consonants. The "s" sound is a burst of noise extending up to 8000 Hz. The fine vertical lines represent the vibrations of the vocal cords.

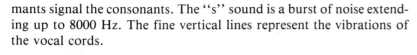

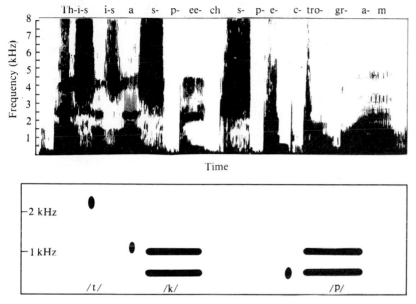

FIG. 16.4
A sound spectrogram of a spoken phrase. The vertical axis is frequency, and the horizontal axis is time.

FIG. 16.5
Stimulus patterns for producing /t/, /k/, and /p/ sounds on the pattern playback machine. A single burst of high-frequency noise is heard as /t/. A noise burst at the frequency of the second formant of a following vowel is heard as /k/; a noise burst below the second formant is heard as /p/. (From data in Cooper et al., 1952).

Experiments with the sound spectrograph have contributed a great deal to our understanding of the recognition of speech sounds, especially of consonants. In some experiments, certain features of speech are altered or eliminated to determine the intelligibility changes. Another type of experiment, however, generates speechlike sounds artificially. In such artificially synthesized speech, it is relatively easy to adjust separately the acoustic features to determine their effect on speech recognition.

Much early knowledge about the recognition of consonants was obtained with the pattern-playback machine built some years ago at the Haskins Laboratories. This machine works like a speech spectrograph in reverse: when a pattern similar to a spectrogram is fed in, it generates a sound with the designated intensity-frequency-time pattern. Arbitrary patterns may be painted on plastic belts in order to study the effects of varying the features of speech one by one.

A dot presented to the pattern-playback machine produces a "pop" that is like a plosive consonant, but difficult to recognize as any particular consonant unless it is followed by a vowel sound. In experiments by Cooper et al. (1952), listeners were presented with 15-ms noise bursts of varying frequencies, followed by two-formant vowel sounds, as shown in Fig. 16.5. High-frequency bursts were heard as

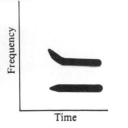

FIG. 16.6
A formant transition, which may produce a /t/, /p/, or /k/ depending on the vowel that follows.

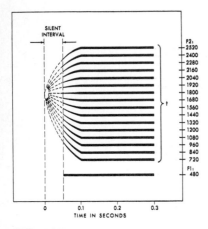

FIG. 16.7
Second-formant transitions perceived as the same plosive consonant "t." (After Delattre, Liberman, and Cooper, 1955).

FIG. 16.8
Spectrographic patterns sufficient for the synthesis of /b/, /d/, and /g/ before vowels. The dashed line at 1800 Hz shows the locus for /d/. (From Delattre, Liberman, and Cooper, 1955. Reprinted by permission of the Amer. Inst. of Physics.)

/t/ for all vowels, but bursts at lower frequencies could be heard as either /p/ or /k/, depending on the vowel sound that followed. Bursts were heard as /k/ when they were on a level with or slightly above the second formant of the vowel; otherwise, they were heard as /p/.

Another way to generate plosive consonants on the pattern-playback machine is by a frequency transition in the second formant, which may be upward or downward. Transitions of the second formant of the type shown in Fig. 16.6 will produce the unvoiced plosives /t/, /p/, or /k/, depending on the vowel formants that follow. A remarkable result emerged from experiments with the pattern-playback machine: All the second-formant transitions perceived as one particular plosive pointed back toward one particular frequency. The transitions in Fig. 16.7, which appear to originate from about 1800 Hz, are all heard as the sound /t/. Similarly, transitions that produce /p/ appear to originate from about 700 Hz, and /k/-producing transitions originate from about 3000 Hz.

The voiced plosives /b/, /d/, and /g/ have associated with their second-formant transition an upward transition in the first formant as well. The first formant is raised from a very low frequency to a level appropriate for the vowel. Figure 16.8 shows patterns that synthesize /b/, /d/, and /g/ sounds before various vowels. Note that although the first formant always moves upward, the second formant can move either upward or downward, depending on the vowel.

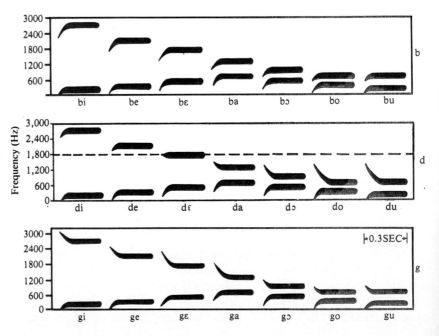

/d/ sounds

In the case of the consonant /d/, the second formants appear to originate from a "d-locus" at about 1800 Hz; the key to distinguishing the voiced /d/ from the unvoiced /t/, therefore, lies in the cue provided by the first-formant transition. It is interesting to note, however, that patterns extending all the way back to the d-locus, as in Fig. 16.9(a), do not always produce a clear /d/. In order to hear a /d/ sound in every case, it is necessary to erase the first part of the transition so that it "points" at the locus but does not actually begin there, as in Fig. 16.9(b). Presumably this is the way we are accustomed to receiving these cues, and major change confuses our speech decoder.

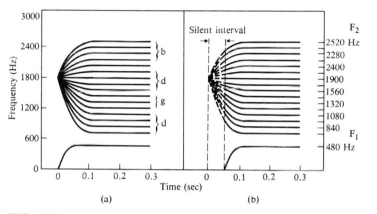

FIG. 16.9

(a) Second-formant transitions that start at the /d/ locus. (b) Comparable transitions that merely "point" at it, as indicated by the dotted lines. Those of (a) produce syllables beginning with /b/, /d/, or /g/, depending on the frequency level of the formant; those of (b) produce only syllables beginning with /d/. (From Delattre, Liberman, and Cooper, 1955).

For the liquids and semivowels /r/, /ℓ/, /w/, and /j/, the second-formant transition begins at the locus, although the exact character of the transition can vary with context.

Fricative consonants can be distinguished from all other sounds by the hissing noise of the turbulent air stream, which appears as a fuzzy area on speech spectra. We may ask, however, "What are the cues for distinguishing one fricative from another?" Experiments on both natural and synthesized speech have indicated that /s/ and /ʃ/

("sh") are distinguished from other fricatives by their greater intensities and from each other by their spectra. In the case of / ʃ /, sound energy is concentrated in the 2000- to 3000-Hz range, whereas for /s/, it is above 4000 Hz. The weaker fricatives /f/ and /θ/ ("th") depend on second-formant transitions in the vowel sound that follows to provide clues about the place of articulation.

The *duration* of a sound may provide an important clue for phoneme recognition. In Chapter 15 we pointed out that a fricative appeared to change into a plosive when its duration was shortened. This can be demonstrated by tape recording a normally pronounced word such as "see"; if the tape is cut to reduce the duration of the initial /s/ from its normal 0.1-s duration to about 0.01 s, the word is heard as "tee."

The effects of third-formant transitions on the perception of consonants are more complicated and less understood than those of second- and first-formant transitions. For example, a third-formant transition provides a clue for the perception of /d/ in "di" but not in "du" (Liberman et al., 1972). Apparently, no simple explanation has been made of this phenomenon, although it may be noted that the second-formant transitions are in opposite directions in the two cases.

16.4 FILTERED SPEECH AND NOISY ENVIRONMENTS

Filters are devices that respond selectively to certain frequencies. The vocal tract acts like a series of filters, each tuned to one of the resonances that we associate with formants; however, electrical filters can be constructed to have a much sharper frequency response than the vocal tract does. Some experiments with electrically filtered speech will be described in this section.

Electrical filters may have high-pass, low-pass, band-pass, or band-reject characteristics (see Section 18.4). A high-pass filter transmits only those frequencies above its cutoff frequency, and a low-pass filter only those frequencies below its cutoff frequency. A band-pass filter has both high and low cutoff frequencies and transmits only frequencies that lie in the band between; a band-reject or "notch" filter rejects only signals between the two cutoff frequencies.

Speech intelligibility is usually measured by *articulation tests* in which a set of words is spoken and a listener or group of listeners is asked to identify them. Articulation tests customarily use lists of specially selected words of one or two syllables (the best known are the test lists published by the Harvard Psychoacoustics Laboratory). The articulation score, which is the percentage of words correctly identified, will be lower for these isolated test words, of course, than for words used in the normal context of speech.

Articulation scores for filtered speech are shown in Fig. 16.10 for both high-pass and low-pass filtering. The curves are seen to cross at 1800 Hz, where the articulation score for both is about 67 percent. Normal conversation, therefore, would be completely intelligible by listening only to components above 1800 Hz, or, equally so, by listening only to components below 1800 Hz. It is also possible to achieve an acceptable level of intelligibility for speech after passage through a band-pass filter with a surprisingly narrow pass-band. The minimum acceptable pass-band is found to vary with frequency; in the range around 1500 Hz, for example, a 1000-Hz band width is sufficient to give a sentence articulation score of about 90 percent (Denes and Pinson, 1973). Using a narrowband filter (⅓-octave bandwidth), intelligibility reached 50 percent around 2000 Hz, but was very low at most frequencies (Chari, Herman, and Danhauer, 1977). Needless to say, speech quality deteriorates more than does intelligibility with filtering. Filtered speech sounds thin and unpleasant even when it is understandable.

The effects of waveform distortion have also been investigated. *Peak clipping* is a type of distortion that often results from overdriving an audio amplifier, but it is sometimes introduced deliberately into speech communication systems in order to reduce the bandwidth required to carry the speech. Figure 16.11 illustrates moderate and severe peak clipping of a speech waveform. Intelligibility of speech is impaired surprisingly little by peak clipping, although the quality of the speech suffers. Even after severe peak clipping, similar to that shown in Fig. 16.11(c), intelligibility remains at 50 to 90 percent, depending on the skill of the listener (Licklider and Pollack, 1948).

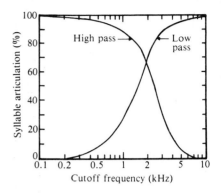

FIG. 16.10
Intelligibility of filtered speech for different cutoff frequencies of both high-pass and low-pass filters. Note that the two curves cross at about 1800 Hz, where the articulation score is 67% for both types of filter. (After French and Steinberg, 1947).

    (a)   (b)   (c)

FIG. 16.11
Peak clipping: (a) waveform of original speech; (b) waveform after peak clipping; (c) waveform after severe peak clipping.

The intelligibility of speech in noisy environments is a timely subject that has been studied at a number of laboratories. The degree of "masking" of speech depends on the intensity and the spectrum of the interfering noise. Using broadband noise or white noise (noise with equal intensity at every audible frequency), the intelligibility of words drops to about 50 percent when the average intensities of the speech and the noise are about equal. The intelligibility of sentences remains

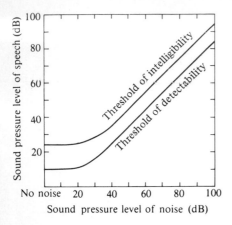

FIG. 16.12
The thresholds of intelligibility and detectability as functions of the intensity of the masking noise. (From J. E. Hawkins, Jr. and S. S. Stevens (1950). "The Masking of Pure Tones and Speech by White Noise," *J. Acoust. Soc. Am.* **19**: 90. Reprinted by permission of the Amer. Inst. of Physics.)

higher, however, because of linguistic and semantic cues. Figure 16.12 indicates how the thresholds of intelligibility and detectability of speech depend on the level of broadband noise.

A tone of lower frequency can mask a tone of higher frequency much more effectively than the converse (see Section 6.7). Thus narrowband noise is most effective in masking speech if its frequency is below the speech band. Potential for speech interference in a noisy environment is sometimes expressed by the speech-interference level, which is the average of the noise level in three appropriate frequency bands (see Section 31.5).

An interesting demonstration illustrating one property of speech interference may be performed using "elliptical speech." As an interfering noise source rises in intensity, one of the first features of speech that is lost is the place of articulation, so "cat" becomes indistinguishable from "tat" and "bed" from "dead," etc. Elliptical speech, in which such substitutions have been made, is difficult to understand under normal conditions, but as the noise level rises, the confusion gradually fades away and linguistic and semantic cues eventually make elliptical speech more understandable.

16.5 THE SYNTHESIS OF SPEECH

"If computers could speak, they could be given many useful tasks. The telephone on one's desk might then serve as a computer terminal, providing automatic access to such things as airline and hotel reservations, selective stock market quotations, inventory reports, medical data, and the balance in one's checking account" (Flanagan, 1972a). Providing computers with the ability to speak has been one target of a substantial amount of research on speech synthesis.

Early efforts to imitate speech sounds resulted in various mechanical "talking machines." One such machine, invented in 1791 by Wolfgang von Kempelen of Vienna and later improved by Sir Charles Wheatstone, is shown in Fig. 16.13. A bellows supplies air to a reed, which serves as the main voice source. A leather "vocal tract" is shaped by the fingers of one hand. Consonants, including nasals, are simulated by four constricted passages controlled by the fingers of the other hand.

During his boyhood in Scotland, Alexander Graham Bell had an opportunity to see the Wheatstone reconstruction of von Kempelen's talking machine. Assisted by his father and his brother, he constructed a talking machine of his own, molding the lips, tongue, palate, pharynx, and velum in guttapercha, wood, and rubber. A larynx box of tin had vocal cords made of rubber sheet. (Flanagan, 1979b).

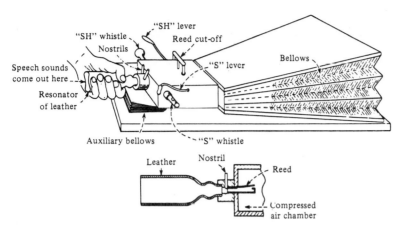

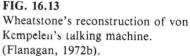

FIG. 16.13
Wheatstone's reconstruction of von
Kempelen's talking machine.
(Flanagan, 1972b).

More recent talking machines have employed electronic rather
than mechanical techniques. A combined speech analyzer and synthesizer is often called a *vocoder* (short for "voice coder"), a term first
applied by inventor Homer Dudley to his pioneering machine (Dudley,
1939). The simple plan of Dudley's machine, shown in Fig. 16.14, is
still the basic plan of most modern vocoders, although the "hardware" has become very much more sophisticated.

One promising application of vocoders is in speech bandwidth
compression systems. The range of frequencies necessary for high-
quality speech transmission is about 100 to 10,000 Hz; telephone-
quality speech contains only frequencies between about 200 and 3400
Hz. By using a vocoder with a speech analyzer on the sending end and
a speech synthesizer at the receiving end, only enough information to
reconstruct the speech would need to be transmitted, and this could
probably be done within a bandwidth of about 300 Hz. A *channel
vocoder* transmits information about the output signal in 16 filters
(plus a seventeenth channel with information about unvoiced consonants, fundamental frequency, etc.). A *resonance vocoder* or *formant
vocoder,* on the other hand, transmits information describing the formants themselves. These and other systems are described in a useful
book by Denes and Pinson (1973).

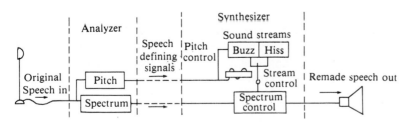

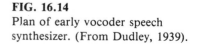

FIG. 16.14
Plan of early vocoder speech
synthesizer. (From Dudley, 1939).

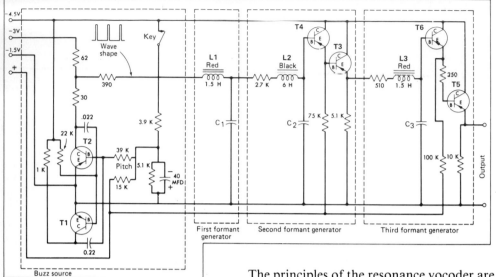

FIG. 16.15
A simple circuit for synthesizing vowel
sounds. (From Coker, Denes, and
Pinson, 1963).

The principles of the resonance vocoder are illustrated by a simple vowel synthesizer developed for the Bell System High School Science Program. The circuit, shown in Fig. 16.15, is capable of synthesizing vowel sounds. The *buzz source* generates a waveform that is rich in harmonics just as the output from the human vocal cords is. The output of the buzz source is filtered by three *formant generators,* whose resonance frequencies are tuned by changing capacitors C_1, C_2, and C_3. The manual accompanying this kit also described more sophisticated speech synthesizers (Coker, Denes, and Pinson, 1963).

16.6   TIME-DOMAIN VS. FREQUENCY-DOMAIN SYNTHESIS

Throughout our discussions of the perception of sound and of musical instruments, we have learned how frequency-domain analysis and time-domain analysis complement each other. The periodic motion of a vibrating system, for example, may be described either by its *waveform* in the time domain or by its *spectrum* in the frequency domain (see Section 7.10). Fourier analysis allows us to go from the time-domain description to the corresponding one in the frequency domain; Fourier synthesis is a means for returning from the frequency domain to the time domain. Another example of this complementarity was encountered in our discussion of pitch perception in Section 7.6, where we learned that time analysis gives us more accurate information about the pitch of low-frequency tones, whereas frequency analysis is more profitable for high-frequency tones.

The formant vocoder and the channel vocoder, discussed in Section 16.5, operate mainly in the frequency domain. They synthesize speech largely on the basis of its spectral content. Time-domain methods, on the other hand, attempt to synthesize speech waveforms. The most successful method for waveform synthesis has been a method called *linear predictive coding* (LPC), which describes a speech waveform in terms of a set of time-varying parameters (about 10 or 12) derived from analysis of speech samples (Atal and Hanauer, 1971). Most "talking chips" found in microcomputers make use of LPC in some way; likewise the remarkable Speak and Spell toy introduced by Texas Instruments in 1978 (Franz and Wiggins, 1982). A detailed description of LPC and other methods of speech synthesis is beyond the scope of this book, but the interested reader is referred to books by Morgan (1984), Lingard (1985) and others.

Although it is now possible to synthesize speech that carries a high degree of intelligibility, achieving a natural voice quality remains quite a challenge. All the characteristics of the human voice must be imitated (fluctuations in source pitch and intensity, fluctuation in formant frequencies, extraneous noises, and so on). Also people have special speaking habits, accents, and inflections that are difficult not only to analyze acoustically but also to reconstruct. Even sophisticated digital computer vocoders have limitations on the quality of sound achievable.

16.7 SPEECH RECOGNITION BY COMPUTERS

Designing a machine that understands language is more difficult than building one that talks. Human listeners have learned to accept a wide range of speech input, including different dialects, accents, voice inflections, and even speech of rather low quality from talking computers. Machines to recognize speech have not yet reached this degree of flexibility, however. Machines that can recognize a limited vocabulary from one speaker will have difficulty recognizing the same words from a different speaker.

Speech recognition may focus either on recognizing individual words or on recognizing connected words in a phrase or sentence. A common strategy for recognizing isolated words is template matching. Templates of appropriate time-varying parameters are created for the words in the desired vocabulary as spoken by selected speakers. These same parameters in a spoken word are then compared to the stored templates, and the closest match is assumed to be the word spoken. Isolated word recognition is practical for such tasks as digit recognition, recognizing simple computer commands, and machine control, but not for general communication.

Continuous speech recognition is much more difficult than isolated word recognition, because it is difficult to recognize the beginning and end of words, syllables, and phonemes. In natural speech, articulatory gestures are made quickly, so that each is modified, to a certain extent, by its neighbors in the spoken sequence. This modification, which can be quite considerable, is known as *coarticulation.* The degree of coarticulation will depend on the rapidity of speech and the mode of speech. Its effect, in some ways, is analogous to the difference between hand-printed letters and handwriting in which the letters are modified as they are connected together.

Much research effort has been devoted to machine recognition of speech, because the potential applications are many. Voice-controlled typewriters and word processors may soon be possible. Voice programming of computers, control of machines, telephone dialing, data entry for materials handling and sorting, financial transactions, etc., while leaving the hands free for other tasks, are of obvious benefit.

16.8   SPEAKER IDENTIFICATION BY SPEECH SPECTROGRAMS: VOICEPRINTS

Can one reliably identify a person by examining the spectrographic patterns of his or her speech? This is a question of considerable legal as well as scientific importance. The Technical Committee on Speech Communication of the Acoustical Society of America asked six distinguished speech scientists to review the matter from a scientific point of view a few years ago (Bolt, et al., 1970). They concluded that "the available results are inadequate to establish the reliability of voice identification by spectrograms." Speech spectrograms, or *voiceprints,* are not analogous to fingerprints, because they do not represent anatomical traits in a direct way. The article by Bolt et al. (1970) does, however, summarize methods used for speaker identification and their validity.

Speech spectrograms portray short-term variations in intensity and frequency in graphical form. Thus they give much useful information about speech articulation. When two persons speak the same word, their articulation is similar but not identical. Thus spectrograms of their speech will show similarities but also differences. However, there are also differences when the same speaker repeats a word, as can be seen in Fig. 16.16.

Our auditory system exhibits an amazing ability to identify speakers, especially if the voices are well known to us, even in the presence of substantial interference. However, wrong identifications are within the experience of all of us. Careful studies, in fact, have shown that listening provides more dependable identification of the speaker than the examination of spectrograms of the same utterances does (Stevens

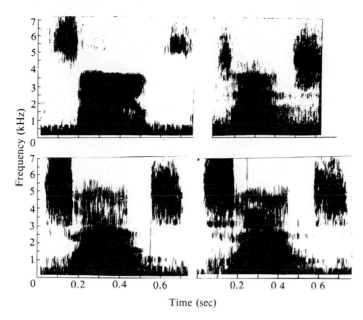

FIG. 16.16
Four spectrograms of the spoken word "science." The vertical scale represents frequency, the horizontal dimension is time, and darkness represents sound level. The two spectrograms at the left are by the same speaker. (Compare similar spectrograms in Bolt et al., 1970.)

et al., 1968). Work is being done on the design and evaluation of methods for objective voice identification using completely automatic procedures but, at this time at least, they do not inspire great confidence in the use of voiceprints for error-free speaker identification.

16.9 SUMMARY

To analyze speech, it is desirable to display sound level as a function of both frequency and time. This can be done on a three-dimensional graph or by a sound spectrograph.

The first two or three formants are usually sufficient for recognition of vowel sounds even in the presence of distortion or interference; The cues for consonant recognition often depend on the vowel sound that follows. The pattern-playback machine, which generates synthesized speech with specified features, has added much to our knowledge about consonant recognition. Filtering speech and masking speech with noise reduce intelligibility.

It is now possible to build machines that synthesize speech of acceptable quality, and machines that can recognize small vocabularies of words. Other machines can identify a speaker by his or her voiceprint, but not with a high degree of reliability. Future research and development will most likely lead to machines that can speak, understand speech, and even identify a speaker.

References

Atal, B. S., and S. L. Hanauer (1971). "Speech Analysis and Synthesis by Linear Predication of the Speech Wave," *J. Acoust. Soc. Am.* **50**: 637.

Bolt, R. H., F. S. Cooper, E. E. David, Jr., P. B. Denes, J. M. Pickett, and K. N. Stevens (1970). "Speaker Identification by Speech Spectrograms: A Scientists' View of Its Reliability for Legal Purposes," *J. Acoust. Soc. Am.* **47**: 369.

Chari, N. C., G. Herman, and J. L. Danhauer (1977). "Perception of One-Third Octave-Band Filtered Speech," *J. Acoust. Soc. Am.* **61**: 576.

Coker, C. H., P. B. Denes, and E. N. Pinson (1963). *Speech Synthesis.* Murray Hill, N.J.: Bell Telephone Laboratories.

Cooper, F. S., P. C. Delattre, A. M. Liberman, J. M. Borst, and L. J. Gerstman (1952). "Some Experiments on the Perception of Synthetic Speech Sounds," *J. Acoust. Soc. Am.* **24**: 597.

David, E. E., Jr., and P. B. Denes (1972). *Human Communication: A Unified View.* New York: McGraw-Hill.

Delattre, P. C., A. M. Liberman, and F. S. Cooper (1955). "Acoustic Loci and Transitional Cues for Consonants," *J. Acoust. Soc. Am.* **27**: 769.

Denes, P. B., and E. N. Pinson (1973). *The Speech Chain.* Garden City, N.Y.: Anchor/Doubleday.

Dudley, H. (1939). "Remaking Speech," *J. Acoust. Soc. Am.* **11**: 169. (This article describes the original vocoder build at the Bell Telephone Laboratories.)

Flanagan, J. L. (1972a). "The Synthesis of Speech," *Sci. Am.* **226**(2): 48.

Flanagan, J. L. (1972b). "Voices of Men and Machines," J. Acoust. Soc. Am. **51**: 1375.

Flanagan, J. L., and L. R. Rabiner (1973). *Speech Synthesis.* Stroudsburg, PA: Dowden, Hutchinson & Ross.

Franz, G. A., and R. H. Wiggins (1982). "Design Case History: Speak and Spell Learns to Talk," *IEEE Spectrum* **19**(4): 45.

French, N. R., and J. C. Steinberg (1947). "Factors Governing the Intelligibility of Speech Sounds," *J. Acoust. Soc. Am.* **19**: 90.

Hawkins, J. E., Jr., and S. S. Stevens (1950). "The Masking of Pure Tones and Speech by White Noise," *J. Acoust. Soc. Am.* **22**: 6.

Levinson, S. E., and M. Y. Liberman (1981). "Speech Recognition by Computer," *Sci. Am.* **244**(4): 64.

Liberman, A. M., F. S. Cooper, D. P. Shankweiler, and M. Studdert-Kennedy (1967). "Perception of the Speech Code," *Psych. Rev.* **74**: 431.

Licklider, J. C. R., and I. Pollack (1948). "Effects of Differentiation, Integration, and Infinite Peak Clipping Upon the Intelligibility of Speech," *J. Acoust. Soc. Am.* **20**: 42.

Lingard, R. (1985). *Electronic Synthesis of Speech.* Cambridge: Cambridge Univ. Press.

Morgan, N. (1984). *Talking Chips.* New York: McGraw-Hill.

Stevens, K. N., C. E. Williams, J. P. Carbonell, and B. Woods (1968). "Speaker Authentication and Identification: A Comparison of Spectrographic and Auditory Presentation of Speech Material," *J. Acoust. Soc. Am.* **44**: 1596.

Stover, W. R. (1966). "Technique for Correcting Helium Speech Distortion," *J. Acoust. Soc. Am.* **41**: 70.

Strong, W. J., and G. R. Plitnik (1983). *Music, Speech and High-Fidelity,* 2nd ed. Provo, Utah: Soundprint.

Glossary

coarticulation Modification of speech sounds when they are connected to other sounds in a spoken sequence.

cues Characteristics of speech sounds that help us to recognize them.

linear predictive coding (LPC) Describing a speech waveform in terms of a set of time-varying parameters derived from speech samples.

masking Obscuring of one sound by another (see Section 6.7)

peak clipping Limiting the amplitude of a waveform so that peaks in the waveform are eliminated; this distorts the waveform.

phonemes Individual units of sound that make up speech.

sound spectrograph An instrument that displays sound level as a function of frequency and time for a brief sample of speech.

spectrogram A graph of sound level vs. frequency and time as recorded on a sound spectrograph or similar instrument.

speech synthesis Creating speechlike sounds artificially.

vocoder A combined speech analyzer and synthesizer ("voice coder").

voiceprints Speech spectrograms from which a speaker's identity may be determined.

Questions for Discussion

1. When you fail to understand an indistinctly spoken word, is it more apt to be the initial consonant, the final consonant, or the vowel that is not recognized? Try to give reasons for your answer.

2. Discuss why a baritone voice played back at a higher speed than that at which it was recorded does not sound like a soprano.

3. Discuss the acoustics of the frequently heard phrase "he projects his voice." Think of other expressions used to describe good speaking techniques and their possible acoustical basis.

4. Would synthesized speech of high intelligibility but unnatural quality be useful in telecommunication? Would it be acceptable to most users of the telephone?

Problems

1. From the spacing of the small striations in the speech spectrogram shown in Fig. 16.3, estimate the fundamental frequency of the speaker. Can you tell whether the speaker was male or female? (The duration of the spectrogram is 1.9 s.)

2. Recognition of vowels requires frequencies from about 200 to 3000 Hz, whereas recognition of certain consonants requires frequencies up to 8000 Hz. While listening to a radio newscast, quickly turn down the treble tone control, and note which consonants are the most difficult to identify.

3. From Fig. 16.8 estimate the frequency change in the first and second formants during articulation of "di," "da," and "du." Estimate also the time over which these formant shifts take place.

4. Time your own speech rates when speaking normally and when speaking as fast as you can. Then count the number of words in a particular paragraph, and time yourself as you read it at each of these rates.

5. Listen to speech recorded at one rate and then played back at both a faster and a slower rate. Describe the speech you hear (quality, pitch, intelligibility, etc.). [Tape recorders usually have speeds of 3¾ and 7½ in/s; 15 in/s is used for high-quality recordings, and 1⅞ in/s is used for cassettes. Phonographs usually have 33⅓, 45, and 78 rpm (revolutions per minute); some turntables have 16 rpm as well.]

CHAPTER 17
Singing

It is somewhat ironic that the oldest musical instrument of all, the human voice, is less well understood than the various instruments we discussed in Part III. This is certainly due, in part, to the inaccessibility of its various components within the human body. Studying the human voice might be likened to studying the violin without being allowed to open the case or, at best, to hearing it played from behind an opaque screen with a small hole through which to peek.

The vocal organ, as shown schematically in Fig. 15.1, consists of the lungs, the larynx, the pharynx, the nose, and the mouth. Air from the lungs is forced through the glottis, a V-shaped opening between the vocal cords or folds, causing them to vibrate and thus modulate the flow of air through the larynx. The output from the vocal cords is characterized as a buzz (a nearly triangular waveform), rich in harmonics that diminish at a rate of about 12 dB per octave (see Fig. 15.7).

The vocal tract, which consists of the larynx, the pharynx, the oral cavity, and the nasal cavity, acts as a filter-resonator to transform this buzz into musical sound, somewhat in the manner of the tubing of a trumpet or oboe (but without a large amount of feedback to the source). Unlike the horns of the orchestra, however, the vocal tract creates its formants (resonances) mainly by changing its cross-sectional area at various points of articulation along its length. (The vocal tract was discussed in Section 15.3.)

17.1 FORMANTS AND PITCH

In both speech and singing, there is a division of labor, so to speak, between the vocal cords and the vocal tract. The vocal cords control the pitch of the sound, whereas the vocal tract determines the vowel sounds through its formants and also articulates the consonants. The pitch and the formant frequencies are virtually independent of each other in speech, but trained singers (especially sopranos) sometimes tune their vowel formants to match one or more harmonics of the sung pitch. The loudness and timbre of the sung sound depend on both the vocal cords and the vocal tract.

Figure 15.16 shows the vocal tract profiles for twelve English vowels, and Table 15.3 tabulates typical formant frequencies for ten of them as spoken by both male and female voices. In Fig. 17.1, these same data are presented on a musical staff for the reader who is more familiar with this notation. It may be surprising to learn that although female voices are pitched about an octave higher than male voices, the formants usually differ by less than a musical third (less than 25 percent in frequency).

FIG. 17.1
Typical formants of male (♩) and female (♪) speakers represented on a musical staff. (Compare with Table 15.3.)

In Table 15.3 the relative formant amplitudes are given. For most spoken vowels, the second formant is considerably weaker than the first; the "ah" and "aw" sounds have the strongest second formants. We have added dynamic markings (pp, p, mp, mf) to Fig. 17.1 to indicate the relative strengths of the second formant. Although the first and second formants contribute equally to vowel sounds (the third formant contributes slightly less), the first formant will usually contribute more to timbre because of its greater amplitude and lower frequency, closer to the fundamental.

Note the position of the formants in Fig. 17.1 in relation to the singing range. In the case of the bass or baritone singer, the fundamental rarely is enhanced by a formant resonance (exceptions are "ee" and "oo" sounds near the top of the singing range). In most cases, the formants enhance higher harmonics of the fundamental; for example, if a bass sings "ah" with a pitch G_2 ($f = 98$ Hz), the first formant gives its greatest boost to the seventh harmonic, the second formant boosts harmonics around the eleventh, and the third formant gives a smaller (but important because of the frequency range in which it lies) boost to the twenty-fourth and twenty-fifth harmonics and their neighbors. The pitch, of course, remains at G_2, since the overtones are harmonics of this "almost-missing" fundamental. A few people have learned to shape their mouths in such a way that harmonics of a sung pitch can be made audible (see Section 15.9, "Single Formants").

Another way to present formant frequencies of vowel sounds is shown in Fig. 17.2. Frequencies of the first formant are plotted on the horizontal scale, and those of the second formant on the vertical scale. The egg-shaped regions represent the approximate limits of formant frequencies that the ear will recognize as a given vowel. Note that there is overlap; that is, certain sounds can be interpreted as more than one vowel, depending on the context.

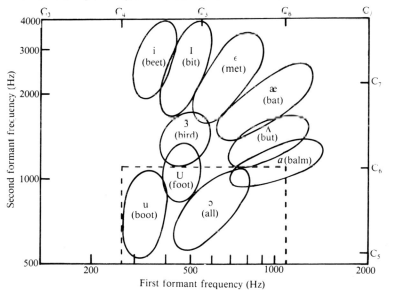

FIG. 17.2
Frequency of first and second formants for ten vowels. The dotted line shows the approximate range of a soprano voice. (After Peterson and Barney, 1952).

17.2   DIFFERENCES BETWEEN SPOKEN AND SUNG VOWELS

Sung vowels are fundamentally the same as spoken vowels, although singers do change a few vowel sounds in order to improve the musical tone. For example, "ee" is often sung like the umlauted "ü" of the German "für," and the short "e" of bed sounds more like the vowel sound in herd.

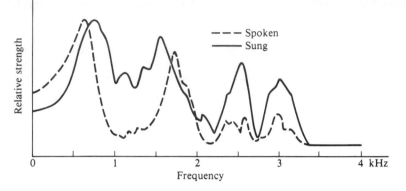

FIG. 17.3

Spectra of vowel sound |ae| as spoken and sung by a professional singer.

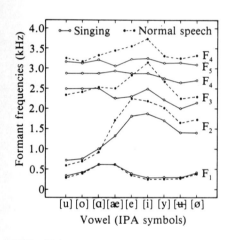

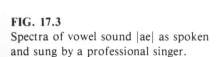

FIG. 17.4

Formant frequencies of long Swedish vowels in normal male speech (dashed lines) and in professional male singing (solid lines). (From J. Sundberg (1974). "Articulatory Interpretation of the 'Singing' Format,'" *J. Acoust. Soc. Am.* **55**: 838. Reprinted by permission from the Amer. Inst. of Physics.)

Analysis of the individual vowel formants, however, reveals changes that may be substantial. Figure 17.3 shows spectra of the vowel |ae| (pronounced "aa") spoken and sung by a professional bass-baritone singer. Note that the first formant is virtually unchanged, but the second formant is lower in frequency in the sung vowel. The third and fourth formants remain at about the same pitch but are markedly stronger in the sung vowel.

Four articulatory differences between spoken and sung vowels were noted by Sundberg (1974) as a result of studying X-ray pictures of the vocal tract and photographs of the lip openings. In singing,

1. the larynx is lowered;

2. the jaw opening is larger;

3. the tongue tip is advanced in the back vowels |u|, |o|, and |ɑ|; and

4. the lips are protruded in the front vowels.

Formant frequencies of nine sung and spoken vowels are shown in Fig. 17.4.

Trained singers, especially male opera singers, show a strong formant somewhere around 2500–3000 Hz. This "singer's formant," which seems to be more or less independent of the particular vowel and the pitch, usually lies between the third and fourth formants and adds brilliance and carrying power to the male singing voice. It is interesting to note that the frequency of this formant is near the resonance frequency of the ear canal, which gives it an additional auditory boost. A formant at 3000 Hz is evident in the spectrograms shown in Fig. 17.5.

Sundberg (1974) attributes the singer's formant to a lowered larynx, which, along with a widened pharynx, forms an additional resonance cavity (about 2 cm long) with a frequency in the range of from 2500 to 3000 Hz. Lowering the larynx also produces the darker vowel sounds favored by most singers. The larynx, which is lowered as much as 30 mm during swallowing, may be lowered up to 15 mm during singing (Shipp, 1977). Untrained singers tend to raise their larynxes as they raise the pitch.

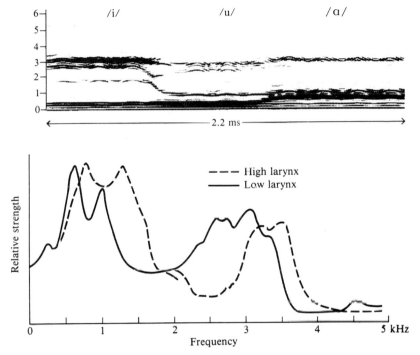

FIG. 17.5
Spectrogram of vowels /i/, /u/, and /ɑ/ (ee, oo, ah). The pitch is E_3 (f = 165 Hz). Note the strong formant at 3 kHz for all three vowels. (From van den Berg and Vennard, 1959).

FIG. 17.6
Spectrum of vowel sound /ɑ/ sung with a high and a low larynx.

Figure 17.6 shows spectra of the vowel sound |ɑ| ("ah") sung with both a high and a low larynx by a professional bass-baritone singer in our laboratory. The broad resonance extending from 2500 to 3000 Hz in the spectrum of the low larynx is a blend of the third vowel formant and the singer's formant.

Since the singer's formant requires a widened pharynx ("open throat"), it is characteristic of good singing in the chest register (see Section 17.4). Professional contraltos usually have such a formant, but sopranos, who sing mainly in the head register, may not. It is not usually present in the falsetto voice of the male singer, either. Figure 17.7 shows how the singer's formant in the voice of operatic tenor Jussi Björling helped him "cut through" a large orchestra.

It is obvious from Fig. 17.2 that the formant frequencies of different speakers (and singers) may vary rather widely, yet still result in understandable vowel sounds. Furthermore, in certain ranges of singing, the vowel formants change substantially from their normal frequencies. Nevertheless, it is instructive to compare the formant frequencies of typical *sung* vowels given in Table 17.1 to the corresponding formant frequencies of the *spoken* vowels given in Table 15.3.

Formant changes that occur throughout the singing range may be roughly described as the gradual substitution of one vowel sound for

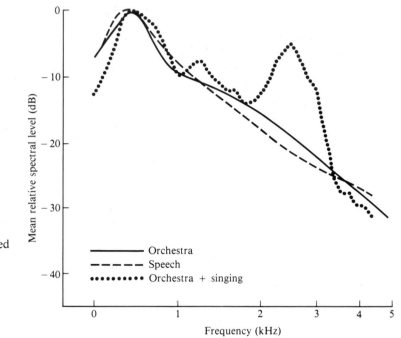

FIG. 17.7
Idealized average spectra of normal speech and orchestra music. The dotted curve shows the average spectrum of Jussi Björling singing with a loud orchestra accompaniment. (From Sundberg, 1977a).

TABLE 17.1 Formant frequencies of basic sung vowels

Formant frequency (Hz)		/i/ (ee)	/ɪ/ (i)	/ɛ/ (e)	/æ/ (aa)	/ɑ/ (ah)	/ɔ/ (aw)	/ʊ/ (ů)	/u/ (oo)	/ʌ/ (u)	/ɜ/ (er)
F_1	M	300	375	530	620	700	610	400	350	500	400
	W	400	475	550	600	700	625	425	400	550	450
F_2	M	1950	1810	1500	1490	1200	1000	720	640	1200	1150
	W	2250	2100	1750	1650	1300	1240	900	800	1300	1350
F_3	M	2750	2500	2500	2250	2600	2600	2500	2550	2675	2500
	W	3300	3450	3250	3000	3250	3250	3375	3250	3250	3050

Source: Appelman (1967).

another. As the pitch rises, for example, many singers find it convenient to make the following substitutions (Appelman, 1967):

Normal range	$|i|$	$|\epsilon|$	$|ae|$	$|a|$	$|\mathfrak{o}|$	$|u|$
High range	$|I|$	$|a|$	$|a|$	$|\Lambda|$	$|\Lambda|$	$|\upsilon|$

The vowel sounds $|I|$, $|u|$, $|\Lambda|$, and $|3|$, which do not change substan-
tially with pitch, are termed *stable*.

17.3 FORMANT TUNING BY SOPRANOS

In low voices, the various formants of the vocal tract emphasize
various harmonics of the source sound from the glottis, as we discussed in Section 17.1. Sopranos, however, do much of their singing
in a range in which the pitch exceeds the frequency of the first formant. Thus they would not receive the benefit of a boost from formant resonance, and their tones would suffer in quality and loudness.
Experienced sopranos have learned how to "tune" their formants
over a reasonable range of frequency in order to make a formant coincide with the fundamental or one of the overtones of the note being
sung.

For example, a soprano singing $|i|$ ("ee") at a pitch of F_5 (698 Hz)
might find her normal first formant at 310 Hz, over an octave below
the sung pitch. She would receive little support from this formant.
However, if she opens her lips somewhat wider than the normal posi-
tion for speaking $|i|$, the formant can be pushed up to the vicinity of
the sung pitch. Or if she were singing $|a|$ ("ah") at a pitch of A_4 (440
Hz), she would find her normal first formant around 700 Hz, between
the fundamental (440 Hz) and the second harmonic (880 Hz) of the
sung note. She would probably find it more convenient, in this case, to
raise the formant to the vicinity of the second harmonic in order to
provide the needed boost. Figure 17.8 shows how formant tuning can
be accomplished by increasing the jaw opening to change the shape of
the vocal tract.

Figure 17.9, also based on the work of Sundberg, shows the extent
to which formant tuning can take place to match one of the harmonics
of various sung pitches. The numbered lines represent the harmonics
of the pitch. Note that formants are usually tuned upward, although
downward tuning is also possible.

Formant tuning might be expected to produce objectionable
distortion of vowel sounds, but this does not seem to be the case. We
are accustomed to recognizing vowels produced at various pitches in
the speech of men, women, and children (see Table 15.3) with vocal
tracts of different lengths. If the pitch is high, we associate it with

relatively high formant frequencies. Recording a vowel sound at one speed and playing it back at another may change it to another vowel sound because the same ratio of f_2/f_1 in the new pitch range is interpreted differently. An "ah" changes to an "oh" when played at half speed (Benade, 1976).

FIG. 17.8
Formant tuning by wider jaw opening: (a) normal first formant lies below the sung pitch; (b) first formant raised to coincide with sung pitch. (From Sundberg, 1977a).

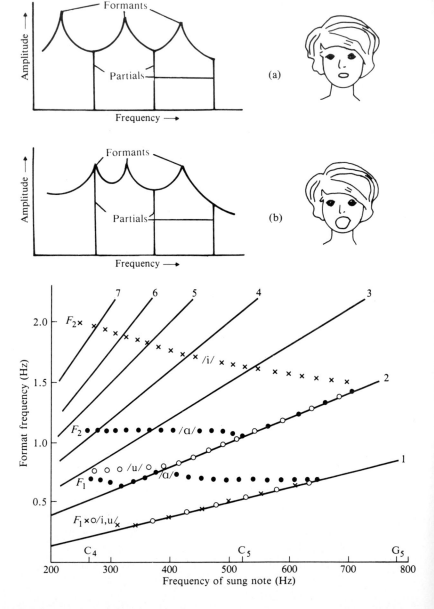

FIG. 17.9
The tuning of formants to match harmonics of the sung note. F_1 and F_2 are the lowest formants of vowels |i|, |a|, and |u|. The solid lines are the first seven harmonics of the sung note. (After Sundberg, 1975).

Near the top of the soprano range, where formant tuning is particularly marked, it is difficult to distinguish one vowel sound from another. (Try listening to a soprano sing various vowels at a high pitch, and see if you can recognize them.) Composers are quite aware of this difficulty in vowel recognition and generally do not present important text at the top of a soprano's range (if they must, they generally repeat the text at a lower pitch).

17.4 BREATHING AND AIR FLOW

Since the lungs have no muscles of their own, breathing is accomplished by changing the size of the chest cavity. There are two basic mechanisms for doing this:

1. Downward and upward movement of the diaphragm to lengthen and shorten the chest cavity;

2. Elevation and depression of the ribs to increase and decrease the front-to-back thickness of the chest cavity.

Normal quiet breathing is accomplished almost entirely by movement of the diaphragm. During maximum breathing, however, increasing the thickness of the chest cavity may account for up to half of the chest cavity enlargement.

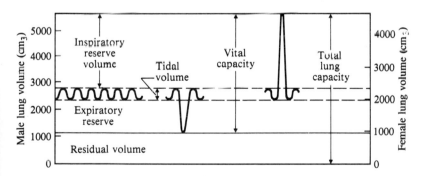

FIG. 17.10
Lung capacity in the normal young adult and its subdivision into functioning volumes. The volume of the male lung is indicated at the left, female at the right.

Figure 17.10 shows how the total lung capacity is divided into four different volumes:

1. The *tidal volume* is the volume of air moved in and out during normal breathing (about 500 cm³ in the normal male adult).

2. The *inspiratory reserve volume* is the volume that can be inspired beyond the normal tidal volume (about 3000 cm³).

3. The *expiratory reserve volume* is the volume that can be expired

by forceful effort at the end of normal tidal expiration (about 1100 cm³).

4. The *residual volume* is the volume of air that remains in the lung after forceful expiration (about 1200 cm³).

Corresponding volumes in the female lung average about 20 to 25 percent less than those given for the male lung.

Vital capacity is the amount of air that can be moved in and out of the lung with maximum effort. The average vital capacity in the young adult male is about 4600 cm³ and in the young adult female is about 3100 cm³. Pathological conditions such as tuberculosis, emphysema, chronic asthma, lung cancer, bronchitis, and pleurisy can greatly decrease vital capacity.

At a normal breathing rate of ten breaths per minute, about 5000 cm³ of air will be moved in and out of the lungs per minute. A young male adult can breath at a rate as high as 2500 cm³/s for a short period of strenuous exercise.

During normal quiet breathing, the air pressure in the lungs will be about 100 N/m² (1 cm H₂O) above and below atmospheric pressure. During maximum expiratory effort with the glottis closed, a pressure of 10,000 N/m² is possible in the strong healthy male lung. Fortissimo singing requires a pressure of 3000 to 4000 N/m² (compare woodwind blowing pressures given in Fig. 12.22). Figure 17.11 shows the subglottal air pressure (essentially equal to lung pressure) required for different sound levels in the case of four singers.

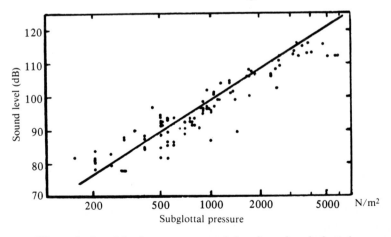

FIG. 17.11
Subglottal pressures and sound levels for many different tones from four singers. (After Bouhuys et al., 1968).

The relationship between sound level and subglottal pressure, shown in Fig. 17.11, tends to be the same for trained and untrained singers. This is not true of the relationship between sound level and air-flow rate, however. In trained singers, the pressure and rate of air flow tend to increase together with sound level, so a trained singer can sustain a soft tone for a long time. Many untrained singers, however, require a fairly large air-flow rate in order to sing softly, the flow rate reaching a minimum for mf or f dynamic. Flow rates range from about 100 to 400 cm^3/s at different dynamic levels of singing (Bouhuys et al., 1968).

Although it would appear that all a singer requires from his or her breathing apparatus is to maintain a stable subglottic pressure, the emphasis put on breathing technique by many singing teachers suggests that the manner of breathing is of some importance.

One recent set of experiments (Leanderson, et al., 1987) set about determining the role of diaphragm activity during singing (some singers use their diaphragms only during inspiration, others contract them during the entire phrase). By simultaneously monitoring pressure across the diaphragm, sound pressure level, and air flow, it was found that the flow rate tended to be higher when the diaphragm was activated, although there were substantial differences in diaphragm use between individual singers. It is probably safe to conclude that use of the diaphragm is not the key to good singing.

17.5 REGISTERS, VOICES, AND MUSCLES

We discussed the larynx and vocal cords and their functions in speech in Section 15.2. Although their functions are essentially the same in singing, there are a few additional features, such as muscular action, which become important when analyzing the singing voice.

The principal muscles internal to the larynx are the thyro-arytenoids, the cricothyroids, and the cricoarytenoids (the names indicate which two cartilages they connect). The *cricoarytenoid muscles* operate the arytenoid cartilages to which the posterior ends of the vocal cords or vocal folds are attached, as shown in Fig. 15.3.

The *cricothyroids* connect the two large cartilages of the larynx, the thyroid and the cricoid (see Fig. 15.2). They can pull the thyroid forward, with respect to the cricoid, and also downward, closer to it. Both of these actions stretch the vocal cords longitudinally, which is one way of increasing their rate of vibration. The action of the cricothyroid muscles can be observed in two ways: One is to press inward on the Adam's apple while singing a note in midrange. Sudden release of the pressure will cause the pitch to go up. A second experiment consists of placing a finger in the small space between the thyroid and cricoid cartilages while singing. Raising the pitch an octave will force the finger outward as the thyroid is pulled down closer to the cricoid.

The *thyroarytenoids*, also called the vocalis or vocal muscles, form the body of the vocal cords themselves. They extend from the notch of the thyroid to the arytenoid cartilages at the rear, and are covered with a membrane that is continuous with the lining of the rest of the larynx. The tension on the vocal cords is a complex balance of forces from all three muscles, and coordination between them is necessary for smooth transition from one pitch to another. In order to hold a steady pitch during a crescendo or diminuendo, these muscles must compensate for the tendency of pitch to rise as the velocity of air flow is increased (due to the Bernoulli force; see Section 11.7).

Much has been said about the various registers used in singing. An idealistic approach is *one register*. The voice, if possible, should produce all the pitches of which it is capable without breaks or radical changes in technique. Some teachers feel that the best way to make this ideal come true is to assume that it *is* true, that it *can* be accomplished.

A more realistic approach is *three registers*. These correspond to the differences in tone caused by different adjustments of the larynx. The registers go by various names, but the most common are "chest," "middle," and "head" (in male voices, they are sometimes labeled "chest," "head," and "falsetto"). The famous teacher Mathilde Marchesi was obviously a proponent of this approach as she wrote, "I most emphatically maintain tha. the female voice possesses *three* registers, and not *two*, and I strongly impress upon my pupils this undeniable fact, which, moreover, their own experience teaches them after a few lessons." According to Marchesi (1970), the highest note in the chest register is about E_4 to F_4 for sopranos and F_4 to $F_4^{\#}$ for mezzo-sopranos and contraltos. The highest note in the middle register is about F_5. This is in agreement with the register ranges shown in Fig. 17.12.

The third approach is *two registers*, which considers that every voice has a potential of roughly two octaves of "heavy" mechanism, with about one octave of overlap. This middle octave can be sung in either laryngeal adjustment, and it is possible to combine some of the best qualities of both. Basses and contraltos sing almost exclusively with the heavy mechanism, mixing in just a bit of the light mechanism at the top of their range. However, the light mechanism is never used exclusively except for comic effects. Lyric and coloratura sopranos, on the other hand, sing with the light mechanism, mixing in just a little of the heavy at the bottom of their range, but never singing in a pure chest voice (Vennard, 1967).

Because of the confusion associated with the use of the term register, it is preferable to refer to the two modes of vocal cord vibration as two mechanisms, heavy and light. We will call them *chest voice* and *head* (falsetto) *voice*. The distinguishing feature seems to be in the

Chest   Middle   Head

FIG. 17.12
Ranges of three registers (according to Mackworth-Young, 1953). Half notes represent the male voice, quarter notes represent the female voice. Male voices use head register when singing falsetto.

state of the thyroarytenoid muscles. In the heavy or chest voice, these muscles are active; in the light or head voice, they are virtually passive. An analogy can be drawn between these two modes of vocal cord vibration and the vibrations of the lips of a trumpet player (with active muscles) as opposed to the (passive) vibrations of a clarinet reed.

In the chest voice, the thyroarytenoids or vocalis muscles are active and hence shortened. At the lowest tones, the muscles are relaxed and the vocal folds are thick. Because of their thickness, the glottis closes firmly and remains closed an appreciable part of each cycle of vibration, as it does during speech (see Fig. 15.5).

As the pitch rises in chest voice, the cricothyroid muscles contract and apply tension to the vocal cords. The cords do not elongate rapidly, however, because the thyroarytenoid muscles come into action, and indeed thicken the vocal folds as they do so. At the top notes of the chest voice, the thyroarytenoids of the inexperienced singer sometimes give way to excessive force from the cricothyroids, and the voice "cracks" into an involuntary head tone (Vennard, 1967).

In the light mechanism or head voice, the thyroarytenoids offer little resistance to the cricothyroids, which can then apply substantial longitudinal tension to the vocal cords, thus elongating them and making them thin. The vocalis muscles fall to the sides, and the vibration takes place almost entirely in the ligaments with much less amplitude of movement than in the chest voice. The glottis closes only briefly, or not at all, and the resulting sound has fewer harmonics than the chest voice does. According to studies by van den Berg (1961) on isolated larynxes, elongations of 30 percent are typical, as shown in the graph of stress vs. strain in Fig. 17.13. Stress is the force applied per unit area and strain is the percentage by which the length increases.

The manner in which the vocal cords vibrate in the chest voice and in the head voice is shown schematically in Fig. 17.14.

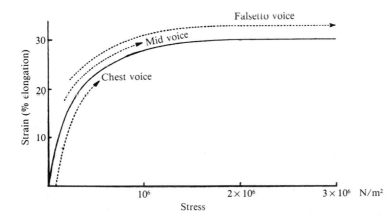

FIG. 17.13
Stress-strain graph for vocal ligaments under passive tension (similar to head voice). The horizontal axis represents applied stress of force. (From van den Berg, 1968).

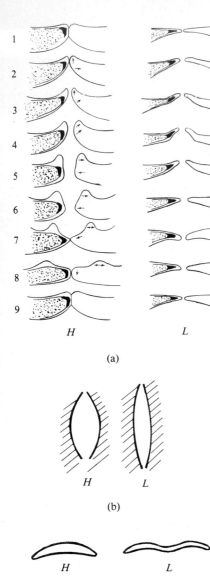

When vibrating in the light mechanism, the vocal cords are up to 30 percent longer, are appreciably thinner, and have a smaller effective mass. The cords do not ordinarily close completely during any part of the cycle (compare the "open phase" speech mode described in Section 15.2). This results in fewer harmonics of the fundamental and also in less efficient conversion of breath power into sound power. The waveforms of the glottal air flow during speech and their spectra of overtones were shown in Figs. 15.6 and 15.7. Waveforms of glottal air flow during various modes of singing are shown in Fig. 17.15.

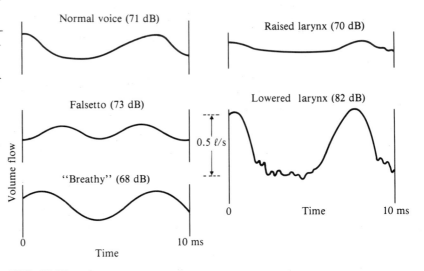

FIG. 17.15
Waveforms of glottal air flow during various modes of singing. (After Sundberg, 1978).

FIG. 17.14
Schematic diagrams of vocal cords vibrating: (a) side view (from Titze, 1973); (b) top view; (c) edge view. In each diagram H denotes the heavy mechanism (chest voice) and L the light mechanism (head voice).

17.6   OTHER FACTORS INFLUENCING THE SPECTRA OF SUNG NOTES

It is characteristic of nearly all musical instruments that raising the dynamic level increases the levels of the higher harmonics more rapidly than that of the fundamental (see, for example, Figs. 11.13(c) and 12.16). The same effect is observed in singing, as can be seen in Fig. 17.16. In loud singing a greater fraction of the total sound energy appears in the higher harmonics as compared to soft singing.

The reason for this gain in energy in the higher harmonics can be seen by comparing the glottal air flow waveforms ("glottograms") in Fig. 17.17. As the loudness of phonation is increased, the rate of closure of the glottis (indicated by the slopes of the heavy lines drawn

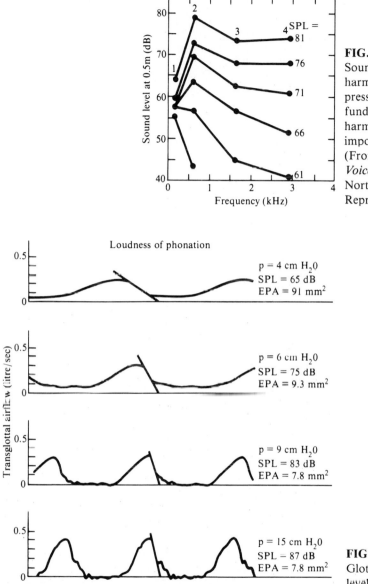

FIG. 17.16
Sound pressure level in the first four harmonics at different total sound pressure levels. In soft phonation, the fundamental dominates, but the higher harmonics take on increasing importance as the loudness increases. (From Sundberg, 1987.
Voice by Johan Sundberg © 1987 by Northern Illinois University Press. Reprinted by permission.)

FIG. 17.17
Glottal waveforms for four different levels of phonation. The rate of glottal closure increases as the phonation level increases. (From Sundberg, 1987).

along the trailing edges of the waveforms) increases. Fourier analysis shows that waveforms with rapid rates of rise or fall have spectra rich in harmonics (compare Fig. 7.11).

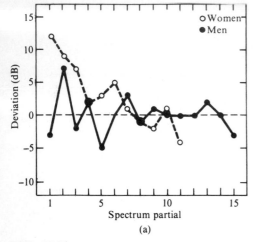

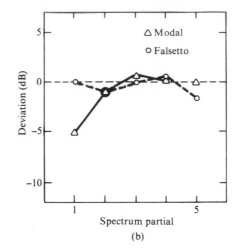

FIG. 17.18
(a) Relative strengths of harmonics in male and female voices. (b) Relative strengths of harmonics in a male voice in the modal and falsetto registers. In both cases the vertical axis shows the deviation from the overall decrease of 12 dB/octave that characterizes voice source. (From Sundberg, 1987).

Normally, a male voice tends to have a weaker fundamental and stronger harmonics than a female voice, as shown in Fig. 17.18(a). Also, a male voice singing falsetto has a stronger fundamental and weaker harmonics than when singing the same note in the modal register, as shown in Fig. 17.18(b). In Fig. 17.18 the vertical axis shows only the deviation from the overall decrease of 12 dB/octave that characterizes the voice source in both speech and singing.

17.7   CHOIR SINGING

Choral singing and solo singing are two distinctly different modes of musical performance, making different demands on the singers. Most research on the acoustics of singing has been directed at solo singing, and so less is known about voice use in choir singing.

The author had the opportunity of participating in some experiments at the Royal Institute of Technology (KTH) in Stockholm, which compared identical passages sung by experienced singers in solo and choir modes. A number of differences were noted, in both male and female singers.

Male singers tended to employ a more prominent singer's formant in the solo mode, as can be seen in Fig. 17.19, while the fundamental is emphasized more in the choir mode, as might be expected (Rossing, et al., 1986). It appeared that this was accomplished through adjustments in both articulation (adjustment of formant frequencies) and phonation (change in the glottal waveform).

Female singers also tend to produce more energy in the 2–4 kHz range in the solo mode, as shown in Fig. 17.20, although different subjects appear to differ substantially in spectral characteristics in this

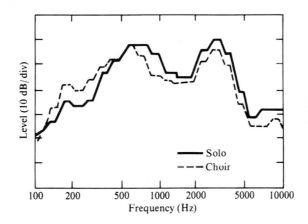

FIG. 17.19
Average spectrum envelopes for a male singer who sang a phrase as a solo singer and as a choral singer. In the latter case his lowest partials are somewhat stronger and his singer's formant is slightly weaker. (From Rossing et al., 1986).

frequency range. It is more difficult to obtain accurate glottal waveforms from female singers, so it is difficult to distinguish changes in articulation from voice source changes. The extent of vibrato appeared to be greater in the solo mode (Rossing et al., 1987).

Other studies by the Stockholm group have observed the degree of unison and accuracy of intervals in choir singing, both under normal conditions and when singers were deprived of feedback from other singers. In a good amateur choir, the standard deviation in the notes sung by members of the bass section was found to be 16 cents (Ternström and Sundberg, 1988). In another experiment, singers were asked to sing a note at constant sound level and with the same pitch, as reference tones presented at different levels. When the reference tone was the vowel /ɑ/, they were able to do this quite well, but when the vowel was /u/, the singers sang about 25 cents sharp with the softest

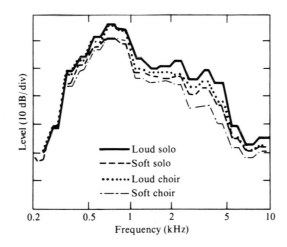

FIG. 17.20
Average spectrum envelopes for a female singer who sang the same phrase at two different sound levels as a solo singer and as a choral singer. Choral singing gives slightly weaker high partials. (From Rossing et al., 1987).

reference tone and about 45 cents flat with the loudest reference tone. Apparently the relatively low number of harmonics in the /u/ tone (due to the low frequencies of the first and second formants) was the cause of this (Sundberg, 1987).

17.8 SUMMARY

In singing, as in speaking, the vocal cords act as a source of sound, which is filtered by the vocal tract. The resonances (formants) of the vocal tract determine the vowel sounds as well as the timbre of the sung tone. Sung vowels and their formants are slightly different than spoken vowels, one of the most important differences being the appearance of a *singer's formant* around 2500–3000 Hz. One of the important results of vocal training is to learn to lower the larynx and open the pharynx to create this extra formant. Sopranos often sing at pitches above their normal formants, and therefore must "tune" these formants if they are to reinforce the sung notes.

There appear to be two mechanisms for singing: In one, the vocalis muscles are active; in the other, they tend to be passive. They can be referred to as *heavy mechanism* (chest voice) and *light mechanism* (head voice). During normal breathing, about 500 cm^3 of air is moved per breath. In a trained singer, both air flow rate and pressure increase with sound level.

In loud singing, a greater fraction of the total sound energy appears in the higher harmonics, partly due to the higher rate of closure of the glottis. Singers tend to concentrate more energy in the 2–4 kHz range in solo singing, while they emphasize the fundamental more in choir singing.

References and Suggested Readings

Appelman, D. R. (1967). *The Science of Vocal Pedagogy.* Bloomington, Indiana: Indiana University Press.

Bartholomew, W. T. (1940). "The Paradox of Voice Teaching," *J. Acoust. Soc. Am.* **11:** 446.

Benade, A. H. (1976). *Fundamentals of Musical Acoustics.* New York: Oxford. (See Chapter 19.)

Bloothooft, G., and R. Plomp (1984, 1985, 1986) "Spectral Analysis of Sung Vowels. I, II, and III," *J. Acoust. Soc. Am.* **75:** 1259; **77:** 1580; **79:** 852.

Bjorklund, A. (1961). "Analyses of Soprano Voices," *J. Acoust. Soc. Am.* **33:** 575.

Bouhuys, A., J. Mead, D. F. Proctor, and K. N. Stevens

(1968). "Pressure-Flow Events During Singing," *Annals N.Y. Acad. Sci.* **155:** 165.

Large, J. (1972). "Towards an Integrated Physiologic-Acoustic Theory of Vocal Registers," *Nat. Assn. Teachers of Singing, Bull.* **28:** 18, 30.

Leanderson, R., J. Sundberg, and C. von Euler (1987). "Role of Diaphragmatic Activity During Singing: A Study of Transdiaphragmatic Pressures," *J. Appl. Physiol.* **62:** 259.

Mackworth-Young, G. (1953). *What Happens in Singing.* London: Neame.

Marchesi, M. (1970). *Bel Canto: A Theoretical and Practi-*

cal Vocal Method. (Dover reproduction of original undated publication by Enoch & Sons, London).

Peterson, G. E., and H. L. Barney (1952). "Control Methods Used in a Study of the Vowels," *J. Acoust. Soc. Am.* **24**: 104.

Rossing, T. D., J. Sundberg, and S. Ternström (1986). "Acoustic Comparison of Voice Use in Solo and Choir Singing," *J. Acoust. Soc. Am.* **79**: 1975.

Rossing, T. D., J. Sundberg, and S. Ternström (1987). "Acoustic Comparison of Soprano Solo and Choir Singing," *J. Acoust. Soc. Am.* **82**: 830.

Seymour, J. (1972). "Acoustic Analysis of Singing Voices, Parts I, II, III," *Acustica* **27**: 203, 209, 218.

Shipp, T. (1977). "Vertical Laryngeal Position in Singing," *J. Research in Singing* **1**: 16. (abstract in *J. Acoust. Soc. Am.* **58**: S95.)

Strong, W. J., and G. R. Plitnik (1977). *Music, Speech and High Fidelity,* Provo, Utah: Brigham Young University Press. (See Chapter 6B.)

Sundberg, J. (1974). "Articulatory Interpretation of the 'Singing Formant,'" *J. Acoust. Soc. Am.* **55**: 838.

Sundberg, J. (1975). "Formant Technique in a Professional Female Singer," *Acustica* **32**: 8.

Sundberg, J. (1977a). "The Acoustics of the Singing Voice," *Sci. Am.* **236**(3).

Sundberg, J. (1977b). "Singing and Timbre," in *Music Room and Acoustics.* Stockholm: Royal Academy of Music.

Sundberg, J. (1978). "Waveform and Spectrum of the Glottal Voice Source," Report STL-QPSR 2–3, 35. Stockholm: Speech Transmission Lab., Royal Institute of Technology.

Sundberg, J. (1987). *The Science of the Singing Voice.* DeKalb, IL: Northern Illinois University Press.

Ternström, S., and J. Sundberg (1988). "Intonation Precision of Choir Singers," *J. Acoust. Soc. Am.* **84**: 59.

Titze, I. R. (1973). "The Human Vocal Cords: A Mathematical Model, Part I," *Phonetica* **28**: 129. (See also Part II, *Phonetica* **29**: 1.)

van den Berg, Jw. (1968). "Register Problems," *Ann. New York Academy of Sciences* **155**: 129.

Vennard, W. (1967). *Singing: The Mechanism and the Technic.* New York: Carl Fischer.

Glossary

chest voice (register) Mode of singing associated with a heavy mechanism or active vocalis muscles.

cricoarytenoids The muscles of the larynx that help to apply tension to the vocal cords.

cricoid Lower cartilage of the larynx.

cricothyroids The muscles of the larynx that determine the relative position of cricoid and thyroid cartilages and thus affect vocal cord tension.

diaphragm The dome-shaped muscle that forms a floor for the chest cavity.

flow rate The volume of air that flows past a point and measured per second.

formant A resonance of the vocal tract.

head voice (register) Mode of singing associated with a light mechanism, passive vocalis muscle, and elongated, thin vocal cords.

heavy mechanism Mode of vocal cord vibration in which the vocalis muscles are active and the vocal folds or cords are thick.

larynx The source of sound for speaking or singing.

light mechanism Mode of vocal cord vibration in which the vocalis muscles are relaxed, and the vocal cords elongated and thin.

middle register A combination of light and heavy mechanism that lies between the chest and head registers.

pharynx The lower part of the vocal tract connecting the larynx and the oral cavity.

singer's formant A resonance around 2500–3000 Hz in male (and low female) voices that adds brilliance to the tone.

thyroarytenoids (vocalis muscles) The muscles that form part of the vocal folds.

thyroid The upper cartilage of the larynx.

tidal volume The volume of air moved in and out of the lungs during a normal breath.

vital capacity The volume of air that can be moved in and out of the lungs with maximum effort during inhalation and exhalation.

vocal tract The tube connecting the larynx to the mouth consisting of the pharynx and the oral cavity.

vocalis muscle The thyroarytenoid muscle.

Questions for Discussion

1. Try to sing as many notes as possible in both chest and head registers. Can you sing in both registers? How much overlap is there in your voice?

2. Is a stress of 100 gm/mm² (see Fig. 17.13) a large stress? What is the breaking stress of a piece of cotton cord? a nylon thread?

3. Normal speaking is done in chest voice. Is it possible to speak in a head voice? Is speech intelligibility affected?

4. Place either a cardboard tube, a length of pipe, or your cupped hands around your lips to extend the vocal tract and lower the formant frequencies. Describe the tone produced. What is often called a "dark" or "covered" tone is produced by extending the vocal tract at the lower end. Is this equivalent to what you have done?

Problems

1. Find the frequencies that correspond to the three singing registers designated in Fig. 17.12.

2. What harmonics of G_2 ($f = 98$ Hz) are enhanced by the formants of /i/? of /u/?

3. Compare the first three formant frequencies in Fig. 17.4 to those in Table 17.1 for the sung vowels /u/, /ɑ/, and /i/.

4. Find the lengths of closed pipes that would resonate at 2500 and at 3000 Hz. Are these reasonable lengths for the cavity formed by the (closed) glottis and the (open) pharynx?

5. The power (in watts) used to move air in or out of the lungs is equal to the pressure (in N/m²) multiplied by the flow rate (in m³/s). Find the power for:

   a) Quiet breathing ($p = 100$ N/m², flow rate = 100 cm³/s);

   b) Soft singing ($p = 1000$ N/m², flow rate = 100 cm³/s);

   c) Loud singing ($p = 4000$ N/m², flow rate = 400 cm³/s).

6. According to Fig. 17.11, a pressure of 4000 N/m² will produce a sound level of about 120 dB.

   a) Find the intensity and sound pressure that correspond to this sound level (see Chapter 6).

   b) Compare the sound pressure at the mouth to the steady subglottal air pressure.

   c) Assuming a mouth opening of 20 cm², calculate the total radiated sound power.

   d) What portion of the total power calculated in Problem 5 is converted into sound? (*Answer*: About 0.1 percent.)

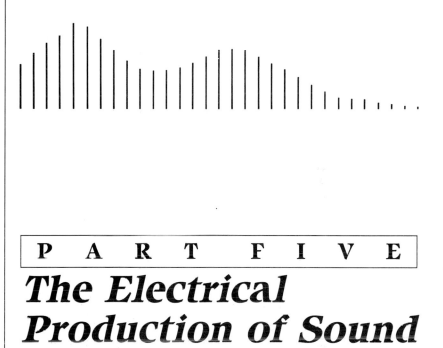

Electroacoustics deals with the conversion of electrical energy into acoustical energy or the reverse. The discipline that deals with practical applications of electroacoustics is called *audio engineering*. Chapters 18–22 present the basic principles of electroacoustics or audio engineering.

Chapter 18 covers the principles of electrical circuits in an elementary way. A student who has taken a basic physics or electronics course will probably skip Chapter 18, skim though Chapter 19, and begin with Chapter 20. Other students will wish to work through Chapters 18 and 19 rather carefully, taking time to work all the problems. The more difficult the problems appear to be, the more important it is to solve them carefully!

One of the most important applications of electroacoustics is the high-fidelity reproduction of sound, which has grown into a multimillion-dollar consumer-oriented industry. Some students may consider themselves audiophiles (lovers of sound), and their reason for taking a course in sound is to gain a better understanding of audio technology or perhaps to help them spend their money more wisely when they purchase audio components.

Although Chapter 22 deals with components for high-fidelity (hi-fi) reproduction of sound, stereo and hi-fi sound systems are discussed in Chapter 25 with

P A R T   F I V E

The Electrical Production of Sound

the study of the acoustics of rooms. This reflects the author's opinion that the listening room is an important part of a high-fidelity sound system.

CHAPTER **18**

A Little About Electricity

Before learning about electronic circuits, we ought to consider a few basic principles of electricity. Although we use electricity every day in countless ways, it is curious how few people learn to understand what electricity is or how it actually performs the useful tasks we call on it to do.

Electricity is everywhere in nature. All the common, everyday materials that we handle are held together by electrical forces. The light that reaches our eyes consists of electric (as well as magnetic) waves, as do the waves that bring us heat from the sun. When we use the term electricity in everyday language, however, we usually mean *electrical energy* or an *electric current*, and that is the way in which we will use the term electricity here.

18.1 ELECTRICAL CIRCUITS

An electric current is a flow of electric charge, and is capable of carrying energy from one place to another. An electric current in a wire consists of a cloud of negatively charged electrons moving through the wire, but in other conducting media (for example, the liquid in a battery), the current may be carried by charged atoms. An early model of electric current, proposed by Benjamin Franklin and others, was that of a charged fluid (practical electricians to this day sometimes refer to electricity as "juice"), and it may be helpful to our understanding of electricity to draw analogies between the flow of electricity in wires and the flow of water in pipes.

The electrical circuit in Fig. 18.1(b) and the water circuit in Fig. 18.1(a) have many similarities. In the water circuit, a pump raises the water pressure, which causes the water to flow around the circuit. It could be made to perform some work (like turning a paddle wheel) as it drops to the lower reservoir. In the electrical circuit, the battery raises the potential (voltage) in one part of the circuit so that an electrical current will flow. The current "drops" through the light bulb and some of its energy is converted to light and heat. The circuit diagram in Fig. 18.1(c) could be used to represent either the water circuit or the electrical circuit; *V* represents the *voltage* supplied by the battery or the pressure supplied by the pump, and *R* represents the *resistance* to the flow of water or electric current on the part of the small pipe or the light bulb.

FIG. 18.1

(a) A water circuit in which a pump raises the water pressure, resulting in a current flow. (b) An electrical circuit in which a battery raises the electrical potential (voltage) so that electric current can flow through the light bulb. (c) A circuit diagram that could represent either of the systems illustrated in (a) and (b).

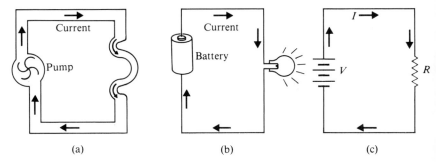

(a)   (b)   (c)

Clearly, the amount of current or water that flows depends on the size of *V* (electrical voltage or water pressure) and *R* (resistance). Large *V* and small *R* leads to a large current. In fact, the relationship between current, voltage, and resistance (referred to as *Ohm's law*) is one of the important laws of nature. Using the symbol *I* for current, it is written

$$I = V/R. \tag{18.1}$$

To use Ohm's law, one needs a consistent set of electrical units. In the international system (SI), *V* it is expressed in *volts*, *R* in *ohms*, and *I* in *amperes* (frequently abbreviated "amps"). The preferred symbols are *V* for volts; Ω for ohms; and *A* for amperes.

If the circuit contains only a single source of voltage (potential) and a single resistance, as in Fig. 18.2, the use of Ohm's law to calculate the current is straightforward. In Fig. 18.2, the current is $I = V/R = 3/6 = 0.5$ A.

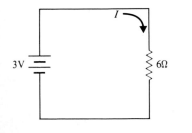

FIG. 18.2

An electrical circuit with a single voltage source and a single resistance.

If the circuit contains two or more voltage sources in series, these can be added to give the total source voltage.* In a two-cell flashlight, two 1.5-volt batteries are connected in series to provide a source of 3 volts, as shown in Fig. 18.3(a). If the resistance of the bulb were 6 ohms, the current would be 0.5 amp, as in the circuit in Fig. 18.2.

Similarly, if two resistances R_1 and R_2 are connected in series, the total resistance is their sum. This can be represented by the equation

$$R = R_1 + R_2. \qquad (18.2)$$

Figure 18.3(b) shows a circuit with two 3-ohm resistances in series in which the current is 0.5 amp, the same as in the circuits shown in Figs. 18.2 and 18.3(a). Note that the total voltage (3 volts) supplied by the battery is equally divided between the two resistances, so that a voltmeter connected across either one of them would read 1.5 volts. Voltages across series elements add together, whether they are the source voltages of two or more batteries connected in series or the voltage "drops" across two resistances in series.

FIG. 18.3
(a) An electrical circuit with two voltage sources in series. (b) A circuit with two resistances in series. In both of these circuits, the current is the same as that in the circuit shown in Fig. 18.2.

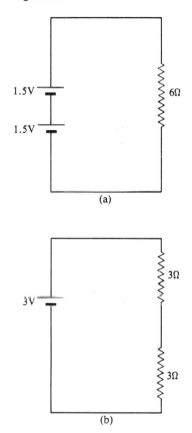

Examples Involving the Use of Ohm's Law to Calculate Current and Voltage

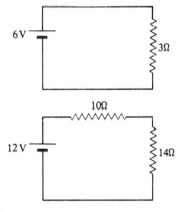

In the first circuit shown in the figure, a six-volt battery causes an electric current to flow through a three-ohm resistor (which could be a small light bulb). The current that flows is easily calculated to be $I = V/R = 6/3 = 2$ amperes.

In the second circuit, two resistors connected in series impede the flow of electrical current. Thus, the total resistance through which current must flow is $R = 10 + 14 = 24\ \Omega$. The current will be $I = V/R = 12/24 = 0.5$ A. If we wish, we could use Ohm's law in the form $V = IR$ to calculate the voltage that appears across either one of the resistors alone. For example, the voltage across the 14-ohm resistor is $V_{14} = IR = (0.5)(14) = 7$ volts. Clearly the voltage across the

---

* If the sources are batteries, they must be connected in the same direction or polarity in order for the total source voltage to be the sum of the individual voltages. If they are connected in opposition, their voltages subtract, as one discovers by inserting a battery into a flashlight in the reverse direction.

other resistor would be the total voltage less V_{14}:

$$V_{10} = V - V_{14} = 12 - 7 = 5 \text{ volts.}$$

The same answer could be arrived at by use of Ohm's law:

$$V_{10} = IR = (0.5)(10) = 5 \text{ volts.}$$

An arrangement of two resistors in series such as this is often called a *voltage divider*, since the entire voltage is divided between the two resistors. Frequently the point of division is made variable, as shown in the figure below. If a voltage V is applied across the resistor R, a fraction R_1/R appears across the variable part R_1. The output voltage, therefore, is VR_1/R. A voltage divider is used for the volume -control in a radio or amplifier. Such a device is also called a *potentiometer,* or simply a "pot."

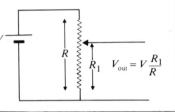

FIG. 18.4
The internal resistance of the battery r has the effect of further limiting the circuit current.

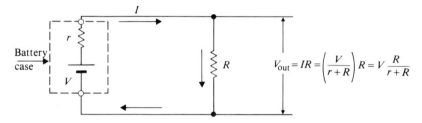

An important consideration we have thus far ignored is the fact that the battery (or pump) itself somewhat impedes the flow of current in addition to providing the pressure. That is to say, the battery has internal resistance. Let the internal resistance be denoted by r (also measured in ohms); thus, a more accurate representation of a real circuit is shown in Fig. 18.4. A new or fully charged battery would have a small internal resistance (0.1 to 0.01 ohm or smaller), whereas a discharged or worn-out battery might have an internal resistance of 10 to 100 ohms or larger.

Another way to connect circuit elements is in parallel, as shown in Fig. 18.5. Connecting two identical voltage sources in parallel does not increase the total voltage, but it does increase the available current, because each source supplies half the current in the circuit as shown in Fig. 18.5(a). Similarly, connecting two resistances in parallel provides two alternate paths for the current, so the total current is the sum of I_1 and I_2, the currents through R_1 and R_2, respectively.

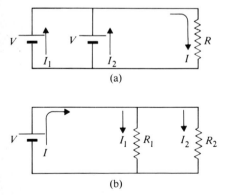

FIG. 18.5
(a) A circuit with two voltage sources in parallel. (b) A circuit with two resistances in parallel. In both circuits, the total current is $I = I_1 + I_2$.

A formula for calculating the total resistance of two or more resistances in parallel can be derived by using Ohm's law. In the circuit shown in Fig. 18.5(b), the voltage across each resistor is V, since they are both connected directly to the voltage source. Thus the currents in the two resistances are $I_1 = V/R_1$ and $I_2 = V/R_2$. The total current is

$$I = I_1 + I_2 = V/R_1 + V/R_2.$$

But we know that if R is the total resistance in the circuit, $I = V/R$. Thus if each term in the equation is divided by V, we obtain an expression for R:

$$\frac{1}{R} = \frac{1}{R_1} + \frac{1}{R_2}. \qquad (18.3)$$

For example, if $R_1 = R_2 = 10$ ohms, then

$$1/R = 1/10 + 1/10 = 2/10$$

so

$$R = 10/2 = 5 \text{ ohms.}$$

The total resistance R of two resistances in parallel will always be less than either individual resistance.

The formula can be extended to any number of resistances in parallel:

$$\frac{1}{R} = \frac{1}{R_1} + \frac{1}{R_2} + \frac{1}{R_3} + \cdots + \frac{1}{R_n}. \qquad (18.4)$$

18.2 ELECTRICAL ENERGY AND POWER

Work, energy, and power were defined in Chapter 1; we recall that power is work or energy divided by time. Electrical power is expressed in watts and electrical energy in joules, just as mechanical power and energy are (one joule = one watt-second). However, when expressing electrical power, the kilowatt-hour (kWh), a much larger unit of energy, is frequently used (1 kWh = 3.6×10^6 joules; see Section 1.8).

When you pay your electric bill, you pay for the total amount of *electrical energy* consumed in all the circuits in your house. The energy depends on the voltage, the average current, and the total time during which the current flows. The "pump," in this case, is the power-generating station, which maintains a constant voltage (usually 120 volts) throughout your house. Electrical energy costs about five cents per kilowatt-hour; a kilowatt-hour would light a 100-watt bulb for 10 hours. In comparison, a clarinetist would have to blow his or her instrument for 20,000 hours (over two years) without stopping in order to generate a kilowatt-hour of acoustic energy.

Calculating Electric Power and Energy

Electric *power* is the product of current and voltage: $\mathcal{P} = IV$. Power is expressed in *watts* when I is in *amperes* and V is in *volts*. A 60-watt light bulb is designed to draw a current of 0.5 ampere. Thus $\mathcal{P} = 0.5 \times 120 = 60$ watts. Power is the rate at which energy is used; it is measured in watts and kilowatts (1 kilowatt = 1000 watts). Energy is the product of power and time. Thus watt-second, watt-hour, and kilowatt-hour (kWh) are units used to measure energy.

The amount of power that can be delivered to a resistance depends on the value of the resistance itself (sometimes called the "load" resistance) and the internal resistance of the source (the battery). Consider a battery of fixed voltage and internal resistance connected to a load resistance R, the resistance of which can be changed. The circuit is that shown in Fig. 18.4. The amount of power delivered to R as it ranges from zero to a large value is shown in the graph in Fig. 18.6. Note that the greatest amount of power is transferred when the load resistance is equal to the internal resistance r of the battery.

Even though these results are presented for DC (direct-current) circuits, they have general applicability and important consequences when it comes to matching a loudspeaker to an amplifier or matching a microphone or phono pickup to an amplifier input.

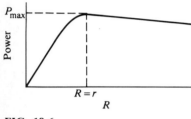

FIG. 18.6
Power delivered to a load resistance R depends on the value of the resistance and is greatest when it is equal to the internal resistance r of the battery.

18.3   ALTERNATING CURRENT

Alternating current (AC) is electrical current that reverses its direction many times each second. Alternating current is generated in a microphone, and alternating current drives the voice coil in a loudspeaker. Nearly all electrical power in the United States is AC with a frequency of 60 cycles per second (60 *hertz*, abbreviated Hz). This means that 120 times each second, the current in a light bulb stops momentarily and reverses its direction.

Ordinarily you are not aware of this interruption, but if you look near the end of a fluorescent lamp, you can sometimes observe a flicker. This slight flicker, incidentally, is often used to check the speed of phonograph turntables at home by viewing a strobe disc with a fluorescent light.

Figure 18.7(a) shows how the current varies with time for a simple alternating current with frequency of 60 Hz, which is the type supplied by power companies in the United States. The waveform closely resembles that of a simple harmonic oscillator, shown in Fig. 2.2. Figure 18.7(b) shows alternating current with a more complex waveform such as one might observe in the output of an audio amplifier.

Ohm's law still holds for AC circuits as well as for DC (direct current) circuits. The same is true for formulas used to calculate power and energy, with one caveat: In the case of AC, we use *effective* voltage, current, and power. An effective AC voltage of 120 volts (such as we are supplied by the power company) means that the voltage rapidly varies from -170 to $+170$ volts, as shown in Fig 18.8. Rarely, however, do we need to concern ourselves with this fact when using AC electricity, because the power available from AC with an effective voltage of 120 volts is the same as would be available from 120-volts DC.

In electronic circuits that use AC, we commonly encounter *capacitors* and *inductors*. They are circuit elements whose impedance to the flow of electricity varies with frequency. The electrical *impedance Z* is defined as the ratio of voltage to current: $Z = V/I$. Impedance in an AC circuit is analogous to resistance in a DC circuit. Like resistance, it is expressed in ohms.

A typical example of an inductor is a coil of wire wound around a core of magnetic material. The buildup of magnetism in the core prevents the current from changing as rapidly as it would otherwise; thus, an inductor impedes the flow of AC even if the resistance of its wire is small. In an inductor, the impedance increases with frequency, so it is more difficult for high-frequency current to flow, as shown in Fig. 18.9(a).

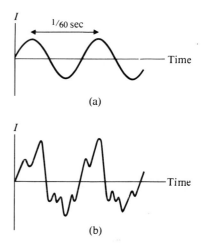

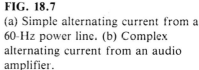

FIG. 18.7
(a) Simple alternating current from a 60-Hz power line. (b) Complex alternating current from an audio amplifier.

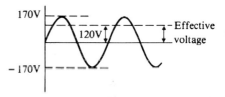

FIG. 18.8
Alternating current with an effective 120 volts has a voltage that actually varies from -170 V to 170 V. It is equivalent in power to a 120-V direct current.

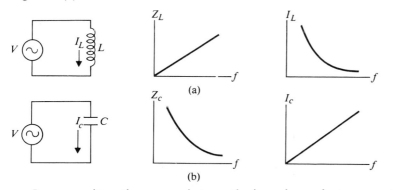

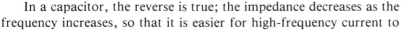

FIG. 18.9
Dependence of impedance Z and current I on frequency f (a) for an inductor; (b) for a capacitor.

In a capacitor, the reverse is true; the impedance decreases as the frequency increases, so that it is easier for high-frequency current to flow, as shown in Fig. 18.9(b). In electrical circuits, we denote inductance by the symbol L and measure it in *henries* (abbreviated H). Capacitance is denoted by C and measured in *farads* (abbreviated F).

The various electrical units are appropriately named after scientists of the eighteenth and nineteenth centuries: Georg Ohm, Andre Amperé, Alessandro Volta, James Watt, James Joule, Joseph Henry, and Michael Faraday.

The formulas that relate V, I, Z, L, and C are as follows:

For an inductor: $Z_L = 2\pi f L$, $\quad I_L = \dfrac{V}{Z} = \dfrac{V}{2\pi f L}$; and

For a capacitance: $Z_C = \dfrac{1}{2\pi f C}$, $\quad I_C = \dfrac{V}{Z} = 2\pi f C V$,

where V = the potential in volts (V), Z = the impedance magnitude in ohms (Ω), I = the current in amperes (A), L = the inductance in henries (H), f = the frequency in hertz (Hz), and C = the capacitance in farads (F). *Note:* In electronic circuits, capacitance is almost always expressed in μF (microfarads) or pF (picofarads), because a farad is a very large amount of capacitance: $1 \ \mu\text{F} = 10^{-6} \ \text{F}$ and $1 \ \text{pF} = 10^{-12} \ \text{F}$.

Examples Involving Alternating Current

1. In Fig. 18.9(a), find the impedance and the current when $L = 5$ henries, $f = 60$ Hz, and $V = 120$ volts.

   Solution: $Z_L = 2\pi f L = 2(3.14)(60)(5) = 1885$ ohms;
   $\qquad\qquad I = V/Z = 120/1885 = 0.064$ amp $= 64$ milliamps.

2. In Fig. 18.9(b), find the impedance and the current when $C = 5 \ \mu$F, $f = 200$ Hz, and $V = 50$ volts.

   Solution: $Z_C = \dfrac{1}{2\pi f C} = \dfrac{1}{2(3.14)(200)(5 \times 10^{-6})} = 159$ ohms;
   Solution: $I = V/Z = 50/159 = 0.312$ amp.

18.4 ELECTRICAL RESONANCE

Mechanical vibrators, such as pendulums, mass-spring systems, open and closed pipes, and Helmholtz resonators, were discussed in

Chapter 2. Each of these vibrators has its own natural frequency and, if acted on by a driving force at this frequency, resonance occurs (see Chapter 4).

It is possible to have resonance in electrical circuits as well. The most common example of an oscillatory circuit is one with inductance and capacitance such as the simple circuit shown in Fig. 18.10. The resonance frequency of this circuit is

$$f_0 = \frac{1}{2\pi\sqrt{LC}}. \tag{18.5}$$

As the frequency of the source is varied, the current shows a maximum at the resonance frequency f_0, which is analogous to the maximum amplitude at resonance in Fig. 4.2. The maximum current and the line width in the electrical circuit are determined by the internal resistance of the inductor, the capacitor, and the driving source.

A slightly different kind of electrical circuit, which also shows resonance, is the one shown in Fig 18.11 in which an inductor and a capacitor are connected in parallel. Now as the frequency of the driving source is increased, the current in the inductor decreases but the current in the capacitor increases (see Fig. 18.9). At the resonance frequency f_0, the two currents are equal but, for reasons that we will not discuss, they are in opposite phase (that is, the currents I_L and I_C at any instant are in opposite directions). Thus the net current from the source drops to a minimum value (which would be zero except for internal resistance in the inductor and capacitor). A graph of current vs. frequency (see Fig. 18.11b) is a sort of "upside-down resonance curve," but if we prefer the more familiar line shape, we can make a graph of impedance Z vs. frequency, as in Fig. 18.11(c).

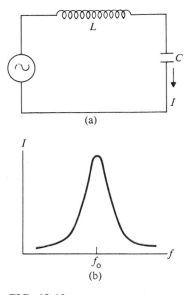

FIG. 18.10
(a) A simple circuit with inductance L and capacitance C in series.
(b) Response of this circuit to a driving voltage; when $f = f_0$ resonance occurs and the current I reaches its maximum value.

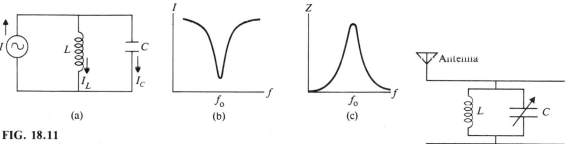

FIG. 18.11
(a) A circuit with inductance L and capacitance C in parallel. (b) The total current reaches a minimum value at the resonance frequency f_0.
(c) Impedance Z reaches a maximum at f_0.

A very common application for a circuit of this type is the tuning circuit in a radio or television receiver, shown in Fig. 18.12. In this case, the tuning capacitor C is variable, so the resonance frequency

FIG. 18.12
Tuning circuit in a radio or television receiver uses a circuit with a variable frequency of resonance.

can be changed in order to select different stations. The AM broadcast band extends from 550 kHz to 1600 kHz, and stations are assigned to frequencies that differ by 10 kHz, so the resonance linewidth should be somewhat less than 10 kHz. (The FM band is somewhat less crowded and greater linewidths can be used for higher fidelity.)

Examples Involving Resonance of Electrical Circuits

1. Find the resonance frequency of a 100-mh inductor and a 0.1-μF capacitor connected in series.

Solution:

$$f = \frac{1}{2\pi\sqrt{LC}} = \frac{1}{2(3.14)\sqrt{(10^{-1})(10^{-7})}} = \frac{10^4}{2(3.14)} = 1592 \text{ Hz.}$$

2. If $L = 0.2$ mh and $C = 200$ pF in a radio tuning circuit, to what frequency is the radio tuned?

Solution:

$$f = \frac{1}{2\pi\sqrt{LC}} = \frac{1}{2(3.14)\sqrt{(2\times10^{-4})(2\times10^{-10})}} = 7.96\times10^5 \text{ Hz}$$
$$= 796 \text{ kHz.}$$

3. Show that the currents I_C and I_L in Fig. 18.11 are equal when

$$f = \frac{1}{2\pi\sqrt{LC}}$$

Solution: Use the formulas in the box in Section 18.3 and set $I_C = I_L$:

$$I_C = 2\pi fCV = I_L = \frac{V}{2\pi fL}.$$

Multiply each side of the equation by f and divide by $2\pi CV$:

$$f^2 = \frac{1}{(2\pi)^2 LC} \quad \text{or} \quad f = \frac{1}{2\pi\sqrt{LC}}.$$

18.5   DIODES

A very common and useful circuit element is the *diode*. Nearly all diodes today are of the solid-state type, which consists of a tiny crystal of germanium or silicon with two regions of carefully selected impurities. In one region (called the *n*-region), there is an excess of electrons; in the other (called the *p*-region), there is a deficiency (see Fig. 18.13). Current (flow of positive charge) flows easily from the *p*-region to the *n*-region but not vice versa; thus, the diode is a *one-way* device for current.

18.6   TRANSFORMERS

A *transformer* consists of a pair of closely coupled inductors. Usually a transformer is constructed by winding two coils of wire on a core of iron or other magnetic material. The coil or winding that is conducted to the voltage source is called the *primary*; the winding that is connected to the load is called the *secondary*. The circuit diagram used to symbolize a transformer is shown in Fig. 18.14.

Transformers, by their nature, function only when the current in the primary is changing. Consequently, they are AC devices. One main function of a transformer is to "transform" the primary voltage up or down into the secondary voltage. If the number of turns of wire in the secondary is less than in the primary, then the secondary voltage will be less than the primary voltage. The converse of this is also true. In fact, the primary-to-secondary voltage ratio is exactly equal to the primary-to-secondary winding turns ratio:

$$\frac{V_p}{V_s} = \frac{N_p}{N_s} \quad \text{or} \quad V_s = V_p \frac{N_s}{N_p}. \tag{18.6}$$

If, for example, the number of turns in the primary is 300 for a given transformer and the number of turns in the secondary is 100, then the secondary voltage will be 40 volts when the primary voltage is 120 volts:

$$V_s = 120 \frac{100}{300} = 40 \text{ volts}.$$

Note that the transformer does not supply any power, so that even though primary and secondary voltages may differ, equal amounts of power are transmitted in the primary and secondary of an ideal transformer: $\mathcal{P}_p = \mathcal{P}_s$.

18.7   POWER SUPPLIES

Nearly all electronic circuits require a source of direct current. This can be supplied by batteries, as it is in portable equipment, but a more economical source of power is the AC power line. By using a transformer and diodes, it is possible to change 120-volt AC power to

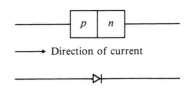

FIG. 18.13
A diode consisting of a *p*-region, an *n*-region, and a junction between them. The symbol for the diode indicates the direction of easy current flow.

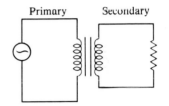

FIG. 18.14
A transformer connected to an AC source of voltage. The inductor connected to the voltage source is called the primary. The inductor connected to the load resistor is called the secondary.

DC power at any voltage desired. This can be done with the simple rectifier circuit shown in Fig. 18.15.

An AC of 120 volts is supplied to the primary of the transformer. By selecting the appropriate ratio N_p/N_s, the desired AC voltage V_s will appear at the secondary. This AC is then *rectified*, or changed, to a pulsating DC by means of the diode, which permits current to flow in one direction only (see Section 18.4). The waveform of the pulsating DC voltage V_R across the load resistor is shown in Fig. 18.15(b).

FIG. 18.15

(a) A simple rectifier circuit using a transformer and a diode to change AC to DC at the desired voltage.
(b) Waveforms at the secondary of the transformer (V_s) and at the load resistor (V_R).

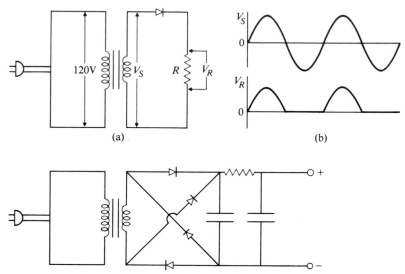

(a)

(b)

FIG. 18.16
Power supply circuit that uses several diodes and capacitors to produce a steady DC voltage by smoothing out the pulsations shown in Fig. 18.15(b).

A more practical power-supply circuit, shown in Fig. 18.16, works on the same principle as the simple rectifier circuit shown in Fig. 18.15, but by using additional diodes and capacitors, the pulsations in the DC voltage are smoothed out so that a steady DC is available.

18.8 SUMMARY

An electric current is the flow of electric charge; an electric current can carry energy from one place to another. An analogy can be drawn between an electrical circuit with a battery and a water circuit with a pump. Alternating current (AC), like direct current (DC), conveys energy or power, but it reverses its direction of flow several times each second. Inductors and capacitors are important elements in AC circuits, and they can be described by an impedance or opposition to the flow or current. A diode is a circuit element that allows current to flow easily in one direction but strongly opposes flow in the opposite direction. An electrical circuit with inductance and capacitance has a resonance frequency at which the current reaches a maximum (series circuit) or a minimum (parallel circuit).

A diode allows current to flow easily in one direction but resists its flow in the reverse direction. A transformer can change the voltage of an alternating current. A power supply generally uses a transformer and one or more diodes to produce DC power from AC power.

Glossary

AC (alternating current) Electric current that reverses its direction of flow several times each second (120 times each second, in the case of electrical power distributed in the United States having a frequency of 60 Hz).

capacitor A device that stores energy by creating an electrical field between two conductors; AC can flow through a capacitor but DC cannot.

current The flow of electrical charge. Current is measured in amperes, often abbreviated "amps."

diode A device that allows easy current flow in one direction only.

DC (direct current) Electric current that flows in one direction only, such as that supplied by a battery.

energy The ability to do work. Various units are used to measure energy, such as joules, BTU's, calories, footpounds, etc.

impedance A measure of the opposition to the flow of electric current by a circuit element such as a resistor, capacitor, or inductor. Impedance is measured in ohms.

inductor A device that stores energy by creating a magnetic field, usually within a coil of wire.

Ohm's law A fundamental law that relates electric current I, voltage V, and resistance R; written $I = V/R$, or $V = IR$.

potential, or **voltage** A measure of the electrical "force" or "pressure" that causes a current to flow; typically supplied by a generator or a battery and measured in volts.

power The rate at which energy is supplied or that rate at which work is done. It is measured in watts; one watt equals one joule per second.

rectifier A diode that is used to change AC into pulsating DC.

resistor A device that converts electrical energy into heat.

resonance The natural frequency of a system, at which its response to a mechanical or electrical force reaches a maximum.

transformer A device that changes AC at one voltage into AC at a higher or lower voltage.

Questions for Discussion

1. Explain why a fluorescent lamp flashes 120 times per second (why not 60 times, for example?). Also, why does it remain lit between flashes?

2. Sketch a simple circuit for a battery charger using a diode as a rectifier.

3. The circuit shown to the right might serve as a treble tone control in an amplifier. Explain how it works. Will more treble sound be heard in position A or in position B?

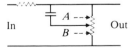

Problems

1. In each of the circuits below, find the current *I*.

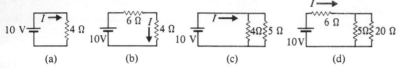

| (a) | (b) | (c) | (d) |

2. Using the figure shown below, determine how much power is consumed if
a) $R = 20 \, \Omega$;
b) $R = 40 \, \Omega$.

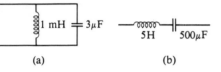

3. Find the resonance frequency of each circuit shown below.

(a)   (b)

4. The circuit shown below represents a flashlight.
a) How much resistance should the bulb have in order for the current to be 2 A?
b) How much power will be supplied to the bulb?

5. What is the total resistance of
a) Resistances of 5 Ω, 6 Ω, and 15 Ω in parallel?
b) Resistances of 20 Ω and 5 Ω in parallel?

6. Show that when resistances R_1 and R_2 are connected in parallel, the total resistance is

$$\frac{R_1 R_2}{R_1 + R_2} \, .$$

Test this out with the resistances given in Problem 5(b).

7. A 5-volt source with an internal resistance $r = 10$ ohms delivers power to a 5-ohm load. How much power is delivered to the load? (*Hint:* Find the current and then use Ohm's law to find the voltage across the load. The product of these two quantities will be the power delivered to the load.) Repeat for a "matched" load, $R = 10$ ohms.

8. A transformer is to furnish an AC of 6 volts from a 120-volt power source.
a) What should the turns ratio be?
b) If the current in the load is 2 A, what is the current in the primary?
c) Verify that power in the primary and the secondary are the same by multiplying current and voltage in the primary and the secondary.

CHAPTER 19
Filters, Amplifiers, and Oscillators

In this chapter we will apply some of the principles of electricity discussed in Chapter 18 to simple electronic circuits, both active and passive. Active circuits include a power source, whereas passive circuits do not. Filter networks are passive in nature; amplifiers and oscillators are active.

19.1   FILTERS

It is frequently desirable to filter out signals of high or low frequency or to alter the balance between them. This can be accomplished by using a passive frequency-selective circuit called a *filter*. An example of a filter is the treble tone control circuit in an audio amplifier (see Question 3 in Chapter 18), which attenuates high-frequency signals by the desired amount. Such a filter is called a *low-pass* filter, because signals of low frequency pass through without attenuation. Other types of filters are called high-pass, band-pass, and band-reject.

A *high-pass* filter passes signals of high frequency but attenuates those of low frequency. A *band-pass* filter passes signals within a certain frequency band but attenuates all others. The reverse of a band-pass filter is a *band-reject*, or *notch*, filter, which attenuates only signals within a certain frequency band and passes all others. The

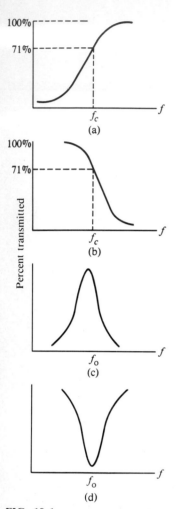

FIG. 19.1

Characteristics of basic filter types: (a) high-pass; (b) low-pass; (c) band-pass; (d) band-reject. The cutoff frequency is f_c and the resonance frequency is f_0.

cutoff frequency of a high-pass or low-pass filter is the frequency at which the response has dropped to 71 percent of its maximum value. Band-pass and band-reject filters are described by a resonance frequency at the center of the frequency band they filter. Characteristics of the four basic filter types are shown in Fig. 19.1

Modern electronic instruments frequently make use of rather sophisticated filter networks. However, rather basic filters can be constructed from simple combinations of a resistor with a capacitor or an inductor, whose impedances, you will recall, depend on frequency.

Either of the networks shown in Fig. 19.2 can act as a high-pass filter. In the first, the low frequencies are attenuated by the series capacitor, whereas high frequencies are allowed to pass. In the second network, the inductor acts as a shunt ("short circuit") for low-frequency signals, whereas high frequencies pass on through the filter.

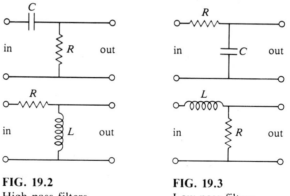

FIG. 19.2
High-pass filters.

FIG. 19.3
Low-pass filters.

Low-pass filters are just the reverse of high-pass filters. In the first example in Fig. 19.3, high frequencies are shunted by the capacitor; in the second example, frequencies above the cutoff are attenuated by the inductor in series.

The cutoff frequency of a basic *RC* high-pass filter is

$$f_c = \frac{1}{2\pi RC}. \tag{19.1}$$

Expressing C in farads (F) and R in ohms (Ω) gives f_c in hertz. The cutoff frequency of a basic *LR* high-pass or low-pass filter is

$$f_c = \frac{R}{2\pi L}; \tag{19.2}$$

inductance L is expressed in henries (H).

Practical filters in electronic instruments frequently combine two or more of the simple networks we have just discussed. A band-pass filter, for example, can be constructed by combining a high-pass filter and a low-pass filter. (The high-pass filter, of course, must have a slightly lower cutoff frequency than the low-pass filter does.) Another way to make a band-pass filter, however, is by using a resonant circuit, as shown in Fig. 19.4. The resonance frequency of the parallel *LC* circuit is

$$f_{\mathrm{o}} = \frac{1}{2\pi\sqrt{LC}}, \tag{19.3}$$

and this becomes the center frequency of the pass band.

19.2 AMPLIFIERS

Basically, an amplifier is an active device in which a small amount of power P_{in} is used to control a larger amount of power P_{out}, as illustrated in Fig. 19.5. The "valve" that controls the flow of current is most often a vacuum tube or a transistor.

In a single transistor amplifier, input power is supplied to the base circuit and the output power P_{out} actually comes from the battery but is *controlled* by P_{in}. In the water circuit, P_{in} turns the handle of a faucet regulating the flow of water. Again the output power is actually supplied by the pressure of the water in the reservoir, but its flow is controlled by the faucet. The power gain of either amplifier is simply $P_{\mathrm{out}}/P_{\mathrm{in}}$.

Prior to 1960 most electronic amplifier circuits used vacuum tubes to control the flow of current. A vacuum tube, shown schematically in Fig. 19.6, consists of a hot cathode that emits electrons, a plate to collect them, and one or more grids that control their rate of flow. A small amount of power P_{in} supplied to the grid controls a large amount of power P_{out} in the plate circuit, hence amplification takes place. The power gain again is simply $P_{\mathrm{out}}/P_{\mathrm{in}}$.

A major revolution in electronics followed the invention of the transistor in 1946. Transistors occupy far less space than do vacuum tubes, and they do not depend on heated cathodes to supply electrons. Radios and amplifiers with transistor circuits can be built incredibly small and are considerably more reliable than their vacuum-tube ancestors. Transistors have replaced vacuum tubes in most ordinary amplifier circuits. We now describe briefly how transistors amplify.

In one common type of transistor, a crystal of silicon has been divided into three regions by the introduction of selected impurities in the three regions. Two of these regions (designated *emitter* and *collector*) have an excess of free electrons, while an intervening layer (called

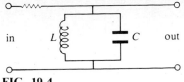

FIG. 19.4
A band-pass filter, which utilizes a parallel combination of an inductor and a capacitor.

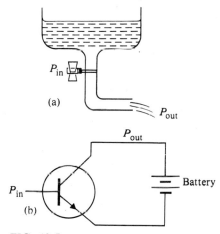

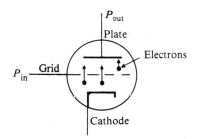

FIG. 19.5
(a) A water "amplifier" circuit in which a small amount of power P_{in} applied to a faucet controls a relatively large amount of power P_{out} in the water from a reservoir. (b) A transistor amplifier circuit in which a small amount of power P_{in} applied to the base controls the flow of a larger power P_{out} in the collector circuit.

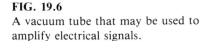

FIG. 19.6
A vacuum tube that may be used to amplify electrical signals.

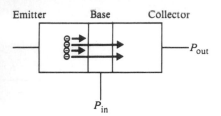

FIG. 19.7
An *npn* junction transistor. The flow of electrons from emitter to collector is controlled by applying an electrical signal to the base.

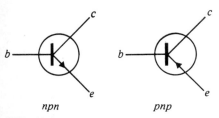

npn   *pnp*

FIG. 19.8
Symbols used for junction transistors in circuit diagrams. The arrow on the emitter designates the direction of flow of the + charge.

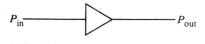

FIG. 19.9
Amplifier symbol used in simplified circuit diagrams.

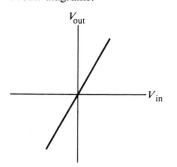

FIG. 19.10
A high-fidelity amplifier has a linear characteristics (that is, the graph of output vs. input is a straight line).

the *base*) has a deficiency of electrons. The flow of a substantial electron current from emitter to collector is controlled by a relatively small amount of power applied to the base, and in this way the transistor amplifies.

A transistor of this type, shown in Fig. 19.7, is called an *npn* junction transistor (*n* denotes that negative electrons abound in the emitter and collector regions; the base region is denoted *p* for positive). Symbols for both *npn* and *pnp* transistors are shown in Fig. 19.8. The base, emitter, and collector are designated *b, e,* and *c,* respectively. The arrow on the emitter indicates the direction of current (flow of the positive charge).

There are other useful types of transistors, given names such as field-effect transistors, unijunction transistors, and metal-oxide-semiconductor devices, etc., which operate in the same general manner as the simple junction transistor described. Some of these devices are described in Chapter 29.

Often in simplified electronic circuit diagrams, an amplifier is merely denoted by a simple triangular symbol with input and output, as shown in Fig. 19.9. The active element may be either a vacuum tube or a transistor. One of the most significant parameters in describing an amplifier is its *voltage gain*, that is, the ratio of the output voltage to the input voltage. If this ratio remains constant, the amplifier is said to have a *linear* characteristic. (See Fig. 19.10.)

Two other important parameters of an amplifier are its current gain and its power gain. *Current gain* is the ratio of output current to input current, and *power gain* is the ratio of output power to input power. Since electrical power is the product of current times voltage, it is not difficult to see that the power gain will be the current gain times the voltage gain.

An amplifier is termed either a *voltage amplifier* or a *power amplifier*, depending on whether it is designed primarily to have a large voltage gain or a large power gain. An example of a voltage amplifier is a preamplifier designed for use with a microphone or a phonograph cartridge. The amplifier that drives a loudspeaker, on the other hand, is a power amplifier. A power amplifier has a low output impedance so that it can deliver a large output current.

Audio amplifiers nearly always incorporate several individual stages, some of which are voltage amplifiers (the early stages) and some of which are power amplifiers (the output stage and the stage that precedes it).

19.3 DISTORTION IN AMPLIFIERS

An ideal amplifier is one in which the output faithfully resembles the input but at a higher level. No amplifier is capable of performing this task perfectly. Real amplifiers invariably produce some distortion, although in high-quality amplifiers, the distortion may be small enough to be unnoticeable.

We will discuss two types of distortion: limited frequency response and nonlinear gain. An audio amplifier should have a "flat" frequency response (that is, gain that is independent of frequency) over the audible range of at least 50 to 20,000 Hz. Nearly all high-fidelity amplifiers meet this criterion. In low-quality amplifiers, however, low-frequency "roll-off" (at point A in Fig. 19.11) may be caused by coupling capacitors that are too small and therefore present too much impedance to the passage of low-frequency signals from one stage to the next. High-frequency roll-off (at point B in the figure) could be caused by stray capacitance in the circuit or by transistors that have insufficient gain at high frequency.

The frequency response of an amplifier is often described by the bandwidth between the "3 dB points," that is, the frequencies at which the voltage gain has dropped to 71 percent of its midfrequency value. At these points, the gain is 3 dB below its midfrequency value.

One way to quickly test the frequency response of an amplifier is by applying a square-wave voltage to the input. A square wave is made up of all the odd-numbered harmonics; the high-numbered ones shape the leading and trailing edges, whereas the low-numbered ones determine the flatness of the top of the wave. Thus, distorted outputs appear as shown in Fig. 19.12 if the gain decreases at either high or low frequency.

In most amplifiers, there is a range of input signal V_{in} for which the output faithfully resembles the input. For larger signals, however, the output may be distorted, as shown in Fig. 19.13. Supplying an amplifier with an input signal that takes it outside its linear region is referred to as "overdriving" the amplifier. The distorted output frequently contains harmonics that were not present in the input signal.

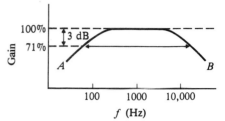

FIG. 19.11
Gain vs. frequency characteristic of an amplifier showing rolloff in gain at high frequency (B) and low frequency (A). The bandwidth is measured between the 3-dB points.

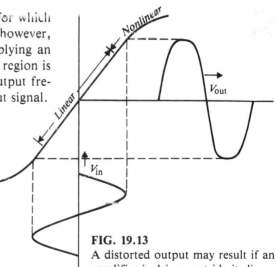

FIG. 19.13
A distorted output may result if an amplifier is driven outside its linear region. The distorted output has a greater harmonic content than the input.

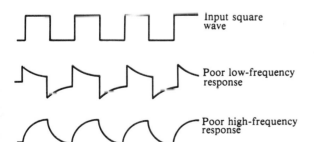

Input square wave

Poor low-frequency response

Poor high-frequency response

FIG. 19.12
Response of an amplifier to a square wave input.

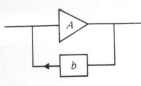

FIG. 19.14
An amplifier circuit with negative feedback. A fraction *b* of the output signal is fed back to the input.

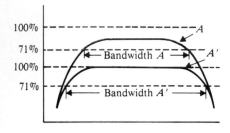

FIG. 19.15
Improvement in frequency response of an amplifier with negative feedback. *A* is the gain without feedback; *A'* is the gain with negative feedback.

19.4   FEEDBACK

One effective way to reduce distortion in amplifiers is to incorporate into the circuit some *negative feedback*. A fraction *b* (see Fig. 19.14) of the output signal is fed back in opposition to the input. This reduces the gain of the amplifier somewhat, but it reduces the distortion much more.

Figure 19.15 illustrates the improvement in frequency response of an amplifier that results from negative feedback. The upper curve gives the gain *A* without feedback, whereas the lower curve gives the gain *A'* with negative feedback. The gain has been reduced, but the frequency range over which the amplifier has a consistent ("flat") response has been extended. The loss in gain is easily made up by additional stages of amplification.

Negative feedback also reduces nonlinear or harmonic distortion, since any part of the output V_{out} that does not appear in the input V_{in} will be subtracted from the input. Thus any harmonic distortion generated within the amplifier will be subtracted from the output. Are we getting something for nothing? Not really, because negative feedback reduces the gain of the amplifier besides minimizing distortion and increasing stability. If the voltage gain of the amplifier itself without feedback is denoted by *A*, the gain with feedback will be

$$A' = \frac{A}{1 + bA} \tag{19.4}$$

19.5   OPERATIONAL AMPLIFIERS

Integrated circuit technology (to be discussed in Chapter 29) has made operational amplifiers, or "op amps," so common that mention should be made of them in this chapter.

An operational amplifier is a high-gain amplifier with a large amount of negative feedback. It actually consists of several transistors, resistors, and other components fabricated on a single semiconductor chip. In the simplified schematic in Fig. 19.16, two inputs are shown: One is called "inverting" ($-$) and one is called "noninverting ($+$). A signal applied to the noninverting input produces an output signal with the same phase as the input, whereas a

FIG. 19.16
A schematic for an operational amplifier (op amp).

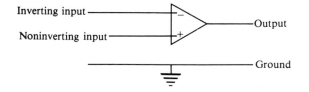

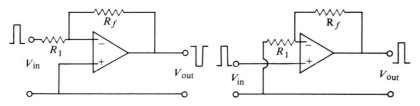

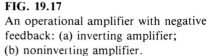

FIG. 19.17
An operational amplifier with negative feedback: (a) inverting amplifier; (b) noninverting amplifier.

signal applied to the inverting input produces an output of opposite phase.

A resistor R_f is usually connected from the output to the inverting input, as shown in Fig. 19.17, to provide negative feedback. The input signal may be applied to either the inverting or the noninverting input, as shown.

The feedback in an operational amplifier is usually so great that the voltage gain is determined by the ratio of the resistances R_f and R_1. In the inverting amplifier, the voltage gain is $A = R_f/R_1$, whereas in the noninverting amplifier, it is $A = (R_f/R_1) + 1$.

19.6   OSCILLATORS

The opposite of negative feedback is positive feedback. Negative feedback decreases the gain of an amplifier but makes it more stable; positive feedback *increases* the gain but may make it *unstable*. An amplifier that becomes sufficiently unstable generates signals of its own. An example of this is acoustic (positive) feedback between a loudspeaker and a microphone that may cause a public address system to go unstable and generate howls and squeals until the operator turns down the gain (see Fig. 19.18).

Amplifier instability is not necessarily bad. An amplifier that is unstable at one frequency, only, can be useful as an *oscillator*, or *generator*. This selective instability can be accomplished by means of positive feedback through a band-pass filter. This is, in fact, how most audio generators operate. The band-pass filter can be made tunable so that the generator can be tuned to the desired frequency (see Fig. 19.19).

Audio generators or oscillators used in the laboratory and studio are usually tunable over a wide frequency range (e.g., 20 Hz to 1 MHz). There is frequently a "range" or "multiplier" switch, which changes the frequency by factors of ten, plus a continuous frequency control for selecting the desired frequency within the range. Audio generators often furnish square waves as well as sine waves (pure tones), and an output amplitude control is usually included. (An audio generator is shown in Fig. 33.4.)

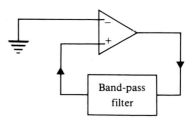

FIG. 19.18
Acoustic feedback path from the loudspeaker to the microphone may cause oscillation.

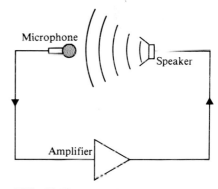

FIG. 19.19
Positive feedback through a band-pass filter causes oscillation at a single frequency determined by the filter.

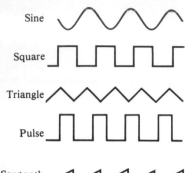

FIG. 19.20
Waveforms available from a function generator.

19.7   FUNCTION GENERATORS

Increasingly popular in the laboratory are function generators, which generate a variety of electrical waveforms or functions (see Fig. 19.20). Sine waves, square waves, and triangle waves are nearly always available, and other waveforms such as the pulse and the sawtooth may be available as well. The frequency of many function generators can be controlled electrically; that is, the frequency can be changed from that indicated on the frequency dial by applying an electrical control voltage. (A function generator is shown in Fig. 33.5.)

19.8   SUMMARY

A filter is a frequency-selective circuit that may have a high-pass, low-pass, band-pass, or band-reject characteristic. Band-pass filters incorporate circuits that have a resonance at the desired frequency.

An amplifier is a device in which a small amount of input power controls a larger amount of output power. Electrical amplifiers accomplish this by using a transistor, vacuum tube, or similar device as a "valve." The main current in a vacuum tube always flows from plate to cathode, but a transistor can be designed to carry current in either direction, and it is designated *npn*- or *pnp*-type, accordingly.

Distortion can be reduced in an amplifier by negative feedback. Positive feedback, on the other hand, causes an amplifier to become unstable and eventually to oscillate. Audio generators or oscillators are amplifiers with positive feedback at a selected frequency. Function generators provide a variety of waveforms and often have provision for voltage control of frequency.

Glossary

amplifier A device in which a small amount of input power controls a larger amount of output power.

base, collector, emitter The three regions of a transistor; in the most common usage, the input signal is applied to the base and the output is taken from the collector.

current gain The ratio of output current to input current.

distortion A measure of the difference between the output and input signals in an amplifier.

feedback An arrangement by which a portion of the output of an amplifier is applied to the input. Negative feedback reduces amplifier gain but also decreases distortion; positive feedback increases the gain and may lead to self-oscillation.

filter An electrical circuit that passes alternating currents of some frequencies and attenuates others. Basic filter types are high-pass, low-pass, band-pass, and band-reject.

function generator An audio generator that provides several different waveforms or functions at the desired frequency.

npn transistor A transistor that conducts primarily by the flow of electrons.

operational amplifier ("op amp") A high-gain amplifier with a large amount of negative feedback and high input impedance.

pnp transistor A transistor that conducts primarily by the flow of positive charge ("holes").

power gain The ratio of output power to input power.

transistor A solid-state amplifying device consisting of a crystal of germanium or silicon with carefully selected impurities.

vacuum tube An amplifying device in which electrons from a hot cathode flow through one or more grids before reaching a plate.

voltage amplifier An amplifier designed mainly for voltage gain.

voltage gain The ratio of output voltage to input voltage.

Questions for Discussion

1. List several advantages of using transistors rather than vacuum tubes. Can you think of any possible advantages of tubes?

2. Is it possible to produce a nearly sinusoidal wave (single frequency) by filtering a square wave? Would you use a high-pass or a low-pass filter? (See Section 7.12, Fig. 7.13 in particular.)

3. a) Will the negative feedback in this amplifier be greater at high or at low frequency? Explain.

b) Will the voltage gain be greater at high or low frequency? Explain.

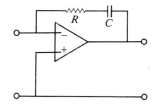

Problems

1. Find the cutoff frequency of each circuit shown below and tell whether it is a high-pass or low-pass filter.

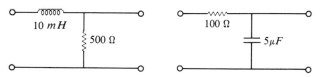

2. a) Find the voltage gain of the amplifier shown below.

b) Is it an inverting or a noninverting amplifier?

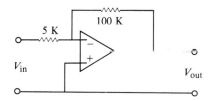

3. A certain amplifier has a voltage gain of 100 and a current gain of 50. What is its power gain? Using the same formula as that used for sound power (see Section 6.4), express the power gain in dB.

4. Two amplifier stages, each having a power gain of 500, are "cascaded," with the output of one connected to the input of the other.

a) Show that the combined amplifier has a power gain of 250 000.

b) Show that the total gain in dB is the sum of the individual amplifiers.

5. A series LC circuit acts as a band-pass filter with

$$f_o = \frac{1}{2\pi\sqrt{LC}}.$$

a) Explain how the operational amplifier circuit shown below functions as an oscillator.

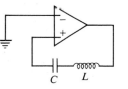

b) At what frequency will it oscillate if $L = 5$ mh and $C = 10 \ \mu$F?

6. An oscillator usually has several ranges. If the

capacitor C in the circuit shown below can be varied from 100 to 500 pF, what frequency range can be tuned with each value of L?

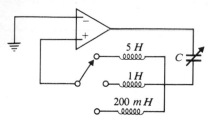

7. A certain amplifier has a voltage gain A as given.

a) Make a graph of A vs. f on semilog graph paper (see Fig. 5.11).

b) Find the gain A' if negative feedback equal to 2 percent of the output is provided.

c) Make a plot of A' vs. f on the graph of part (a).

d) Find the bandwidth with and without feedback.

f	A	A'
20	50	
50	120	
100	140	
200	150	
2,000	150	
5,000	140	
10,000	110	
20,000	50	

CHAPTER 20
Microphones and Loudspeakers

It is virtually impossible to amplify sound waves (that is, to add energy to them as sound waves). Electrical signals, on the other hand, are relatively easy to amplify. Thus, a practical system for amplifying sound includes input and output *transducers*, together with an electronic amplifier. The input and output transducers are usually a microphone and a loudspeaker, respectively. A microphone converts a sound signal to an electrical signal, and a loudspeaker converts the amplified electrical signal to a much louder sound, as shown in Fig. 20.1.

Similarly, to record music, whether on tape or disc, it is necessary to convert sound signals to electrical signals, which can be amplified in order to drive a recording stylus or recording head, as in Fig. 20.2. The playback system includes an amplifier and a loudspeaker to convert electrical signals into sound. The record and playback transducers will be discussed in Chapter 21.

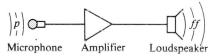

FIG. 20.1
Public address system consisting of a microphone, an amplifier, and a loudspeaker. Sound is converted to an electrical signal, which is amplified and used to generate a louder sound.

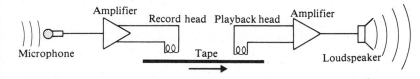

FIG. 20.2
Tape recording system. To the system shown in Fig. 20.1 are added record and playback heads and a second amplifier.

393

20.1   MICROPHONES AS TRANSDUCERS

A microphone is a transducer that produces an electrical signal when actuated by sound waves. Microphones may be designed to respond to variations in air pressure due to the sound wave or to variations in particle velocity as the sound wave propagates. Most microphones in common use are *pressure* microphones.

A pressure microphone has a thin diaphragm, which moves back and forth with the rapid pressure changes in a sound wave. The diaphragm is connected to some type of electrical generator, which may be a piezoelectric crystal (crystal microphone), a moving coil (dynamic or magnetic microphone), or a variable capacitor (condenser microphone). Each of these microphone types will be discussed briefly.

One can also classify microphones according to their ability to pick up sound arriving from different directions. Microphones may be omnidirectional (that is, nearly the same sensitivity in all directions), unidirectional (that is, considerably more sensitive in one direction), or bidirectional in their response pattern. The directionality of a microphone is usually determined by the acoustical design of its housing.

Crystal Microphones

Crystal microphones (see Fig. 20.3) use piezoelectric crystals that generate a voltage when deformed by a mechanical pressure. Rochelle salt crystals generate a large electric signal and are used in some inexpensive microphones. Better-quality crystal microphones generally use a ceramic material, however, which is much less sensitive to humidity, temperature, and mechanical shock. Ceramics commonly used include barium titanate, lead titanate, and lead zirconate. The piezoelectric effect is described in the box below.

Crystal microphones have a comparatively large electrical output voltage, which makes them convenient for use in portable sound equipment and tape recorders. On the other hand, the high-frequency response of most crystal microphones is less "flat" than that of condenser microphones due in part to the greater mass that must move. Therefore, they are not often used in the professional recording of sound.

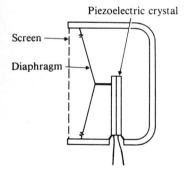

FIG. 20.3

A crystal microphone. Sound pressure on the diaphragm causes a deformation of the crystal, generating an electrical signal.

The *piezoelectric* (pronounced "pah-ee-zo-electric") effect was discovered in 1880 by the brothers Pierre and Jacques Curie, who observed that certain crystals acquire electric charge when they are compressed or distorted. The amount of charge that appears is proportional to the distortion and disappears when the force is removed. If leads are attached to the crystal at the appropriate places, an electrical output signal is obtained.

The piezoelectric effect is reversible; if an electrical voltage is applied to a piezoelectric crystal, it changes its shape. This has led to the use of piezoelectric crystals as ultrasonic generators and even as loudspeakers for high-frequency sound. Quartz, tourmaline, topaz, and Rochelle salt are examples of natural piezoelectric crystals.

Dynamic Microphones

In a *dynamic* or magnetic microphone, an electrical signal is generated by the motion of a conductor in a magnetic field. This is an example of *electromagnetic induction*, or the dynamo principle, discussed in the box below. In the most common type of dynamic microphone, sound pressure on a diaphragm causes the attached voice coil to move in the field of a magnet, as shown in Fig. 20.4. Movement of the voice coil thus generates an output voltage.

The electrical voltage generated by a dynamic microphone is quite small, but the impedance of the voice coil is also small. Because of the low source impedance, the power generated by a dynamic microphone is not necessarily small, even though the voltage output is small. The low source impedance is a distinct advantage when the microphone is located a substantial distance from the amplifier, as will be discussed later (see Section 20.4). If a higher output voltage is desired, a transformer can be used, preferably at the amplifier end of the transmission line. It is essential that the mass of the moving coil be kept very small in order to yield good response at high frequency.

A dynamic microphone is similar in principle to a small loudspeaker, and, indeed, a loudspeaker will serve as a crude microphone. This is frequently done in two-way intercommunication units. The mass of a loudspeaker cone plus voice coil is too large to give satisfactory response at high frequency, however, and so its fidelity as a microphone is generally low.

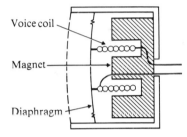

FIG. 20.4
A dynamic (moving coil) microphone. Sound pressure on the diaphragm causes the voice coil to move in a magnetic field.

Electromagnetic induction was discovered by Michael Faraday in 1831. When a wire moves in a magnetic field B (see the figure on p. 360), an electrical voltage is generated. When the wire moves downward, the voltage is as shown; when the wire moves upward, the voltage is reversed.

Electromagnetic induction has a wide range of applica-

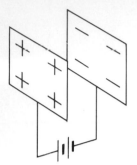

FIG. 20.5
A simple capacitor (condenser) consisting of two conducting plates connected to a battery. One plate acquires a positive charge, the other a negative charge. The amount of charge changes as the spacing between the plates changes.

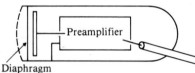

FIG. 20.6
A condenser microphone. The diaphragm constitutes one plate of a capacitor and it moves with sound pressure.

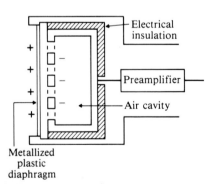

FIG. 20.7
An electret condenser microphone. A thin metalized plastic diaphragm is tightly stretched across a perforated backing plate. The holes in the back plate couple to an air cavity.

tions, from electrical generators at power plants to magnetic phonograph cartridges.

The reverse effect forms the basis of electric motors as well as loudspeakers. If an electric current flows in a conductor that is in a magnetic field, the conductor experiences a force that depends on the strengths of the current and the magnetic field.

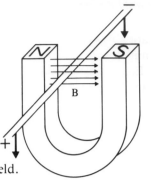

Dynamic microphones are rugged, have a broad frequency response, and are able to withstand the high sound levels that occur in popular music. Thus they are widely used in both live performance and recording.

Condenser Microphones

When two metal plates that face each other are connected to a battery, they acquire and store electric charge (see Fig. 20.5). Such an arrangement of charged plates is called a *capacitor*, or *condenser*. The amount of electric charge that can be stored in a condenser depends on the size of the plates and on their spacing. Thus if one of the plates moves, an electric current will flow in the circuit as charge arrives and departs from the plate. This is the principle of the condenser, or electrostatic, microphone.

In a condenser microphone, one plate is usually the thin movable diaphragm, and the other plate is the fixed backing plate, as shown in Fig. 20.6. As the diaphragm is moved by the pressure of a sound wave, a small current flows in the circuit.

A condenser microphone has a very high source impedance, and for this reason, a preamplifier is usually incorporated into the microphone itself. The diaphragm can be made very light; therefore, the condenser microphone is capable of excellent response at high frequencies. A major disadvantage is the need for a relatively high voltage source to maintain electrical charge (bias) on the plates of the capacitor.

Electret-Condenser Microphones

A type of microphone that retains most of the advantages of the condenser microphone but eliminates the need for a high-voltage bias supply is the *foil-electret*, or *electret-condenser*, microphone, shown in Fig. 20.7. The diaphragm consists of a plastic foil less than one-thousandth of an inch thick with an even thinner layer of metal attached. The foil has been given a permanent electrical charge by a combination of heat and high voltage or by electron bombardment during manufacture.

Electret-condenser microphones often incorporate a built-in preamplifier that is usually a field-effect transistor. Because of their high input impedance, field-effect transistors are a better match for condenser microphones than the junction transistors we discussed in Chapter 19.

20.2 VELOCITY MICROPHONES

A *ribbon* microphone is a magnetic microphone in which the lightweight ribbon diaphragm is also the moving conductor (see Fig. 20.8). The ribbon responds to the particle velocity rather than the pressure of the sound wave. A ribbon microphone responds readily to sound waves arriving from the front or back but is insensitive to sound arriving from the sides. Thus it has a bidirectional response.

Ribbon microphones were popular in the early days of radio broadcasting, because their bidirectional nature allowed performers to stand on opposite sides facing each other. They are sensitive to moving air currents, however, and therefore are not satisfactory for outdoor use. Their use is now mainly restricted to special situations, such as a vocal soloist who wants to sing very close to the microphone and minimize the danger of "popping" when plosive consonants (p, t, d, etc.) are sung.

20.3 UNIDIRECTIONAL MICROPHONES

In many applications, it is desirable to have a microphone with maximum sensitivity in one direction only. One such application is in sound systems in which acoustic feedback may lead to oscillation if the microphone picks up sound from the loudspeaker (see Chapter 24).

The most popular unidirectional microphone is the *cardioid* microphone, which has a heart-shaped pickup pattern (see Fig. 20.9). A cardioid microphone can be made by adding a hole in the rear of the case of a sealed omnidirectional microphone. When a sound wave arrives from the rear, the pressure maxima and minima reach both front and back of the diaphragm at the same time, and its deflection is thus minimal. A sound wave arriving from the front produces normal deflection, however.

It is relatively difficult to obtain a cardioid response pattern over a wide range of frequencies; thus, many cardioid microphones have an

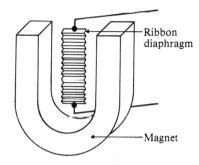

FIG. 20.8
A ribbon microphone. A lightweight ribbon diaphragm moves in a magnetic field, thus generating an electrical signal.

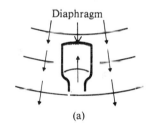

(a)

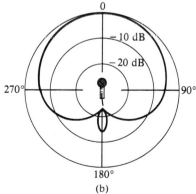

FIG. 20.9
A simple cardioid microphone (a) and its pickup pattern (b).

uneven frequency response, especially a roll-off at bass frequencies. Off-axis response, in particular, will be colored by a variation in the directional response with frequency. This is offset, however, by the unidirectional characteristic of the cardioid microphone, which reduces the pickup of background noise.

For even greater directivity, line microphones or reflector microphones can be used. The line microphone shown in Fig. 20.10(a) (sometimes called a "shotgun" microphone) has a tube with side holes through which sound can enter. Sound waves approaching from nearly head-on will leak into the tube at each side hole, building up into a traveling wave within. Sound waves from other directions, however, will enter the tube in varying phases through the side holes, and these contributions will tend to cancel each other.

A reflector microphone consists of a microphone element at the focus of a parabolic reflector, as shown in Fig. 20.10(b). Parabolic reflectors are also used in spotlights and telescopes; they are shaped so that all the sound (or light) arriving from straight ahead is reflected toward the focus. A reflector microphone has even greater directivity than a line microphone does.

FIG. 20.10
Directional microphones: (a) line microphone; (b) reflector microphone.

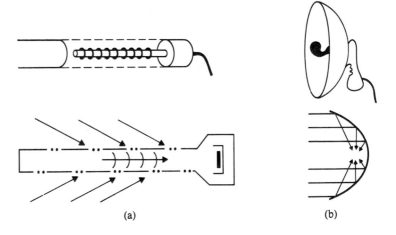

(a) (b)

20.4 MICROPHONE IMPEDANCE

Microphones are generally classified as having *high impedance* or *low impedance* (see Section 18.3). Crystal, condenser, and electret-condenser microphones have high output impedances from 50,000 ohms and up. Dynamic microphones, on the other hand, have low impedances, typically from 50 to 600 ohms.

For best results, the output impedance of a microphone should closely match the input impedance of the amplifier into which it operates. If they do not match, a step-up or a step-down transformer should be used.

If a high-impedance microphone is used with a long microphone cable, high-frequency signals will be attenuated (a 20-foot cable, for example, may result in a 6-dB loss at 10,000 Hz). This loss may be prevented by using a line-matching transformer or by locating a pre-amplifier near the microphone. Low-impedance microphones can be used with long cables without much loss of signal at high frequency. Low-impedance (600-ohm) cables are frequently used in sound systems.

20.5 MICROPHONE SENSITIVITY

There are basically two ways to express the sensitivity of a microphone: in terms of voltage output or in terms of power output for a given sound pressure. Sensitivity is usually expressed in decibels, however, so a reference level is used, but the reference level often is not stated in microphone specifications found in catalogs. All of this makes it somewhat difficult to select and compare microphones.

The three most commonly used methods for rating microphone sensitivity and the formulas for calculating them are as follows:

1. Open-circuit output voltage for a sound pressure of one microbar ($=0.1$ N/m^2). The voltage sensitivity S_v is expressed in dBV, or simply in dB:

$$S_v(\text{dBV}) = 20 \log V/p - 20, \qquad (20.1)$$

where V is the open-circuit output voltage and p is the sound pressure in N/m^2.

2. Maximum output power in milliwatts for a sound pressure of ten microbars ($=1$ N/m^2). The power sensitivity S_p is expressed in dBm, where "m" means that 0 dBm is at one milliwatt. (Unfortunately, dBm is sometimes written as merely dB.)

$$S_p(\text{dBm}) = 20 \log V/p - 10 \log Z + 24, \qquad (20.2)$$

where V is the open-circuit output voltage, Z is the output impedance in ohms, and p is the sound pressure in N/m^2.

3. Maximum output power in milliwatts for a sound pressure of 2×10^{-5} N/m^2 (the same reference used in expressing sound-pressure level). The result, called G_m, is the standard recommended by the Electronic Industries Association (EIA). (Unfortunately, G_m is also expressed in dB.)

$$G_m(\text{dB}) = 20 \log V/p - 10 \log Z - 70, \qquad (20.3)$$

where V is the open-circuit output voltage, Z is the output impedance in ohms, and p is the sound pressure in N/m².

In each case, the open-circuit output voltage is given. The output voltage into a matched load (load resistance equal to Z) will be one-half as great.

Of primary interest to the microphone user is the output voltage that can be expected at a given sound level. The open-circuit output voltage is expressed by the following formulas:

$$20 \log V = S_v + L_p - 74 \qquad \text{(Method 1)}, \qquad (20.4)$$
$$20 \log V = S_p + 10 \log Z + L_p - 118 \qquad \text{(Method 2)}, \qquad (20.5)$$
$$20 \log V = G_m + 10 \log Z + L_p - 24 \qquad \text{(Method 3)}. \qquad (20.6)$$

To determine the microphone output voltage at a given sound-pressure level L_p, use one of the above formulas, the choice depending on how the microphone sensitivity is expressed.

Examples of Computing Microphone Output Voltage

1. A microphone has a voltage sensitivity of -66 dBV re one microbar. What is its output voltage when $L_p = 90$ dB?

$$20 \log V = S_v + L_p - 74 = -66 + 90 - 74 = -50$$
$$V = 3.16 \times 10^{-3} \text{ volts} \approx 3.2 \text{ millivolts.}$$

2. A microphone has a power sensitivity of -50 dBm re ten microbars and an output impedance of 600 ohms. When $L_p = 90$ dB, find its output voltage (a) into a load whose resistance is much larger than 600 ohms, and (b) into a matched 600-ohm load.

a) $20 \log V = S_p + 10 \log Z + L_p - 118$
$$= -50 + 28 + 90 - 118 = -50$$
$$V = 3.16 \times 10^{-3} \text{ volts} \approx 3.2 \text{ millivolts.}$$

b) $V = \frac{1}{2}V \approx 1.6$ millivolts.

3. A microphone has an EIA sensitivity rating $G_m = -144$ dB and a rated impedance of 600 ohms. Find its open-circuit output voltage when the sound level is 90 dB.

$$20 \log V = G_m + 10 \log Z + L_p - 24$$
$$= -144 + 28 + 90 - 24 = -50$$
$$V = 3.16 \times 10^{-3} \text{ volts} \approx 3.2 \text{ millivolts.}$$

> It is clear that these three examples deal with the same microphone but with the sensitivity rated in three different ways ($S_v = -66$ dBV re 1 μbar; $S_p = -50$ dBm re 10 μbar; and $G_m = -144$ dB).

Figure 20.11 can be used to determine the microphone output voltage for different sound-pressure levels and different values of voltage sensitivity.

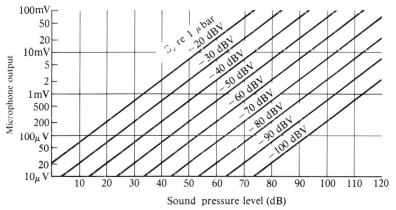

FIG. 20.11
Microphone output voltage at different sound levels and sensitivities.

20.6   LOUDSPEAKERS AS TRANSDUCERS

A loudspeaker is a transducer that converts electrical energy into acoustic energy. Actually, this conversion takes place in two steps: The electrical signal causes mechanical motion of the speaker cone or diaphragm, which in turn causes the pressure waves in the air that we call sound.

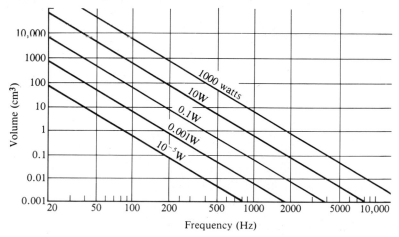

FIG. 20.12
Volume of air moved by a loudspeaker as a function of sound power and frequency.

How much air do we actually need to move? That depends on the sound power and the frequency, as shown in Fig. 20.12. It is clear from this graph that loudspeakers for low-frequency reproduction (*woofers*) must be considerably larger than high-frequency transducers (*tweeters*).

20.7   STRUCTURE OF DYNAMIC LOUDSPEAKERS

A *dynamic* loudspeaker is an electromagnetic transducer consisting of a voice coil, a cone diaphragm, and a magnet structure with a small gap in which the voice coil moves (see Fig. 20.13). The magnet structure is designed to provide a strong magnetic field across the gap, so that when a current flows in the voice coil it will experience a force (see box, p. 396). Although early dynamic speakers used electromagnets, modern loudspeakers have high-efficiency permanent magnets, thus eliminating the need to supply a current to the electromagnets.

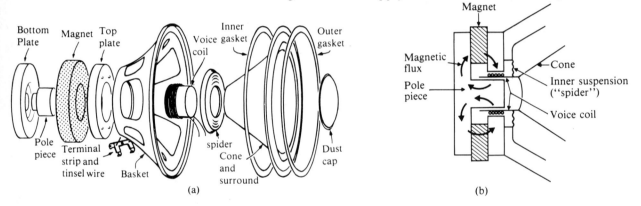

FIG. 20.13

(a) An exploded view of a dynamic loudspeaker. (b) A detailed view of magnet and voice coil structures; note the path of magnetic flux. (Part (a) courtesy of Speakerlab, Inc., Seattle, Washington).

An alternating electric current in the voice coil causes it to move in and out in the magnetic field, thus radiating sound. In order to move a larger quantity of air and thereby radiate more sound, a cone diaphragm is attached to the voice coil and moves in and out with the coil. In general, the larger the cone the better the radiation at low frequencies because of the greater volume of air moved by the cone. On the other hand, a large cone has a large mass, and it is difficult to move at high frequencies. Thus it is difficult (although not impossible) to build a single loudspeaker that radiates efficiently at both high and low frequencies. Many schemes have been devised to overcome this difficulty, but the most common solution is to use two or more speakers with an electrical or mechanical crossover to separate low-frequency and high-frequency signals, as will be discussed in Section 20.15.

Several different magnet systems are used in loudspeakers. Three of the most common types are illustrated in Fig. 20.14. Each one consists of a cast magnet of a special magnetic alloy (usually Alnico V, an alloy of aluminum, nickel, and cobalt) or a ceramic material. In general, the greater the magnetic flux, the greater the efficiency of the speaker. However, efficiency is not the only criterion in magnet design. For high-fidelity reproduction of sound, the magnetic field in the gap must be uniform throughout the maximum distance that the voice coil moves.

The magnet is usually one of the most costly components in loudspeaker manufacture, so there is a tendency to make the magnets as small as practical. High-quality speakers, however, require substantial magnet systems. Some manufacturers of loudspeakers customarily specify the magnet weight in the loudspeaker specifications, whereas other manufacturers specify the magnetic flux across the gap.

An important consideration in loudspeaker design is the suspension of the cone and the voice coil. In addition to supporting the cone, the suspension system must keep the voice coil accurately centered in the magnet gap, provide springlike action to bring the coil back to its equilibrium position with no input signal, and also to provide some degree of mechanical damping. This is quite an order!

The *compliance* of a speaker indicates how flexible its suspension is. Compliance is expressed in units of distance divided by force (m/N or cm/dyne). Most suspension systems use a flexible inner suspension to keep the voice coil centered in the magnet gap plus an outer suspension to support the cone. The inner suspension may be a plastic or thin metal "spider" with long, thin legs (see Fig. 20.15a) or a plastic-impregnated corrugated cloth. The outer suspension may consist of soft folds in the cone material itself, or a high-compliance cloth may be used to attach the cone to the frame (see Fig. 20.15b). The outer suspension is usually designed to provide edge damping to inhibit standing waves of short wavelength in the cone.

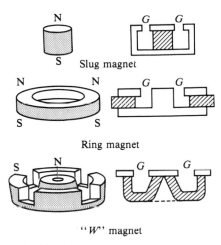

FIG. 20.14
Three types of permanent magnets commonly used in loudspeakers. At the right are shown the complete magnet systems, consisting of the magnet itself (shaded) plus a yoke of soft iron to complete the flux path and provide a circular gap in which the voice coil moves (indicated by *G*).

FIG. 20.15
(a) Inner suspension of a loudspeaker consists of a flexible "spider" or cloth corrugations, which keep the voice coil centered in the magnet gap. (b) Outer suspension may consist of high-compliance cloth or soft folds in the cone material itself.

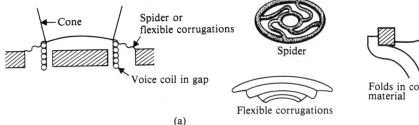

(a)

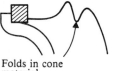

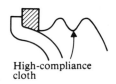

(b)

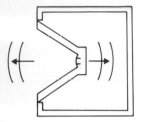

FIG. 20.16
An air-suspension speaker system consists of one or more high-compliance speaker cones mounted in an airtight box.

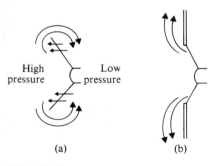

(a)   (b)

FIG. 20.17
(a) Without a baffle there is a short-circuit air path from the front side of the cone to the back side. (b) A baffle eliminates this path or greatly increases the path length.

20.8 AIR SUSPENSION

Air-suspension speaker systems, consisting of a loudspeaker with very high compliance mounted in an airtight box, are quite popular (see Fig. 20.16). Much of the restoring force that returns the cone and the voice coil to their rest position is furnished by the pressure of the air in the box.

Air-suspension systems are designed to operate with large excursions of the voice coil and the cone. This allows a relatively small speaker cone to radiate bass tones satisfactorily, since the radiation of low frequencies requires the movement of a relatively large volume of air (see Fig. 20.12). At the same time, large excursions of the voice coil require a magnetic field extending over a great distance or a voice coil that overhangs the magnet, either of which tends to reduce loudspeaker efficiency. Thus air-suspension speakers are usually low-efficiency speakers requiring fairly large amounts of electrical power from the amplifier. If space is an important consideration, however, air suspension makes it possible to achieve very good performance from small speakers.

20.9 BAFFLES AND ENCLOSURES

In order for a loudspeaker to radiate sound efficiently, especially at low frequency, it must be mounted on some sort of baffle. To understand why a baffle is important, note that when the loudspeaker cone moves forward, a wave of increased pressure is generated at the front, but a wave of decreased pressure is generated on the rear side of the cone. If there is no baffle, there will be a short-circuit path around the edge of the cone from front to back, as shown in Fig. 20.17. When this path is less than a quarter wavelength, the radiation efficiency drops off rapidly. Thus the baffle must be large enough to eliminate paths shorter than a quarter wavelength at the lowest desired frequency.

Mounting the loudspeaker in the wall of a large enclosure or room (effectively an "infinite" baffle) eliminates all short-circuit paths from back to front, but no use is made of the sound radiated from the back of the cone. Greater efficiency is obtained if some use is made of this back wave, especially at low frequency where it is difficult to achieve efficient radiation of sound. Therefore, many modern loudspeaker systems have a type of speaker enclosure that not only serves as a baffle but also as an acoustic circuit to transmit the back wave to the front of the enclosure in the proper phase to reinforce the sound radiated from the front of the cone. Three common types are the bass-reflex, drone-cone, and rear-horn enclosures.

High pressure   Low pressure

Two simple experiments demonstrate the important effect of a baffle on sound radiation. If a tuning fork is set into vibration and slowly passed through a slot in a sheet of paper, as shown in Fig. 20.18(b), the loudness increases noticeably as each tine passes through the slot. The paper, acting as a crude baffle, partially blocks the short-circuit path from one side of the vibrating tine to the other.

In the second experiment, a small speaker (two or three inches in diameter) with no baffle radiates very little sound at low frequency. Music thus sounds distorted. Placing the same speaker against a piece of plywood with an appropriate hole greatly increases the radiation of sound at low frequency, as illustrated in Fig. 20.18(d).

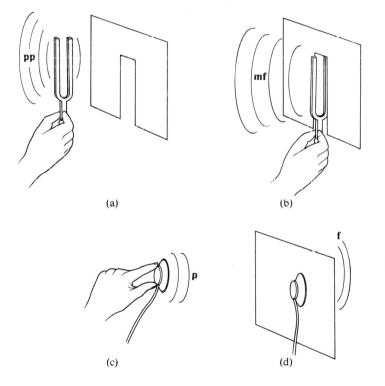

(a) (b)

(c) (d)

FIG. 20.18
Two experiments that demonstrate the effect of a baffle: (a) tuning fork with no baffle; (b) tuning fork with one tine in a slot cut in a sheet of paper; (c) small loudspeaker with no baffle; (d) same loudspeaker with a plywood baffle.

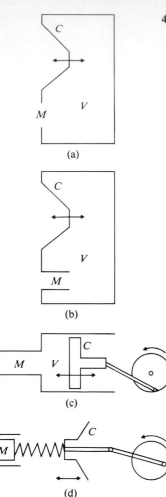

FIG. 20.19

Bass-reflex speaker enclosures (a) with a port; (b) with a duct; (c) represented as a Helmholtz resonator driven by a piston; (d) a mechanical analogue.

FIG. 20.20

Output from a loudspeaker in a bass-reflex enclosure compared with the same speaker in an infinite baffle and a closed box. f_0 is the resonance frequency of the loudspeaker and also the bass-reflex enclosure in this design.

The Bass-Reflex Enclosure

A *bass-reflex enclosure* has a port or a duct through which the back wave is radiated to the front. The enclosure is a sort of Helmholtz resonator with a piston driver at the closed end, as shown in Fig. 20.19(c). A mechanical analogue is shown in Fig. 20.19(d). As the speaker cone C is driven in and out by the voice coil, the air in the enclosure is compressed in the same manner as the spring in Fig. 20.19(d). The mass of the air in the port or duct is represented by mass M.

All four systems shown in Fig. 20.19 behave in the manner described in Section 2.4. At frequencies below resonance, the mass M moves in phase with the driver C (c.f. Fig. 2.7a), which in (a) and (b) means that sound radiated from the port or duct will be out of phase with that radiated from the speaker (in phase with the back side of the speaker means out of phase with the front side). At frequencies above resonance, however, the mass in each case moves out of phase with the driver (c.f. Fig. 2.7b), which in (a) and (b) means that sound radiated from the port or duct will be in phase with radiation from the front side of the speaker cone. Although the area of the port or duct is smaller than the speaker, the amplitude of motion may be considerably greater near resonance, so that the sound radiated from the port exceeds that radiated from the speaker cone. This is the "reflex action" from which the enclosure derives its name.

Figure 20.20 shows the output from a loudspeaker in a bass-reflex cabinet compared to the same loudspeaker in an infinite baffle and in a closed box. The bass-reflex cabinet and the loudspeaker have the same resonance frequency f_0. Note that the response at frequencies between $0.5f_0$ and f_0 is increased by the bass-reflex enclosure at the expense of a sharper cutoff below $0.5f_0$.

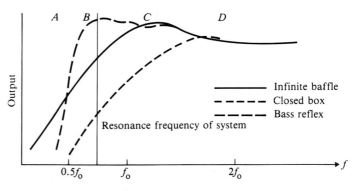

The acoustic behavior of the bass-reflex enclosure can be understood by comparing four different frequency ranges in Fig.

20.20. In range *A*, the sound radiated from the port or duct is out of phase with that radiated by the loudspeaker, which results in cancellation. In range *B*, just below resonance, sound radiation from the port exceeds that from the speaker cone, and so even though the two are out of phase, the total radiation is increased. In range *C*, above resonance, the two are in phase, and radiation is enhanced, but this enhancement grows small in range *D*, well above resonance. Note that the resonance frequency of the loudspeaker in the bass-reflex enclosure is different from the resonance frequency f_0 of the same speaker when it is not enclosed.

In addition to increasing the amount of sound radiated at low frequency, the bass-reflex enclosure decreases the speaker cone excursions near resonance; it therefore decreases distortion and increases the power-handling capability. The manner in which it accomplishes this can be demonstrated by suspending a load on the end of a long rubber band, the other end of which is held in your hand. Applying an up-and-down force with your hand causes the load to move up and down, and you soon become aware of the fact that resonance occurs at a particular frequency. At resonance, two things can be noticed: First, the amplitude of motion of the load reaches a maximum; and second, the free end of the rubber band offers substantial resistance to being moved by your hand.

As an analogy, your hand represents the speaker cone, the rubber band represents the enclosed air, and the load represents the air in the port of the enclosure. At resonance, the enclosure offers considerable resistance to the motion of the speaker cone and limits its amplitude of motion. At the same time, the motion of the air in the port reaches a maximum, so the radiation of sound is mainly from the port at resonance.

Generally the enclosure is tuned to the loudspeaker cone resonance frequency, although if the enclosure volume is large, it may be tuned substantially lower. The use of a duct (see Fig. 20.19b) rather than a port makes it possible to achieve a low resonance frequency with an enclosure of smaller volume.

Fairly detailed analyses of bass reflex enclosures are found in a number of books and papers (Beranek, 1954; Thiele, 1961; Small, 1973), some of which are summarized briefly in a paper by the author (Rossing, 1980).

Design of a Bass-Reflex Enclosure

A bass-reflex enclosure that will perform satisfactorily with a given loudspeaker can be designed by the use of charts and graphs readily available (see, for example, Cohen, 1968; or Tremaine, 1969), although it may not perform quite as well as one in which all the parameters have been carefully selected in accordance with one of the alignments given by Thiele (1961). The easiest procedure is to tune the enclosure to the loudspeaker resonance frequency.

The resonance frequency of the speaker can be determined by using the circuit shown in Fig. 20.21(a). Resistor R should be 200 ohms or more so that the current supplied to the loudspeaker does not change appreciably when the frequency is varied. The voltage across the speaker, which is then proportional to the input impedance, will show a rather sharp maximum at the speaker resonance, as shown in Fig. 20.21(b). When the speaker is mounted in a properly tuned bass-reflex enclosure, its input impedance will have two peaks, also shown in Fig. 20.21(b).

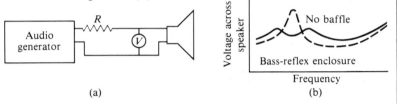

(a)

(b)

FIG. 20.21
(a) Circuit for determining the resonance frequency of a speaker.
(b) Voltage across the speaker (input impedance) when unenclosed and when mounted in a bass-reflex enclosure.

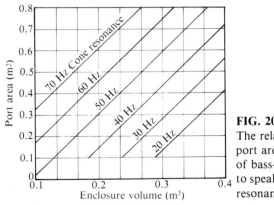

FIG. 20.22
The relationship of port area and volume of bass-reflex enclosure to speaker-cone resonance frequency.

Figure 20.22 shows possible combinations of port area and enclosure volume for speakers with different resonance frequencies.

Drone Cones

A speaker enclosure with a passive speaker, or *drone cone*, is similar in principle to a bass-reflex enclosure (see Fig. 20.23). The drone cone, which is a loudspeaker cone without a voice coil or a magnet, is driven by the back wave from the main speaker. Like the air in the port or duct of a bass-reflex speaker, it radiates a substantial part of the total sound near the resonance frequency of the enclosure. The drone cone is usually either the same size as the main speaker or slightly larger.

Because the drone cone has a substantial amount of mass, the volume of the cabinet can be modest in size and yet achieve a low resonance frequency. An additional advantage over the bass-reflex enclosure with a port is that large air velocities in the latter can lead to turbulence, whereas the larger area of the drone cone reduces the air velocity at high power levels.

Rear Horn and Acoustic Labyrinth Enclosures

An acoustic horn is a very effective acoustic radiator. (The horn loudspeaker will be discussed in Section 20.10.) It is possible to use a horn to radiate the back wave from a cone loudspeaker, as shown in Figs. 20.24(a) and (b). The cone loudspeaker in each of these arrangements radiates directly from the front side of its cone, but the back side drives a modified horn that radiates indirectly.

The acoustic labyrinth in Fig. 20.24(c) incorporates an open pipe that is approximately half a wavelength in length for low frequency sound, so that sound from the back of the speaker cone arrives at the front in phase with the direct-radiated sound.

20.10   HORN LOUDSPEAKERS

A *horn* loudspeaker consists of an electrically-driven diaphragm and a voice coil coupled acoustically to a horn. The acoustic horn transforms high pressure at the throat to lower pressure distributed over the large mouth of the horn. The acoustic horn is the acoustical analogue of an electrical step-down transformer, as shown in Fig. 20.25.

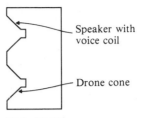

FIG. 20.23
A drone cone enclosure substitutes a speaker without a voice coil for the port of duct of a conventional bass-reflex enclosure.

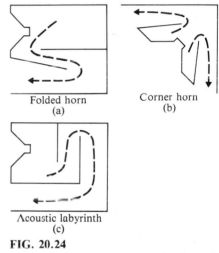

Folded horn
(a)

Corner horn
(b)

Acoustic labyrinth
(c)

FIG. 20.24
Three arrangements for rear loading of a cone speaker with a horn or labyrinth.

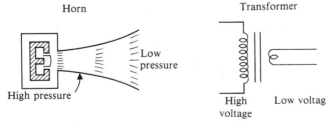

Horn

Transformer

FIG. 20.25
An acoustical horn is analogous to an electrical transformer.

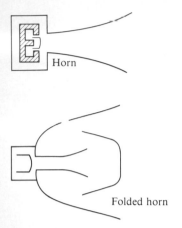

FIG. 20.26
Straight horn and folded horn.

A well-designed horn loudspeaker is a very efficient radiator of sound, with efficiency reaching as high as 40 to 50 percent compared to a typical 3 to 5 percent for a cone-type loudspeaker. Therefore, it is widely used in public address systems (especially outdoors) and theater sound systems, where a large amount of power is desired. A horn capable of radiating low frequencies becomes large and cumbersome, however, unless a folded-horn arrangement such as the one shown in Fig. 20.26 is used.

One type of low-frequency folded horn, first designed in 1941 by Paul Klipsch, is still in use today. It is called the Klipschorn and is shown in Fig. 20.27. Designed to stand in the corner of a room, it uses the walls of the room as an important part of the horn. The efficiency (of converting electrical energy to acoustical energy) is about 30 percent over a frequency range of 50 to 350 Hz (Klipsch, 1941).

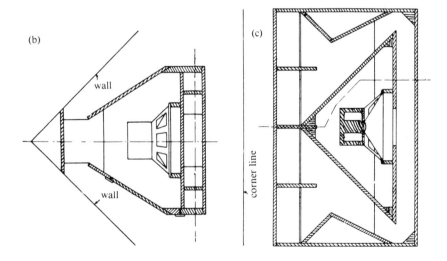

FIG. 20.27
A low-frequency horn design: (a) the Klipschorn; (b) sectional top view; (c) sectional side view. (Courtesy of Klipsch and Associates).

20.11   MULTISPEAKER SYSTEMS

Because of the difficulty of obtaining high performance at both high and low frequencies from the same speaker, nearly all high-fidelity loudspeaker systems have two or more speakers. A typical speaker system will consist of a large-cone low-frequency *woofer*, a small high-frequency *tweeter*, and an electrical crossover network that feeds low-frequency signals to the woofer and high-frequency signals to the tweeter, as shown in Fig. 20.28(a). More elaborate systems add a midrange speaker and a second crossover network to direct a band of

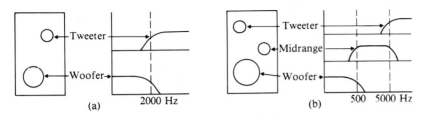

FIG. 20.28
Two-way and three-way loudspeaker systems along with typical division of frequencies by the crossover networks. Crossover frequencies vary among manufacturers.

frequencies to this speaker, as shown in Fig. 20.28(b). Crossover frequencies vary widely among different manufacturers.

A woofer usually has a large voice coil, a deep cone, and a fairly soft suspension in order to have a low frequency of resonance. A tweeter, on the other hand, has a small lightweight voice coil and a relatively stiff suspension. Horn tweeters have become very popular, because *if properly oriented*, they provide good angular dispersion of high-frequency sounds that tend to be quite directional. A dome tweeter substitutes a lightweight diaphragm (typically aluminum) for the cone in order to decrease the mass that moves with the voice coil, and thus to improve the response at high frequency. Dome tweeters are popular as "super tweeters" in multiple-speaker systems. Several types of tweeters are illustrated in Fig. 20.29.

Several manufacturers supply coaxial two-way and even three-way (triaxial) speaker systems with two or three different speakers mounted on the same frame. This is a convenience in designing the enclosure. Figure 20.30 illustrates multispeaker systems, including two-way, three-way (see Fig. 20.28), and coaxial three-way. The coax-

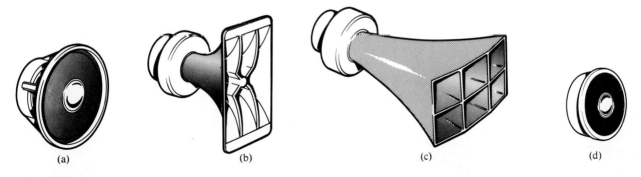

FIG. 20.29
Several types of tweeters: (a) cone type; (b) diffraction horn; (c) multicellular horn; (d) dome tweeter. Types (b) and (c) includes special horns for good angular dispersion of high-frequency sound.

ial system shown has cones attached to the woofer voice coil; other triaxial systems have three independent voice coils.

FIG. 20.30
Multispeaker systems: (a) two-way system; (b) three-way system; (c) coaxial three-way system. (Courtesy of Electro-Voice).

The crossover network may consist of simple high-pass and low-pass filters, as shown in Fig. 20.31. The purpose of the crossover network is to send high frequencies to the tweeter and low frequencies to the woofer. If a separate midrange speaker is included, it may be fed through a band-pass filter. If the midrange speaker is a second cone attached to the voice coil of the woofer, the crossover will be mechanical rather than electrical.

FIG. 20.31
Crossover network consisting of high-pass and low-pass filters.

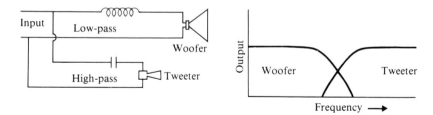

20.12 OTHER LOUDSPEAKER TYPES

New speaker designs are constantly being marketed. Some of them are variations of more familiar designs; others are quite different. We mention only a few of them.

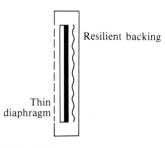

FIG. 20.32
An electrostatic speaker.

Electrostatic Speakers The electrostatic speaker, shown in Fig. 20.32, is similar in construction to the condenser microphone. A perforated screen serves as one electrode, and a metal film on a plastic diaphragm forms the other. The plastic diaphragm carries a polarizing charge that is attracted or repelled by the charge in the screen.

Electrostatic speakers have smooth response at high frequency, but they are inefficient at low frequency due to the difficulty in obtaining a large excursion of the diaphragm. They have the added disadvantage of a high input impedance, which makes them somewhat difficult to drive efficiently with a transistor amplifier.

Planar Speakers   Several planar- or flat-speaker systems, both of magnetic and electrostatic type, have appeared, but they tend to be inefficient and have not become very popular.

Cylindrical Radiators   Several loudspeakers designed to radiate equally in all directions have appeared on the market. In some cases, the loudspeaker is pointed toward a conical reflector. The Ohm F speaker features sideways radiation from the outside of a vertical cone. Cylindrical radiators can be placed almost anywhere in the room, but interference effects can distort the sound field in the room when they are used as components in a stereophonic system.

Air-motion Transformer   This speaker uses a flexible pleated diaphragm to which strips of conductive aluminum have been bonded. Motion of the aluminum strips in a magnetic field causes the pleats to open and close and thus to pump air in and out.

20.13   LOUDSPEAKER EFFICIENCY

The purpose of a loudspeaker is to convert electrical energy into mechanical energy and to radiate it as acoustic energy. Unfortunately, only a small part of the electrical energy is converted to sound; in most home loudspeaker systems, 90 to 99 percent of it is wasted as heat. The percentage of electrical power radiated as sound determines the *efficiency* of the loudspeaker. In spite of the fact that it is now possible to buy high-fidelity amplifiers that deliver enormous amounts of electrical power, loudspeaker efficiency remains an important consideration.

Loudspeaker manufacturers tend not to be very informative about speaker efficiency. Almost never are efficiencies stated as simple ratios or percentages. The term "power-handling capability" is quite meaningless; it merely tells how much electrical power a speaker system can absorb without damage. Obviously, a low-efficiency loudspeaker system will be able to "handle" a larger amount of power (and convert it to heat) than can a comparable system with a high efficiency. But loudspeakers are not intended to be space heaters.

The efficiency of a loudspeaker depends on many design parameters, such as magnet strength, cone area, type of enclosure, resistance of the voice coil, mass of cone and coil, etc. It may change with position of the loudspeaker in the room. The efficiency increases with the square of the magnetic field strength and also with the square of the cone diameter, so these tend to be the most important parameters. The magnetic field strength depends on the weight of the magnets (sometimes stated), but also on other parameters, such as magnet materials, air gap size, etc. (see Section 20.7). Horn loudspeakers are much more efficient than cone speakers, but low-frequency horns are considered too large for most home sound systems. Air-suspension speaker systems tend to have very low efficiencies, but are popular because of their compactness.

Some loudspeaker manufacturers state the sound level that will be obtained per watt of input, on the speaker axis at one meter (or some other distance) from the speaker. This depends on both the efficiency of the loudspeaker and its directivity (how much of its total power is radiated straight ahead). The directivity factor of a loudspeaker varies markedly with frequency, so the on-axis sound level will also vary and, to be meaningful, it should be stated for both high and low frequencies. The on-axis sound level should be expressed as, for example, "89 dB at 1 m with 1W input at 500 Hz." It certainly would be a great help to the intelligent buyer of high-fidelity sound systems if manufacturers would specify the efficiency (preferably in percent) of their loudspeaker systems at several different frequencies!

The subject of loudspeaker efficiency and also distortion in loudspeakers is discussed further in Section 22.10.

20.14 SUMMARY

A microphone produces an electrical signal when actuated by sound waves; most microphones respond to variations in sound pressure. There are many different microphone types, including crystal, dynamic, condenser, electret-condenser, and ribbon. The most popular directional characteristics are omnidirectional and cardioid (unidirectional). Important microphone specifications include sensitivity and output impedance. High-impedance microphones require transformers or in situ preamplifiers if long microphone cables are to be used.

A loudspeaker converts electrical power into sound. Most loudspeakers use cone-type direct radiators, but a wide variety of loudspeaker enclosures are in use. In an air-suspension system, the air in the sealed enclosure provides most of the restoring force for the speaker cone. In a bass-reflex speaker, the back wave from the speaker cone causes a substantial amount of low-frequency sound to be radiated from a duct or port. Horn loudspeakers have very high efficiencies, but low-frequency horns are very large. Although speaker efficiency is a very important consideration, it is rarely specified in the case of systems designed for home use.

References and Suggested Readings

Beranek, L. L. (1954). *Acoustics.* New York: McGraw-Hill. Reprinted by Acoust. Soc. Am., 1986.

Cohen, A. B. (1968). *Hi-Fi Loudspeakers and Enclosures,* 2nd ed. New York: Hayden.

Institute of High Fidelity (1974). *Guide to High Fidelity.* Indianapolis: Howard Sams.

Klipsch, P. W. (1941). "A Low-Frequency Horn of Small Dimensions," *J. Acoust. Soc. Am.* **13**: 137.

Long, J. (1972). "A Microphone Primer: Basic Construction, Performance and Applications," *Audio* **56**(12): 18.

Long, J. (1973). "A Microphone Primer: Basic Construction, Performance and Applications," *Audio* **57**(1): 34.

Olson, H. F. (1972). *Modern Sound Reproduction.* New York: Van Nostrand-Reinhold. (See Chapters 2 and 4.)

Rossing, T. D. (1980). "Physics and Psychophysics of High-Fidelity Sound, Part III," *Physics Teach.* **18**: 426.

Small, R. H. (1973). "Vented-Box Loudspeaker Systems," *J. Audio Eng. Soc.* **21**, 363, 438, 549, and 635.

Thiele, A. N. (1961). "Loudspeakers in Vented Boxes," *Proc. I.R.E. (Australia)* **22**: 487. (Reprinted in *J. Audio Eng. Soc.* **19**: 382, 471 (1971).)

Tremaine, H. M. (1969). *Audio Cyclopedia,* 2nd ed. Indianapolis: Howard Sams.

Glossary

air-suspension speaker A loudspeaker mounted in the front of an airtight box so that the pressure of the enclosed air furnishes a major part of the force that restores the speaker cone to its equilibrium position.

baffle An arrangement that reduces interference between sound radiated from the front and rear of a speaker by increasing the path length from front to back.

bass-reflex enclosure A speaker enclosure in which the back wave from the speaker is radiated through a port or duct in the front.

compliance A measure of flexibility; it is expressed in units of distance divided by force (m/N or cm/dyne).

condenser microphone A microphone in which the diaphragm serves as one plate of a small capacitor or condenser. As the diaphragm moves, the electrical charge on the condenser varies.

crossover network A network designed to send high frequencies to the tweeter and low frequencies to the woofer. The crossover frequency is the approximate point of division between high and low frequencies.

diffraction horn A horn that has one broad dimension and one narrow one, designed to spread sound (by means of diffraction) in the direction of the narrow dimension.

drone cone A passive loudspeaker that has no voice coil or magnet.

dynamic microphone A microphone that generates an electrical voltage by the movement of a coil of wire in a magnetic field.

electret-condenser microphone A condenser microphone that has an electrified foil as a dielectric, thus eliminating the need for a polarizing voltage as required in an air-dielectric condenser microphone.

electromagnetic induction Generation of an electrical voltage in a wire that moves across a magnetic field.

horn loudspeaker A system that uses an acoustic horn to enhance sound radiation from a moving diaphragm.

impedance The ratio of voltage to current. In the case of source impedance or output impedance, it is the current that the device can deliver; in the case of input impedance, it is the current that the device draws from the source.

infinite baffle A large baffle or an enclosure that prevents interference between sound radiated from the front and back of the speaker cone.

piezoelectric crystal A crystal that generates an electric voltage when it is bent or otherwise distorted in shape.

sensitivity (microphone) Voltage or power generated in a microphone at a given sound pressure level.

transducer A device that converts one form of energy into another; in this chapter, we discussed the conversion from acoustic energy to electrical energy (microphone) or vice versa (loudspeaker).

tweeter A loudspeaker designed to produce high-frequency sound.

velocity microphone A microphone that responds to particle velocity rather than to sound pressure.

woofer A loudspeaker designed to produce low-frequency sound.

Questions for Discussion

1. What are the advantages and disadvantages of using microphones of low impedance as compared to those of high impedance?

2. Describe the directional pattern of a cardioid microphone. For what applications are cardioid microphones superior?

3. What are the advantages and disadvantages of air-suspension loudspeakers?

4. Why does a horn loudspeaker have a higher efficiency than a cone loudspeaker?

5. If the baffle in which a loudspeaker is mounted is insufficient in size, which will suffer most: the bass response or the treble response? Why?

6. Why does a woofer have a large-diameter cone? Why does a tweeter have a small cone?

Problems

1. Compare the sound power level of a loudspeaker with an efficiency of one percent to one with an efficiency of ten percent supplied with the same electrical power.

2. Compare the cone areas of speakers having diameters of 20 cm, 30 cm, and 38 cm (8 in, 12 in, and 15 in).

3. A loudspeaker 20 cm in diameter is mounted at the center of a one-meter square baffle board.

   a) Determine the path length from the center of the front side to the center of the back side of the speaker.

   b) At what frequency will this path be equal to one-half wavelength of sound?

4. a) From the graph shown in Fig. 20.12, determine the volume of air that must be moved in order to generate a sound power of 0.1 watt at 100 Hz.

   b) How far must a speaker cone move in order to

move this volume of air if the cone has a diameter of 8 inches?

5. From the graph shown in Fig. 20.11, determine the output voltage from a microphone having a sensitivity of -60 dBV (re 1 microbar) when $L_p = 60$ dB.

6. Compare the voltage outputs from two microphones: Microphone A has a sensitivity -60 dB re 1 volt per 1 N/m^2 and microphone B a sensitivity of -66 dB compared to the same reference.

7. Calculate the actual voltage output of microphone A at a sound pressure of 1 N/m^2. To what sound pressure level does this correspond (see Chapter 6)?

8. For satisfactory operation down to a cutoff frequency f_c, (a) a horn loudspeaker should have a mouth diameter at least one-fourth of the wavelength at f_c,

and (b) its diameter should double no more rapidly than every one-ninth of the wavelength at f_c. Determine the mouth diameter and the total length of a horn loudspeaker with $f_c = 100$ Hz if the throat diameter is 5.4 cm.

9. If a loudspeaker can produce a sound pressure level of 92 dB for one watt of input power at a distance of one meter, estimate its efficiency in converting electrical power to acoustic power. (*Hint*: Assume that the sound is uniformly radiated into a hemisphere so that the sound level at one meter is 8 dB less than the sound power level; see Section 6.2.)

10. How large a volume of air must be moved by a loudspeaker cone radiating 0.1 W of acoustical power at 50 Hz? How far would the cone of a 12-inch loudspeaker (actual cone diameter is 25 cm) have to move to displace this volume?

11. A certain loudspeaker has a compliance of 8.7×10^{-4} m/N and a mass (cone plus voice coil) of 71 g. Estimate its resonance frequency. (See Section 2.1; the compliance is the reciprocal of stiffness or spring constant: $C = 1/K$.)

12. Express the compliance of the loudspeaker in Problem 11 in cm/dyne (1 dyne $= 10^{-5}$ newton). If you placed the speaker face up and added a 100-g mass, how far would the speaker deflect? Is this a practical method for measuring compliance?

CHAPTER 21

The Recording and Reproduction of Sound

Spectacular advancements in the technology of sound recording in recent years have had a profound effect on the performance of music, both classical and popular. Millions of people enjoy high-fidelity reproduction of music in their own homes. The production and distribution of phonograph records and tapes has become a major industry.

The basic principles of disc and tape recording and playback will be discussed in this chapter. Chapter 22 deals with the components used for high-fidelity sound reproduction. In Chapter 28, we will discuss digital techniques for recording sound (including Compact Discs).

21.1 DISC RECORDING

Most modern phonograph records are vinyl plastic discs with circular grooves having an average density of about 100 grooves per cm (250 per inch) and designed to revolve at either 33-1/3 or 45 revolutions per minute. When a record is played, the stylus rides in the groove, making contact with the sides of the groove and following its undulations. On a monophonic recording, the motion of the stylus is purely lateral; on a stereophonic recording, there is both horizontal and vertical motion. Early recordings, including the cylinders of Edison and others, used a vertical ("hill and dale") motion of the stylus, but dis-

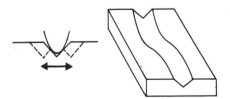

FIG. 21.1
Lateral modulation of groove in monophonic disc recording.

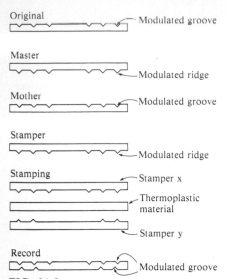

FIG. 21.2
The steps in the mass production of phonographic records from the lacquer original. (From *Modern Sound Reproduction* by Harry F. Olson, © 1972 by Litton Educational Publishing, Inc. Reprinted by permission of Van Nostrand Reinhold Company.)

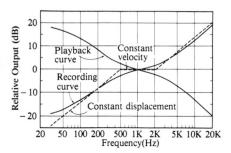

FIG. 21.3
Recording and reproducing characteristics recommended by the RIAA. Dotted lines show constant displacement and constant velocity characteristics for comparison. The playback curve is the inverse of the recording curve.

tortion and noise are greater for vertical motion of the stylus than for lateral motion.

The first step in producing commercial disc recordings is to cut the original lacquer disc. The cutting stylus is driven by an electromagnetic arrangement not unlike the voice coil of a small loudspeaker. It has sharp edges, which cut the grooves as it moves back and forth. The original record is coated with metal for hardness, and this metal coating is separated from the lacquer base to make the "master" record. The master record is used to make "mother" records and "stampers," the hard metal dies that stamp the vinyl discs for mass distribution. The convenience of mass-producing copies of the original recording gives disc recordings a distinct economic advantage over tape recordings.

21.2 RECORDING CHARACTERISTICS

All modern recordings are made with some degree of bass attenuation and treble emphasis. Nearly all record manufacturers have adopted the recording and reproducing characteristics recommended by the Record Industry Association of America (RIAA) and shown in Fig. 21.3. Note that the recording characteristic calls for reducing the amplitude of signals with frequencies below 1000 Hz and raising the amplitude of signals above 1000 Hz. The reproducing characteristic (also called the "playback equalization curve") is just the inverse of the recording curve, so that the final response curve will be "flat."

Reduction of low-frequency (bass) signals is done to prevent the grooves from interfering with each other. If when vibrating at 100 Hz the stylus were to move at the same velocity as when vibrating at 200 Hz, for example, it would travel twice as far laterally during one cycle. This means that the grooves would need to be spaced farther apart to allow for the occasional large stylus excursions for low bass notes. Attenuation of the bass during recording avoids these large excursions, and the playback equalization network restores the bass tones to their original strength.

Treble preemphasis during recording is used to override surface noise, which is strongest at the higher frequencies. In the playback characteristic, the treble is attenuated in the right proportion to bring it back into proper balance with the bass and midrange signals.

21.3 DYNAMIC RANGE

The dynamic range of a recording is the power ratio (expressed in dB) between the loudest and softest sounds that can be reproduced satisfactorily. On a disc, the upper limit of recorded sound is determined by the allowable excursion of the groove, and the lower limit by the surface noise. Disc recordings of high quality will have dynamic ranges of from 40 to 55 dB. The use of the recording characteristics described in Section 21.2 allows a wide dynamic range, but it can be expanded further by using *variable-pitch* recording.

To use variable-pitch recording, one first records the music on tape and then transfers it to the master disc (recording studios make use of this technique for several other reasons as well). An extra playback head monitors the tape before it reaches the main playback head, and when the monitor detects a loud passage coming, it "warns" the stylus to increase the spacing between grooves in order to avoid overcut into a neighboring groove. During soft passages, the groove spacing may be made small to compensate for the wider grooves cut during loud passages.

21.4 PHONOGRAPH PICKUPS

As the stylus follows the groove in the phonograph record, it is forced to vibrate back and forth by the lateral variations in the groove. The pickup must generate an electrical signal that conforms to these mechanical vibrations and that can be amplified electrically. The required signal is generated by using a piezoelectric crystal or by electromagnetic induction.

Magnetic pickups are the most common type used in high-fidelity reproducing systems; they may be one of the three types shown in Fig. 21.4: moving coil, moving magnet, or variable reluctance. The *moving-coil* pickup is similar to a moving-coil microphone. The stylus is attached to a small coil, which moves in a magnetic field as the stylus moves in the record groove. An electrical voltage, which is proportional to the stylus velocity, is thus induced in the coil.

In a *moving-magnet* pickup, the coils are stationary and a tiny magnet is attached to the stylus. Since electromagnetic induction requires only relative motion of the coil and the magnet, a voltage is induced in the coils as the magnet moves. The induced voltage is again proportional to the velocity of the stylus.

In a moving-iron or *variable-reluctance* pickup, both the coil and the magnet are fixed, but a piece of magnetic material attached to the stylus moves in such a way as to vary the amount of magnetic flux that passes through each coil, thereby inducing a voltage in the coils.

Piezoelectric or "crystal" pickups, like crystal microphones, generate their voltages by the bending of a piezoelectric material as the stylus moves in the record groove. Crystal pickups usually have a

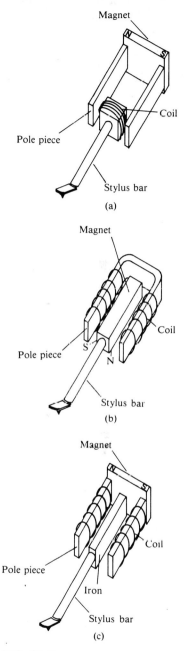

(a)

(b)

(c)

FIG. 21.4
A simplified diagram of three types of magnetic phonograph pickup: (a) moving coil; (b) moving magnet; (c) variable reluctance.

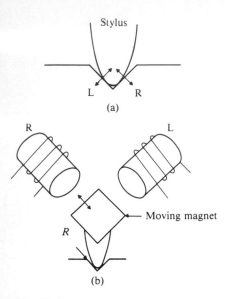

FIG. 21.5
(a) Two channels of stereo information L and R recorded on two sides of groove. (b) A simplified diagram of moving-magnet stereo pickup. The motion of the stylus and magnet in direction R induces a voltage in coil R but almost no voltage in coil L.

larger electrical output voltage than do magnetic pickups, and they are less costly. However, it is generally not possible to achieve as wide a frequency response as that obtained from magnetic pickups.

21.5   STEREOPHONIC DISCS

Stereophonic recording requires two independent channels of recorded information. On stereo discs, this is accomplished by recording the "right" and "left" channels along two sides of the same groove, as shown in Fig. 21.5. The sides of the groove are at right angles to each other, so that motion of the stylus in the direction marked L does not affect its motion in the direction marked R.

An important consideration in the recording and reproduction of sound is the compatibility of different types of records with different kinds of playback equipment. Monophonic and stereophonic disc records are compatible as follows:

1. When a two-channel stereo record is played on a monophonic reproducing system, the output signal will be a combination of both recorded channels. This is because variation in either side of the groove moves the stylus laterally.

2. When a single-channel monophonic record is played on a stereophonic reproducing system, lateral motion of the stylus will induce the same voltage in both stereo pickup coils, so the monophonic sound will appear in both channels.

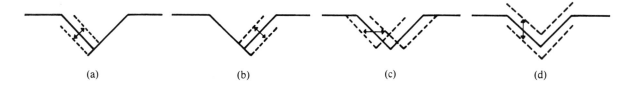

| (a) | (b) | (c) | (d) |

FIG. 21.6
Groove cross section and stylus motion for four types of recorded program material: (a) left channel only; (b) right channel only; (c) both channels, in phase (horizontal movement); (d) both channels, out of phase (vertical movement).

The manner in which two channels of stereo information are recorded in a single groove may become clearer by referring to Fig. 21.6. In (a) and (b), information is recorded on the left or right channel only. In (c), both channels receive the same signal (in phase), and the net result is a vertical motion of the stylus. In (d), both channels receive the same signal but with the phase in one channel reversed, so that when the groove moves inward on one side it moves outward on the other side, resulting in a horizontal motion of the stylus. The lateral motion of the stylus in Fig. 21.6(d) is identical to the motion of the stylus in a monophonic recording. A photograph of a phonograph disc recorded with stereo material is shown in Fig. 21.7.

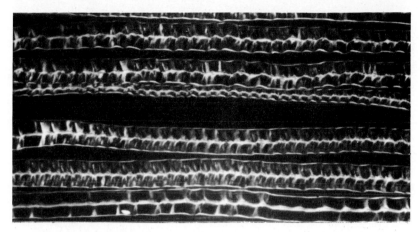

FIG. 21.7
Photomicrograph of grooves of a phonograph disc recorded with stereo program material. (Courtesy of Shure Brothers, Inc.)

21.6 MAGNETIC TAPE RECORDING

Most sound recording is done initially on magnetic tape. Modern recording tape consists of a thin coating of magnetic particles on a plastic base. Iron oxide (Fe_2O_3 and Fe_3O_4) powders are still the most widely used, although chromium dioxide (CrO_2) and metal powders have recently become very popular because of the improved signal-to-noise ratio that can be achieved through their use.

The preferred base material is polyester film (which has replaced the older cellulose acetate). Polyester film has high strength even when as thin as 0.5 mil (0.01 mm), which has made ultralong playing times possible. It is possible to wind as much as 3600 feet (1100 m) of 0.5-mil tape on a standard 7-inch (18 cm) reel, which provides a playing time of over three hours (each direction) at a recording speed of $3\frac{3}{4}$ in./sec (9.5 cm/sec). (The user should be warned, however, that handling and, especially, splicing these extremely thin tapes is a challenge.)

The magnetic coating on the tape is made up of very tiny particles (typically 0.5 μm in length) held in a plastic binder. Particles this small have desirable "single-domain" properties, which means that their magnetization remains stably oriented in one of two opposite directions. (Thus a recorded signal remains on the tape until it is erased.)

The magnetic coating may be regarded as having a large number of tiny bar magnets in random alignment in the unmagnetized state. When a magnetic field is applied by means of the record head, the magnets align themselves in the direction of the field, as shown in Fig. 21.8. This model should not be carried to the point of thinking that the magnetic particles move or rotate, however. They do not. The "motion" involved in alignment of the magnets is inside the atoms themselves.

FIG. 21.8
Magnetic domains within the oxide coating shown as tiny bar magnets. As the magnetic field increases, the magnets align along the field.

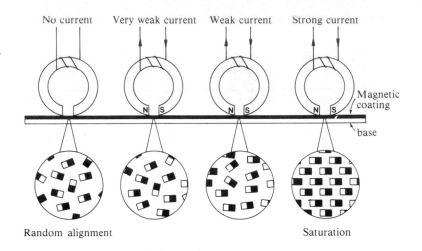

FIG. 21.9
Magnetic hysteresis loop. (See the text for an explanation of the symbols.)

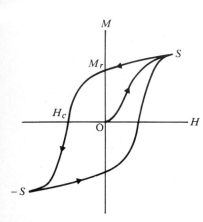

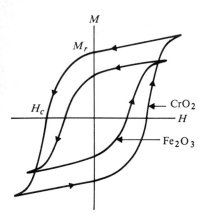

FIG. 21.10
Magnetic hysteresis loops for chromium dioxide and one type of iron oxide commonly used in recording tape. Note that CrO_2 has a higher remanent magnetization M_r and a higher coercivity H_c.

The magnetic properties of recording tape can be represented by a *hysteresis loop* of the type shown in Fig. 21.9. The magnetization M is plotted vertically and the magnetic field H is plotted horizontally. As the field H increases from the unmagnetized state O, the magnetization M increases (as the magnets align with the field) until magnetic *saturation* is reached at point S. When the field is now reduced to zero, the magnetization does not return to zero, but rather to the magnetic *remanence* M_r. In order to reduce the magnetization to zero, a field must be applied in the reverse direction. This reverse field H_c is called the *coercivity*. Further increase in the reverse field saturates the magnetization in this direction, indicated by $-S$ on the hysteresis curve. Note that the hysteresis loop is traversed in one direction only.

Different magnetic oxides have different magnetic properties. For example, CrO_2 has greater magnetization than Fe_2O_3, and thus it is capable of a greater signal-to-noise ratio (which is particularly important in cassette recorders). However, it also has a larger coercivity, which means that the record bias must be increased. Most recorders can adjust to either type of material by changing the bias to the recording head. Hysteresis loops of CrO_2 and Fe_2O_3 are compared in Fig. 21.10. Bias will be discussed further in Section 21.7.

In a typical tape recorder, the tape passes three recording heads in succession. First the *erase head* applies a rapidly oscillating magnetic field that erases the old information (demagnetizes the tape); then the *record head* magnetizes the tape in the desired pattern; and finally a *playback head* reads the recorded pattern and generates an output voltage as the tape passes. Many tape recorders use a single head for the record and playback functions, but this is slightly less convenient, because the tape cannot be monitored during recording.

The basic principles of magnetic tape recording are simple: The record head generates an alternating magnetic field across its gap, and this field leaves a pattern to the magnetic remanence of the tape passing by, as illustrated in Figs. 21.11 and 21.12. When the tape arrives at the gap in the recording head, it sets up a magnetic "disturbance" that induces an electrical signal in the coil windings. Although the basic principles are simple enough, there are several practical considerations.

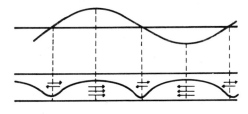

FIG. 21.11
Magnetization of a magnetic tape on which a pure tone has been recorded. Arrows represent the magnetization.

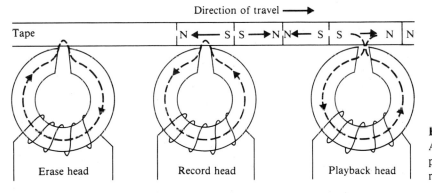

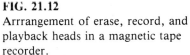

FIG. 21.12
Arrrangement of erase, record, and playback heads in a magnetic tape recorder.

21.7 TAPE SPEED AND FREQUENCY RESPONSE

Magnetic tape recorders usually operate at one of the following speeds: 15, 7½, 3¾, or 1⅞ inches per second (ips); the corresponding metric speeds are 38, 19, 9.5, and 4.8 cm/s, respectively. The choice of speed is a compromise between tape economy and performance. Professional recording is usually done at 15 ips, whereas cassette rcorders use 1⅞ ips.

The high-frequency limit usually occurs when the wavelength of the recorded signal approaches the size of the gap in the playback head. A simple calculation will indicate the frequency at which this might be expected.

The recorded wavelength for a pure tone is the tape velocity divided by the frequency of the tone. Thus, for a 10-kHz tone recorded at 3¾ ips,

$$\text{Wavelength} = \frac{3.75\ \text{ips}}{10,000\ \text{Hz}} = 3.75 \times 10^{-4}\ \text{in.} = 9.5\mu\text{m}.$$

This is only a few times larger than the gap size in a typical playback head, and therefore near the upper limit that can be reproduced with fidelity.

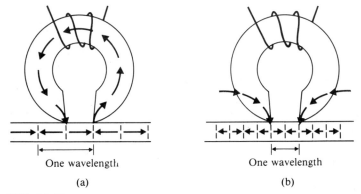

(a)   (b)

FIG. 21.13
High-frequency cutoff occurs when a wavelength of recorded signal approaches the head gap size. (a) The wavelength is twice the gap size; magnetic flux travels through the head. (b) The wavelength equals the gap size; no flux reaches the coil windings.

FIG. 21.14
The frequency characteristic for an unequalized tape at four different tape speeds.

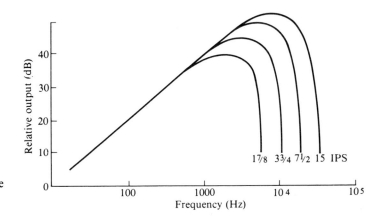

The graph of relative output as a function of frequency for different tape speeds (see Fig. 21.14) shows the frequency characteristic for a tape without equalization. Note that the output rises with frequency (since the rate of flux change through the head increases) until the wavelength of the recorded signal on the tape approaches the gap width and then falls off rapidly.

21.8 BIAS AND EQUALIZATION

In order for the magnetization to be a linear function of the recording field, it is necessary to use a magnetic *bias*. Although older tape recorders occasionally used a constant field as bias (and even a constant erase field), high-frequency erase and bias currents are used almost exclusively in modern equipment. A high-frequency current of 40 to 150 kHz is supplied to the erase head and also superimposed on the signal at the record head.

High-frequency bias has two very important advantages over "DC" bias: (1) the recorded magnetization is a linear function of the sound signal over a larger dynamic range, and (2) when the sound signal dips to zero, the tape remains demagnetized and therefore generates less noise in the playback head. The relationship between the current in the recording head and the recorded magnetization in the tape is shown in Fig. 21.15 (compare with the inner curve OS in Fig. 21.9). The bias amplitude is selected to make use of the linear portion of the curve. A physical description of high-frequency bias is given in the box below.

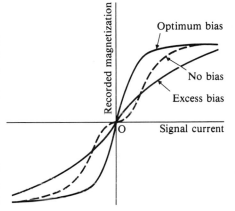

FIG. 21.15

The recorded magnetization as a function of signal current for varying amounts of high-frequency bias. With no bias (dashed curve) the distortion is large. (After Westmijze, 1953.)

High-frequency bias can be understood by referring to the waveforms shown in Fig. 21.16. The current supplied to the recording head (c) is the sum of the bias (a) plus the sound signal (b); the bias current must be substantially larger than the signal current. During the time that the tape is next to the recording gap, the bias field is large enough to drive the magnetization rapidly around the hysteresis loop from one magnetic state to the other. As the tape moves away from the gap, however, the field diminishes, and the magnetization traverses a steadily diminishing "minor" hysteresis loop (e). In the first case shown, the magnetization diminishes toward zero; in

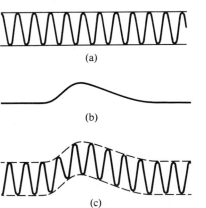

the second case, the excursions are unequal, and a remanent magnetization results as the tape leaves the field of the gap.

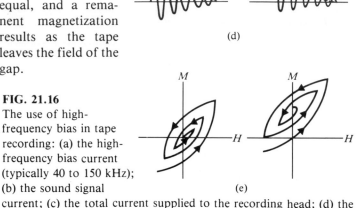

FIG. 21.16
The use of high-frequency bias in tape recording: (a) the high-frequency bias current (typically 40 to 150 kHz); (b) the sound signal current; (c) the total current supplied to the recording head; (d) the field "seen" by the tape as it leaves the record gap; (e) the magnetic states of the tape as it leaves the gap field. (From Rossing, 1980. Reprinted by permission from the Amer. Assoc. of Physics Teachers.)

The equalization of response in tape recording is a more complex problem than in disc recording. The voltage induced in the playback coils is proportional to the *rate of change* of magnetic flux through the playback head. Thus, for a given amplitude of variation of the tape magnetization, the output signal increases linearly with the frequency of the recorded signal and the tape speed (the response rises with frequency at the rate of 6 dB per octave). There are other complicating factors, however (see Fig. 21.14, for example), so that it is not enough to merely attenuate the high frequencies by 6 dB/octave. In fact, it is common practice to provide equalization during both recording and playback in order to reduce tape noise. Nearly all tape recorders provide for different equalization networks for different tape speeds.

21.9   TAPE NOISE

Unrecorded tape will produce a hiss when it is played back. This is due to microscopic random variation in the magnetization M. If the tape has been carefully erased with a high-frequency field, the average magnetization at any point will be zero, but M may vary in small regions the size of the gap. Erasing with the steady field of a permanent magnet (called "D.C. erase"), used on a few inexpensive tape recorders, results in a much higher level of tape noise than does high frequency erase. Some tapes have a lower level of tape noise than others.

Quality tape recorders are capable of recording signals that will play back 50 dB or more above the level of the tape noise. An important tape recorder specification is its *signal-to-noise ratio.* Unfortunately, there are several different ways of measuring this ratio (e.g., weighted or unweighted signal, how much distortion is allowed in the signal, etc.), so that a comparison is difficult.

Dolby noise-reduction systems have become very popular lately, especially in cassette tape recorders. These systems change the equalization with the level of the recorded signal. In the Dolby B-system, low-level high-frequency signals are boosted 10 dB when recorded and attenuated by the same amount during playback. This has the effect of reducing tape noise during soft passages, and yet loud sounds will not overdrive the tape (the extra noise reduction is not needed during loud passages). The Dolby C-system provides about 20dB of noise reduction at high frequency by using two compressors in series (Dolby, 1983). Figure 21.17 shows the preemphasis used in the Dolby B-system encoder.

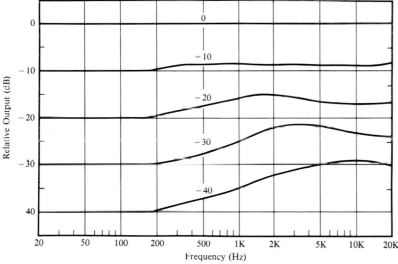

FIG. 21.17
Dolby encoder characteristics. Low-level high-frequency signals are boosted during recording and attenuated during playback.

There are other noise-reduction systems in use. The Dolby A-system, slightly more sophisticated than the B-system, is widely used in professional recording. Another popular noise-reduction system is the dbx system. Any of these systems should improve the signal-to-noise ratio by at least 5 to 10 dB, which is especially important in the case of low-speed cassette recorders.

21.10   MULTIPLE TRACKS

Advances in tape recorder technology have made it possible to record several tracks of information on the same tape. For example, the standard ¼-inch-wide tape can easily handle four tracks, and eight tracks are possible with a slight decrease in the signal-to-noise ratio. In general, the narrower the recorder track, the poorer the signal-to-noise ratio will be. Magnetic tape is usually recorded on alternate tracks by passing it through the recorder in both directions. Formats for multitrack recording are shown in Fig. 21.18.

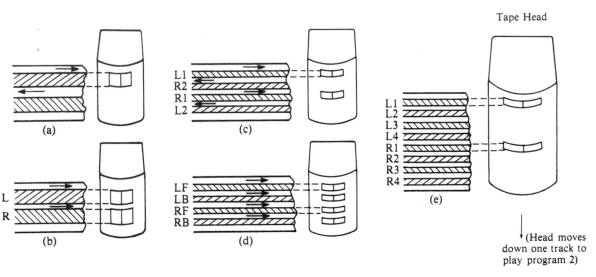

FIG. 21.18
Formats for multitrack recording on tape: (a) two-track monophonic; (b) two-track stereo; (c) four-track stereo; (d) four-track quadraphonic; (e) eight-track stereo.

Note that a two-track stereo tape can be played on a monophonic playback, and two-track mono or stereo tape can generally be played on a four-track machine. However, a four-track recording cannot be played on a two-track machine, because one head will play both the L1 and R2 tracks (which were recorded in opposite directions).

Stereo cassette recorders manage rather remarkably to record four-track stereo on ⅛-inch-wide tape at a speed of only 1⅞ ips. By using Dolby noise reduction, signal-to-noise ratios of 50 dB are obtained, which compares favorably to many good open reel recorders without a noise-reduction system.

In professional recording, where a high signal-to-noise ratio is a more important consideration than tape economy, wider tracks are used (e.g., two tracks on ¼-inch tape, four tracks on ½-inch tape, eight tracks on 1-inch tape) as shown in Fig. 21.18(a).

21.11   DIGITAL TAPE RECORDING

The digital recording of sound offers major advantages over conventional tape recording. Among them are wide dynamic range, low distortion, flat frequency response, freedom from wow and flutter, and no cross talk between channels. Digital tape recording requires a very high frequency response, such as that of video tape recorders. Although only a few digital sound recorders are commercially available, a number of systems have been demonstrated successfully. Digital recording (sometimes called "pulse-code modulation") will be described in Chapter 28.

21.12 SUMMARY

Modern phonograph records are vinyl discs with approximately 100 grooves per centimeter, designed to revolve at either 33⅓ or 45 rpm. The pickup stylus senses the motion of the grooves, and an electrical signal is generated in a magnetic coil or a piezoelectric crystal. Recordings are made with treble emphasis and bass attenuation and played back with the opposite characteristic. In stereophonic discs, the two channels of information are carried by the two side walls of the groove.

Magnetic recording tape consists of a thin layer of magnetic oxide coated on a plastic tape. As the tape passes the recording head, its magnetization is caused to vary in accordance with the magnetic field at the head gap, and it retains this pattern of magnetization until erased. In order that the magnetization be a linear function of the recording field, it is necessary to use a magnetic bias, preferably one that varies at a frequency well above the audible range. Tape noise, due to random magnetization, can be reduced by using a Dolby or similar system. Digital tape recording appears attractive for the future.

References and Suggested Readings

Ballou, G., ed. (1987). *Handbook for Sound Engineers.* Indianapolis: H. W. Sams. Chaps. 23 and 24.

Bauer, B. B. (1977). "The High-Fidelity Phonograph Transducer," *J. Audio Eng. Soc.* **25:** 729. (This and other review articles appear in the Oct/Nov 1977 issue celebrating the centennial of the phonograph.)

Camras, M. (1988). *Magnetic Recording Handbook.* New York: Van Nostrand Reinhold.

Dolby, R. (1983). "A 20 dB Audio Noise Reduction System for Consumer Applications," *J. Audio Eng. Soc.* **31:** 98.

Eargle, J. (1976). *Sound Recording.* New York: Van Nostrand Reinhold.

Gravereaux, D. W., A. J. Gust, and B. B. Bauer (1970). "The Dynamic Range of Disc and Tape Records," *J. Audio Eng. Soc.* **18:** 530.

Institute of High Fidelity (1974). *Guide to High Fidelity.* Indianapolis: Howard Sams. (See Chapters 6 and 7.)

Mallinson, J. C. (1987). *The Foundations of Magnetic Recording.* San Diego: Academic Press.

Mee, C. D. (1964). *The Physics of Magnetic Recording.* Amsterdam: North-Holland.

Meyer, E., and E-G. Neumann (1972). *Physical and Applied Acoustics.* New York: Academic. (See Chapter 9.)

Olson, H. F. (1972). *Modern Sound Reproduction.* New York: McGraw-Hill. (See Chapter 8.)

Rossing, T. D. (1980). "Physics and Psychophysics of High-Fidelity Sound, Part II," *Physics Teach.* **18**: 278.

Rossing, T. D. (1981). "Anhysteretic Magnetization and Tape Recorder Bias," *Am. J. Phys.* **49**: 655.

Westmijze, W. K. (1953). "Studies on Magnetic Recording. III The Recording Process," *Philips Res. Rep.* **8**: 245.

Glossary

bias That which is added to the desired signal to produce a composite. In the case of magnetic tape recording, for example, the bias may be either a constant magnetic field or a field that varies at a high frequency.

coercivity The magnetic field that must be applied in the "reverse" direction to reduce the magnetization to zero.

demagnetization "Erasing" a magnetically recorded signal by applying a rapidly varying magnetic field of large amplitude; in a demagnetized tape, the tiny magnetic "domains" are oriented in nearly random directions.

Dolby system A widely used noise-reduction system that boosts low-level high-frequency signals when recorded and reduces them in playback.

domain (magnetic) A tiny region in which the atomic magnets point in the same direction.

equalization Boosting some frequencies and attenuating others to meet some prescribed recipe. Equalization is usually applied during both recording and playback.

hysteresis (magnetic) The effect of past history on the magnetic state of a material; when a magnetic field is applied momentarily and removed, the material does not revert to its original state. Magnetic hysteresis may be described by a "hysteresis loop," a graph of magnetization vs. applied magnetic field.

matrix An array of numbers that indicates the recipe for mixing inputs to achieve various outputs. In the case of matrix quadraphonic sound, it indicates the manner in which four channels of sound are to be mixed to be recorded as two channels on a phonograph disc.

piezoelectric Materials that generate an electrical voltage when placed under stress. Crystal pickups on phonographs consist of a stylus attached to a piezoelectric crystal.

remanence (magnetic) The net magnetization that remains after the magnetic material has been saturated and the field has been removed.

signal-to-noise ratio The ratio (usually expressed in dB) of the average recorded signal to the background noise.

variable reluctance A type of magnetic cartridge in which the magnetic flux in a circuit is changed by the movement of a small piece of iron attached to the stylus.

variable-pitch recording The spacing of the grooves on a disc according to the amplitude of the recorded material.

Questions for Discussion

1. In very early phonograph records, the stylus moved up and down in the groove. What are the advantages of the lateral movement of the stylus used in today's records over the vertical movement of early records?

2. Why is the signal-to-noise ratio in a cassette recorder usually lower than it is in a reel-to-reel recorder?

3. What is the purpose of tape bias?

4. What are several troublesome effects of the dust that collects on records?

5. Why does a high-frequency erase field result in a lower level of noise than does a unidirectional (DC) erase field, as in older tape recorders?

6. Why does a high tape speed give better high-frequency response?

7. Would decreasing the particle size in the oxide coating on a tape be expected to improve the high-frequency response of the tape? Why?

Problems

1. Measure the maximum and minimum radii of the grooves on a phonograph record.

    a) Assuming an average of 100 grooves per cm, determine how many grooves there are.

    b) Actually, of course, there is only one long groove. What is its total length? (Multiply the number of grooves times the average circumference.)

    c) Assuming $33\frac{1}{3}$ revolutions per minute, determine how long the record will play.

    d) Divide the total groove length by the time to find the average velocity of the stylus in the groove.

2. What is the wavelength of a 8000-Hz signal recorded on cassette tape moving at $1\frac{7}{8}$ ips?

3. A record intended to be played at $33\frac{1}{3}$ rpm is played at 45 rpm by mistake. By what ratio are the frequencies of all tones raised? Approximately what musical interval is this?

4. What is the wavelength of a 100-kHz bias field recorded on tape moving at a speed of $7\frac{1}{2}$ ips? Would you expect to see a very strong signal recorded at this frequency? Explain why.

5. Estimate the spacing between track centers for two-track, four-track, and eight-track formats recorded on $\frac{1}{4}$-inch-wide tape (see Fig. 21.18).

CHAPTER 22

Components for High-Fidelity Sound

One of the most rapidly growing consumer-oriented industries in the United States today is the sale of components for high-fidelity sound reproduction. Most homes now have one or more systems for the reproduction of sound; many of these systems represent substantial financial investments. Many systems combine components from several different manufacturers.

Although many purchasers of high-fidelity sound systems are interested only in the end product, sound, a growing number of users (especially young people) are becoming interested in the scientific and technical aspects of sound reproduction, and this chapter, like the preceding ones, is intended for them. Unfortunately, many popular books on the subject include misleading statements and technical explanations of dubious quality.

In the preceding chapters, we have discussed the principles of amplifiers, microphones, loudspeakers, and recording media. In this chapter, we will apply these principles to specific high-fidelity components. High-fidelity sound systems, including the all-important listening room, are discussed in Chapter 25.

22.1 THE COMPONENTS OF A HIGH-FIDELITY SOUND SYSTEM

Besides the listening room itself, a high-fidelity sound system includes an amplifier, one or more loudspeakers, and program sources, such as a record player, a tape player, a compact disc player, and an FM tuner. The program sources supply electrical signals of low amplitude to the amplifier, which replicates them at much higher levels of power and feeds them to loudspeakers. The loudspeakers convert the electrical signals to sound, part of which is transmitted directly to the listener, and part of which interacts with surfaces in the room before reaching the ear of the listener. A high-fidelity system is shown schematically in Fig. 22.1.

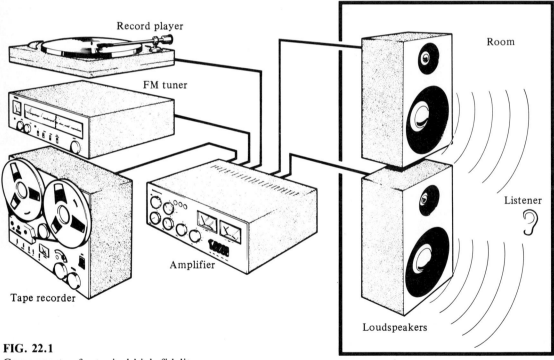

FIG. 22.1
Components of a typical high-fidelity sound system.

Some systems separate the preamplifier and power amplifier into two components. Others combine the amplifiers, the FM tuner, and an AM tuner into a single component called a receiver. Some systems include two different tape recorders: a reel-to-reel type and a cassette type. Some include earphones for private listening.

22.2 RECORD PLAYERS

Three types of record players are in common use: manual turntables, automatic turntables, and automatic record changers. They are further distinguished by the particular turntable drive system—rim-drive, belt-drive, and direct-drive—and by the tone arm and cartridge design.

On a *manual turntable*, the tone arm must be raised and lowered by the user. On an *automatic turntable*, the tone arm cycles automatically. Some automatic turntables can be programmed, so that the tone arm will play individual bands or selections in any desired sequence. A *record changer* will automatically play a stack of records one at at time, thus providing several hours of program material without need of attention.

Although automatic record changers were once considered incapable of high-fidelity performance, this is not necessarily true today. Because manual turntables have fewer parts and linkages, they tend to be more rugged and require less servicing than automatic turntables or record changers, but even so it is impossible to generalize. The choice is largely governed by the listening habits of the user.

At one time, most home record players used a *rim-drive* system, in which a small induction motor was coupled to the rim of the turntable platter by rubber idler wheels. A different diameter of idler wheel was used for each turntable speed.

Record players often suffer from three mechanical defects, called "wow," "flutter," and "rumble." *Rumble* is the term applied to low-frequency vibrations and noise transmitted to the pickup and eventually heard as sound. Variations in turntable speed are called *wow* when they occur at a slow rate or *flutter* when they occur at a more rapid rate. Variations in speed can be minimized by using a hysteresis synchronous motor, a cousin to the familiar electric clock motor, which closely follows the 60-Hz frequency of the electric power line.

Rumble can be reduced by substituting a *belt-drive* system for a rim-drive system; in a belt drive, there is no idler wheel to transmit vibrations from motor to turntable. Another effective means of reducing vibration is to use a motor of slower speed. Rim and belt drives are shown in Fig. 22.2.

Direct-drive motors turn at the same slow speed as the turntable, and thus minimize vibration. The motor and turntable platter, in fact, have the same center shaft. The rotating component of the motor has a large number of small permanent magnets that are attracted by small magnetic coils surrounding it. The current in these coils, and thus the speed of rotation, is controlled electronically.

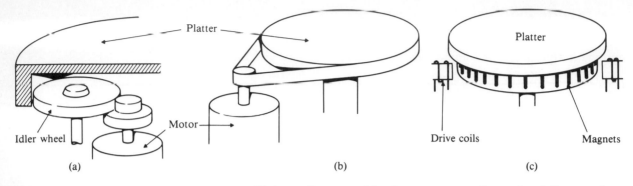

(a)   (b)   (c)

FIG. 22.2
Turntable drives: (a) rim drive; (b) belt drive; (c) direct drive.

High-quality turntable platters are usually made of die-cast aluminum, carefully machined for accuracy and balance. Heavy platters will have large rotational inertia, which opposes variations in speed.

Unfortunately, there are no universal standards for specifying rumble, wow, and flutter. Rumble, for example, may be expressed by comparing the total (unweighted) rumble noise to the signal (the NAB* method) or by using a suitable weighting network (called the DIN, JIS, and RIAA* methods after the various organizations recommending them). The amount of rumble described by the weighted DIN method may be 20 to 30 dB lower than the amount described by the NAB unweighted method for the same turntable. A good turntable should have a rumble of −50 dB or below, described by the DIN method.

Wow and flutter are usually expressed in percent deviation from true speed. A good turntable should have less than 0.2 percent wow and flutter.

22.3 TONE ARMS AND CARTRIDGES

The phono cartridge tracks the record groove and transforms the undulations into an electrical signal. (Various types of cartridges were described in Section 21.4.) Nearly all high-fidelity record players use magnetic cartridges.

The stylus in the cartridge is made of either sapphire or diamond. Diamond is much to be preferred, because it maintains its tip shape much longer, and thus the risk of damage to records is lessened. The tip of the stylus is machined to extreme precision, and may be either *spherical*, with a tip radius of about 0.01 mm (4×10^{-4} inch), or *elliptical*. Elliptical styli have their long axis at a right angle to the groove,

* NAB: National Association of Broadcasters; DIN: Deutsche Industrie Normen (German standards); JIS: Japanese standards; RIAA: Record Industry Association of America.

and are better able to follow rapid groove undulations. A special type of stylus, called the *Shibata* stylus after its inventor, is designed to track the very high frequencies associated with quadraphonic records (see Chapter 25).

A good cartridge should have a smooth frequency response throughout the audio range and, if it is a stereo cartridge, the separation between channels must be sufficient to prevent cross talk. The response of a good cartridge should vary no more than ± 3 dB over the frequency range of 20 to 20,000 Hz. (Merely stating that the "frequency response is 20–20,000 Hz" is a rather meaningless specification.) The actual output voltage of a cartridge is specified in millivolts for a certain stylus velocity, but this is not particularly significant since the preamplifier usually has sufficient gain for almost any cartridge. The channel separation should be at least 20 dB (at 1000 Hz).

The *tracking force* is related to the force necessary to keep the stylus in the groove even when loud passages cause it to move rapidly. It is customarily expressed in grams, although a gram is *not* a unit of force but rather of mass (see Section 1.5). To state that the "tracking force is 1–3 grams" really means that the tracking force is equal to the force of gravity on a mass of from 1 to 3 grams, which is approximately 0.01 to 0.03 newtons. Since the mass of the cartridge is almost certainly greater than 1 to 3 grams, the tone arm has a counter-weight, usually adjustable to provide the desired amount of tracking force.

An important specification of a cartridge is its *compliance,* which is a measure of how easily it moves or "complies" with force from the groove wall. A compliance of 10^{-5} cm/dyne (10^{-2} m/N) or greater generally means that the stylus will not wear the sides of the groove unduly. The compliance and the mass of the stylus plus the moving magnet (or coil) determines the resonance frequency of the cartridge, which is typically in the vicinity of 2000 Hz. At frequencies well below this resonance, the mechanical response of the stylus is largely determined by the compliance; at high frequency, the mass dominates; whereas near resonance, damping is most important.

Important design parameters in a tone arm are the *offset angle* and the *overhang*, shown in Fig. 22.3. Together they determine the *tracking error* of the cartridge, that angle by which the cartridge axis differs from the tangent to the groove. High-quality tone arms are usually designed so that the tracking error will be zero at two points on the record and will never exceed 3° at any other point. Excessive tracking error produces distortion.

The resonance frequency of the tone arm should be less than 10 Hz in order to avoid being excited to resonance by motion of the car-

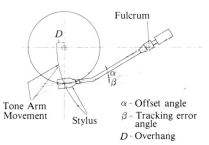

α - Offset angle
β - Tracking error angle
D - Overhang

FIG. 22.3
Tracking error, offset angle, and stylus overhang. (From the Institute of High Fidelity, © 1974. Reprinted by permission of the Electronic Industries Association.)

tridge in the record groove. Many tone arms include an antiskating device, which counteracts the inward force on the tone arm and prevents it from causing the stylus to jump out of the groove and "skate" across the record.

22.4   TAPE DECKS

The term "tape deck" has come to mean a tape recorder without a power amplifier and speaker, designed to serve as a program source in a high-fidelity sound system. Most reel-to-reel or open-reel decks operate at tape speeds of 7½ in./s and 3¾ in./s. Some include the professional speed of 15 in./s or the slow speed of 1⅞ in./s as well. (The corresponding metric speeds are 38, 19, 9.5, and 4.8 cm/s.)

The principles of tape recording and playback were presented in Chapter 21. Many high-quality recorders use three separate heads: one for erase, one for recording, and one for playback. This allows the newly recorded signal to be monitored for quality. Some multichannel tape recorders have a provision for using the record head as a playback monitor on one or more channels. Known as "sel-sync," this feature facilitates the synchronization of new program material with material already recorded on another track, a feature that makes it very useful in recording electronic music.

Tape decks or recorders may have one, two, or three motors. Three-motor machines have separate motors to drive each reel and a third to drive the *capstan*, which moves the tape past the heads. In two-motor machines, both reels are driven by one motor; in one-motor machines, both reels and the capstan are driven from one motor by the use of belts.

Becoming very popular in recent years, *cassette decks* record on narrow ⁵⁄₃₂-inch tape at a slow speed of 1⅞ in./s. The arrangement of tracks in stereo recording on cassettes, shown in Fig. 22.4, is different from that used for stereo recording on open reels, shown in Fig. 21.18. The left and right tracks are adjacent to each other so that a stereo recording can be played on a mono machine and vice versa.

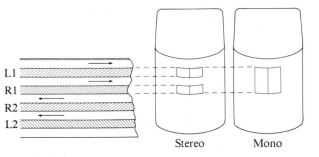

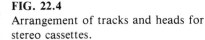

FIG. 22.4
Arrangement of tracks and heads for stereo cassettes.

22.5   AM/FM TUNERS

Radio stations are assigned a frequency in one of two bands: 540–1600 kHz (5.4–16×10^5 Hz) and 88–108 MHz (8.8–10.8×10^7 Hz). Broadcasts in the upper band use *frequency modulation* (FM), whereas broadcasts in the lower band use *amplitude modulation* (AM). At the lower frequency of the AM band, signals can be transmitted over long distances, but because the AM band is crowded, stations are allowed to transmit audio frequencies up to 5000 Hz only. This severely limits the fidelity of sound reproduction. Furthermore, AM signals are subject to interference from electrical noise, both natural (atmospheric electricity) and human-made (car ignition, motors, etc.). On the other hand, FM stations are allowed to transmit a much greater range of audio frequency, but the high-frequency waves in the FM band are normally limited in range to "line of sight" or slightly more. Within metropolitan areas, however, FM radio has become the major source of high-fidelity program material.

Figure 22.5 illustrates the difference between amplitude modulation and frequency modulation. In both cases, audio information is superimposed on a radio carrier wave of high frequency. In AM broadcasting, the amplitude is varied (modulated) in accordance with the audio signal; in FM broadcasting, it is the frequency that varies.

The AM band, which has a total frequency range of 1060 kHz, is divided into broadcast channels that are 10 kHz wide. On the FM band, on the other hand, with a total frequency range of 20 MHz, channels can be 200 kHz wide. This bandwidth, 20 times greater than that of AM stations, allows not only a wider range of audio frequency but a wider dynamic range as well; furthermore, it permits the broadcasting of stereophonic sound.

The tuner selects the AM or FM signal from the desired station, amplifies it, and then "demodulates" or "detects" it (that is, extracts the audio signal). (The technical details of how this is accomplished will not be described.) Many modern AM/FM tuners are finely

FIG. 22.5
A schematic diagram showing
(a) amplitude modulation (AM);
(b) frequency modulation (FM).

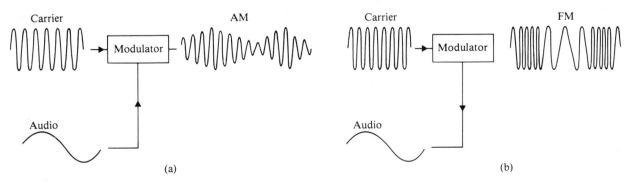

(a)   (b)

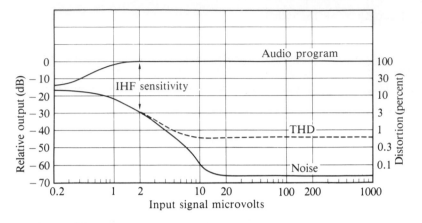

FIG. 22.6
Typical noise and distortion in an FM tuner. (From the Institute of High Fidelity, © 1974. Reprinted by permission of the Electronic Industries Association.)

engineered instruments, capable of high-fidelity sound reproduction. Furthermore, the performance specifications are much more accurate and complete than for other components in an audio system. A few of the more important specifications are as follows.

1. *Sensitivity* describes the ability of the tuner to pick up weak signals. In accordance with the recommendation of the Institute of High Fidelity (IHF), the sensitivity is usually expressed as the electrical signal (in microvolts) at the antenna, which will result in an audio signal 30 dB greater than the noise level.

2. *Signal-to-noise (S/N) ratio* expresses the ratio of audio signal to noise for a strong input signal. As the input signal increases, both noise and distortion usually decrease, as shown in Fig. 22.6. The S/N ratio for stereo reception is generally somewhat less than for mono reception.

3. *Total harmonic distortion (THD)*, also shown in Fig. 22.6, compares (in percent) the harmonic distortion generated in the tuner to the undistorted signal.

4. *Selectivity* measures the ability of the tuner to reject stations on nearby channels. It is usually expressed in decibels of attenuation for a station two channels (400 kHz) away.

5. *Capture ratio* expresses the ability of a receiver to reject the weaker of two signals on the same channel. The capture ratio (in decibels) tells how much smaller the one signal must be to be rejected; a capture ratio below 3 dB is desirable.

6. *Stereo separation*, the separation between stereo channels, should be at least 30 dB at all audio frequencies.

Other tuner specifications may include *image rejection*, *i-f rejection*, and *spurious-response rejection*, all of which describe the ability to reject some type of undesired signal.

22.6 STEREO BROADCASTING

Much of the program material of FM stations these days is broadcast in stereo. However, it is broadcast in a compatible format, so that it can be received on either stereophonic or monophonic tuners. This is done by mixing the left (L) and right (R) signals together to create sum and difference signals, which are then broadcast.

The sum (L + R) is broadcast as a conventional FM signal, as described in Section 22.5, and is received by both monophonic and stereophonic receivers. At the same time, a difference signal (L − R) is used to modulate a 38-kHz subcarrier signal, and this subcarrier, in turn, modulates the main carrier wave. A monophonic FM radio, of course, receives this subcarrier along with the regular audio signal, but it is not heard because of its high frequency. A stereophonic receiver, on the other hand, reconstructs the desired (L) and (R) signals from the (L + R) and (L − R) signals received. Because the signal-to-noise ratio of a tuner is usually greater when operated in the mono mode, it is sometimes desirable to use this mode when listening to a weak or distant FM station.

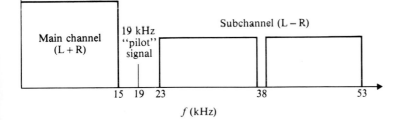

FIG. 22.7
Composition of a stereo broadcast signal.

To conserve channel space, the 38-kHz subcarrier itself is not actually transmitted, but only the modulation "sidebands" (which contain sum and difference frequencies from combining the subcarrier and the audio signal) actually go out over the air. In order to reconstruct the entire signal at the FM tuner, a 19-kHz "pilot" signal is broadcast. (Many receivers also use this 19-kHz pilot signal to trigger a stereo indicator light.) The composition of a stereo broadcast signal is shown in Fig. 22.7. This signal modulates the frequency of the main carrier wave, which lies in the FM band (88–108 MHz).

22.7 AMPLIFIERS

High-fidelity amplifiers consist of two parts: the preamplifier and the power amplifier. Although they may be packaged as separate components, more often they are sections of an integrated amplifier or combined with the tuner to form an integrated receiver. In the days of vacuum-tube electronics, separate components usually resulted in less hum and better heat dissipation, but these are not particular problems with transistor circuits, and integration of components results in a substantial reduction in cost.

The preamplifier amplifies signals of a millivolt or two from the phonograph cartridge, and also "equalizes" the signals in accordance with the RIAA equalization curve shown in Fig. 21.2. The phono input impedance of the preamplifier should match the cartridge used, which is usually 47,000 ohms for stereo cartridges in common use. (some cartridges designed for quadraphonic reproduction have a higher impedance).

The preamplifier has controls to select the desired program source, adjust the desired signal level ("volume"), and adjust the tonal balance between high, middle, and low frequencies. The tonal balance is usually adjusted by means of bass and treble controls, although some amplifiers provide equalizers which adjust the gain in several frequency bands individually. The action of typical treble and bass tone controls is illustrated by Fig. 22.8.

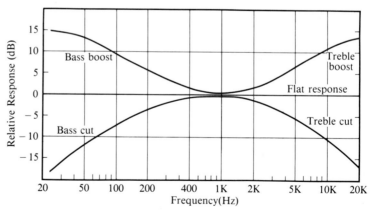

FIG. 22.8
Action of typical bass and treble controls.

Many amplifiers have a loudness control. With the loudness switch turned on, the frequency response of the amplifier will automatically change with the volume level. At high level, the response will be essentially flat, but as the volume level is reduced, the bass and treble are increased to compensate for the reduced sensitivity of the ear to low-level sounds of high and especially low frequency (see Fig. 6.1).

Amplifiers are sometimes equipped with filters designated "low" and "high" or "rumble" and "scratch," which reduce gain at frequencies below about 100 Hz and above about 5 kHz, respectively.

The power amplifier section is designed to supply substantial amounts of audio power to the loudspeakers with as little distortion as possible. Because large amounts of power are involved, efficiency becomes an important consideration. The large output transistors are supplied DC power from the power supply; some of it is converted into audio output, the rest into waste heat. Since this heat must be carried away by heat "sinks," designers try to minimize its production.

One way to minimize waste heat in the output transistors is to have them conduct electricity only when an audio signal is present, and then only as much as necessary. A circuit so designed is called a class-B amplifier, as opposed to a class-A amplifier in which the transistors conduct nearly all the time. Unfortunately, a class-B amplifier generates substantially more distortion than does a class-A amplifier. The solution arrived at in most high-fidelity power amplifiers is to operate in an intermediate condition known as "class-AB."

Actual power amplifier circuits are substantially more complicated than the simplified circuit shown in Fig. 22.9. Some type of negative feedback is included (see Section 19.3), as is temperature compensation to prevent the amplifier characteristics from changing with operating temperature. A protective fuse or circuit breaker is nearly always included in the output line, and often a large capacitor is also, which allows AC to flow but prevents DC from reaching the speaker. Sometimes a more elaborate protection circuit involving several transistors is provided.

Virtually all vacuum-tube power amplifiers include an output transformer, but only a few transistor power amplifiers have them. Although expensive and bulky, output transformers offer several advantages. With a transformer it is possible for all the output transistors to be of one type, usually *npn*, by using an arrangement called "push-pull." Also the loudspeaker impedance can be more nearly matched to the amplifier, allowing for efficient power transfer to the speaker.

Waste heat is usually carried away from the output transistors by attaching them to large aluminum or copper heat sinks, which are good conductors of heat. The heat sinks in turn are cooled by air

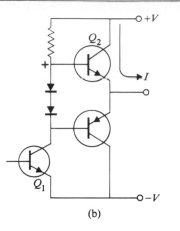

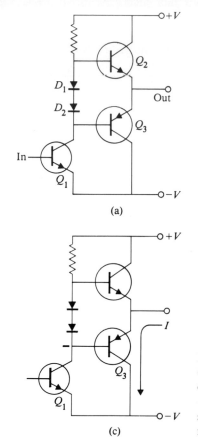

FIG. 22.9

(a) A simplified amplifier circuit employing a complementary pair of transistors. (b) The positive signal turns Q_2 on. (c) The negative voltage turns Q_3 on.

Figure 22.9 is a simplified schematic diagram of a power amplifier. The output transisters Q_2 and Q_3 are a "complementary pair," consisting of one *npn* and one *pnp* transistor with nearly identical characteristics otherwise. The *npn* transistor Q_2 is turned on by a positive voltage on its base, the *pnp* transistor Q_3 by a negative voltage. Both transistors conduct in the direction shown by the arrows, so if they were both turned on at the same time, a direct current (DC) would flow through both of them from $+V$ to $-V$. The power carried by such a direct current would convert to waste heat, however.

Since both transistors are supplied their input signals by transistor Q_1, they will not both be turned on at the same time. If Q_1 supplies a positive voltage, transistor Q_2 turns on, and a current I flows to the loudspeaker in the direction shown in Fig. 22.9(b). If Q_1 supplies a negative voltage, however, transistor Q_3 turns on, and the current flows in the opposite direction, as shown in Fig. 22.9(c). Without the diodes D_1 and D_2, transistors Q_2 and Q_3 would essentially operate class B. However the small voltage drop across D_1 and D_2 is sufficient to maintain a small amount of current in both transistors to operate class AB and thus prevent what is known as *crossover distortion*.

motion or convection. For this reason, it is important to comply with the manufacturer's instructions regarding the free circulation of air past the output circuitry. Transistors, which normally last for years, can be destroyed in a wink by overheating or overloading electrically!

22.8 DISTORTION

Distortion refers to any signal not in the original recording or program source that appears in the reproduced sound. Distortion can be generated in any component in the audio system. Once generated, it will most likely be reproduced by the remaining components along with the desired sound.

The three types of distortion that are most significant in a high-fidelity system are harmonic distortion, intermodulation distortion, and transient distortion. Harmonic and intermodulation distortion usually result from a nonlinear transfer characteristic in some component (see Section 19.2). Transient distortion usually results from insufficient damping.

Harmonic distortion results in the creation of harmonics of the fundamental frequency that are not present in the original signal. "Clipping" of waveform peaks, which results in substantial harmonic distortion, can occur because of overloading some component (see Fig. 19.9). Small amounts of harmonic distortion often escape notice, since musical tones already contain several harmonics. Harmonic distortion is expressed as a percentage of the total signal.

Intermodulation distortion may result when tones of two or more different frequencies are reproduced at the same time (almost always the case in music). A nonlinear characteristic in some component results in the creation of sum and difference frequencies. These new frequencies are harmonically and musically unrelated to the desired tones, and are therefore more noticeable and objectionable than harmonic distortion products. Intermodulation distortion is also expressed as a percentage of the total signal.

Transient distortion occurs when some component cannot respond quickly enough to a rapidly changing signal. In a loudspeaker, it may occur because of insufficient damping of a mechanical resonance. The loudspeaker depends on the amplifier to damp its motion, and an amplifier with a damping factor of at least 20 is desirable. (The *damping factor* expresses the ratio of speaker impedance to "internal resistance" in the amplifier output.)

Transient intermodulation distortion (TIM) can occur in an amplifier because time delay in the feedback loop causes momentary overloading of some circuit element. Thus the percussive sound of a cymbal is distorted, although the steady sound of a violin is not.

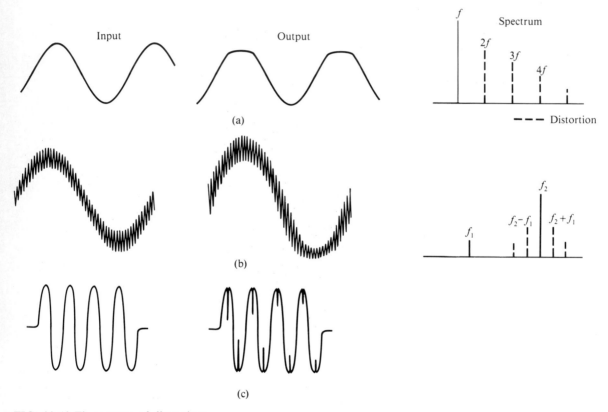

FIG. 22.10 Three types of distortion: (a) harmonic; (b) intermodulation; (c) transient. In each case, input and output waveforms are shown, and spectra of (a) and (b) are shown as well.

Figure 22.10 illustrates the three types of distortion just discussed. If the input signal is a pure tone of frequency f, harmonic distortion will create distortion products with frequencies $2f$, $3f$, etc. Intermodulation distortion may appear as amplitude modulation (AM) or frequency modulation (FM). AM distortion, shown in Fig. 22.10(b), produces distortion products at the sum and difference frequencies. (FM distortion, not shown, produces additional frequencies as well.) One type of transient distortion is shown in Fig. 22.10(c).

22.9 AMPLIFIER POWER AND DISTORTION

The most familiar specification of amplifiers is the output power. Until recently it was also one of the most misleading. In an electronics catalog from the early 1970s, one can find an assortment of expressions such as "music power," "IHF power," "peak power," "rms power," etc. In 1974, the Federal Trade Commission (FTC) ruled that the continuous average power per channel, the power frequency response, and the total harmonic distortion at full rated power must be stated "clearly, conspicuously, and more prominently than other representations." Furthermore, the test conditions for measuring power were clearly defined, including a preconditioning at one-third of rated output (the heating of the output transistors is usually greater at one-third of rated output than at full power). A few manufacturers still use the incorrect term "rms power" to refer to continuous average power.

How important is power? It is not easy to answer this question, because room conditions, loudspeaker efficiencies, and listening habits differ widely. A typical average listening level of 65 dB in a room may require from 0.01 to 0.03 mW of acoustic power. Even with a low-efficiency (one percent) loudspeaker, this requires only 0.001 to 0.003 W of electrical power. However, at an orchestra concert, you would hear peaks to 105 dB or even more. To reproduce such peaks, 40 dB above the average level, would require a 10,000-fold increase in power! This would require 10 to 30 W of power from the amplifier.

Table 22.1 is a rough guide to the amplifier power requirements recommended in a publication by high-fidelity equipment manufac-

TABLE 22.1 A rough guide to power-amplifier requirements for various speaker efficiencies and room sizes

Highest sound pressure level (in dB) possible for a room of the indicated volume (in cu ft)

Amplifier power (continuous watts per channel)	Low-efficiency systems			Medium-efficiency systems			High-efficiency systems		
	2000 cu ft	3000 cu ft	4000 cu ft	2000 cu ft	3000 cu ft	4000 cu ft	2000 cu ft	3000 cu ft	4000 cu ft
10	94 dB	92 dB	91 dB	97 dB	95 dB	93 dB	102 dB	101 dB	100 dB
20	97 dB	95 dB	94 dB	100 dB	98 dB	96 dB	105 dB	104 dB	103 dB
35	99.5 dB	97.5 dB	96.5 dB	102.5 dB	101.5 dB	98 dB	107 dB	106 dB	105 dB
50	101 dB	99 dB	98 dB	104 dB	102 dB	100 dB	109 dB	108 dB	107 dB
75	103 dB	101 dB	100 dB	105 dB	103.5 dB	101.5 dB	110.5 dB	109.5 dB	108.5 dB
100	104 dB	102 dB	101 dB	107 dB	105 dB	103 dB	112 dB	111 dB	110 dB
125	105 dB	103 dB	102 dB	108 dB	106 dB	104 dB	113 dB	112 dB	111 dB

Notes: Ninety to ninety-five dB of typical sound pressure level would be about as loud as what you would hear when sitting in the balcony seats at a live concert. One hundred dB is fairly loud, about what you could expect at midorchestra seating in a concert hall. One hundred ten dB is reached by an orchestra playing full blast, with the listener in the front rows of the hall. One hundred twenty-six dB is considered to be the threshold of physical pain. (From the Institute of High Fidelity, © 1974. Reprinted by permission.)

turers (Institute of High Fidelity, 1974). The figures appear to pertain to living rooms with large amounts of absorption. In most dormitory rooms, ten watts per channel even into low-efficiency speakers will reproduce peaks well in excess of 100 dB.

Salespeople have been known to recommend the purchase of high-power amplifiers because the large amount of reserve power will supposedly assure less distortion. This is not normally the case. In most modern amplifiers, harmonic and intermodulation distortion remain low until one approaches the rated output power. Rarely is the amplifier the main source of distortion in a high-fidelity system, in any case.

The harmonic and intermodulation distortion of a typical amplifier are shown in Fig. 22.11. Intermodulation is generally measured by the application of two signals with widely spaced frequencies (e.g., 60 Hz and 7000 Hz) in a 4:1 ratio of amplitudes. In both cases, distortion is expressed as a percent of the output signal.

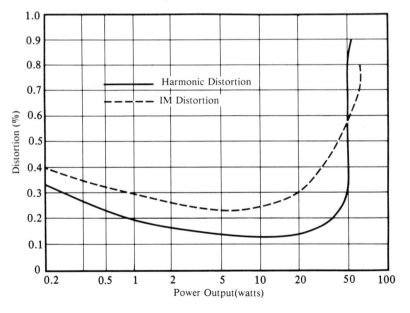

FIG. 22.11
Typical amplifier distortion curves.
(From the Institute of High Fidelity,
© 1974. Reprinted by permission of
the Electronic Industries Association.)

22.10   LOUDSPEAKER EFFICIENCY AND DISTORTION

Loudspeakers tend to be the weakest component in most high-fidelity sound-reproducing systems. Their frequency-response curves show peaks and dips much greater than those of amplifiers or phono cartridges; they are also the component most likely to produce noticeable distortion. Rarely are complete specifications of their performance available from either manufacturer or retailer.

The trend in recent years has been toward the compact use of low-efficiency speaker systems. In order to produce bass sound from a compact speaker system, long-throw air-suspension drivers are used. Radiation of low-frequency sound requires moving a substantial quantity of air (see Fig. 20.12); this can be done by a speaker cone of small diameter moving through a large amplitude or by a large-diameter cone moving through a much smaller amplitude. As a general rule, the distortion will be less in the latter case, as the curves in Fig. 22.12 illustrate. These curves should not be regarded as descriptive of all speakers of those sizes, of course. The amount of distortion will depend on many design parameters, such as magnet size, voice-coil design, suspension, etc. Certainly there are speaker systems with eight-inch and twelve-inch drivers with lower distortion. Nevertheless, minimizing cone travel by using large-diameter high-efficiency drivers tends to reduce distortion as well as to conserve amplifier power.

One source of distortion in loudspeakers, which cannot be eliminated even by careful design, is intermodulation due to the Doppler effect (see Section 3.7). If the loudspeaker cone is making large excursions at low frequency, it will act as a moving source for high-frequency sound. The amount of distortion depends on the amplitude of the low-frequency excursions as well as the frequencies of the two interacting sounds (Klipsch, 1972).

The crossover frequency in two-way speaker systems is typically at 2000–4000 Hz; even in three-way systems, the woofer often receives signals up to 1000 Hz. Thus, many pairs of frequencies can interact to produce a wide assortment of intermodulation distortion products resulting in a "muddy" sound at high levels of power. Fortunately, average power levels are quite low, and noticeable distortion due to the Doppler effect occurs mostly during loud peaks.

In the future, hopefully, accurate specifications of the efficiency, directivity, distortion, and frequency response of loudspeakers will be made available to prospective users.

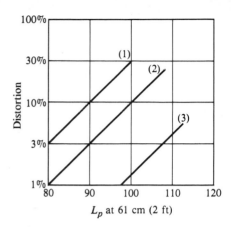

FIG. 22.12

Distortion as a function of output for three loudspeakers: (1) eight-inch diameter cone; (2) twelve-inch diameter cone; (3) large horn. Total distortion is largely intermodulation (IM) from a mixture of the 41-Hz and 350-Hz test signals. (From Klipsch, 1972. Reprinted by permission of AUDIO Engineering Society.)

22.11   EARPHONES (HEADPHONES)

Earphones have been used in electronic communications for many years, but more recently they have become popular for the individual listening to stereophonic music. Several different types of earphones or headphones are in use, including *dynamic (moving coil)*, *electrostatic*, *piezoelectric*, and the *magnetic (metal-diaphragm)* type used in telephone receivers. Dynamic, electrostatic, and piezoelectric earphones are shown in Fig. 22.13.

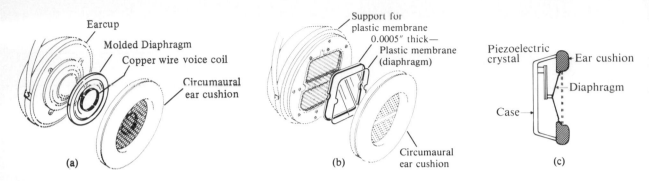

FIG. 22.13
(a) Dynamic earphone. (b) Electrostatic earphone. (c) Piezo-electric earphone. (Parts (a) and (b) courtesy of Koss SAC Training Lesson No. 9.)

Dynamic earphones resemble small loudspeakers in their construction and operation. The diaphragm is analogous to the cone of the loudspeaker but usually has no inner suspension or spider. The electrostatic earphone with its light plastic diaphragm is analogous to a condenser microphone. Some electrostatic earphones have a permanently charged electret diaphragm as do electret-condenser microphones. Piezoelectric earphones are similar to crystal (piezoelectric) microphones.

Earphones differ in the way they couple sound to the ear. *Circumaural* earphones have a liquid- or foam-filled doughnut-shaped pad that makes a nearly airtight seal against the head. Circumaural earphones have high efficiency at low frequencies, because the moving diaphragm produces pressure changes directly in the ear canal. Efficiency is lost if the seal is not tight, however.

Open-air, or *supra-aural*, earphones do not depend on an airtight seal against the wearer's head. The diaphragm is spaced away from the ear by a foam pad that leaks air in a controlled way. Open-air earphones are lighter in weight than the sealed type, but tend to be somewhat less efficient at low frequency.

Recently miniature dynamic earphones that extend into the ear like a stethoscope have become available. Their advantage is the fact that they are very lightweight, one popular model having a mass of only 33 grams.

Listening to stereophonic music with earphones is quite a different experience from listening in a normal room environment; earphone listening will be discussed further in Chapter 25.

22.12   SUMMARY

High-fidelity sound systems usually include an amplifier, loudspeakers, a record player, an FM tuner, and often a tape deck and earphones. Record players, which may be manual or automatic, include a turntable, a tone arm, and a cartridge. All of these should be designed for low noise and distortion. Tape decks may use several different tape speeds and tape formats, including tape cassettes and open reels. Up to eight tracks may be recorded on ¼-inch-wide tape. Radio stations transmit in two frequency bands; the greater bandwidth allocated to stations in the FM band makes high-fidelity stereophonic sound transmission possible. The performance of FM tuners and receivers is usually well described by rather elaborate specifications.

Integrated amplifiers include a preamplifier and a control section as well as a power amplifier. Power amplifiers produce large audio signals with very small amounts of noise and distortion. Although very much less a problem with transistors than with vacuum tubes, waste heat must be removed from power amplifiers. Distortion is of three main types: harmonic, intermodulation, and transient. Distortion in loudspeakers is more prevalent than in amplifiers, but is seldom in the manufacturer's specifications. High-efficiency, large-cone speakers tend to have less distortion at high power levels than do low-efficiency, long-throw speaker systems.

References and Suggested Readings

Ballou, G., ed. (1987). *Handbook for Sound Engineers.* Indianapolis: H. W. Sams.

Institute of High Fidelity (1974). *Guide to High Fidelity.* Indianapolis: Howard Sams.

Klipsch, P. W. (1972). "Modulation Distortion in Loud-speakers: Part III," *J. Audio Eng. Soc.* **20**: 827.

Rossing, T. D. (1980). "Physics and Psychophysics of High-Fidelity Sound, Part II," *The Physics Teacher* **18**: 278.

Glossary

amplitude modulation (AM) The method of radio broadcasting in which the amplitude of a carrier wave is determined by the audio signal. The AM band extends from 540 to 1600 kHz.

carrier wave A high-frequency electromagnetic wave capable of being modulated to transmit a signal of lower frequency.

class AB A power amplifier circuit in which current flows in each transistor for slightly more than one-half of the cycle.

complementary symmetry An *npn* and a *pnp* transistor that otherwise have identical electrical characteristics; such pairs of transistors are used as audio power amplifiers.

compliance A measure of how easily the stylus in a phono cartridge moves; it is measured in units of distance/force (e.g., m/N or cm/dyne).

cross-field bias Supplying the tape recording bias by a separate head placed adjacent to or directly opposite the record head.

crossover frequency A separation point for the distribution of low-frequency signals to the woofer and high-frequency signals to the tweeter.

distortion Signals that appear in the output of a sound reproduction system that were not present in the original program material.

elliptical stylus A phono stylus with an elliptical cross section designed to track very rapid undulations in the record groove.

flutter Rapid changes in the speed of a phono turntable or tape transport that can cause a wavering of the musical pitch.

frequency modulation (FM) The method of radio broadcasting in which the frequency of a carrier wave is determined by the audio signal. The FM band, which extends from 88 to 108 MHz, allows stations sufficient bandwidth to transmit high-fidelity stereophonic, and even quadraphonic, sound.

harmonic distortion The creation of harmonics (frequency multiples) of the original signal by some type of non-linearity in the system (the most common cause is over-driving some component).

intermodulation (IM) distortion The creation of sum and difference frequencies from signals of two different frequencies.

push-pull An arrangement used in power amplifiers; a positive voltage in the input causes the current to rise in one transistor, whereas a negative voltage does the same for another transistor. Push-pull amplifiers usually have an output transformer, but do not require complementary transistors.

rumble Low-frequency noise from a turntable or tape transport.

Shibata stylus A stylus with unusually good high-frequency response, which makes it capable of playing discrete (CD-4) quadraphonic, as well as stereophonic, discs.

sidebands Sum and difference frequencies created in the modulation process.

tracking error The angle between the axis of a phono cartridge and a tangent to the groove.

tracking force The vertical force that keeps the phonograph stylus in the groove; it is usually stated in grams, although the gram is a unit of mass, not force.

transient distortion Overshoot, or other unprogrammed response, that results from the inability of some component to follow a very rapid change in signal.

wow Slow periodic variations in the speed of a turntable or tape transport.

Questions for Discussion

1. The RIAA phono recording curve includes treble boost and bass cut.

   a) What is the reason for the treble boost?

   b) What is the reason for the bass cut?

2. Describe the operation of a Dolby noise-reduction system. Could the system be made to work on disc records as well as on tape?

3. Would it be possible to broadcast stereophonic programs on AM radio? Why?

4. What characteristics of loudspeakers should be described in the specifications?

5. Why is intermodulation distortion generally more objectionable to the ear than an equal amount of harmonic distortion?

Problems

1. If two eight-ohm speakers are connected to the same output terminals of the amplifier, what net impedance will be "seen" by the amplifier?

2. An amplifier has a damping factor of 30 when used with an eight-ohm speaker. What will the damping factor be if used with a sixteen-ohm speaker?

3. An eight-ohm loudspeaker is capable of handling 30 watts of audio power. How much current will flow when it is receiving 30 watts of power? What size of fuse or circuit breaker should be used to protect it from overload?

4. What is the wavelength at the center of the FM band?

(See Problem 1 in Chapter 3.) How long should the horizontal loop of an FM antenna be, if it is to be one-half of a wavelength (in order to be resonant)?

5. Translate the phrase "tracking force of 1–3 grams" into correct units of force (dynes or newtons).

6. A certain phono cartridge has an inductance $L - 600$ mH. At what frequency will its internal impedance equal the recommended load resistance $R = 47,000$ ohms?

7. A certain phono cartridge has a moving mass of 0.5 mg and a compliance of 10^{-2} m/N. Find its resonance frequency using Equation 2.3 (remember that the spring constant K is the reciprocal of the compliance).

In Chapter 3, we defined sound waves as longitudinal waves that travel in a solid, liquid, or gas. The most common path from source to receiver is through air, but not necessarily a *direct* path through the air. In a room, most of the sound waves that reach the listener's ear have been reflected by one or more surfaces of the room or by objects within the room. In a typical room, sound waves undergo dozens of reflections before they become inaudible.

It is not surprising, therefore, that the acoustical properties of the room play an important role in determining the nature of the sound heard by a listener. Performers may not be able to change the acoustics of a concert hall, but they can (consciously or unconsciously) adapt their performance to the particular hall, so that listeners receive the optimum quality of sound. To do this, it is important to understand some of the principles of room acoustics.

Chapter 24 discusses systems for the electronic reinforcement of sound. Sound systems may be necessary to provide an adequate level of sound in a very large auditorium, or they may be designed to compensate partially for some acoustical defect (such as inadequate or excessive reverberation). Chapter 25 discusses the properties of small rooms and high-fidelity sound reproducing systems.

PART SIX
The Acoustics of Rooms

CHAPTER 23
Auditorium Acoustics

The subject of concert hall acoustics is almost certain to provoke a lively discussion by both performers and serious listeners. Musicians recognize the importance of the concert hall in communication between performer and listener. The opening of a new concert hall invokes a flurry of reviews, opinions, and criticisms of its acoustical qualities. Amateur and professional critics try to compare a piece of music performed in the new hall with how they remember it sounding in other halls on other occasions (by other performers perhaps). Opinions of new halls tend to polarize toward the extremes of "very good" or "very bad." In spite of the extensive research on the acoustics of concert halls, the opinion still prevails in some circles that their acoustic design is "black magic."

23.1 SOUND PROPAGATION OUTDOORS AND INDOORS

Before discussing the acoustics of auditoriums, concert halls, and other large rooms, let us briefly review how sound propagates in various environments.

An environment in which the sound pressure varies as $1/r$ is called a *free field*. When a source of sound is small enough to be considered a point source and is located outdoors away from reflecting objects, a

459

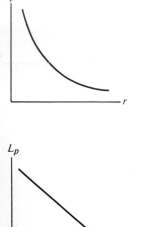

FIG. 23.1
The way in which sound pressure p and sound pressure level L_p decrease with distance r in a free field.

FIG. 23.2
Reflection of sound by various surfaces: (a) flat surface acts like a mirror; (b) concave surface concentrates sound in the region S'; (c) convex surface scatters sound; (d) rough surface leads to diffuse reflection.

free field results. Sound waves travel away from the source in all directions, the wave fronts having the shape of spheres, as shown in Fig. 6.1. The sound pressure in a free field is proportional to $1/r$, as shown in Fig. 23.1 (r is the distance from the source). Thus the pressure p is halved when the distance r doubles. From the definition of the decibel, given in Chapter 6, we see that the sound pressure level decreases 6 dB each time the distance r is doubled. Free-field conditions rarely occur indoors, except in reflection-free *anechoic* rooms. (Anechoic means "echo-free"; this is generally achieved by covering the walls, ceiling, and floor with wedges of sound-absorbing material.)

Indoors, sound travels only short distances before encountering walls and other obstacles. These obstacles reflect and absorb sound in ways that largely determine the acoustic properties of the room. Figure 23.2 illustrates the reflection of sound by flat, curved, and rough surfaces ("rough" in this case means that irregularities have dimensions comparable to the sound wavelength).

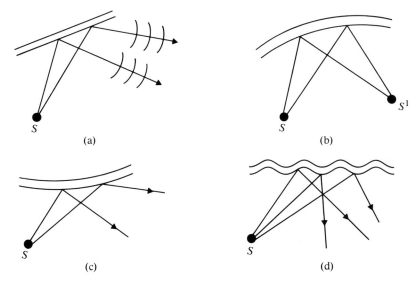

23.2   DIRECT, EARLY, AND REVERBERANT SOUND

Let us examine how sound travels from the source to the listener in an auditorium. Sound waves travel at 344 m/s (1130 ft/s); the direct sound, therefore, may reach the listener after a time of anywhere from 0.02 to 0.2 second (20 to 200 ms), depending on the distance from source to listener. A short time later, the same sound will reach the listener from various reflecting surfaces, mainly the walls and ceiling. In Fig. 23.3, these reflections are shown arriving with various time delays t_1, t_2, t_3, etc. The first group of reflections, reaching the listener within about 50 ms of the direct sound, is often called the *early* sound.

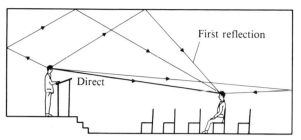

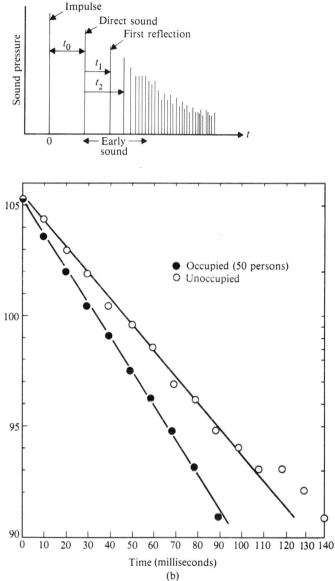

FIG. 23.3

Paths of direct and reflected sound from source to listener with corresponding time delays for a sound impulse. (From *Music, Acoustics, and Architecture.* © 1988 Leo Beranek. Used with author's permission.)

FIG. 23.4

Sound decay curves in a 400-m³ classroom: (a) sound pressure as a function of time (room unoccupied); (b) sound pressure level as a function of time for the same room occupied and unoccupied. An exponential decay of sound pressure corresponds to a linear decay of sound pressure level. (From Jesse, 1980. Reprinted by permission from the Amer. Assoc. of Physics Teachers.)

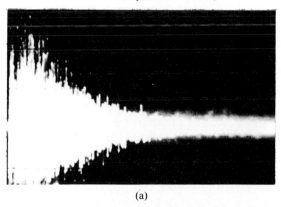

(a)

(b)

After the first group of reflections, the reflected sounds arrive thick and fast from all directions. These reflections become smaller and closer together, merging into what is called *reverberant* sound. If the source emits a continuous sound, the reverberant sound builds up until it reaches an equilibrium level. When the sound stops, the sound level decreases at a more or less constant rate until it reaches inaudibility. In the case of the impulsive sound shown in Fig. 23.3, the decay of the reverberant sound begins immediately, and there is no equilibrium level. This is illustrated by the decay curve of sound pressure shown in Fig. 23.4(a) for a 400-m³ classroom occupied by 50 students. The sound-pressure level decay is shown for the same room when occupied and unoccupied. Note that in the occupied room the sound dies out slightly faster due to absorption by the occupants (Jesse, 1980).

A reasonably simple but accurate analysis of the acoustics of an auditorium can be obtained from a careful study and comparison of *direct, early*, and *reverberant* sound. In Section 23.7 we will consider the criteria for good concert hall acoustics and how these criteria relate to direct, early, and reverberant sound. In the intervening sections, the character of these three types of sound will be discussed in more detail.

23.3   DIRECT AND EARLY SOUND: THE PRECEDENCE EFFECT

Most sound sources heard in auditoriums and concert halls are non-directional (that is, they radiate essentially the same intensity of sound in all directions). Thus the level of *direct* sound depends only on the distance from the source, decreasing by 6 dB for each doubling of the distance. Minor exceptions occur in the case of some percussion instruments and brass instruments toward the upper end of their playing range (where the size of the bell is comparable to the sound wavelength).

Our auditory system has an uncanny ability to determine the direction of a sound source, even in the presence of many distracting sounds. The manner in which we localize the source of the direct sound was discussed in Section 5.5. For sounds of low frequency, localization depends mainly on the observation of a very slight difference in the time of arrival (or the phase of steady sounds) at our two ears. For sounds of high frequency (above about 1000 Hz), the difference in sound level at our two ears, due to the shadow cast by our head, provides the main clue.

The arrival of the reflected sound from many directions adds complications. When an orator articulates a phoneme or a musician attacks a note, our ears are provided with several reflections that closely follow the direct sound. The spectrum and time envelope of these reflected sounds will be more or less identical to those of the direct sound and if they arrive within about 35 ms of the direct sound, the ear does not hear them as separate sounds. Rather, they tend to reinforce the direct sound, a fact that is especially important to listeners located quite a distance from the source. Most remarkably, however, the auditory processor continues to deduce the direction of the source from the first sound reaching the ears, which it interprets as following the direct path.

This remarkable ability of our auditory system, called the *precedence effect,** has considerable significance in our perception of stereophonic and quadraphonic sound. The source is perceived to be in the direction from which the first sound arrives provided that (1) successive sounds arrive within 35 ms, (2) the successive sounds have spectra and time envelopes reasonably similar to the first sound, and (3) the successive sounds are not too much louder than the first.

As a result of studying 54 of the world's leading concert halls, Beranek (1962) concluded that a concert hall is considered "intimate" if the delay time between direct and first reflected sound is less than 20 ms. If the auditorium has the traditional rectangular shape, this first reflection for most listeners will come from the nearest side wall, although listeners located near the center may receive their first reflection from the ceiling. In some large concert halls, a portion of the audience will be too far removed from both ceiling and side walls to receive early reflections within the desirable time interval; in those cases, reflecting surfaces of some type are often suspended from the ceiling.

More recent studies have shown, however, that early reflections from side walls are not equivalent to early reflections from the ceiling or from an overhead reflector. One study showed a high preference for concert halls with ceilings sufficiently high that the first lateral reflection reaches the listener before the first overhead reflection (West, 1966). Others have found that if the total energy from lateral reflections is greater than the energy from overhead reflections, the hall takes on a desirable "spatial responsiveness" or "spatial impression" (Jordan, 1975).

23.4 REVERBERANT SOUND

The characteristic of auditoriums familiar to most people is the reverberation time. Although the behavior of reverberant sound is too complicated to be described by a single number, the reverberation time at mid-frequency (usually 500 Hz) does give quite a fair indication of the "liveness" of the auditorium or concert hall.

Instead of the impulsive sound source used in Figs. 23.3 and 23.4, let us switch on a steady source, leave it on for a time interval T, and then switch it off. The growth and decay of the reverberant sound are shown in Fig. 23.5.

---

* Sometimes it is called the *Haas effect,* because of a paper by H. Haas in *Acustica* 1: 49 (1951), which describes it in some detail. Similar effects had apparently been noted by others, including Joseph Henry, as early as 1856.

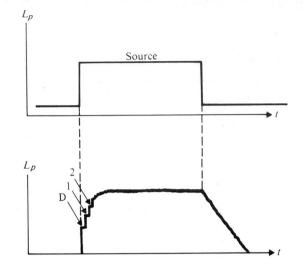

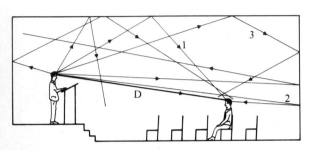

FIG. 23.5
Growth and decay of reverberant sound in a room: *D* represents direct sound; 1, 2, 3, etc., are early reflections.

The rate of growth, the rate of decay, and the reverberant sound level can be deduced by considering the sound energy. The source supplies the energy, which is stored in the air space of the auditorium and eventually absorbed by the walls, the ceiling, the objects within the auditorium, and, under certain conditions, the air itself. The reverberant level is reached when the rate at which energy is supplied by the source (that is, the source power) is equal to the rate at which sound is absorbed.

Although the reverberant sound, like the early sound, reinforces the direct sound and adds to the overall loudness (which is an important consideration in a large auditorium), too great a level of reverberant sound may result in a loss of clarity. In a good concert hall the direct sound should be substantially louder than the background noise at all locations. In large auditoriums, this may call for electronic reinforcement of the direct sound. (In Chapter 24, electronic reinforcement of both direct and reverberant sound will be discussed.)

When the sound source is a full-size symphony orchestra, most listeners find the level of direct sound to be optimum when they are seated about 20 meters (60 feet) from the source (Beranek, 1962). This distance would place one at roughly the center of many of the world's best concert halls.

In principle, it is easy to determine the theoretical reverberation time of a room. The sound energy stored in the room depends on the power of the source and the volume of the room; the rate at which that energy is absorbed depends on the area of all surfaces and objects in the room and their absorption coefficients. In a bare room, where all surfaces absorb the same fraction of the sound that reaches them, the reverberation time is thus proportional to the ratio of volume to surface (Sabine, 1922):

$$RT = K \frac{\text{Volume}}{\text{Area}}. \qquad (23.1)$$

In general, large rooms have longer reverberation times than do small rooms.

If the sound energy were uniformly distributed throughout the room, its decay would follow a curve like that shown in Fig. 23.6(a), which is called an exponential curve. The logarithm of such a curve is a straight line, and so the decay of sound level is a straight line, as shown in Fig. 23.6(b). It is customary to define the reverberation time (abbreviated RT or T_{60}), as the time required for the sound level to decrease by 60 dB. Often it is not convenient to measure the time for the entire 60-dB decay; assuming that the decay is a straight line, however, one can double the time required for a 30-dB decay.

Decay curves similar to those shown in Figs. 23.6(c) and (d) are often observed in auditoriums. In Fig. 23.6(c), there are two different reverberation times, one describing the initial and one the final portion of the decay. This may indicate an insufficient distribution of sound, and can lead to a feeling of "dryness" in a hall even though the final reverberation time falls within acceptable limits. Spikes on the decay curve of the type shown in Fig. 23.6(d) result from the storage of sound energy in the form of room resonances.

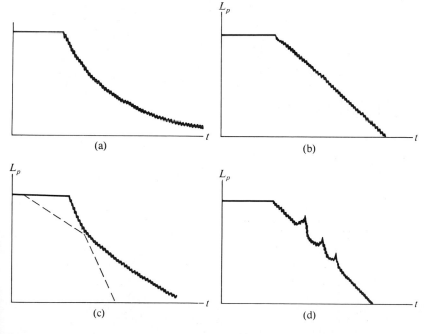

FIG. 23.6
Decay of reverberant sound: (a) and (b) are decay curves of sound pressure and sound level in a somewhat idealized room with uniform energy distribution; (c) and (d) illustrate curves often observed for auditoriums; (c) shows different initial and final reverberation times; (d) shows peaks due to prominent standing waves or room resonances.

23.5   CALCULATION OF THE REVERBERATION TIME

Let us consider a hypothetical room of volume V with hard walls that absorb very little sound but with a window of area A through which sound can escape to the outside. The reverberation time now depends on the ratio of the room volume to the area of the absorbing window. When V is in cubic meters (m³) and A in square meters (m²), the constant K equals 0.16 s/m, so the formula for reverberation becomes

$$\text{RT} = 0.161 \frac{V}{A} \quad (V, \text{ in m}^3; A \text{ in m}^2). \tag{23.2}$$

If room dimensions are given in feet, the formula may be written

$$\text{RT} = 0.049 \frac{V}{A} \quad (V, \text{ in ft}^3; A \text{ in ft}^2). \tag{23.3}$$

In a real room, we can compare the absorbing power of each surface to that of the window in our hypothetical room. The window is assumed to "absorb" all the sound that reaches it; thus we assign it an absorption coefficient $a = 1$. A surface that absorbs half of the sound will have an absorption coefficient $a = 0.5$. Two square meters of this material have an absorption equivalent to one square meter of open window. A surface having an area S and an absorption coefficient a thus has an absorption

$$A = Sa \tag{23.4}$$

equivalent to an open window of area A.

TABLE 23.1   Absorption coefficients for various materials

Material	Frequency (Hz)					
	125	250	500	1000	2000	4000
Concrete block, unpainted	0.36	0.44	0.31	0.29	0.39	0.25
Concrete block, painted	0.10	0.05	0.06	0.07	0.09	0.08
Glass, window	0.35	0.25	0.18	0.12	0.07	0.04
Plaster on lath	0.14	0.10	0.06	0.05	0.04	0.03
Plywood paneling	0.28	0.22	0.17	0.09	0.10	0.11
Drapery, lightweight	0.03	0.04	0.11	0.17	0.24	0.35
Drapery, heavyweight	0.14	0.35	0.55	0.72	0.70	0.65
Terrazzo floor	0.01	0.01	0.02	0.02	0.02	0.02
Wood floor	0.15	0.11	0.10	0.07	0.06	0.07
Carpet, on concrete	0.02	0.06	0.14	0.37	0.60	0.65
Carpet, on pad	0.08	0.24	0.57	0.69	0.71	0.73
Acoustical tile, suspended	0.76	0.93	0.83	0.99	0.99	0.94
Acoustical tile, on concrete	0.14	0.20	0.76	0.79	0.58	0.37
Gypsum board, one-half inch	0.29	0.10	0.05	0.04	0.07	0.09

The total absorption in the room is found by adding up the contributions from each surface exposed to the reverberant sound:

$$A = S_1 a_1 + S_2 a_2 + S_3 a_3 + \cdots. \tag{23.5}$$

The absorption is sometimes expressed in sabins or metric sabins, units named in honor of Wallace Sabine, a pioneer in the study of room acoustics. One sabin is the absorption of one square foot of open window, and one metric sabin is the absorption of one square meter of open window. It is less confusing, however, to express absorption A in square meters (or square feet). Table 23.1 gives the absorption coefficients a for a number of materials at different frequencies.

Example 1

Let us calculate the reverberation time for a room 20 m $\times$ 15 m and 8 m high. The walls are painted concrete block, the ceiling plaster on lath, the floor carpet on concrete. We neglect all furnishings and do the calculation at $f = 500$ Hz:

Walls: $A_1 = (2 \times 15 \times 8 + 2 \times 20 \times 8)(0.06) = 34;$
Ceiling: $A_2 = (15 \times 20)(0.06) = 18;$
Floor: $A_3 = (15 \times 20)(0.14) = 42;$
$A = A_1 + A_2 + A_3 = 94$ m^2;
$V = 20 \times 15 \times 8 = 2400$ m^3;
$RT = 0.161 V/A = (0.161)(2400)/(94) = 4.1$ s.

TABLE 23.2   Sound absorption by people and seats, and air absorption

Material	125	250	500	1000	2000	4000	8000	Unit*
Wood or metal seats, unoccupied	0.014	0.018	0.020	0.036	0.035	0.028		m²
Upholstered seats, unoccupied	0.13	0.26	0.39	0.46	0.43	0.41		m²
Audience in upholstered seats	0.27	0.40	0.56	0.65	0.64	0.56		m²
Air absorption (per m³):								
20°C, 30%	—	—	—	—	0.012	0.038	0.136	
20°C, 50%	—	—	—	—	0.010	0.024	0.086	

Note: Values of sound absorption are given in m²; to convert to ft², multiply by 10.8. Values of air absorption are given in m⁻¹; to convert to ft⁻¹, divide by 3.3.

Neglecting the furnishings, as in Example 1, leads to an unrealistically long reverberation time. In a typical room, the furnishings will contribute a substantial fraction of the total absorption. People are also good absorbers of sound, each member of the audience contributing 0.2 to 0.6 m² of absorption, depending on frequency. The values of sound absorption of various types of seats, occupied and unoccupied, are given in Table 23.2.

23.6   AIR ABSORPTION

In a large auditorium, the air itself contributes a substantial amount to the absorption of sound at high frequencies. The absorption of air depends on the temperature and relative humidity, and an additional term, mV, proportional to the volume should be added to the absorption A. The constant m is given in the last two lines of Table 23.2. The reverberation time for a large auditorium is

$$\text{RT} = 0.161 \, \frac{V}{A + mV} \quad \text{(large auditorium).} \qquad (23.6)$$

Example 2

Let us add 200 upholstered seats to the room used in Example 1. We assume half of them occupied and calculate the reverberation time at 2000 Hz. We also consider the absorption of the air at 20°C, 30% relative humidity.

Walls: $A_1 = (2 \times 15 \times 8 + 2 \times 20 \times 8)(0.09) = 50;$
Ceiling: $A_2 = (15 \times 20)(0.04) = 12;$
Floor: $A_3 = (15 \times 20)(0.60) = 180;$
Empty seats: $A_4 = (100)(0.43) = 43;$
Occupied seats: $A_5 = (100)(0.64) = 64;$
Air absorption: $mV = (0.012)(2400) = 29;$
$A = 50 + 12 + 180 + 43 + 64 = 349 \text{ m}^2;$
$V = 20 \times 15 \times 8 = 2400 \text{ m}^3;$
$$\text{RT} = 0.161 \, \frac{2400}{349 + 29} = 1.02 \text{ s.}$$

Note that in Example 2, nearly one-third of the total absorption results from the seats and the audience. This is not unusual behavior at this frequency. The formulas we have used assume that the reverberant sound is distributed uniformly throughout the room, which is not always the case. Near large absorbers, the sound energy may be less than at other locations within the room; more complicated formulas take this and other factors into account. Nevertheless, the reverberation times calculated from the formulas given in this chapter will be reasonably close to the measured values.

23.7 CRITERIA FOR GOOD ACOUSTICS

From our discussion of reverberation, the interdependence of reverberant sound level and reverberation time should be clear. Increasing the absorption decreases both the reverberant level and the reverberation time. The optimum reverberation time is thus a compromise between clarity (requiring a short reverberation time), sound intensity (requiring a high reverberant level), and liveness (requiring a long reverberation time). The optimum reverberation time will depend on the size of the auditorium and the use for which it is designed. An auditorium intended primarily for speech should have a shorter reverberation time than one intended for music. Figure 23.7 indicates reverberation times considered desirable for auditoriums of various sizes and functions.

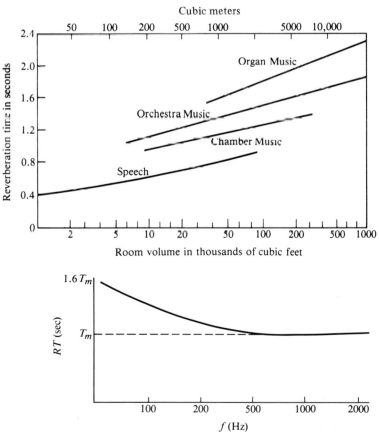

FIG. 23.7
Desirable reverberation times for auditoriums of various sizes and for various functions.

FIG. 23.8
Variation of reverberation time with frequency in a good concert hall.

A feeling of liveness or reverberance is especially important at low frequency to give support to bass notes. Fortunately, many building materials have lower absorption coefficients at low frequencies. Figure 23.8 shows how the reverberation time may vary with frequency in a good concert hall.

Recent work has stressed the importance of having sufficient reflected sound arriving from the sides. Such lateral reflections arriving with time delays of from 25 to 80 ms add to the reverberance, whereas overhead reflections during the same period add mainly to the early sound (Reichardt et al., 1975).

A number of studies have been made of various concert halls in order to determine criteria for "good" and "bad" acoustics. Beranek (1962) found 18 subjective attributes of musical-acoustic quality that can be related to concert hall acoustics. Among the more important are these:

1. *Intimacy.* A hall has acoustical intimacy when music sounds as if it were being played in a small hall. The time delay between the direct and first reflected sound should be less than 20 ms in order for a hall to be intimate.

2. *Liveness.* This is related primarily to the reverberation time for middle and high frequencies. The optimum reverberation time depends on size and function, as shown in Fig. 23.7. A hall with insufficient reverberation is termed "dry."

3. *Warmth.* This is related to liveness and fullness of bass tone. Reverberation time at 250 Hz and below should be somewhat longer than at middle and high frequencies.

4. *Loudness of direct sound.* The auditorium should be designed so that no listener is seated too far from the sound source (since direct sound decreases with distance). If the hall is too large, sound amplification may be necessary.

5. *Reverberant sound level.* The level of the reverberant sound, which will be the same throughout the hall, depends on the power of the source and the reverberation time.

6. *Definition or clarity.* The level of the early plus direct sound should be greater than the reverberant sound level at all locations.

7. *Diffusion or uniformity.* Good spatial distribution of the sound is achieved by diffuse or irregular reflecting surfaces, and by the avoidance of focused sound or sound shadows.

8. *Balance and blend.* This depends on the stage design. If the stage is wider than about 15 meters, the ceiling should be low (10 meters or less) and irregular in shape.

9. *Ensemble.* There should be ample reflecting surfaces to the sides
   and above the orchestra so that players can hear each other.

10. *Freedom from noise.* In order to allow sufficient dynamic range,
    the noise level should be less than NC-20, preferably less than the
    NC-15 curve shown in Fig. 23.9.

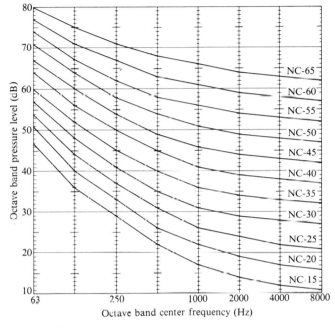

FIG. 23.9
Noise criteria (NC) curves. (From
Noise and Vibration Control,
revised. © 1988 Leo Beranek. Used
with author's permission.

TABLE 23.3 Acoustical characteristics of concert halls

	Year built	Volume (m³)	Area (m²)	Number of seats	t_1 (ms) Floor	t_1 (ms) Balc.	RT (s) 125	RT (s) 500	RT (s) 2000 Hz
Symphony Hall, Boston	1900	18,740	1550	2630	15	7	2.2	1.8	1.7
Orchestra Hall, Chicago	1905	15,170	1855	2580	40	24	—	1.3	—
Severence Hall, Cleveland	1930	15,700	1395	1890	20	13	—	1.7	1.6
Carnegie Hall, New York	1891	24,250	1985	2760	23	16	1.8	1.8	1.6
Opera House, San Francisco	1932	21,800	2165	3250	51	30	—	1.7	—
Arie Crown Theatre, Chicago	1961	36,500	3265	5080	36	14	2.2	1.7	1.4
Royal Festival Hall, London	1951	22,000	2145	3000	34	14	1.4	1.5	1.4
Royal Albert Hall, London	1871	86,600	3715	6080	65	70	3.4	2.6	2.2
Concertgebouw, Amsterdam	1887	18,700	1285	2200	21	9	2.2	2.1	1.8
Kennedy Center, Washington	1971	19,800	1220	2760	—	—	2.5	2.2	1.9

Source: After Beranek (1962).

Table 23.3 gives acoustical data for several concert halls. The time
delay t_1 is given for selected locations near the center of the main floor
and the middle of the balcony. Reverberation times RT in seconds are

given at frequencies of 125, 500, and 2000 Hz. The two halls in London are nearly at the extremes of acceptability. The Royal Festival Hall in London is considered "dry," excellent for chamber music and music from the Baroque period. The Royal Albert Hall is not very popular with musicians; the type of orchestral music that can be performed there is limited by its long reverberation, which also adds excitement to certain types of music, such as Tchaikovsky's *1812 Overture*. The Royal Festival Hall has been improved by the addition of "assisted resonance," which will be discussed in Chapter 24.

Recent work on concert halls has emphasized criteria in addition to those listed earlier:

Spatial impression. This depends on the contributions to the early sound from side reflections and overhead reflections. So long as a sufficient portion of the early sound arrives from the side, the exact time delay does not appear to be important.

Early decay time. The initial rate of decay of reverberant sound appears to be more important than the total reverberation time. A rapid initial decay is apparently interpreted by the ear as meaning that the reverberation time is short.

Things to be avoided in auditorium design include the following:

1. *Echoes.* An echo is a strong reflected sound that is sufficiently delayed (over 50 ms, usually) from the direct sound that it can be heard as a separate entity rather than as a continuation of the original sound. When echoes are heard in an auditorium, the most likely culprit is the rear wall.

2. *Flutter echoes.* Flutter echoes are a series of echoes that occur in rapid succession; they usually result from reflections between two parallel surfaces that are highly reflective.

3. *Sound focusing.* Focusing of sound can be caused by reflection from large concave surfaces (see Fig. 23.2). Certain sounds will be heard too loudly near the focus of a curved surface.

4. *Sound shadows.* Under balconies at the rear of the auditorium, there may be insufficient early sound, since most of the reflections from the side walls and ceiling do not reach this area even though they are in a direct line of sight to the performer and therefore receive the direct sound.

5. *Background noise.* This will be discussed in Section 23.8.

A study of 22 European concert halls by Schroeder, Gottlob, and Siebrasse (1974) showed the following:

1. The greater the early decay time, the greater the preference for the

hall, up to a reverberation time (determined from the early decay time) of two seconds. Above two seconds, the preference for the hall decreased with increasing reverberation time.

2. Narrow halls were generally preferred to wide ones.

3. Considerable preference was shown for halls having a high "binaural dissimilarity"; that is, listeners preferred dissimilarity of sound at their two ears, such as might result from a high degree of sound diffusion.

4. Halls with less "definition" were preferred. Definition represents the ratio of energy in the first 50 milliseconds to the total energy.

23.8 BACKGROUND NOISE

Nothing is more disappointing than to have a performance of music spoiled by background noise from an air-conditioning system or from nearby traffic. Although this does not often happen in major concert halls, it is all too common in churches, school auditoriums, and smaller concert halls where a qualified acoustician has not been consulted.

Background noise can be of internal or external origin. The common source of internal noise is an improperly designed ventilating system. Low-frequency noise from the ventilating equipment itself may be carried through ducts into the auditorium, and a broad band of noise is often generated by air flow in the ducts and grills. In addition, low-frequency vibrations from machinery may be transmitted by the building structure into an auditorium. Other noise originates from noisy doors, inadequate isolation from corridor noise, and so on.

External noise is somewhat more difficult to eliminate. Depending on the hall's location, external noise may come from traffic on a nearby street, aircraft overhead, etc. Possible solutions include placing the auditorium at the center of a building complex, constructing heavy walls, making certain no windows face the offending street, and so forth.

A family of noise criteria curves is shown in Fig. 23.9. The NC curves specify the maximum noise level permissible in each octave band. For example, if architectural specifications call for NC-20 or below, the sound pressures in all octave bands must be below those shown in Fig. 23.9. A concert hall should meet at least the NC-20 curve and preferably the NC-15 standard.

23.9 AVERY FISHER HALL: A CASE STUDY

One of the most famous concert halls in the world is Avery Fisher Hall, originally called Philharmonic Hall, in New York; its history illustrates the tribulations of acoustic design. Since its opening in 1962, it has undergone several renovations, ending with a complete rebuilding in 1976. An apparent success at last, its metamorphosis has taught us a great deal about auditorium acoustics.

Philharmonic Hall was designed by architect Max Abramovitz after an extensive research study by acoustician Leo Beranek of the world's concert halls. Acoustical expertise was furnished by members of Beranek's firm (Bolt Beranek and Newman) as well as by Hope Bagenal of England. The hall was intended to rank with the best in the world.

When the hall opened in 1962, most musicians and listeners were disappointed. The major defects were (1) weak bass; (2) a lack of liveness; (3) echoes from the rear wall; (4) inadequate sound diffusion; (5) poor hearing conditions for the musicians on stage (see Schroeder et al., 1966). Scientists from the Bell Telephone Laboratories were called on to evaluate the acoustics, and a distinguished committee of acoustical consultants headed by V. O. Knudsen was asked to recommend improvements. Changes made during the summers of 1963, 1964, 1965, 1969, and 1972, costing over two million dollars, improved the hall, but criticisms continued. In 1975, the hall was completely redesigned by architect Philip Johnson and acoustician Cyril Harris. The reconstruction, following somewhat the proven lines of Boston's Symphony Hall and the more recent Kennedy Center (Washington) and Orchestra Hall (Minneapolis), cost over five million dollars.

What went wrong with the original Philharmonic Hall, the product of so much careful planning? Beranek (1975) maintains that the final plans were expanded and modified without his consent or that of the orchestra. To enlarge the seating capacity from 2400 to 2600, the side walls were spread out into a more "modern" fan shape. The adjustable ceiling was eliminated for reasons of economy, as were some irregularities on the side walls, which would have acted as diffusers of sound.

The designers of the hall placed a great emphasis on acoustical intimacy, and suspended 136 panels ("clouds") from the ceiling to provide some of the early reflections that would come from the side walls in a smaller hall. It is now known that early sounds arriving from above and from the sides are not nearly so equivalent as was thought. The designers apparently counted rather heavily on last-minute adjustments during "tuning week" to optimize several parameters.

Several of the problems in the hall resulted from insufficient diffusion of sound. This resulted in a decay curve with two slopes. Although the total reverberation time was about right, the initial slope was too steep ("early decay time" too short), thus giving the hall the impression of dryness or lack of reverberance. The story behind the redesign of Avery Fisher Hall (renamed in honor of its chief benefactor) is told in a dramatic account by Bliven (1976).

23.10 VARIABLE ACOUSTICS

Quite a number of attempts have been made to design auditoriums with variable acoustics. They have met with varying success. Usually it is the reverberation time that is varied by changing the amount of absorption in the room. One technique is the use of wall panels that can be rotated, one side with a high absorption coefficient and one with a low absorption coefficient. Another technique is the use of extended or retracted absorbing panels or blankets to change the absorption. In a large hall, the amount of absorption that must be added or subtracted in order to produce a noticeable change is very large.

Another technique for varying the acoustics, which will be discussed in Chapter 24, is to enhance the reverberation and/or the direct sound by electronic means. There are distinct limits to the range of variability, however, and the baseline from which these variations take place should be one of good initial acoustical design.

23.11 SUMMARY

In a free field, sound level decreases by 6 dB for each doubling of the distance. Inside rooms, however, in which sound reflects from many surfaces, the sound level is greater than the free-field level. In analyzing the acoustics of an auditorium, attention should be given to the direct, early, and reverberant sounds. The ability of our ears to locate a source by analyzing the direct sound and virtually ignoring the early sound is called the precedence effect. Reverberant sound builds up and decays at a rate characterized by the reverberation time RT, which depends on the volume of the room and the absorption of all the surfaces in it.

Criteria for good concert hall acoustics (e.g., intimacy, liveness, warmth, clarity, etc.) can be related to measurable parameters. Optimum values for some parameters, such as reverberation time, depend on the size and use of the auditorium. Background noise, due to internal and external sources, should be kept at a minimum.

References

Beranek, L. L. (1962). *Music, Acoustics and Architecture*. New York: Wiley.

Beranek, L. L. (1971). *Noise and Vibration Control*. New York: McGraw-Hill.

Beranek, L. L. (1975). "Changing Role of the Expert," *J. Acoust. Soc. Am.* **58**: 547.

Bliven, B., Jr. (1976). "Annals of Architecture," *New Yorker* (Nov. 8): 51.

Cremer, L., H. A. Müller, and T. J. Schultz (1982). *Principles and Applications of Room Acoustics,* vols 1 and 2. London: Applied Science.

Doelle, L. L. (1972). *Environmental Acoustics.* New York: McGraw-Hill.

Jesse, K. E. (1980). "A Classroom Acoustical Absorption Experiment," *The Physics Teacher* **18**: 41.

Jordan, V. L. (1975). "Auditoria Acoustics: Developments in Recent Years," *Applied Acoustics* **8**: 217.

Knudsen, V. O. (1963). "Architectural Acoustics," *Sci. Am.* **209**(5): 78.

Knudsen, V. O., and C. M. Harris (1950). *Acoustical Designing in Architecture.* New York: J. Wiley. Reprinted by The Acoustical Society of America, 1978.

Marshall, A. H., D. Gottlob, and H. Altruz (1978). "Architectural Conditions Preferred for Ensemble," *J. Acoust. Soc. Am.* **64**: 1437.

Reichardt, W., O. Abdel Alim, and W. Schmidt (1975). "Definition and Basis of Making an Objective Evaluation Between Useful and Useless Clarity Defining Musical Performances," *Acustica* **32**: 126.

Sabine, W. C. (1922). *Collected Papers on Acoustics.* Reprinted by Dover, New York, 1964.

Schroeder, M. R. (1979). "Binaural Dissimilarity and Optimum Ceilings for Concert Halls: More Lateral Sound Diffusion," *J. Acoust. Soc. Am.* **65**: 958.

Schroeder, M. R., B. S. Atal, G. M. Sessler, and J. E. West (1966). "Acoustical Measurements in Philharmonic Hall," *J. Acoust. Soc. Am.* **40**: 434.

Schroeder, M. R., D. Gottlob, and K. F. Siebrasse (1974). "Comparative Study of European Concert Halls: Correlation of Subjective Preferences with Geometric and Acoustic Parameters," *J. Acoust. Soc. Am.* **56**: 1195.

West, J. E. (1966). "Possible Subjective Significance of the Ratio of Height to Width of Concert Halls," *J. Acoust. Soc. Am.* **40**: 1245.

Glossary

anechoic Echo-free; a term applied to a specially designed room with highly absorbing walls.

direct sound Sound that reaches the listener without being reflected.

early sound Sound that reaches the listener within a short time (about 50 ms) after the direct sound.

free field A reflection-free environment, such as exists outdoors or in an anechoic room, in which sound pressure varies inversely with distance ($p \propto 1/r$).

noise criteria (NC) curves A family of curves defining levels of room noise in several octave bands.

precedence effect The ability of the ear to determine the direction of a sound source from the direct sound without being confused by the early sound that follows.

reverberant sound Sound that builds up and decays gradually and can be "stored" in a room for an appreciable time.

reverberation time The time required for the stored or reverberant sound to decrease by 60 dB.

sabin, metric sabin Units for measuring absorption of sound; the sabin is equivalent to one square foot of open window, the metric sabin to one square meter.

Questions for Discussion

1. Why does the use of cushioned seats help to make the reverberation time of an auditorium independent of audience size?

2. Which is easier to correct, a reverberation time that is too long or one that is too short?

3. What are desirable reverberation times for speech and for orchestral music in an auditorium with $V = 1000$ m^3?

4. An auditorium is thought to have excessive reverberation, especially at low frequency. It is proposed that the ceiling be covered with acoustic tile to reduce this. What do you think of this solution? Is there a better one?

Problems

1. Compare the absorption of 100 square meters of plastered wall with that of 100 square meters of carpeted floor at

   a) 125 Hz;

   b) 2000 Hz.

2. An auditorium has dimensions 40 m × 20 m and a ceiling height of 15 m. The front and back walls are covered with plywood paneling; the side walls and ceiling are plaster. The floor is wood. There are 1100 wooden seats. Estimate the reverberation time (500 Hz) when

   a) the hall is empty;

   b) the seats are half filled;

   c) all the seats are occupied.

3. Estimate the time delay t_1 of the first reflected sound for a person seated near the center of the auditorium described in Problem 2. Does the first reflection arrive from the side or from overhead?

4. If the ceiling in this auditorium were covered with acoustical tile, by how much would the reverberation time be decreased?

5. Specify reasonable values for the reverberation times at 100, 200, 500, and 1000 Hz for a 2000-m³ concert hall to be used primarily for orchestral music.

6. If two hard parallel walls are spaced 30 m apart, calculate the repetition rate of the flutter echo that might result. What efforts might be made to prevent its occurrence?

7. Repeat a historic experiment done by Joseph Henry over 130 years ago. Clap your hands periodically as you move away from a large flat wall. Determine how far away you have to be in order to distinguish the echo from the original sound. Divide twice this distance by the speed of sound to obtain the "limit of perceptibility," as Henry called it [see *J. Acoust. Soc. Am.* **61** 250 (1977)]. Compare your result to that given in Section 23.3.

8. Find the reverberation time at 8000 Hz for a very "live" room having a volume of 1000 m³ when the temperature is 20°C and the relative humidity is 30 percent. Assume that absorption by the walls is negligibly small. Would your answer be different if $V = 100$ m³ instead?

CHAPTER 24

Electronic Reinforcement of Sound

A small- to medium-sized room with good acoustical design should not require electronic reinforcement for either speech or music, and an unnecessary sound system is an annoying distraction. Nevertheless, in many large auditoriums, sound systems are necessary to obtain adequate loudness and good distribution of sound. This is especially true in athletic arenas or other enclosures with high levels of background noise. The sound system should be integrated into both the visual and acoustical design of the room. A sound reinforcement system is optimum when most listeners are hardly aware of its presence.

Electronic reinforcement of either the direct, early, or reverberant sound (see Section 23.2) is possible, although it is generally the direct sound that most needs to be reinforced. Reinforcement of the reverberant sound in dry auditoriums, however, can increase the reverberation time and create a feeling of liveness. If excessive reverberation or background noise interferes with clarity of speech, selective reinforcement of the direct sound over a selected range of frequency can improve the situation. It should be remembered, however, that electronic reinforcement of the direct sound will increase the reverberant level as well.

Sections 24.1–24.3, along with the example in the box on p. 487, discuss sound sources and sound fields in a more quantitative way

479

than is found throughout the rest of the chapter. This is necessary for a full understanding of the subject. However, it is possible to skim these sections and still gain some appreciation for the electronic reinforcement of sound.

24.1   SOUND SOURCES IN A ROOM

For the purposes of this discussion, we characterize sound sources by their *power* and *directivity*. Both of these parameters will vary with frequency, of course. A source tends to be more directional at high frequencies where the wavelength of sound is comparable to the dimensions of the source. A person speaking is a virtually nondirectional source, since the wavelength of spoken sound exceeds the size of the mouth opening.

The power of a source is expressed in watts. The sound power level L_w compares the power W of a source to the reference power $W_0 = 10^{-12}$ watt (see Eq. 6.2). The average source power of a person speaking at a conversational level is about 10^{-5} watt ($L_w = 70$ dB). (It has been said that it would take fifty speakers five years to generate enough energy to boil a cup of tea.)

The directivity factor Q is defined as the ratio of the sound intensity at a distance r in front of a source to the sound intensity averaged over all directions. A source that radiates equally in all directions (a spherical source) has a directivity factor $Q = 1$. A hemispherical source has $Q = 2$; a source in a corner (which radiates into one-quarter of a sphere) has $Q = 4$; a source at a three-surface corner (in the corner of a room at the floor or ceiling, for example) has $Q = 8$.

The sound pressure level L_p depends on the power and the directivity of the source, its distance, and the strength of the reflected sound. In a free field (away from reflecting surfaces), the sound pressure level at a distance r meters from the source is

$$L_p = L_w + 10 \log \frac{Q}{4\pi r^2} . \tag{24.1}$$

24.2   SOUND FIELDS

We speak of the distribution of sound in space as a sound field. The character of the field due to a sound source varies with the distance from the source and the acoustic environment. Figure 24.1 illustrates how sound level varies with distance from the source in a room. When the distance is small compared to the dimensions of the source, the sound level varies with location, because some parts of the source may radiate more strongly than others. This part of the sound field is called the *near field*, and is cross-hatched in Fig. 24.1.

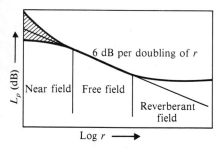

FIG. 24.1
Variation of sound pressure level L_p with distance from the source r in a typical large room. In the free-field region, intensity varies as $1/r^2$, so L_p decreases by 6 dB each time r is doubled. In the near field, L_p depends on the geometry of the source. (Compare Fig. 23.1.)

Farther away from the source, the sound pressure varies as $1/r$; this is called the *free field*. In the free field, the sound level decreases by 6 dB each time the distance r is doubled. As the distance r increases, the contribution from reflected sound takes on increasing importance, and we enter the *reverberant field*. The sound level in a large room eventually reaches the reverberant level at which it no longer decreases with increasing distance.

The reverberant level depends on the absorption of various surfaces in the room as well as on the power of the source. If the total absorption A is given in square meters, the reverberant sound level is $L_p(R) = L_w + 10 \log 4/A$. The sound level due to both direct and reverberant sound is

$$L_p = L_w + 10 \log \left(\frac{Q}{4\pi r^2} + \frac{4}{A} \right). \qquad (24.2)$$

(A more precise value of L_p results from using $4/R$ in place of $4/A$; the "room constant" R is given by $A/(1-\bar{a})$, where $\bar{a} = A/S$ is the average absorption coefficient when S is the total surface area of the room.) Let us consider a hypothetical room.

Example 1 Suppose an auditorium has dimensions $20 \times 30 \times 10$ meters and an average absorption coefficient $\bar{a} = 0.15$. A speaker standing in the corner radiates 10^{-5} watt of acoustical power. Thus, $A = 0.15 \times 2(20 \times 30 + 20 \times 10 + 30 \times 10) = 330$ m^2. For a corner source, $Q = 4$. At a distance of 5 meters from the source,

$$L_p = 10 \log 10^7 + 10 \log \left(\frac{4}{4\pi(25)} + \frac{4}{330} \right) = 70 - 16 = 54 \text{ dB}.$$

The reverberant level in this rather live room will be

$$L_p = 10 \log 10^7 + 10 \log 4/330 = 70 - 18 = 52 \text{ dB}.$$

Provided the room noise is not too great, it should be possible to hear the speaker without sound reinforcement. On the other hand, a listener seated 30 meters away, who receives only 35 dB of direct sound, 17 dB below the reverberant level, will have trouble understanding the speaker. Reinforcement of the direct sound would improve the quality of sound this listener hears.

Studies made in the Netherlands on the intelligibility of speech have led to the following formulas for percentage articulation loss for consonants (percentage of sounds identified incorrectly) in a reverberant room (Peutz, 1971):

$$\% \text{ AL} = \frac{200r^2(\text{RT})^2}{V} + k, \qquad (24.3)$$

where $RT =$ the reverberation time (s), $r =$ the distance from the source (m), $V =$ the volume (m³), and $k =$ the constant for each listener that indicates listening ability (1.5% for the best listener to 12.5% for the poorest). As the distance from the source increases, a distance D is reached beyond which the articulation loss remains constant at a value

$$\% \ AL = 9RT + k. \tag{24.4}$$

This critical distance D is given by

$$D = 0.20\sqrt{V/RT}. \tag{24.5}$$

For skilled speakers and listeners, a % AL of 25 to 30 as calculated from these formulas may be acceptable, since speech includes a fair amount of redundancy (that is, it is possible to "guess" a few lost consonants on the basis of the text). A better strategy, however, is to reduce % AL to 15 or less in order to accommodate the "average" speaker and listener.

Example 2 Determine the % AL in the room described in Example 1, where

$$RT = 0.161 V/A = 0.161(6000)/330 = 2.9 \ s.$$

Assume $k = 7$ for a typical listener.

$$\% \ AL \quad \frac{200(2.9)^2 r^2}{6000} + 7 = 0.28 r^2 + 7.$$

Thus, % AL = 15 for $r = 5.4$ m.

$$D = 0.20 \ \sqrt{6000/2.9} = 9.1 \ m.$$

$$\% \ AL = 9RT + a = 9(2.9) + 7 = 33 \quad \text{for } r > D.$$

The % AL is excessive in much of the room because of the reverberation. The direct sound should probably be reinforced.

Equations 24.3 and 24.4 for articulation loss assumed nondirectional sources ($Q = 1$). When a directional source is used, Davis and Davis (1975) recommend dividing Eq. 24.3 by the directivity factor Q, since the ratio of direct sound to reverberant sound increases by that factor. Also D increases by the factor Q. Thus,

$$\% \ AL = \frac{200 r^2 T^2}{QV} + a; \tag{24.3a}$$

$$D = 0.20\sqrt{QV/T}. \tag{24.5a}$$

If the main problem in a room is background noise, the intelligibility can obviously be improved with electronic reinforcement of sound. The sound level should be raised at least 25 dB above the background noise for all listeners, if possible. However, if the problem is with a poor ratio of direct sound to reverberant sound (such as one finds in old cathedrals, for example), the sound system must use speakers with high directivity factors Q. Otherwise, the reverberant sound, as well as the direct sound, will be reinforced. Speech clarity can be improved by reinforcement of only midrange frequencies (500–2000 Hz), which carry most of the speech intelligence.

24.3 POWER CONSIDERATIONS

Equation (24.2) can be used to determine the power capacity of a sound system to be used for sound reinforcement, since it works for any source. What must be determined are the desired levels of direct sound and reverberant sound. The direct-sound level is given by Eq. (24.1), and the reverberant sound is represented by Eq. (24.6) below:

$$L_p \text{ (direct)} = L_w + 10 \log Q/4\pi r^2, \qquad (24.1)$$

$$L_p \text{ (reverb)} = L_w + 10 \log 4/A. \qquad (24.6)$$

[Note that the combined direct and reverberant sound, given by Eq. (24.2), is not simply the sum of L_p (direct) and L_p (reverb).] In the above equations, A and r^2 are in square meters; if they are expressed in square feet, add 10 dB to each equation.*

For speech, a sound pressure level of 65 to 70 dB will be adequate, provided that it is at least 25 dB above the noise level in the room. Music, however, spans a wide dynamic range, and peak levels of 90 to 100 dB or more may be required. Systems used to amplify musical instruments in rock groups deliver 110 to 120 dB peaks, although usually not without some distortion.

Figure 24.2 is a useful chart for determining the acoustic power needed in a room. Each curve represents a single value for the room constant R. The vertical axis at the right gives the sound pressure level for a source power W of one watt, and the vertical axis on the left can be used to determine the sound pressure level with a source of any sound power level.

Example 3 Determine the acoustical power necessary to reach a level of 100 dB in a room with a total absorption $A = 400$ m² and an average absorption coefficient $\bar{a} = 0.2$.

---

* Since 1 m² = 10.76 ft², 10 log 10.76 = 10.32 dB is the more precise figure; however, adding 10 dB is usually accurate enough.

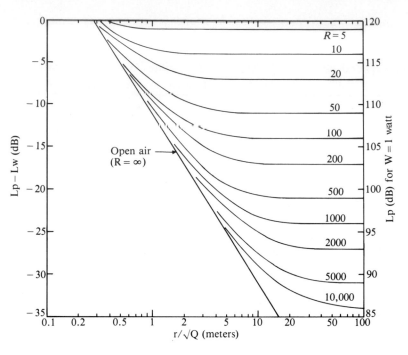

FIG. 24.2
Chart for determining sound pressure level L_p in a room, where r is the distance from the source, Q is the directivity of the source, W is its power, L_w is its sound power level, and R is the room constant (in m²).

The room constant is

$$R = \frac{A}{1-a} = \frac{400}{1-0.2} = 500 \text{ m}^2.$$

At a large distance ($r/\sqrt{Q} > 10$ m), $L_p - L_w \simeq -21$ dB (see the chart). Thus, $L_w = L_p + 21 = 121$ dB, which corresponds to $W = 1.26$ watts of acoustical power that must be delivered by the sound reinforcement system.

24.4   LOUDSPEAKER PLACEMENT

One of the most important considerations in designing a sound system is the placement of loudspeakers in order to give the best coverage over the listening area. Most systems use a large single source or a distribution of small sources throughout the room. In most auditoriums, single-source systems are preferred because they preserve best the spatial pattern of the sound field.

A single source generally consists of a cluster of loudspeakers with directivity factors Q selected to give the best coverage of the audience. The preferred location of a single source is on the centerline of the room, near the front, over the speaker's head. Vertical displacement of the source is not particularly distracting, because of our inability to localize sound in a vertical plane. The loudspeakers should be aimed toward listeners at the rear of the auditorium, as shown in Fig. 24.3(a).

Another arrangement, which provides satisfactory coverage in a long room with a low ceiling, is the distributed-speaker system, shown in Fig. 24.3(b). Each unit mounted in the ceiling covers 60 to 90 degrees. If the room is long, it is important to have an electronic time delay for the rear speakers; otherwise, the direct sound arrives after the sound from the loudspeakers, so that it appears to be a distracting echo. Some time-delay systems use digital delay networks; others use magnetic tape or disc. Loudspeakers should not be placed along the side walls of an auditorium, where their crossfire will cause the listener to hear sound from several loudspeakers at the same time.

Another loudspeaker arrangement to be avoided is the all-too-common practice of putting one speaker on each side of the stage area or front wall, as shown in Fig. 24.4. Listeners seated at points *A* or *B* hear sound from one of the speakers before they hear the direct sound. If this arrangement of speakers is necessary, sound to both loudspeakers should be delayed electronically. However, the single loudspeaker shown in Fig. 24.3(a) is much to be preferred.

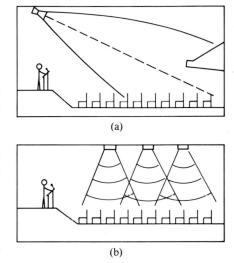

(a)

(b)

FIG. 24.3

(a) Central loudspeaker system in a large auditorium. (b) Distributed loudspeaker system in a long, low room.

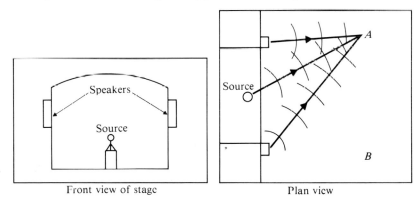

Front view of stage

Plan view

FIG. 24.4

Unsatisfactory arrangement of loudspeakers. Listeners on side *A* hear the sound from one speaker before the direct sound.

24.5 LOUDSPEAKER DIRECTIVITY

Three types of loudspeaker arrays are commonly used in sound systems: (1) cone radiators, (2) line or column radiators, and (3) horn radiators. An ordinary cone-type loudspeaker has a directivity factor Q that depends on frequency. When the wavelength of sound is much larger than the cone size, the radiation is quite uniform in the forward hemisphere; even directly backwards, the sound level is only 10 or 15 dB less than in the forward direction, depending on the design of the speaker housing. The directivity factor Q of the speaker will be

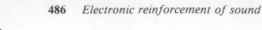

FIG. 24.5
Radiation pattern and directivity factor Q for a typical eight-inch cone-type loudspeaker. (After Davis and Davis, 1975).

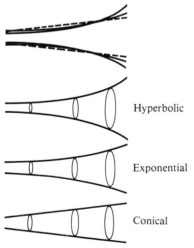

FIG. 24.6
A comparison of the rates of expansion of conical, exponential, and hyperbolic horns.

about 2. At higher frequencies, the radiation is mainly in the forward direction, and Q increases to 20 or more depending on speaker size. The radiation pattern and Q of a typical eight-inch loudspeaker are shown in Fig. 24.5. The polar radiation pattern indicates the relative sound level at each angle compared to the level at the same distance directly in front of the speaker.

Combining two or more elements changes the directivity. A vertical column of speakers is equivalent to increasing the speaker size in the vertical direction but not the horizontal. Thus the beam spreads out much less in the vertical direction than in the horizontal. Line or column radiators are frequently used in sound systems because they increase Q while maintaining the broad distribution of a single speaker in the horizontal direction.

Horn loudspeakers nearly always have greater efficiency in converting electrical energy to acoustical energy than do cone-type loudspeakers. They also can be designed to have greater directivity. Thus they are well suited for use in sound systems. Their main disadvantages lie in their larger size and the difficulty in achieving good low-frequency response. To achieve smooth response down to 100 Hz, for example, the horn mouth should have an area of at least eight square feet. Many systems use horn loudspeakers at middle and high frequencies, and cone-type loudspeakers at low frequency.

Horns are called conical, exponential, or hyperbolic according to the way in which their area expands with distance from the driver. Figure 24.6 compares the rates of expansion. The two most common horn designs in use are the *multicellular horn* and the *radial/sectoral horn*. The radial/sectoral horn has straight sides on two boundaries and curved sides on the other two boundaries, as shown in Fig. 24.7(a). The multicellular horn consists of several exponential horns with axes passing through a common point. At the lower frequencies, the entire unit radiates as one horn, but at high frequencies, each horn radiates its own narrowing beam. Figure 24.7(b) shows a multicellular horn.

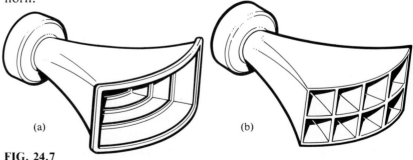

(a)   (b)

FIG. 24.7
(a) A radial/sectoral horn. (b) A multicellular horn.

24.6 ACOUSTIC FEEDBACK

You have probably been in an auditorium when the sound system begins to howl or squeal loudly. This phenomenon results from *acoustic feedback*, an example of positive feedback, which, as we noted in Chapter 19, can cause an amplifier to act as an oscillator. Acoustic feedback occurs when the microphone picks up sound from the loudspeaker and sends it to the amplifier to be reamplified. If the electrical gain is greater than the acoustical loss, the signal continues to build up and the system goes into oscillation. This process is illustrated in Fig. 24.8. If the gain is not quite large enough to send the system into oscillation, acoustic feedback may still cause speech to sound "tinny" due to a long decay time at certain frequencies.

In most auditoriums, acoustic feedback limits the amount of gain that can be achieved from an amplifying system. Microphones will always be in the reverberant sound field, of course, and sometimes they receive a fairly large amount of direct sound as well. By turning down the amplifier gain, one will find that greater sound levels can be tolerated without acoustic feedback, but the sensitivity to the desired sound will decrease as well.

The useful gain without feedback oscillation can be increased by using microphones of high directivity (see Chapter 20) and placing them as far from the loudspeakers as possible and well off the axis of directional loudspeakers. Since feedback oscillation tends to occur at the frequencies of prominent room resonances (peaks in the reverberant sound field), room equalization will raise the oscillation level substantially.

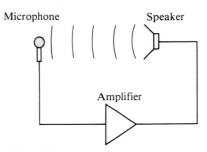

FIG. 24.8
Acoustic feedback from loudspeaker to microphone to amplifier can cause a sound system to become unstable and go into oscillation.

> **Example 4** Consider the case illustrated in Fig. 24.9. We define three distances, and also factors $G(\theta)$ to express the off-axis output of the loudspeaker and $F(\phi)$ to describe the directivity of the microphone. Thus, $r =$ the distance from the loudspeaker to the listener, $d_1 =$ the distance from the microphone, $d_2 =$ the distance from the source to the microphone, $G(\theta) =$ the sound level at angle θ compared to the sound level on the loudspeaker axis
>
>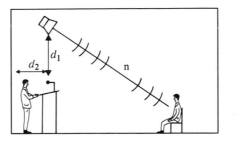
>
> **FIG. 24.9**
> Parameters of sound systems with single microphone and loudspeaker.

at the same distance, and $F(\phi)$ = the directivity of the microphone. The sound level at a listener (on the axis of the loudspeaker) due to direct sound from the loudspeaker will be

$$L_p(1) = L_w + 10 \log \frac{Q}{4\pi r^2}.$$

The sound fed back to the microphone directly from the loudspeaker and from the reverberant field in the room will be

$$L_p(m) = L_w + 10 \log \left(\frac{Q}{4\pi d_1^2} + \frac{4}{A} \right) - G(\theta) - F(\phi).$$

The sound picked up by the microphone from a nondirectional source with power level $L_w(s)$ will be

$$L_p(s) = L_w(s) + 10 \log \frac{1}{4\pi d_2^2}.$$

It is up to the sound engineer to design the system such that $L_p(1)$ will be sufficiently large to provide clarity to the listener, yet $L_p(m)$ will be small enough to prevent feedback oscillation. The directivity factors $G(\theta)$ and $F(\phi)$, as well as the Q of the loudspeaker, are within the engineer's control. Obviously if d_2 (source-to-microphone distance) can be kept small, the task is much easier.

24.7 EQUALIZATION

One of the most important steps in the installation of a sound system is the adjustment of its frequency response to complement that of the room in which it is installed. This is called *equalization*; unfortunately, it is a step that is too often neglected. Equalization serves two important functions: (1) it results in a more natural sound by compensating for predominant room resonances, and (2) by suppressing acoustic feedback oscillation, it permits the sound system to be operated at a higher level.

Pipe organ builders have long recognized the importance of room resonances in churches and auditoriums, and they regulate or voice individual pipes accordingly. It is unthinkable for an organ builder not to make this final adjustment to the room, in spite of the labor involved. Many expensive sound systems, however, are installed without any provision for equalizing their response to the auditorium in which they will be used.

Very crude equalization can be obtained by means of the bass and treble tone controls, which boost or cut the low and high frequencies.

This is usually not sufficient to accomplish the goals of equalization, however. Some type of selective filtering is generally necessary. The most economical type of filter set has eight to ten band-pass filters (see Chapter 19) with pass bands one octave wide, each of which controls one octave of sound. More flexibility is obtained by using ⅓-octave filters, but this adds to the cost, because three times as many filters are required. Filter sets with octave and ⅓-octave filters are shown in Fig. 24.10.

FIG. 24.10
Filter sets for equalization of a sound system (a) set with eight octave-band filters; (b) set with 36 ⅓-octave filters. (Photograph (a) courtesy of Shure Brothers, Inc.; (b) courtesy of United Recording Electronics Industries).

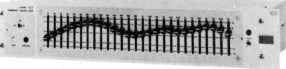

Figure 24.11 illustrates how equalization can increase the usable gain of a sound system. The sound system before equalization will generally show peaks in gain at certain frequencies. These correspond to room resonances, enhanced perhaps by the sound system itself (the character of this enhancement depends on the placement of the microphones and loudspeaker). As the amplifier gain is increased, acoustic feedback causes the system to oscillate at the highest of these peaks.

After the system is equalized, the response curve may look like the dashed line in Fig. 24.11. It is clear that the level may now be safely raised several decibels without danger of oscillation due to feedback. Sometimes tunable band-reject filters with narrow bands (⅙-octave or ¹⁄₁₀-octave) are used to reduce prominent resonances. Such a filter set is shown in Fig. 24.12.

Another useful device for suppressing acoustic feedback oscillation in a sound system is a *frequency shifter,* which shifts the frequencies in the microphone signal by an inaudible amount, typically from 3 to 5 Hz. This small shift in frequency is often enough to avoid feedback due to room resonances that are narrow but prominent.

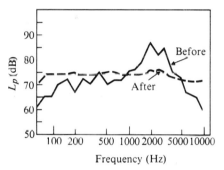

FIG. 24.11
Frequency response of a typical auditorium with a sound system before and after equalization. The solid curve indicates the sound level in each ⅓-octave band of frequency for an input with equal level in each band ("pink noise"). (After Davis and Davis, 1975).

FIG. 24.12
A feedback suppressor that includes four tunable band-reject filters with ⅙-octave bandwidths. Each filter can be tuned to the frequency of a room or sound system resonance. (Courtesy of United Recording Electronics Industries).

24.8   TIME DELAY

A sound system with loudspeakers at several positions in a large auditorium generally requires some type of time delay for best results. As we pointed out in Section 23.3, sound that arrives up to 35 ms after the direct sound (provided, of course, that it is similar in spectrum and time envelope) will reinforce the direct sound and yet preserve the apparent direction of the sound source. Time delay is especially important in the case of supplementary speakers positioned in problem areas, such as underneath a balcony.

The best time-delay systems use digital time delay. In a digital system, the audio waveform is sampled at regular intervals, and the samples are converted to digital numbers by a device called an *analog-to-digital converter* (ADC). The resulting series of numbers is circulated in a digital memory and read out after the desired delay time has passed. The numbers are then converted back to voltages in order to reconstruct the original audio waveform. This conversion of numbers back to voltages is done by a *digital-to-analog converter* (DAC). Digital time-delay systems are still relatively expensive, although their cost has been greatly reduced by the development of microelectronics.

Another type of delay system uses a loop of magnetic tape or a magnetic disc with record and playback heads located close to each other. The time delay is the time required for the magnetic medium to pass from the record head to the playback head. In the case of magnetic tape, a rather high tape speed is required for short time delays.

An acoustical time-delay system may consist of a microphone and loudspeaker at opposite ends of a pipe. The delay will be the length of the pipe divided by the speed of sound.

24.9   ENHANCEMENT OF REVERBERATION

The fact that some halls are used for varied purposes makes the adjustment of reverberation time desirable. Maximum clarity of speech demands a short reverberation time; a pipe organ sounds best in a reverberant room (see Fig. 23.6). Some control of reverberation time is possible by using movable wall panels, etc., but these techniques are generally expensive. A solution that is becoming increasingly popular is the use of "assisted resonance" or the electronic enhancement of reverberation.

One method used for reverberation enhancement is to place a loudspeaker and a microphone in a *reverberation chamber*, a small room of 300 to 3000 m^3 with highly reflecting walls, ceiling, and floor.

By amplifying the reverberant sound in this chamber and feeding it back into the main auditorium through loudspeakers, the reverberation time may be increased. The reverberation chamber usually has nonparallel walls to increase the number of resonances, but some type of equalization is necessary to smooth out the frequency response curve of the chamber.

A simple means of providing reverberation, often used in electronic organs and music sythesizers, is a spring with a transducer at each end. The transducers are usually piezoelectric crystals or magnetic pickups similar to phonograph pickups. One transducer generates a sound wave, which propagates down the spring, stimulating a second transducer at the other end. The sound wave travels back and forth many times on the spring before dying out, and this approximates the decay of reverberant sound. The effect tends to resemble the decay of sound in a one-dimensional room or tunnel, however; units with several springs of varying length minimize this effect.

A more pleasing two-dimensional effect is obtained by placing a number of transducers on a thin plate (Kuhl plate) or foil. By careful design and provision for damping the plate, a pleasing effect can be obtained. A reverberation unit of high quality using a gold foil is available commercially.

One of the most elegant systems for enhancing the reverberation of a large concert hall is the *assisted resonance* system developed by acoustician P. H. Parkin and his associates for the Royal Festival Hall in London. The Royal Festival Hall, opened in 1951, had some desirable acoustical characteristics but was found to be deficient in low-frequency reverberation. The system developed for this hall consists of 172 independent channels, each with its own tuned resonator spaced over the frequency range 58–700 Hz. At the lower frequencies, each channel uses a microphone placed inside a tuned Helmholtz resonator; above 300 Hz, tuned pipes are used. The gain of each channel is adjustable individually to obtain satisfactory reverberation at present and perhaps in the future in the event that changes are made in the hall. Figure 24.13 shows the results of tests made using music as a source.

In order to ensure acceptance of the system, the reverberation time was increased gradually over several concert seasons. As Parkin and Morgan (1970) describe it, "the policy has been to keep discussions to a minimum, because of the passions likely to be aroused in some breasts by the thought of loudspeakers in the RFH." After conducting two concerts with the Berlin Philharmonic Orchestra, Herbert von Karajan commented that the acoustics of the Royal Festival Hall

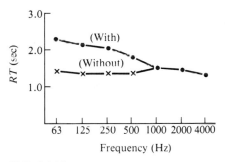

FIG. 24.13
Reverberation time in the Royal Festival Hall with and without assisted resonance. (From P. H. Parkin and K. Morgan (1970). '"Assisted Resonance' in the Royal Festival Hall, London: 1965–1969," *J. Acoust. Soc. Am.* **48**: 1025. Reprinted by permission from the Amer. Inst. of Physics.)

are probably now among the finest in the world. He and several members of the Orchestra were interested in what had been done to the hall since they had last performed in it two years earlier. Recent work on the system has been aimed at slightly increasing the reverberation time in the full auditorium to the same value as the empty auditorium, so that artists can rehearse and perform under nearly identical conditions.

24.10   OUTDOOR SOUND SYSTEMS

The discussion in this chapter has focused on sound systems for use indoors in auditoriums and large rooms. For purposes of comparison, it is interesting to consider an outdoor system. In the past few years there have been a number of rock music concerts held outdoors with thousands in attendance. Portable sound systems generating kilowatts of acoustic power have been set up to provide coverage to large areas having high levels of background noise.

Clearly, sound power takes on substantially more importance outdoors, where there is no reverberant sound field for reinforcement. Thus, large power amplifiers and arrays of loudspeakers with high efficiency are needed.

An example of a large outdoor system is that used to ''broadcast'' the concerts-in-the-park series to audiences as large as 100,000 people in New York's Central Park. The system, described by Rosner and King (1977), is designed to cover an area of about 80,000 m^2 and to provide peak levels of 105 dB at 45 m and 99 dB at 90 m from the stage. The frequency response is flat to $\pm$ 3 dB over the range 63–1000 Hz, and has a downward slope of 2 dB per octave above 1000 Hz. The loudspeakers are built into two 40-foot portable towers. Each tower has 8 woofers in 4 bass horn cabinets, 6 midrange radial horns, and 12 tweeters. Sixteen 80-watt amplifiers in each tower supply the electrical power. The system includes a mechanical reverberation unit to provide liveness for the outdoor environment.

24.11   SUMMARY

Electronic enhancement of either direct, early, or reverberant sound is possible, although it is generally the direct sound that must be reinforced. Sound sources are characterized by their power (in watts or decibels) and their directivity, which can be expressed by a factor Q. At various distances from the source, the sound field in a room is characterized as the near field, the free field, or the reverberant field. The clarity of speech, described by a percentage of articulation loss

(% AL) of consonants, can be improved by using a well-designed sound system. Loudspeaker design and placement are important factors in the performance of a sound system. Horn loudspeakers usually have the highest efficiency and directivity.

Acoustic feedback can cause oscillation to occur in a sound system at a frequency determined by the room resonances or the distance from the loudspeakers to the microphones. This annoyance can be prevented by equalizing the system to smooth out peaks in the gain.

A sound system that includes loudspeakers at several locations in a large auditorium may require time delay. Digital time-delay systems provide the best performance but are expensive. Enhancement of reverberation can improve the sound of music in a dry concert hall.

References and Suggested Readings

Beranek, L. L. (1954). *Acoustics.* New York: McGraw-Hill. Reprinted by the Acoustical Society of America, 1986.

Boner, C. P., and C. R. Boner (1968). "Sound-Reinforcing System Design," *Appl. Acoust.* **1**: 115.

Davis, D., and C. Davis (1975). *Sound System Engineering.* Indianapolis: Howard Sams.

Everest, F. A. (1978). *The Complete Handbook of Public Address Sound Systems.* Blue Ridge Summit, PA: TAB Books.

Klepper, D. L. (1970). "Sound Systems in Reverberant Rooms for Worship," *J. Audio Eng. Soc.* **18**(4): 391.

Parkin, P. H., and K. Morgan (1970). " 'Assisted Resonance' in the Royal Festival Hall, London: 1965–1969," *J. Acoust. Soc. Am.* **48**: 1025.

Peutz, V. M. A. (1971). "Articulation Loss of Consonants as Criterion for Speech Transmission in a Room," *J. Audio Eng. Soc.* **19**: 915.

Rosner, A., and L. S. King (1977). "New Mobile Sound Reinforcement System for the Metropolitan Opera/New York Philharmonic Orchestra Park Concerts," *J. Audio Eng. Soc.* **25**: 566.

Glossary

acoustic feedback Sound from a loudspeaker picked up by a microphone (either in the direct field or the reverberant field) and reamplified.

analog-to-digital converter (ADC) A circuit that converts (analog) voltages to a digital or numerical representation.

articulation loss of consonants (% AL) A measure of speech intelligibility; the percentage of consonants heard incorrectly (strongly influenced by noise or excessive reverberation).

digital-to-analog converter (DAC) A circuit that generates a voltage proportional to a digital number.

equalization Changing the gain of a sound system at certain frequencies to compensate for room resonances and other peaks in the response curve.

free field That part of the sound field where the sound level decreases by 6 dB for each doubling of distance.

L_p Sound pressure level = $10 \log p/p_0$ (see Chapter 6).

L_w Sound power level = $10 \log W/W_0$ (see Chapter 6).

near field That part of the sound field where the sound level varies from point to point because of the radiation pattern of the source.

Q (directivity factor) Comparison of the sound power radiated directly ahead of a sound source to that radiated in all directions.

reverberant field That part of the sound field in which sound level is independent of distance from the source.

room constant A quantity that describes the absorption in a room; it is slightly greater than the total absorption A used to calculate reverberation time (see Section 23.5).

Questions for Discussion

1. Draw a cross section (to scale) of the auditorium in Example 1 on p. 445 (30 m long, 10 m high). Indicate the correct placement of a single-source loudspeaker cluster, and measure the angle of the beam that would give coverage to the rear two-thirds of the auditorium.

2. What are the advantages and the disadvantages of using horn loudspeakers? Why are they not often used in home hi-fi systems?

3. Explain the advantages and disadvantages of using directional microphones (e.g., cardioid type) in auditorium sound systems.

Problems

1. At what frequency does the wavelength of sound equal the diameter of a 15-inch woofer? a 2-inch tweeter?

2. Assuming the outdoor sound system described in Section 24.10 were to convert 2.5 kW of electrical power to acoustic power with an average efficiency of 20 percent and to radiate it into a 90° angle ($Q=4$), determine the sound level at 90 m. (First determine L_w for the source.) Compare this to the design level at this distance.

3. What are the upper and lower frequencies passed by an octave-band filter tuned to 500 Hz? (In other words, what frequencies are one-half octave above and below 500 Hz?)

4. If we assume the record and playback heads are spaced 1 cm apart, what tape speed is required to give a delay time of 20 ms? Is this practical?

5. If a microphone is five meters in front of a loudspeaker, at what frequency might oscillation take place due to acoustic feedback? Could it also occur at harmonics of this frequency? Explain.

6. Check the statement in Section 24.1 concerning the 50 speakers and the cup of tea. (It takes about 1050 joules to raise the temperature of a ¼-liter cup of water one degree Celsius; a joule is equal to a watt-second.)

CHAPTER 25

High-Fidelity Sound-Reproducing Systems

Circuits and components for high-fidelity sound were discussed in Chapter 22. However, there is more to the high-fidelity reproduction of sound than merely assembling high-quality components. Although it is certainly desirable to have components with low distortion and noise, ample power-handling capacity, and so on, the importance of the listening room itself is often underestimated. Just as in the case of live performance, the room has much to do with the quality of the sound we hear. The listening room is the weakest link with many high-fidelity sound systems.

In this chapter, we will consider the acoustics of small rooms. We will also discuss some of the psychoacoustical effects that result from having multiple sound sources and their significance in stereophonic and quadraphonic sound systems. Finally, we will glance into the future of high-fidelity sound reproduction.

25.1 ACOUSTICS OF SMALL ROOMS

In Section 23.2 we defined three sound fields: direct sound, early reflected sound, and reverberant sound. Much of the acoustical character of an auditorium or concert hall depends on the balance and the time relationship between these three types of sound. The first reflected sound typically reaches a listener from 10 to 30 milliseconds after the direct sound. Reverberation times of one to three seconds are common.

In a small room, such as a home listening room, quite a different time relationship exists. Walls and ceiling are so close to the listener that many reflections arrive within a few milliseconds after the direct sound. It is therefore not important to distinguish between early reflected sound and reverberant sound. Achieving "intimacy" (which depends on a short time delay between direct and first reflected sound—see Section 23.7) is no problem at all in a small room.

Home listening rooms seldom sound reverberant even though the level of the reverberant sound may exceed the level of the direct sound ($4/A > Q/4\pi r^2$ in Eq. 24.2) except for listeners seated very close to the source. That is because the reverberation time of a small room is generally short, typically less than half a second for a home listening room. To produce a "concert-hall" sound, some reverberant sound must be included on the recording or some type of time-delay device ("room expander") used.

According to the Sabine formula (Eq. 23.1), the reverberation time is proportional to the room volume V and inversely proportional to the absorption A. Thus the reverberation time of a small room that has limited volume can only be increased by making the absorption small. This tends to produce a "shower room" type of reverberation, however, because the individual resonances of the room become prominent. The individual resonances of a small room with little absorption are similar to those of a rectangular box, as described below.

Consider the resonances of a rectangular box with dimensions a, b, and c. The frequencies of the various resonances are

$$f_{lmn} = \frac{v}{2} \sqrt{ \left(\frac{l}{a}\right)^2 + \left(\frac{m}{b}\right)^2 + \left(\frac{n}{c}\right)^2 }. \qquad (25.1)$$

where v is the speed of sound and l, m, and n are integers $0, 1, 2, \ldots$.

Example 1 Find the lowest resonances of a box with dimensions of 0.8, 1.0, and 2.0 meters. The lowest resonance occurs when $l = 0$, $m = 0$, and $n = 1$; we label this the 001 mode, and its frequency is

$$f_{001} = \frac{344}{2} \sqrt{ \left(\frac{0}{0.80}\right)^2 + \left(\frac{0}{1.0}\right)^2 + \left(\frac{1}{2.0}\right)^2 } = 86 \text{ Hz}.$$

The next few resonances are found by letting l, m, and n be other small integers. The results are:

$$f_{001} = 86 \text{ Hz}, \qquad f_{010} = 172 \text{ Hz}, \qquad f_{100} = 215 \text{ Hz},$$
$$f_{002} = 172 \text{ Hz}, \qquad f_{011} = 192 \text{ Hz}, \qquad f_{101} = 232 \text{ Hz},$$
$$f_{111} = 289 \text{ Hz}, \qquad f_{020} = 344 \text{ Hz}, \qquad f_{200} = 430 \text{ Hz}.$$

Suppose that you stand in a shower room with hard, smooth walls that reflect sound well. The dimensions may not be too different from those of the box in Example 1. As you sing up and down the scale, you may hear strong resonances near the calculated frequencies. The resonances are close enough together in frequency to give a "bathroom baritone" full-bodied, resonant sound.

Unfortunately, dormitory rooms may have acoustical behavior that is more like the shower room just described than like a living room amply endowed with carpet, upholstered furniture, and other sound absorbers. Many purchasers of high-fidelity sound systems have experienced disappointment that the sound they hear at home is "colored" by room resonances that didn't exist in the dealer's listening room. For this reason, some dealers allow potential buyers to audition high-fidelity systems at home.

The point of equality between the direct and the reverberant field can be estimated from the graph shown in Fig. 24.2. First the room constant R is calculated; then the point at which the appropriate R-curve would intersect the direct-field ($R = \infty$) curve, if extended horizontally to the left, gives the value of $r/\sqrt{Q}$ at which the two fields contribute equally. If the directivity factor Q of the source is known, the distance r can be determined.

Example 2 A home listening room has a room constant $R = 20$. The source, a loudspeaker placed along one wall, has a directivity factor ranging from $Q = 2$ at low frequency to $Q = 10$ at high frequency. At what distance r_c will the direct and reverberant sound fields be equal in level?

Solution. The $R = 20$ curve in Fig. 24.2, if extended horizontally to the left, would intersect the direct-field line at $r/\sqrt{Q} = 0.5$. Thus at low frequency, $r_c = (0.5)\sqrt{2} = 0.7$ meter; at high frequency, $r_c = (0.5)\sqrt{10} = 1.6$ m.

Most listeners will be at least two meters from the loudspeakers, and hence will be well into the reverberant field for low-frequency sound. At higher frequencies, the room constant R will probably increase as will the directivity factor Q of the loudspeaker, so the point of equality between the two fields moves to a greater distance. A distance of three to four meters from the source may be required to get well into the reverberant field. Nevertheless, even in a "dead" room, a substantial portion of the sound is reverberant sound that arrives from all directions.

An important difference between the reverberant fields of large and small rooms is that in a large room, the reverberant sound builds up over an appreciable time interval, whereas in a small room the buildup and decay times are short. This provides an important auditory clue to room size.

As the dimensions of the room increase, the number of resonances within the audible range multiplies rapidly. In a large room there are so many resonances that the room tends to take on a fairly smooth frequency response, although there may be an emphasis or a deficiency in some particular range of frequency. Not so in a small room with hard walls, however; room resonances may give emphasis to certain frequency ranges, and may result in considerable coloration of sound from a high-fidelity system.

25.2   SOUND IMAGES FROM MULTIPLE SOURCES

In order to better understand stereophonic and quadraphonic sound, as well as room acoustics, let us consider various ways in which our auditory system perceives sounds from multiple sources. First we consider monophonic two-speaker arrangements (that is, two loudspeakers receiving signals from the same source). Recall that at low frequencies the main clue to the direction of a source is the difference in the arrival times of the sound at our two ears, whereas at high frequencies it is the interaural intensity difference that dominates (see Section 5.5).

In Fig. 25.1 the listener is located on the median plane equidistant from both speakers. If both speakers receive the same signal at the same strength, an "image" will be created at location A on the median plane, as shown in Fig. 25.1(a). If the signal strengths at the two speakers are different, however, the image will shift toward the speaker receiving the stronger signal, as indicated by B in Fig. 25.1(b).

The angle of the image θ_I with respect to the median plane can be calculated from the following equation:

$$\frac{\sin \theta_I}{\sin \theta_A} = \frac{p_L - p_R}{p_L + p_R}, \tag{25.2}$$

where θ_A is the angle of each speaker with the midplane, and p_L and p_R are the signal strengths (that is, the sound pressures at the listening point due to sound from the two speakers).

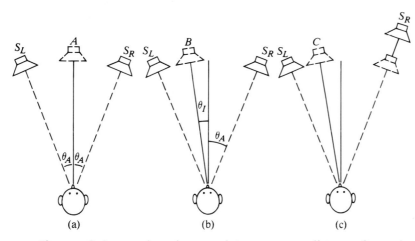

FIG. 25.1
Location of sound images from two sources S_L and S_R with the same program material: (a) identical sources; (b) same signal at different levels; (c) one source delayed by increasing the distance from source to listener.

If one of the speakers is moved to a greater distance from the listener than the other, the sound image moves toward the nearer source (position C). If S_R is farther away by one-third meter or more (so that the sound delay is more than 1 ms), the image coincides with S_L. This is another example of the precedence effect (see Section 23.3). However, if the sound received from S_R is made greater than that from S_L, the image moves back toward the median plane. Thus, it is possible to trade off amplitude for time delay within certain limits.

The extent to which this trade-off works and the effective "trading ratios" (the difference in arrival time divided by the equivalent difference in level) have not been completely established, although they have been the subjects of recent experiments. The following results are representative of several experiments:

1. The trading ratio is frequency-dependent, as would be expected from the fact (see Section 5.5) that localization of the sound source at low frequency is mainly by comparison of the arrival times (or phases) of the sound at our two ears, whereas intensity differences dominate at high frequency.

2. The trade-off is not complete. Although at low and mid frequencies a large fraction of the image shift due to a change in distance from the source can be compensated

for by a change in level, the image (position *C* in Fig. 25.1(c)) cannot be restored completely to the midplane (Gilliom and Sorkin, 1972).

3. There is disagreement among the results of different experiments. At 200 Hz, for example, trading ratios ranging from 60 to 147 $\mu s/dB$ are reported; at 500 Hz, the range is all the way from 10 to 200 $\mu s/dB$.

4. At high frequency, where the ear is insensitive to phase, no trade-off is noted for steady tones. In one experiment, localization of a 2400-Hz tone was found to be entirely based on interaural level. However, when the 2400-Hz tone was modulated at 200 Hz, the trade-off was comparable to that obtained with a 200-Hz tone by itself (Young and Carhart, 1974).

You may wish to perform some simple experiments with your own stereo system on the roles of time and intensity in sound localization. By switching the amplifier to its monophonic mode, apply the same signal from an audio generator to both loudspeakers. While seated at equal distances from the two speakers, adjust the balance control until the sound appears to originate from an image midway between the speakers. Now move closer to one speaker and try to restore the image to the midplane by manipulating the balance control. Try the experiment with tones of high and low frequency, music and speech.

Figure 25.2 illustrates time/intensity trading and also the approximate range of time and intensity differences over which the precedence effect applies. The horizontal axis gives the time by which a

FIG. 25.2
Range of time delay and intensity difference over which time/intensity trading takes place, and also the limits of applicability of the precedence effect. (After Madsen, 1970)

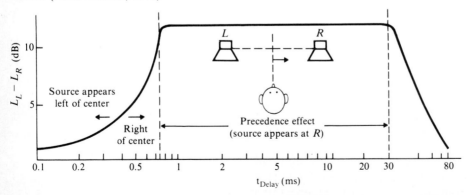

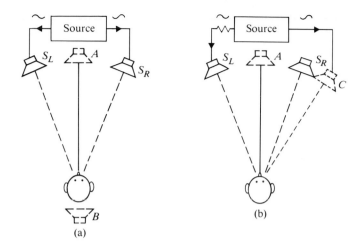

FIG. 25.3
Location of sound images from two out-of-phase sources. (After Rossing, 1981).

pulse to the speaker on the left (*L*) is delayed with respect to the speaker on the right (*R*). The vertical axis is the amount by which the sound level due to speaker *L* exceeds that from speaker *R*. The ascending curve at the left end indicates the approximate combination of time and intensity differences that will center the source image between the speakers. Note that when $L_l - L_R$ exceeds about 15 dB, it is impossible to compensate completely with a time delay. Also, when the time delay exceeds about one millisecond, the precedence effect (Section 23.3) defeats time/intensity trading. A listener seated only about 0.33 m (one foot) closer to the speaker on the right would experience such a one-millisecond time delay.

Now consider what happens when the phase of the signal to one of the speakers is reversed (sometimes hi-fi sets are inadvertently installed this way!). Provided that the listener is located on the median plane, the image appears to move from position *A* to a position inside or in back of the listener's head (position *B* in Fig. 25.3(a). If the signal level at one of the speakers is reduced sufficiently, however, the sound image may shift to position *C* beyond the other speaker.

In the examples thus far considered, the same signal has been presented to both source speakers, although with different intensities and phases. If the spectra of the two sources are different, the image appears to be broadened. In particular, if one speaker is given a signal with a high-frequency emphasis and one a signal with a low-frequency emphasis, as shown in Fig. 25.4(a), both speakers will appear to de-

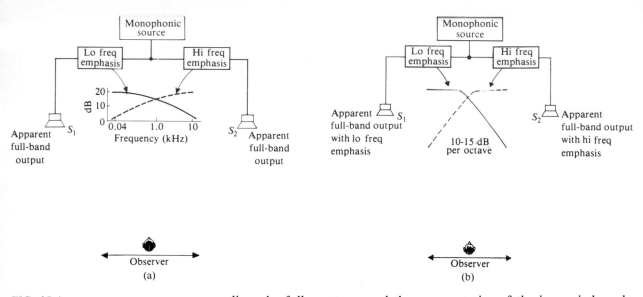

FIG. 25.4

Broadening of sound images by complementary frequency emphasis at two sources. (a) Broad crossover range (approximately eight octaves) results in an apparent full-band output from both sources. (b) Narrow crossover range results in different source timbre. (From Gardner, 1973. Reprinted by permission of AUDIO Engineering Society.)

liver the full spectrum and the apparent size of the image is broad. Furthermore, the listener can shift away from the median position without losing the effect. If, however, the frequency crossover is abrupt, as shown in fig. 25.4(b), there will be a noticeable difference in timbre between the two sources (Gardner, 1973).

Other differences between two sources, other than spectrum shape, are found to broaden the sound image. One effective way to achieve image broadening is to add reverberation to one source but not to the other.

In summary, three important properties of sound from multiple speakers are (1) the degree of *fusion* into a single image, (2) the *broadening* of the fused image, and (3) the *displacement* of the image in space. These are strongly influenced by differences in *level, spectrum, phase, time of arrival,* and *reverberant sound* between the two sources.

You can perform a simple experiment that illustrates sound localization in a "live" room (with little absorption). Let the source be a steady tone (provided by an audio generator and loudspeaker) and move around the room. You will find it very difficult to locate the source. When you are near an antinode (maximum), for example, the apparent direction of the source may change with a small movement of your head. But now substitute a percussive sound source for the steady one; have a friend clap his hands or strike a bell. If you close your eyes and point to the source as it moves around the room, you will

> make few errors. This is a good illustration of the precedence
> effect; you derive the clues needed to localize the source from
> the direct sound, even though the reflected sound, which
> follows in a few milliseconds, may be much louder.

25.3   WHAT IS HIGH-FIDELITY SOUND?

Before we describe the various types of sound-reproducing systems,
let us attempt to define high-fidelity sound.

To achieve realism in reproduced sound, five conditions should
be satisfied:

1. The frequency range of the reproduced sound should be sufficient
   to retain all the audible components in the source sound, and the
   sound spectrum of the reproduced sound should be identical to
   that of the source.

2. The reproduced sound should be free of distortion and noise.

3. The reproduced sound should have loudness and dynamic range
   comparable to the original sound.

4. The spatial sound pattern of the original sound should be reproduced.

5. The reverberation characteristics (in space and time) of the original sound should be preserved in the reproduced sound (Rossing,
   1979).

No sound-reproducing system is able to satisfy all five of these
criteria completely. The extent to which a given system is able to satisfy them determines its fidelity. A high-fidelity system should satisfy
them to the degree desired by the discerning listener.

High-fidelity components were discussed in Chapter 22. Most
quality components manufactured by reputable companies are capable of fulfilling the first two criteria reasonably well (assuming proper
installation and optimum adjustment). The most common exceptions
are the noise and distortion one encounters in phonograph records
and record players and the distortion that may be present in loudspeakers (for example, harmonic distortion, intermodulation distortion, and transient distortion). The distortion in most modern amplifiers is acceptably low when they are operated within their rated output.

Many high-fidelity systems are capable of reproducing sound at
the level heard in a concert hall, as required to fulfill the third criterion. However, it is doubtful whether such a level would be pleasing
in a small room. Sound levels in a concert hall reach 100 dB and more,

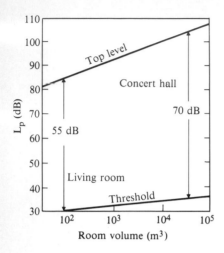

FIG. 25.5
Top level, threshold level, and dynamic range for music in rooms of various sizes.

whereas in a home listening room, music played at a level that reaches 85 dB will sound loud. The dynamic range in a room is limited by the tolerable top level and by the threshold that can be heard above the background noise, which may be about 25 to 30 dB in a home listening room and 30 to 35 dB in a concert hall (see Section 23.8).

Figure 25.5 gives a general idea of the loudness and dynamic range at which music may be heard in rooms of various sizes. Attempts to reproduce the full dynamic range of a concert hall in a small listening room would not create a very pleasing effect. The amount of sound power needed to achieve various sound levels is discussed in Section 24.3.

Criteria (4) and (5) also are requirements of the system rather than the individual components, and often tend to be overlooked by high-fidelity sound enthusiasts ("audiophiles"). Attempts to improve the ambience or spatial characteristics of reproduced sound have led to the development of stereophonic and quadraphonic sound systems, direct-reflecting speaker systems, "room expanders," stereophonic "spreaders" and "shifters," etc.

25.4 SINGLE- AND MULTI-CHANNEL SOUND-REPRODUCING SYSTEMS

Monophonic system   A monophonic system consists of one microphone, one amplifier, and a single loudspeaker in the listening room. Ambience is provided by the acoustic characteristics of the listening room. The performance of a high-quality monophonic system in a room with good acoustics should not be underestimated. Not many good monophonic systems exist, however. Figure 25.6(a) illustrates monophonic sound.

Monaural system   A monaural system differs from a monophonic system in that sound is fed to only one ear of the listener. This type of sound reproduction, used in the telephone and in some psychoacoustics experiments, would definitely not be called high-fidelity. Figure 25.6(b) illustrates monaural sound.

Binaural System   A binaural system, which uses two microphones to feed two earphones, is capable of satisfying all five criteria for sound realism. The microphones are usually placed in a "dummy head" to reproduce the directionality of the human auditory system at all frequencies. One disadvantage of binaural reproduction is the failure of the sound to change as the head of the listener is moved; this robs the listener of some important clues about the spatial characteristics of the sound, and tends to give the impression that the sound

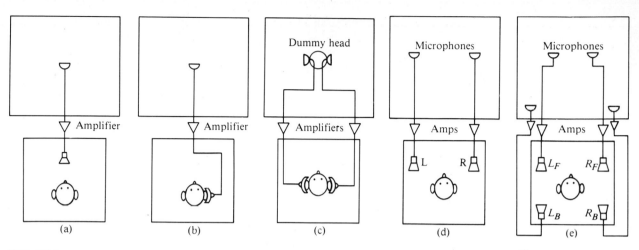

FIG. 25.6

(a) Monophonic sound reproduction: A single microphone feeds a single loudspeaker in the listening room. (b) Monaural sound reproduction: Sound is fed to one ear only by means of an earphone. (c) Binaural sound reproduction: Two microphones, placed in a "dummy head," feed two earphones. (d) Stereophonic sound reproduction: Two microphones feed two loudspeakers in the listening room. (e) Quadraphonic sound reproduction: Four microphones feed four loudspeakers.

source is located "inside the head" of the listener. Partly because of this problem, few binaural recordings have been made available commercially.

Listening to stereophonic music with stereo earphones produces a greatly exaggerated stereo effect that is interesting but not realistic; the source image usually appears to be inside or above the head. This is not true binaural reproduction however, because the recording microphones were probably not positioned in a pattern resembling that of the auditory system. A binaural system is illustrated in Fig. 25.6(c).

Stereophonic system   Sound picked up by two microphones is fed to two loudspeakers in the listening room. This popular system, shown in Fig. 25.6(d), will be discussed in Section 25.5.

Quadraphonic system   Sound picked up by four microphones feeds four loudspeakers in the listening room. This system, shown in Fig. 25.6(e) will be described in Section 25.8.

25.5 STEREOPHONIC SOUND

The most popular, and thus far the most successful, system for reproducing sound with a spatial dimension is the stereophonic system. In the basic stereophonic (or stereo) system, sound from the source is picked up by two microphones, recorded or transmitted in two separate channels, and reproduced by two loudspeakers. In theory, there are many ways to do this. In practice, several different arrangements of microphones and loudspeakers have been used with varying degrees of success.

Although experiments with stereophonic sound took place in the 1930s at the Bell Telephone Laboratories in the USA and at Electric and Musical Industries Ltd. (EMI) in England, stereophonic sound recording did not develop commercially until the mid-1950s. The Bell Labs experiments are described in a series of papers by Fletcher, et al. (1934), while the EMI work by Alan Blumlein and his colleagues resulted in a series of patents (Lipshitz, 1986). Early stereophonic recordings of the Philadelphia Orchestra (conducted by Leopold Stokowski), made by Bell Labs engineers in 1931–32, are of remarkably high quality.

Blumlein's preferred technique for recording stereophonic sound was to use a pair of bidirectional (figure-of-eight) microphones with their axes of maximum response at an angle of 90° to each other. This arrangement, still used by many recording engineers, is referred to as a stereosonic system or a "Blumlein pair."

Stereo listening tests were performed at a convention of the Audio Engineering Society, using a musical selection recorded in a concert hall with six different microphone arrangements used for stereophonic recording. In the listening tests, 34 listeners were seated in a favorable stereophonic listening area, and 30 listeners were seated in unfavorable positions. Although the listening position had little influence on some parameters such as intimacy, dynamic range, and brilliance, it was found that impressions of liveness, warmth, and source width were very much dependent on listening position (Ceoen, 1972). A recording system that used a pair of cardioid microphones spaced 17 cm apart with their axes of maximum response at an angle of 110° was judged to give the best results overall.

Many different microphone arrangements have been used for stereophonic recording. Most of them are based on one or more of the following systems, however, which are described for the sake of the technically minded reader (Ceoen, 1972):

1. *XY system.* A coincident pair of cardioid microphones with their axes of maximum response at an angle of 135°.

2. *Stereosonic system.* A coincident pair of velocity microphones with their axes of maximum response at an angle of 90° (a "Blumlein pair").

3. *MS system*. A coincident pair consisting of a forward-pointing cardioid microphone and a sideways-pointing velocity microphone. The sum and difference of the signals from these two microphones are recorded as the right and left stereo channels. (This system is also described in Blumlein's early patents).

4. *ORTF system*. A pair of cardioid microphones spaced 17 cm apart with their axes of maximum response at an angle of 110°. This system is widely used by the French Broadcasting System (ORTF).

5. *NOS system*. A pair of cardioid microphones spaced 30 cm apart with their axes of maximum response at an angle of 90°. This system is widely used by the Dutch Broadcasting System (NOS).

6. *Pan-potted system*. Five cardioid microphones spaced 2.85 m apart feed appropriate levels of sound into the two stereo channels.

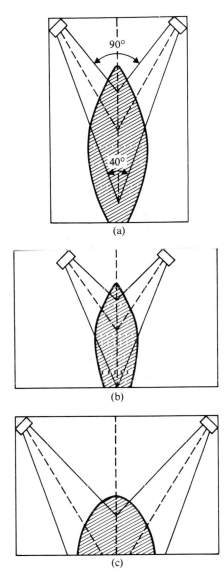

(a)

(b)

(c)

FIG. 25.7
Favorable stereo listening areas for three different loudspeaker arrangements in a rectangular room with dimension in the ratio 3:2. (After Rossing, 1981).

One of the important criteria for realism in reproduced sound is that the spatial sound pattern of the original sound should be reproduced. Stereophonic sound systems accomplish this to a much greater degree than do monophonic systems, provided the listener is seated in a favorable stereophonic listening area. What is a "favorable" listening area? It means being fairly close to the median plane and at a distance such that the two speakers are separated by an angle of about 40 to 90 degrees. This frequently presents difficulties in arranging the home living room for good stereo.

    In a rectangular room, the best location for the speakers is usually in the corners of the end wall. Corner placement provides good room coverage and also enhances the radiation of low-frequency sound. Figure 25.7 shows three loudspeaker arrangements in a room having dimensions in the ratio 3:2. Note that the most favorable stereo listening area in Fig. 25.7(a) is substantially larger than those in Figs. 25.7(b) and (c). The favorable listening area may be moved to one side or the other of the median plane by changing the balance control of the amplifier to give one speaker a greater gain. There are practical limits to this, however, because only the relative loudness of the speakers, and not the relative phase, is changed.

    Although the limiting angles (40° to 90°) suggested above are somewhat arbitrary, it is well known that the spatial effect of stereophonic sound disappears outside a certain range of angle. If the angle is too narrow, the source appears to be monophonic. If the angle is too great, the listener hears two distinct sources with a distracting "hole" in the center.

If you are installing a stereophonic sound system in a home listening room, considerable experimentation with speaker placement is recommended. The results vary somewhat from room to room due to the influence of absorbing and reflecting surfaces in the sound field.

25.6   THE SOUND FIELD IN LISTENING ROOMS

Stereophonic sound-reproducing systems can be quite successful in satisfying four of the five criteria for realism in reproduced sound, including the reproduction of the spatial sound pattern, when the loudspeakers and the listeners are in favorable locations. However, most listening rooms do not begin to produce the ambience experienced in a concert hall.

We can walk into a strange room blindfolded, and rather quickly and accurately estimate its size by listening to the sound in the room. We do not have to make any special effort to do this; our auditory processor has been programmed by experience. Presumably, when reflected sounds follow the direct sound with little delay, an auditory impression of smallness is created. Conversely, when the reflections arrive with a distribution in time and space that is characteristic of a large concert hall, a feeling of spaciousness is created. Various attempts have been made to introduce some of the features of concert hall sound into home listening rooms. Placing additional speakers in the room creates some feeling of spaciousness, because the sound arrives from several directions. Sometimes high- and low-frequency emphases are given to certain speakers, in the manner shown in Fig. 25.4. This is sometimes referred to as "pseudo-quadraphonic" or "simulated quadraphonic" sound, although this may not be an accurate description. Other ways of creating the illusion of spaciousness include those shown in Fig. 25.8. In Fig. 25.8(a), the rear speakers are supplied with a signal similar to that of the front speaker on the opposite side. In Fig. 25.8(b), the difference between the front channels is

FIG. 25.8

Three methods to enhance the feeling of spaciousness from stereophonic sound.

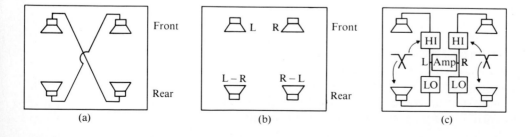

(a)   (b)   (c)

Front view

Back view

FIG. 25.9
A direct-reflecting speaker system
designed to radiate 90 percent of its
energy into the reverberant sound field
of the room. (From Bose, A. G.,
"Sound Recordings & Reproduction,"
Technology Review (MIT) 75 (7,8),
June 1973. Reprinted with permission
of the author.)

used to create two back channels, the two rear channels being in opposite phase ($L_B = L - R$ and $R_B = R - L$, where L_B and R_B are the new channels).

Adding reverberation to the recorded sound can transfer some of the feeling of space to the listening room. Ideally, the reverberant sound would be recorded on separate channels, and reproduced with a different spatial distribution from the direct sound. A device called a stereophonic spreader and shifter, developed at the CBS Laboratories, is an example of devices designed to introduce reverberation into stereophonic recordings (Bauer, 1969).

One loudspeaker design that has been a commercial success creates ambience in a room by radiating 90 percent of the energy toward the wall and only 10 percent directly toward the listener. It is claimed that the large ratio of reflected to direct sound helps compensate for the fact that the listener is much closer to a loudspeaker in the listening room than to the live performers in a concert hall (Bose, 1973). The direct-reflecting speaker design shown in Fig. 25.9 uses a driver at the front of the cabinet to provide direct sound for proper localization and eight drivers aimed at a 30° angle toward the wall to supply sound to the reverberant field.

In the room described in Example 2 (see Section 25.1), nearly all listeners would find themselves well into the reverberant field for sounds of low frequency, but for sounds of high frequency, the effect would depend on where they were seated. This situation is amplified by the fact that the absorption in most listening rooms is greater at high frequency. Although the direction of a source is determined by the direct sound (precedence effect), the tonal balance appears to be derived from the total sound reaching the ear. This strongly suggests that the frequency response of a loudspeaker should be described by the total *power* radiated at each frequency rather than the sound pressure level on its axis (which depends on the directivity factor Q at that frequency).

In one experiment, the spectral distributions of sound from a high-quality loudspeaker, placed at 22 locations in 8 listening rooms, were measured (Allison and Berkovitz, 1972). When these measurements were compared to similar curves made at favorable seat locations in concert halls, it was found that in the important midband range (250–2500 Hz), the curves were similar. Below 250 Hz, however, the sound levels of the combined loudspeaker-room systems were considerably below those in the concert halls. This loss of low-frequency sound in the smaller rooms can be attributed to inadequate stiffness of walls, windows, etc., which causes them to absorb low-frequency sound by vibrating sympathetically. At high frequency, on the other hand, the average sound level in listening rooms was higher than in concert halls, relative to the low frequencies. This experiment suggests that to emulate the tonal balance heard in a concert hall, the tone controls of a high-fidelity system should be adjusted for bass boost and treble cut, rather than for flat response.

Another factor to be considered is the loading effect of nearby walls on a loudspeaker radiating sound of long wavelength. Placing a loudspeaker in a corner of a room can increase its total radiated power at low frequency by as much as 9 dB. What happens, in effect, is that the walls and floor of the room form a sort of pyramidal horn that increases the efficiency of radiation (see Section 20.14). The walls also cause an unevenness in frequency response, however, which can be minimized by placing the loudspeaker so that the distances from the woofer cone to each of the nearby reflecting surfaces differ by at least a factor of two (Allison, 1979).

25.7   TESTING AND EQUALIZING THE LISTENING ROOM

The best method of testing a high-fidelity sound-reproducing system, especially the loudspeaker/listening room combination, is with well-trained ears. Irrespective of what is measured, if a combination of speaker placement, furniture placement, and tone control settings sounds good, then it should be considered good. Often it is advantageous to obtain the opinions of several listeners who are accustomed to hearing live music in concert halls. Objective testing with instruments can be a great help and time-saver, however.

The objective test that seems to correlate best with subjective judgment is a measurement of the room response in ⅓-octave bands using pink-weighted random noise as the sound source. Pink-weighted

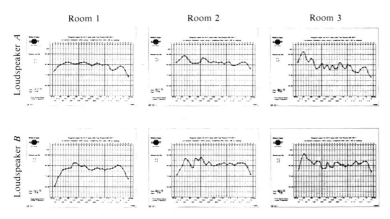

FIG. 25.10
Response of two different loudspeakers in three different rooms. Source material consisted of ⅓-octave bands of random noise. (From ''Relevant Loudspeaker Tests . . . Random Noise,'' 47th Audio Eng. Sol. Convention, Copenhagen. Reprinted as B & K Application Note 17-197 © 1974 by B & K Corporation. Reprinted by permission.)

noise is noise that is weighted so that the average power in each octave band or ⅓-octave band is the same (as opposed to white noise, which has the same average power in a given frequency band Δf throughout the audible range). If a real-time ⅓-octave analyzer is available, the entire room response curve can be displayed at once. Alternatively, the response curve can be plotted point by point by using a pink-noise test record (which presents each ⅓-octave band of noise in turn, one at a time) plus a sound level meter.

Figure 25.10 shows the response curves for two different loudspeakers in three different rooms. In listening tests that included several different types of music, speaker A was preferred in rooms 1 and 2, but speaker B was preferred in room 3 (Møller, 1974). Room 1 was fairly large (140 m³), whereas rooms 2 and 3 were substantially smaller (49 m³ and 32 m³, respectively).

A certain amount of equalization is possible by using the treble and bass tone controls on the amplifier. However, an equalizer that includes a set of octave-band or ⅓-octave-band filters with adjustable gain allows more flexibility. Some amplifiers have built-in octave-band filters.

25.8   QUADRAPHONIC SOUND

The spatial and time distributions of sound in a listening room can be improved by adding channels of sound. A four-channel system, called *quadraphonic sound,* appeared in the 1970s. In most quadraphonic systems, the two front loudspeakers are placed as they would be for good stereophonic reproduction, and the other two loudspeakers are placed near the rear corners of the room (as in Fig. 25.5(e)) or along the sides of the room.

When a listener seated in a favorable location between the four loudspeakers hears quadraphonic sound, skillfully recorded on four channels of magnetic tape and played back on quality equipment, the result can be spectacular! Unfortunately, few persons were ever afforded this privilege. The technical problems associated with reproducing quadraphonic sound in a home listening room are formidable, although not insurmountable. Because of these problems, quadraphonic sound was a great disappointment, even to the point of being described as a "quadrafizzle," and it has all but disappeared. The main problems with quadraphonic sound appear to be (1) recording four channels of information on phonograph discs, and (2) reproducing the sound in a way that will sound realistic in a variety of listening rooms.

Methods of recording quadraphonic material on discs are classed as discrete or matrix systems. In a *discrete* system, four completely independent channels are recorded, two of them at ultrasonic frequencies. In a *matrix* system, the four channels are combined into two by a suitable code. These systems are described in an anthology of articles on quadraphonic sound published by the Audio Engineering Society (1975).

25.9   TIME DELAY AND AMBIENCE SYNTHESIZERS

Digital and analog time-delay circuits, designed to supply ambience to small rooms, are now available under various descriptive titles such as "ambience synthesizer," "room expander," "sound space control," etc. Their function is to delay the normal stereo program material (usually by several different amounts) to approximate the effect of multiple reflections in a hall. The delayed signals go to a second set of amplifiers and speakers placed in positions similar to those of the rear speakers in a quadraphonic system. Sometimes the delayed signals are mixed into the main channels as well.

A variety of delay circuits are used in the various ambience synthesizers available. Some units use integrated electronic circuits known as "charge-coupled devices" to develop a controlled time delay. Digital delay lines offer the greatest amount of flexibility for time-

FIG. 25.11
An ambience synthesizer. (Courtesy of Advent Corp).

delay circuits. In addition to providing a wide range of time delay, digital delay lines feature high signal-to-noise ratios and low distortion. The ambience synthesizer shown in Fig. 25.11 uses a digital delay line.

One technique for "expanding the listening room," suggested by Hafler several years ago, is shown in Fig. 25.12. The out-of-phase (L-R) portion of the stereo signal, which is rich in recorded ambience, is fed to a pair of rear speakers, so that it arrives at the ears from a different direction than the direct, in-phase sound (L and R).

Several commercially available systems (e.g., the Carver Sonic Hologram Generator and the Sound Concepts Image Restoration System) apply the principles of crosstalk cancellation developed by Schroeder and Atal (1963) at the Bell Telephone Laboratories some years ago. To understand this principle, consider what happens when we listen to stereophonic sound (Fig. 25.13). A signal LL from the left speaker reaches the left ear about 100 μs before it reaches the right ear (LR). This interaural time difference is one of the major cues by which we locate the loudspeaker. Adding a second speaker with a different signal provides additional cues that are used to construct an image of the sound source. But we are always aware of the location of each of our speakers.

Suppose that we now add a delayed signal $-aLR$, proportional to the negative of LR into the right speaker to cancel signal LR and a delayed signal $-aRL$ into the left speaker to cancel RL. The cancellation signals $-aLR$ and $-aRL$ are delayed by just the right amount (100 to 200 μs, depending upon the angle of the speakers). In this way only signal LL reaches the left ear and only signal RR the right ear. This essentially cancels out the information about the location of the speakers, and enhances the cues that are used to reconstruct the source image.

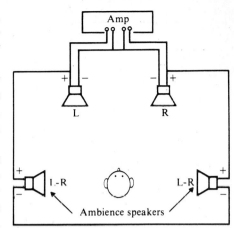

FIG. 25.12

A simple circuit for adding ambience by feeding the out-of-phase (L-R) portion of the stereo signal to a pair of rear speakers.

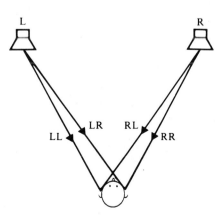

FIG. 25.13

In stereophonic sound reproduction the right ear hears a signal RR from the right speaker plus crosstalk LR from the left speaker. The left ear hears LL plus RL.

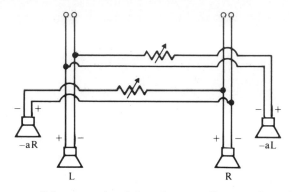

FIG. 25.14
Crosstalk cancellation with two auxiliary speakers.

Of course the delayed crosstalk-canceling signals can come from two auxiliary speakers connected as shown in Fig. 25.14. The amplitudes of the cancellation signals are adjusted with two series resistors, and the time delay is determined by the positions of the two auxiliary speakers.

25.10 LISTENING WITH EARPHONES

Most high-fidelity amplifiers have a headphone jack, and most readers have listened to music with a set of stereophonic earphones or headphones. Opinions will vary, no doubt, concerning the end result. Some people find the unnatural sound exciting; others find it unnerving. Nearly all will agree that it is very different from the sound radiated from loudspeakers that one hears in the listening room.

Listening to stereophonic recordings with earphones is quite different from listening to binaural reproduction (described in Section 25.5). Not only does the source appear to be "inside the head," but it appears to be a greatly enlarged source. The inside-the-head feeling comes partly from the failure of the sound to change as the listener's head is moved, but a more important factor is the spectral change brought about by setting up standing waves in the outer ear. With an earphone in place, the resonances of the outer ear change, more or less, from those of an open pipe to those of a closed pipe. Placing absorbing material between the earphone and ear reduces the inside-the-head impression of the source (Schroeder, 1975). Earphones of open design, which are becoming popular, appear to accomplish this to some extent, also.

25.11 LET THE EAR JUDGE!

The final judgment as to whether or not sound reproduction is high-fidelity must be made by the trained ear itself. We say "trained ear," because we tend to judge sound by what we are accustomed to hearing. A listener who regularly hears good music in the concert hall or a performing musician will detect shortcomings in a sound-reproducing system that elude a person who listens mostly to reproduced sound (and vice versa, perhaps). Some listeners have so tuned their ears to omnipresent stereophonic sound in small rooms that live performance in a concert hall sounds strange to them.

Needless to say, this point of view has been challenged. One student articulated the opposing point of view: "Musicians frequently could care less about quality of sound in recording and will listen to anything if it's a performance they want to hear, while it is the audiophiles who can discern . . . high fidelity in recorded music."

Perhaps the best advice that can be given to a prospective buyer auditioning high-fidelity equipment is to take along a favorite recording of a type of music that best suits your taste. It is wise to steer clear of the special demonstration records that many dealers use to show off their equipment. I tend to favor solo piano music as program material for auditioning equipment, because its transparency makes distortion easier to detect, and at the same time the percussive nature of piano sounds provides useful information about damping in the system. It is a good idea to do your listening with a knowledgcable friend with whom to share your impressions, but in the final analysis, let *your* ears judge. If at all possible, arrange to hear the system you tentatively select in your own listening room; many dealers allow this.

Olson (1969) emphasizes the physiological and psychological aspects of sound reproduction, which he divides into two parts: the *sensorial* process and the *synthetic emotional* process. Both are subject to training, the emotional process especially so. The emotional process depends on current tastes in music, which can change rather rapidly.

25.12 SUMMARY

High-fidelity sound reproduces much of the spectrum, dynamic range, and spatial characteristics of the original sound with minimal addition of distortion and noise. The listening room is probably the weakest link in most high-fidelity sound systems. Important considerations in understanding the sound field in a listening room include (1) room resonances; (2) the relationship of reverberant to direct sound; (3) the variation of loudspeaker power level L_W and directivity factor Q with frequency; (4) sound images from multiple sound sources; (5) placement of loudspeakers and furniture; and (6) program material.

Efforts to reproduce the spatial character of the sound source and to provide ambience in the listening room have led to stereophonic, quadraphonic, and other multichannel techniques for sound reproduction, as well as to a variety of ambience synthesizers that attempt to "expand the listening room."

References and Suggested Readings

Allison, R. F. (1979). "Influence of Listening Rooms on Loudspeaker Systems," *Audio* 63(8): 37.

Allison, R. F., and R. Berkovitz (1972). "The Sound Field in Home Listening Rooms," *J. Audio Eng. Soc.* **20**: 459.

Audio Engineering Society (1975). *Quadraphony*, a collection of 34 technical articles on quadraphonic sound that appeared in the *J. Audio Eng. Soc.* from 1969 to 1975.

Bauer, B. B. (1969). "Some Techniques Toward Better Stereophonic Perspective," *J. Audio Eng. Soc.* **17**: 410.

Bose, A. G. (1973). "Sound Recording and Reproduction," *Technology Review (MIT)* **75**(7, 8).

Ceoen, C. (1972). "Comparative Stereo Listening Tests," *J. Audio Eng. Soc.* **20**: 19.

Fletcher, H., et al. (1934). Symposium on Wire Transmission of Symphonic Music and its Reproduction in Auditory Perspective. *Bell System Tech. J.* **13**: 239.

Gardner, M. B. (1973). "Some Single- and Multiple-Source Localization Effects," *J. Audio Eng. Soc.* **21**: 430.

Gilliom, J. D., and R. D. Sorkin (1972). "Discrimination of Aural Time and Intensity," *J. Acoust. Soc. Am.* **52**: 1635.

Hirsch, J. (1979). "Audio Listening-Room Expanders," *Popular Electronics* **15**(2): 41.

Lipshitz, S. P. (1986). "Stereo Microphone Techniques . . . Are the Purists Wrong?" *J. Audio Eng. Soc.* **34**: 716.

Madsen, E. R. (1970). "The Disclosure of Hidden Information in Sound Recording," (Lecture given May 1970).

Møller, H. (1974). "Relevant Loudspeaker Tests in Studios, in HiFi Dealers' Demo Rooms, in the Home, etc.,—Using ⅓-Octave, Pink-Weighted, Random Noise," 47th Audio Eng. Soc. Convention, Copenhagen. Reprinted as B & K Application Note 17-197 (B & K Instruments, Marlborough, Mass.).

Olson, H. F. (1969). "Home Entertainment: Audio 1988," *J. Audio Eng. Soc.* **17**: 390.

Olson, H. F. (1972). *Modern Sound Reproduction*. New York: McGraw-Hill.

Rossing, T. D. (1979). "Physics and Psychophysics of High-Fidelity Sound, Part I," *Physics Teach.* **17**: 563.

Rossing, T. D. (1981). "Physics and Psychophysics of High-Fidelity Sound, Part IV," *Physics Teach.* **19**: 293.

Schroeder, M. R. (1975). "Models of Hearing," *Proc. IEEE* **63**: 1332.

Schroeder, M. R., and B. S. Atal (1963). "Computer Simulation of Sound Transmission in Rooms," *IEEE Int. Conv. Record,* Pt. 7: 150. See also *Am. J. Phys.* **41**: 461 (1963).

Young, L. L., Jr., and R. Carhart (1974). "Time-Intensity Trading Functions for Pure Tones and a High-Frequency AM Signal," *J. Acoust. Soc. Am.* **56**: 605.

Glossary

ambience Spaciousness; the degree to which sound appears to come from many directions.

binaural Sound reproduction using two microphones (usually in a "dummy" head) feeding two headphones, so that the listener hears the sound he or she would have heard at the recording location.

cardioid microphone A microphone with a heart-shaped directivity pattern designed to pick up sound in one direction preferentially.

CD-4 (discrete) A system for recording quadraphonic sound on phonograph discs that preserves four separate channels by recording two of them at ultrasonic frequencies.

high-fidelity sound Sound that reproduces much of the spectrum, dynamic range, and spatial characteristics of the original sound and adds minimal distortion and noise.

matrix A mathematical array that indicates how a set of signals can be combined in an orderly way to produce a new set.

monaural Sound reproduction using one microphone to feed a single headphone, such as is used in telephone communication.

monophonic Sound reproduction using one microphone to feed one or more loudspeakers with one signal.

pink noise Random noise that has the same power in each octave or ⅓-octave band.

quadraphonic Sound reproduction using four microphones to feed four loudspeakers; usually two are in front of the listener and two are behind or to the sides.

reverberant sound Sound that reaches the listener after a large number of reflections; as one moves away from a sound source, the sound level reaches a steady value called the reverberant level.

SQ (matrix) A system for recording quadraphonic sound on phonograph discs that combines the four signals into two channels.

stereophonic Sound reproduction using two microphones to feed two loudspeakers.

Questions for Discussion

1. In binaural sound reproduction, why are the two microphones placed in a "dummy head" rather than merely given that same separation in air?

2. Comment on the practice of connecting a third speaker between the two stereo channels to eliminate the "hole" in the middle.

3. Is the restriction on listener location in the room likely to be a serious deterrent to the acceptance of quadraphonic sound?

4. Is it possible to have realistic stereophonic sound in an automobile? What about quadraphonic sound?

5. What is the difference between a binaural system and a stereophonic system for recording and reproducing sound?

Problems

1. Calculate the frequencies of the first three resonances of a room with dimensions 5 m × 10 m × 2.5 m. Do they have any significance acoustically?

2. Suppose that two loudspeakers, each 30° from the median plane, carry the same program material, but the loudspeaker on the left has twice the signal strength of the one on the right. Describe the location of the image. (Calculate $\sin \theta_I$, and look in a set of tables or use a pocket calculator to determine the angle θ_I corresponding to your value of $\sin \theta_I$; $\sin 30° = \frac{1}{2}$.)

3. Draw a diagram of a room with dimensions in the ratio 4:3 and indicate the favorite listening areas for two different placements of a pair of speakers in a stereo system.

4. From the chart given in Fig. 24.2, determine the distance from a loudspeaker with $Q = 5$ in a room with $R = 500$ at which direct and reverberant sound are equal in level. (*Hint:* $L_p - L_w$ should be 3 dB above the reverberant level.)

5. The sketch below represents the room described in Problem 1. Indicate the location of the maximum sound pressure level when the room is excited in each of its first three resonances.

6. Suppose the time/amplitude "trading ratio" were found to be 100 μs/dB. If one speaker is 1 m farther away than a second, how much greater must its sound power level be in order that the sound image will appear to be on the median plane?

Electronic music has been called the most important development in music during the twentieth century. Although experiments with electrically generated sound date back to the early years of this century, the technology as we know it today began with two inventions: the *tape recorder* and the *transistor*. Two subsequent inventions, the *digital computer* and the *electronic music synthesizer*, triggered a rapid growth in the art. More recently the impact of another development, microelectronics, has had a substantial effect on electronic music.

Among the various musical endeavors that have been described as types of electronic music are:

Musique concrète: real sounds that have been recorded on tape and altered to create new sounds;

Music played on an electronic musical instrument;

Music by animation: a time-consuming method for creating complex sounds by drawing and filming their waveforms;

Music composed using a digital computer;

Music synthesized with a digital computer;

Music from an electronic music synthesizer.

Chapters 26–29 describe some of the principles of electronic music with emphasis on

PART SEVEN
Electronic Music

the technology rather than the techniques. The reader should have an understanding of basic electronic circuits, such as that presented in Chapters 18 and 19; a reading of all five chapters in Part 5 is recommended. Specific circuits and more technical information are enclosed in boxes, as has been the practice in earlier parts of the book.

CHAPTER 26

Electronic Organs and Other Musical Instruments

A musical instrument may be considered a system that consists of three essential elements: a vibrator, a tone modifier, and a radiator. In traditional musical instruments, the vibrator is an air column, string, membrane, rod, or plate that is set into vibration with a percussive blow or maintained in steady vibration by an external driving force. Acoustical resonators serve both as tone modifiers and as sound radiators in wind instruments, as do wood plates in most string instruments. Many examples of such systems are described in Chapters 10–14.

An electronic musical instrument incorporates the same three elements described above, but the radiator is a loudspeaker driven by an electronic power amplifier. In some electronic instruments, the vibrator is mechanical; in others, it is an electrical circuit. The tone modifiers may include acoustical resonators, electronic circuits, or both.

In Chapter 10, we described the electric guitar, which uses magnetic pickups to convert the vibrations of a string into an appropriate electrical signal that can be amplified and radiated through a loudspeaker. In this chapter, we will describe other electronic musical instruments, including the electronic organ, the electropiano, and the electronic carillon. Electronic tone modifiers will also be discussed.

(a)

(b)

FIG. 26.1
Two early electronic musical
instruments: (a) Cahill's
Teleharmonium; (b) an early
Theremin. (Courtesy of Electronic Arts
Foundation.)

26.1 EARLY INSTRUMENTS

About the beginning of the twentieth century, a number of experimenters constructed musical instruments that used some type of electrical generator and loudspeakers to produce sound. One of the more successful inventors was Thaddeus Cahill, whose 1897 patent incorporated the idea of multiple alternating current generators on a single rotating shaft. He called his instrument the *Teleharmonium,* and this 200-ton monster was heard in several performances (see Fig. 26.1(a)). Since electronic amplifiers had not yet been invented, the electromechanical tone generators had to supply enough electrical power to drive the loudspeakers directly.

One of the first instruments to use electronic tone generators was the *Theremin,* invented by Leon Theremin in 1919. It uses two electronic feedback oscillators (see Section 19.6) designed to oscillate at frequencies substantially above the audible range. The outputs from these two oscillators are combined to produce an audible difference tone (see Section 8.6), which is amplified and used to drive a loudspeaker. One oscillator is fixed in frequency, but the frequency of the other can be varied by changing the position of the player's hand near an antenna (see Fig. 26.1(b)). The presence of the player's hand adds capacitance to the feedback circuit (see Section 19.6). A small change in frequency of the variable-frequency oscillator can cause a relatively large change in the pitch of the output tone, which is determined by the difference in frequency of the two oscillators.

A second antenna on the Theremin makes it possible to change the amplitude. Bringing the hand near this second electrode changes the capacitance in a carefully balanced circuit, so that the performer can control pitch with one hand and loudness with the other.

Other electronic instruments followed the Theremin. In France, Maurice Martenot built the Onde Martenot in 1928, and the same year Frederick Trautwein completed his Trautonium in Germany. The Hammond organ was invented in the United States in 1929.

26.2 ELECTRONIC ORGANS

Although interest in electronic organs began in the 1930s, it was in the decade following World War II that practical electronic organs really developed. These early instruments, while ingenious, left much to be desired musically. In no way could they compete tonally with the pipe organ, which had two thousand years of research and development behind it. Nevertheless, these early instruments form the basis on which many people judge electronic organs as "cheap imitations" of pipe organs.

Modern electronic organs have developed along three rather different lines: First, there are the compact single-manual organs used in

(a)   (b)   (c)

FIG. 26.2
Different kinds of electronic organs: (a) single-manual portable organ;
(b) two-manual home-type organ; (c) full-size console organ. (Photograph
(a) courtesy of Farfisa Music, Inc.; (b) and (c) courtesy of Baldwin Piano
and Organ Co.).

popular bands. These portable instruments are used with a separate
amplifier and loudspeakers.

Second, there are two-manual organs with amplifier and loudspeaker built into the console, which have been quite popular for
home use. In order to keep the cost down, these organs usually have
two short keyboards, staggered for the left and right hands, and an
abbreviated set of pedals, as shown in Fig. 26.2(b).

Finally, there are organs with a full-size console with pedal board,
a wide selection of stops, and several loudspeakers. These larger instruments are sometimes classified as either "concert" or "theatre"
organs, depending on their tonal design.

Within these three general lines, there is an almost infinite variety
of sizes, models, special features, and tonal designs. Only a brief look
at some of the principles of tone generation and modification is possible in this chapter.

26.3   ELECTROMECHANICAL TONE GENERATORS

Many clever electromechanical devices were used to generate sound in
earlier organs, although they have gradually given way to all-elec-
tronic and digital tone generators. One of the most successful electromechanical generators was the magnetic tone wheel, introduced by
Hammond in 1935. A magnetic tone wheel generator is illustrated in
Fig. 26.3. The steel tone wheel rotates close to a magnetized rod, and
the fluctuating magnetic field induces an electrical voltage in the coil.
The frequency of the electrical voltage induced in the coil depends on

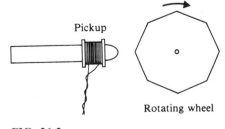

FIG. 26.3
Magnetic tone wheel used in
Hammond organs.

the number of "teeth" on the tone wheel and on its speed of rotation. Tone wheels may have one hundred teeth or more, and may be mounted in pairs rotating in opposite directions in order to generate tones of high frequency.

The harmonic tones generated by the different tone wheel generators can be mixed in a wide variety of ways by "drawbars," which select desired amounts of the various harmonics. One advantage of the magnetic tone generator is that the individual notes need not be tuned, since the tone generators are all driven by the same motor. Of course, such absolute tuning tends to produce a rather dry sound, as we will discuss later.

Other interesting types of electromechanical oscillators have included the vibrating reed oscillator used by Wurlitzer for a number of years, the photoelectric tone generator used by Baldwin and Kimball, and the rotating electrostatic electrone used by Compton in England and by Dereux in France. The electrone produced tone only when a polarizing voltage was applied, thus eliminating the key clicks from "keying" the actual sound. The Wurlitzer reeds vibrated independently, and avoided the dry sound from too perfect a pitch relationship between notes.

26.4   ELECTRONIC TONE GENERATORS

Most electronic organs built in recent years use electronic oscillators. Some of them use a single master oscillator together with frequency dividers; others have twelve separate oscillators, one for each note of the scale; still others use an independent oscillator for each note of the keyboard. Some organs begin with simple waveforms or even pure tones, and build up various tones by *addition*. Others begin with signals that are rich in harmonics and then construct the desired tone by *subtraction*.

Large electronic organs which use additive synthesis of sound have hundreds of independent oscillators, and build up tones by the addition of the desired harmonics. Such organs may be designated as "classic" or "theatre" designs depending on their tonal design. One company that specializes in large electronic organs of this type is Rodgers (Hillsboro, Oregon).

The basic Rodgers tone generator is an LC feedback oscillator that is tuned by changing the inductance of the ferrite core L in Fig. 26.4. The oscillator is keyed on or off by changing the bias on transistor Q_1, and there is provision for introducing tremulant. One medium-size organ has 399 independent oscillators arranged in five banks, three main and two celeste (celeste is generated by keying two oscillators tuned to slightly different frequencies). Rodgers uses from two to six output channels in their classic organs. Each channel, in general, handles one division of the organ; 80- to 100-watt amplifiers in each channel drive several speakers (woofers, mid-range speakers, and tweeters).

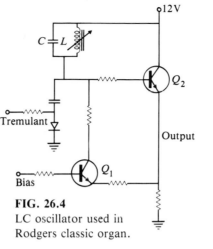

FIG. 26.4

LC oscillator used in Rodgers classic organ.

Saville organs use independently tuned RC oscillators in each voice (400 to 500 oscillators in a large instrument). A feature of the Saville design is the use of twelve independent output amplifiers that are assigned to the various voices and drive loudspeakers in cylindrical horn-loaded enclosures. These speakers are usually placed in an arrangement resembling that of a typical pipe organ. Exceptionally low intermodulation distortion is claimed for this type of amplifier-speaker design.

26.5 FREQUENCY DIVIDERS

Most electronic organs do not have a separate oscillator for each note. Instead, they use frequency division by electronic circuits to derive several frequencies from a master oscillator. There are basically two different ways in which this is accomplished.

In one system, a set of twelve master oscillators generates the frequencies for the top octave (usually 4435–8372 Hz), and each of these frequencies is divided by a series of bistable dividers to obtain all the other octaves. Each such circuit divides the frequency by two, which corresponds to one octave change in pitch. For example, from the 8372-Hz (C_9) master oscillator would be derived 4186 Hz (C_8), 2093 Hz (C_7), 1047 Hz (C_6), etc., right down to the pedal note of 32.7 Hz (C_1).

The principle of frequency division is shown in Fig. 26.5. Each divider consists of a bistable circuit ("flip-flop"), which switches from one stable state to the other for each cycle of the incoming square wave. Thus, the output square wave has exactly half the frequency of the input.

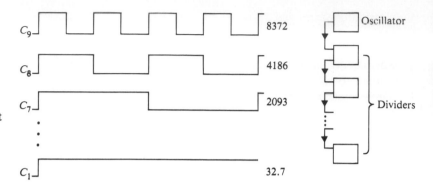

FIG. 26.5
The principle of frequency division.
An oscillator furnishes square waves at
the frequency of a note in the top
octave, which is then divided by eight
bistable circuits to generate that same
note in the other eight octaves.

To generate the 97 notes in 8 octaves requires 12 oscillators and
84 frequency dividers. The dividers, however, have no tuning element,
which makes them easy to fabricate on an integrated circuit (see Chapter 29), and tuning the organ is simplified by having only 12 independent oscillators. Voicing is more difficult, however, since only one
wave shape (a square wave) is generated, and these cannot be added
together to construct other useful tones.

One system, for example, uses a master oscillator frequency
of 3872 kHz; the master oscillator is crystal-controlled for
great stability in frequency. A series of divider circuits
generate the 12 frequencies of the upper octave indicated in
Table 26.1.

TABLE 26.1   Divisors used to generate the upper octave from the
master oscillator at 3872 kHz

Note	Divisor	True Scale Frequency (Hz)	Digital Frequency (Hz)	Error (cents)
C^8	925	4186.01	4186.0	0
B^7	980	3951.07	3951.1	0
$A^{\#7}$	1038	3729.31	3730.4	+0.5
A^7	1100	3520.00	3520.0	0
$G^{\#7}$	1165	3322.44	3323.5	+0.55
G^7	1235	3135.96	3135.3	−0.3
$F^{\#7}$	1306	2959.96	2960.3	+0.1
F^7	1386	2793.83	2793.6	−0.14
E^7	1468	2637.02	2637.0	0
$D^{\#7}$	1555	2489.02	2490.0	+0.5
D^7	1648	2349.32	2349.7	+0.2
$C^{\#7}$	1746	2217.46	2217.8	+0.2

Source: Douglas (1976).

The other system for tone generation by frequency dividers uses a single master oscillator at a very high frequency (1 to 4 MHz). This frequency is divided by a series of circuits with large divider ratios to generate the upper octave. Division by 2 to generate the lower octaves then proceeds as described above.

Deriving all the frequencies from a single crystal-controlled master oscillator has both advantages and disadvantages. Its main advantage is a saving in cost, and it also virtually eliminates the need for tuning. However, an instrument too precisely in tune tends to sound dry. The chorus effect resulting from combining the tones from independent oscillators is now lost.

26.6   VOICING AND KEYING

An organ tone may be characterized by the following properties:

1. *Attack:* (a) rate of attack; (b) change in harmonic structure during attack; (c) noise during attack (chiff, puff, jitter, etc.).

2. *Sustained tone:* (a) pitch; (b) harmonic structure; (c) pitch and amplitude vibrato (frequency and amplitude modulation); (d) noise.

3. *Decay:* (a) rate of decay; (b) change in harmonic structure during decay; (c) noise during decay.

To equal the sound of a fine pipe organ in an electronic instrument, careful attention must be paid to each of these properties. Some very ingenious circuits and techniques have been developed by various organ companies that accomplish this goal with varying degrees of success; space will allow us to describe but a few of them.

In an organ using additive synthesis, tone forming is accomplished by mixing together the various partials in the right proportion. This can be done by *drawbars,* as in some Hammond organs, or by *stop* switches.

Each of the nine drawbars shown in Fig. 26.6 sets the desired amplitude of one of the harmonics of a complex tone by tapping a particular winding on a transformer (the more turns, the greater the voltage; see Section 18.5). The subfundamental is an octave below the fundamental, and the subthird is an octave below the third.

More commonly used on organs are stop tablets. On an organ with tone synthesis by addition, each stop tablet selects a combination of harmonics that will give a particular sound. Stops are more convenient to use than are drawbars; they are also more familiar to organists because of their similarity to pipe organ stops.

In an organ with tone synthesis by subtraction, setting a stop switches in various filter circuits and mixer circuits to modify the wave

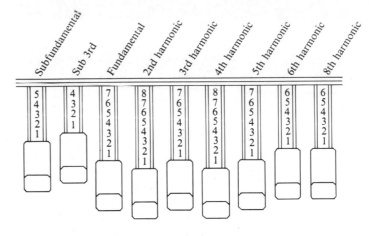

FIG. 26.6
The Hammond drawbar system for selecting partials of a tone.

shapes supplied by the tone generators. The designs of the filter circuits used tend to vary considerably from one builder to another. However, most of them are based on the basic filter types described in Section 19.1. Passing a square wave through a low-pass filter, for example, gives a rounded square wave with a clarinet-like sound, while passing it through a resonant *LC* filter produces a stringlike tone.

The function of the keys and pedals of an electronic organ is to close switches that will route the outputs of the selected tone generators to the mixing and tone-forming circuits that create the desired waveforms. In practice, this can be more difficult than it sounds. A simple switch creates rather prominent transients known as *key clicks*. To prevent the occurrence of these extraneous sounds, most organs have special keying circuits or devices that reduce undesired switching noises below audibility.

At the same time that undesired switching noises are eliminated, however, others may deliberately be added. Many electronic organs have "chiff" generators, which generate transients similar to those that occur during the onset of sound in organ pipes (especially flute pipes), which our ears are accustomed to hearing.

Keying circuits also shape the attack of each tone, including the rate of attack and changes in harmonic structure during the attack.

26.7 OTHER FEATURES OF ELECTRONIC ORGANS

Rotating speakers or sound reflectors are often used in organs to produce a type of vibrato by means of the Doppler effect (Section 3.7). The most common example is in the Leslie speaker, shown in Fig. 26.7, which has a rotating reflector facing the loudspeaker.

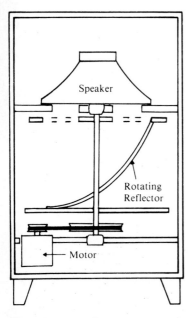

FIG. 26.7
Leslie speaker.

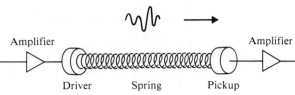

FIG. 26.8
A spring reverberation
unit. Generally, more than
one spring is used.

Many electronic organs provide artificial reverberation to compensate for the lack of reverberation, a common deficiency in small churches and auditoria. Reverberation is usually generated either by an electronic delay line (see Section 25.9) or by an electromechanical system consisting of springs on which mechanical waves can bounce back and forth (see Fig. 26.8). A magnetic transducer at one end of the spring acts as a driver, and a similar one at the other end acts as a pickup, so that the signal is repeated several times as it dies away.

Organs designed for home or entertainment use frequently include a rhythm generator, which can generate a wide variety of percussive sounds and tempos. A ring counter and an adjustable clock can be programmed to trigger the percussion sound generators in the desired rhythmic pattern and tempo.

An entirely different concept in electronic organ design is the digital computer organ now used by the Allen Organ Company and by Nippon Gakki (Yamaha) in Japan. The specifications for the various tones are stored in digital memories and processed in digital form before being converted to analog signals and sent to the output amplifiers and speaker. The Allen digital organ will be discussed further in Chapter 28, "Digital Computers and Digital Sound."

An interesting feature included on many electronic organs is the ability to transpose the entire keyboard up or down by some selected musical interval. Some Rodgers organs can transpose up or down by four semitones, for example, and the Allen and Baldwin organs can transpose up by five semitones or down by seven semitones. The large Rodgers organ in Carnegie Hall can tune from $A_4 = 435$–445 Hz to match the tuning of the various orchestras that play there (Bender, 1974).

It is quite clear that the electronic organ, which some early designers intended to imitate the pipe organ as nearly as possible, has some features that are not possible on a pipe organ. It has developed a personality of its own, and although it competes with pipe organs, it will probably never replace them. For a given cost, it usually offers a wider variety of sound than does the pipe organ, and it will undoubtedly continue to develop as the technology of electronic circuits develops.

Several attempts have been made to build combination pipe/electronic organs. Because of the expense of fabricating pipes for 32-foot

FIG. 26.9
Digital electronic piano. (Photo
courtesy of Yamaha)

and even 16-foot ranks, pipe organs occasionally have one or two
ranks of electronic stops. The Rodgers Organ Company and Ruffatti
Pipe Organ Builders of Italy developed a combination pipe/electronic
organ that they named the "Gemini." The Great division of the Gemini organ has 9 to 15 ranks of pipes, while the other divisions are
electronic. Several design features serve to make the pipe and electronic divisions compatible. The Ruffatti windchest is meant to respond rapidly in the manner of electronic tone generators, whereas
the electronic divisions of the organ have a special control to make
their pitch change with temperature as do organ pipes.

26.8   ELECTRONIC PIANOS

Electronic pianos produce pianolike sounds but substitute loudspeakers for the sounding boards of conventional pianos. The tones
may be produced in one of several ways.

One method of tone generation is to use magnetic pickups to sense
the vibrations of piano strings. Another is to strike rods or reeds in
order to produce percussive sounds that can be filtered to produce
pianolike tones. Still another method is to use electronic tone generators along with suitable keying circuits, which give the sound the fast
rise and slow decay of a piano.

Digital electronic pianos are digital synthesizers that produce piano tones of high quality; they will be discussed in Chapter 29. Most
of them rely on stored samples of real piano tones to "emulate" piano
sounds. Among the more successful piano emulators are instruments
by Kurzweil, Roland, and Yamaha.

Because of the their portability and ability to be amplified to any level of loudness, electronic pianos are widely used in popular music. They are also used in schools to teach piano playing, since the student can listen through earphones without disturbing others in the room.

26.9   ELECTRONIC CARILLONS

Because of the high cost of casting and tuning bronze bells, most carillons in recent years have used electromechanical tone generators. These are usually metal bars or rods, clamped at one end and carefully shaped to have bell-like overtones. One or more magnetic or electrostatic pickups are placed along the bar to generate electrical signals as the bar vibrates.

The frequencies of vibration for a bar clamped at one end have the ratios $1.0:6.27:17.55:34.39$, etc. These are somewhat greater than the corresponding ratios for a bar with free ends (see Section 13.1) and considerably greater than the frequencies of the overtones of a tuned carillon bell (see Section 13.12). However, if the lower two modes are overlooked, the mode frequencies are proportional to 5^2, 7^2, 9^2, 11^2, etc., which are in the ratios $1.0:1.96:3.64:4.84$, etc. By shaping the bar, these can be brought into the desired ratio to give a bell-like sound. If the bar is rectangular with a width-to-thickness ratio of 1.2, the tuning of the bar to bell-like modes is facilitated (Slaymaker, 1956).

26.10   AMPLIFICATION OF CONVENTIONAL INSTRUMENTS

Electronic amplification of conventional string, wind, and percussion instruments is becoming increasingly common. Sometimes the original sound is amplified with high fidelity; sometimes the sound is modified as well as amplified.

Three different methods are used to pick up sound from the instrument to be amplified. The first uses a pressure microphone. In the case of a trumpet or trombone, the microphone can be attached to the bell; in the case of an acoustic guitar, the sound hole is a convenient location. Woodwind instruments present a more difficult problem, however, since sound is radiated primarily from the open tone holes (see Section 12.2), and the best location for a microphone is inside the bore. A microphone can be inserted conveniently into the cork plug of a flute to create an "electric head joint," but the bore of a clarinet or saxophone is less accessible. Sometimes a small hole is bored just below the mouthpiece of a clarinet and a tiny microphone is inserted, a practice sometimes referred to as "bugging" the instrument.

A second method of picking up sound from an instrument is to attach a vibration pickup or contact microphone to the instrument. In the case of a violin or guitar, such a device can be attached conveniently to the bridge. Contact microphones have been attached to the outside of clarinet and saxophone mouthpieces and even to the reeds.

The third method is to use an electrostatic or electromagnetic pickup near a vibrating member. The most familiar example of this is the electric guitar pickup (see Section 10.14). Electrostatic pickups are similar in principle to the condenser microphone in that the varying capacitance due to vibration is used to generate an electrical signal.

One experimental electronic instrument that deserves special mention is the electronic violin, shown in Fig. 26.10, which was constructed at the Bell Telephone Laboratories (Mathews and Kohut, 1973). Metal strings vibrate in a magnetic field to produce the original signals, and a bank of resonant (band-pass) filters are used to simulate the resonances of a violin body. It was found that in order to achieve a rich tone quality approaching that of a violin, the resonance frequencies of the filters should be somewhat irregularly spaced and their response curves should have rather narrow peaks (large Q).

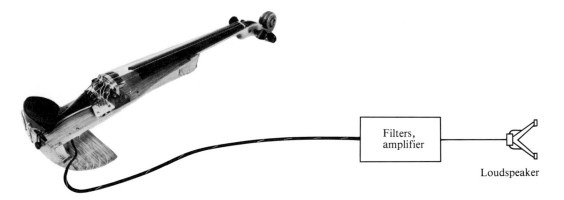

FIG. 26.10
Electronic violin with magnetic pickup and without a body. (From M. V. Mathews and J. Kohut (1973). "Electronic Simulation of Violin Resonances," *J. Acoust. Soc. Am.* **53**: 1620. Reprinted by permission from the Amer. Inst. of Physics.)

26.11   TONE-MODIFICATION CIRCUITS

A wide variety of tone-modifying circuits have been used to modify the sounds of both conventional and electronic musical instruments.

One form of electronic modification is the addition of *vibrato.* Electronic organs with separate oscillators for each note usually generate vibrato (or tremulant, as it is often called in organs) right at the source (see Figs. 26.4 and 26.5). In other elecronic instruments, vibrato is added as a modification.

Amplitude vibrato can be added by means of a voltage-controlled amplifier, whose gain is varied by application of an AC control voltage at the desired vibrato frequency. However, most musical vibratos combine both amplitude and pitch variations (see Section 7.13), and so a more pleasing vibrato is obtained by varying both. Although it is difficult to change the actual frequency of a tone, it is possible to modulate the phase to achieve a similar effect.

Controlled distortion of the electrical waveforms has been used to modify sound. One of the best known distorters is the *fuzz box*, which clips the waveform peaks and thus adds harmonic distortion. The resulting sound is described as "fuzzy" and conveys the feeling of loudness even at moderate sound levels. *Phasers* and *flangers* are rather sophisticated signal modifiers sometimes used to introduce a spacious or ethereal quality to recorded sounds. Flangers mix the signal with a time-delayed signal (the time delay also varies with time), whereas phasers mix the signal with a phase-shifted signal (Hartmann, 1978). For a pure tone these are equivalent, but not for a complex tone, because different harmonics will undergo different phase shifts when a signal is delayed in time.

Flanging can be done by means of a tape recorder with sound-on-sound capability. A signal is re-recorded almost in synchronism with the original signal (a suitable delay time is 25 milliseconds). Another simple method is to record the same material on two tape recorders and sum the outputs from the playback heads. If the second tape recorder is made to run slower than the first (by pressing a thumb on the flange of the tape reel, for example), then the signal from the second recorder is time-delayed with respect to the first. The two signals are then added together in a mixer (Hartmann, 1978). A phaser is a series of all-pass filters that have flat amplitude response but introduce a phase shift.

We discussed artificial reverberation in electronic organs in Section 26.7. Reverberation can also be added to other musical instruments to create special effects. Various combinations of filters can be used to modify the sounds of electronic and conventional instruments, just as they are used in the tone-forming circuits of electronic organs.

One type of tone modifier generates harmonics and subharmonics of the input tone, so that varying amplitudes of tone one or two octaves lower or higher than the played pitch will sound. Modifiers that perform several of these functions have included the Gibson Maestro, the Selmer Varitone, the Conn Multi-Vider, the Warwick Ampliphonic, and others.

26.12 SUMMARY

Early electronic musical instruments included the Teleharmonium, the Theremin, and others. Advances in electronics in the past thirty years have made sophisticated electronic organs possible. The electromechanical tone generators used in early organs have been replaced largely by electronic tone generators. Some organs synthesize tones by the addition of pure tones; others begin with complex tones and do subtractive synthesis by means of filters. Many organs derive all frequencies from a master oscillator by means of frequency dividers. Special features found on electronic organs include devices to produce artificial reverberation, rotating speakers, rhythm generators, and transposers.

References and Suggested Readings

Bender, W. (1974). "Carnegie Goes Electronic," *Time,* October 14, 1974, p. 83.

Hartmann, W. M. (1978). "Flanging and Phasers," *J. Audio Eng. Soc.* **26**: 439.

Mathews, M. V., and J. Kohut (1973). "Electronic Simulation of Violin Resonances," *J. Acoust. Soc. Am.* **53**: 1620.

Slaymaker, F. H. (1956). "Bells, Electronic Carillons, and Chimes," *IRE Trans. Audio* **AU-4**: 24.

Towers, T. D. (1976). *Electronics in Music,* London: Butterworth.

Trythall, G. (1973). *Principles and Practice of Electronic Music.* New York: Grosset and Dunlap.

Glossary

celeste (voix celeste) An organ stop that uses two different tones of slightly different frequency to produce beats.

drawbar A multiposition switch that selects the amount of a particular harmonic that will be used to synthesize a tone.

electromechanical generator A device that uses mechanical motion to produce an oscillating electrical tone.

flanging Mixing a signal with a time-delayed copy in order to produce a new signal.

flip-flop A bistable circuit that can be triggered from one state to the other by an electrical signal; extensively used as a frequency divider.

fuzz-box A device that adds harmonic distortion to a signal in order to create a fuzzy tone.

phaser A series of all-pass filters that have flat amplitude response but introduce a phase shift.

Theremin An electronic instrument in which pitch and loudness are controlled by the proximity of the player's hands to two antennae.

tremulant A term used to denote vibrato in organs; it may refer specifically to pitch vibrato or to both pitch and amplitude vibrato.

vibrato Periodic variation in one or more characteristics of a musical sound, such as frequency, amplitude, and phase. (The terms "pitch vibrato," "amplitude vibrato," etc., are used to describe variations in one of the individual characteristics of a sound.)

Questions for Discussion

1. What are some possible advantages of tone generation by addition and subtraction?

2. Why does the use of multiple output channels, each with its own loudspeaker, decrease intermodulation distortion (see Section 22.10)?

3. Why does the Theremin use the difference tone between two high-frequency oscillators as a tone generator rather than, say, using the capacitance of the hand to control the frequency of an audio-frequency oscillator directly?

4. Compare the oscillator circuit in Fig. 26.4 to the basic oscillator circuit in Fig. 19.20. Trace the path of the feedback in each case.

5. Many reverberation units use two or more springs with different lengths to spread out the resonances. Can you explain how this works?

6. How is the sound picked up inside the bore of a flute different from the sound that is radiated to the room? (See Section 12.2.)

7. A clarinet tone is played through a modifier that shifts the pitch down one octave. Will the tone sound substantially different from a tone played an octave lower on the same instrument? Explain.

Problems

1. Show that magnetic tone wheels with 196 and 185 teeth, rotating at the same speed, generate tones approximately one semitone different in pitch.

2. Express 3872 kHz in Hz and MHz.

3. A 100-Hz square wave is time-delayed 25 ms. Calculate the phase shift of the fundamental, the third, and the fifth harmonics.

4. Suppose that a small loudspeaker is mounted on a phonograph turntable so that its center moves in a circle of 10-cm radius at 33⅓ r.p.m.

   a) What is its speed as it moves around the circle?

   b) What is the maximum shift in frequency (f/f_s) due to the Doppler effect?

   c) To what musical interval does this correspond?

   d) Would it be noticeable to a listener?

CHAPTER 27
Electronic Music Synthesizers

Electronic music synthesizers create complex sounds by generating, altering, and mixing various electrical waveforms. Theoretically, any musical or nonmusical sound can be synthesized electronically, but in a given synthesizer, the number of electronic circuit "building blocks" available (plus, of course, the skill and imagination of the programmer) determines the range of sounds that can be synthesized.

With respect to the type of circuits used to generate sound, synthesizers can be classified as analog, digital, or hybrid (i.e., employing both analog and digital circuits). In this chapter, we will discuss analog synthesizers, while digital synthesizers will be discussed in Chapter 29. Although more and more synthesizers employ digital circuitry exclusively, some familiarity with analog synthesizers aids in understanding how digital synthesizers function.

Although several of the experiments with electronic musical instruments that we described in Chapter 26 date back to the beginning of this century, the rapid growth in electronic music occurred after the development of electronic music synthesizers in the 1950s and 1960s. We will begin with a brief history of their development.

27.1 THE DEVELOPMENT OF SYNTHESIZERS

An electronic organ is an electronic synthesizer with a rather limited program in that it generates only discrete pitches and a limited number of harmonically related overtones. Its great advantages, of course, are that many notes can be executed simultaneously and that it can be very easily "programmed" by changing the stops, so that it can be used for live performances. (Most electronic synthesizers, on the other hand, are not designed for live performance of music.)

The RCA Music Synthesizer, designed by Harry Olson and Herbert Belar at the RCA Laboratories in Princeton, New Jersey, was completed in 1955 after about six years of construction. It was designed to produce tones "with any frequency, intensity, growth, duration, decay, portamento, timbre, vibrato, and variation" (Olson and Belar, 1955). This landmark instrument is shown in Fig. 27.1.

A modified and expanded version, the Mark II, was built in 1959 for the Columbia-Princeton Music Center (Olson, 1967). Seven racks full of oscillators, relays, modulators, amplifiers, and so on, were controlled by a 40-channel paper-tape reader. This large synthesizer was used by Milton Babbitt, Vladimir Ussachevsky, Otto Luening, and others who pioneered in the composition of electronic music. Unfortunately, instruments of this magnitude and complexity were not available to most composers.

In Köln (Cologne), Germany, an electronic music studio was completed in 1953 by composer Herbert Eimert, physicist Dr. Meyer-Eppler, and their associates. Composer Karlheinz Stockhausen used the facilities of this studio for several of his pioneering works.

The "Model-T Ford" of music synthesizers, which put electronic music into the hands of many composers and performers, was the first Moog (rhymes with "vogue") synthesizer, developed by Robert A. Moog in the 1960s. He became interested in electronic music while he was a graduate student in physics at Cornell University, where he built and sold Theremins in his spare time. Moog pioneered in the use of voltage-controlled modules, which are used in most modern analog synthesizers. Many listeners were introduced to synthesized music by hearing the best-selling record "Switched-on Bach," performed by Walter Carlos on a custom-built Moog synthesizer.

FIG. 27.1
RCA Music Synthesizer: schematic diagram. (From H. F. Olson and H. Belar (1955). "Electronic Music Synthesizer," *J. Acoust. Soc. Am.* **27**: 595. Reprinted by permission from the Amer. Inst. of Physics.)

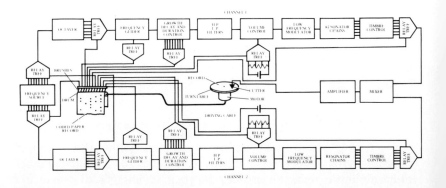

 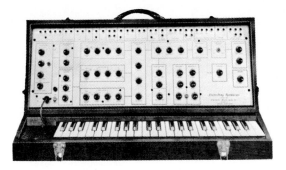

FIG. 27.2
Two early synthesizers:
(a) Mini Moog Sonic Six.
(b) Electrocomp Model 101.
(Photograph (a) courtesy of Moog
Music, Inc.; (b) courtesy of Electronic
Music Laboratories, Inc.).

About the same time, across the country at the San Francisco Tape Music Center, Donald Buchla was working closely with composers Morton Subotnik and Ramon Sender to develop the first Buchla Electric Music Box. Buchla pioneered in the use of sequencers for repeating patterns of sound, and in the use of touch plates for controlling pitch.

The Electronic Music Studios (EMS) of London were created in 1970 through a partnership between composer Peter Zinovieff and engineer David Corkerell. The EMS Synthi AKS, with a 256-event sequencer and other innovations, became one of the first successful portable synthesizers.

Widespread use of synthesizers for live performance led to the widespread development of portable synthesizers with many of the features of the more flexible studio units. Portable synthesizers generally use internal interconnections (patches) between modules, with switches or sliders to select the most widely used patch configurations. Among the more popular synthesizers have been the ARP Odyssey, the Mini-Moog and Sonic Six, the Ionic Performer, the Synthi AKS, and the Electrocomp 101, two of which are shown in Fig. 27.2.

27.2   VOLTAGE-CONTROLLED OSCILLATORS
AND AMPLIFIERS

Since the most important modules in an analog synthesizer are voltage controlled, it is important to understand this principle. In any oscillator, filter, or amplifier that is voltage controlled, *applying a voltage to the control input will have exactly the same effect as adjustment of the corresponding manual control.* This principle of voltage control makes it possible for one module to control the function of another. For example, the output of one oscillator may be used to control the frequency of another oscillator to produce frequency modulation.

A *voltage-controlled oscillator (VCO)* generates a variety of waveforms with a frequency that is controlled by (a) a manual control, (b)

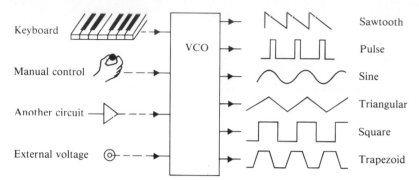

FIG. 27.3
Schematic of voltage-controlled oscillator (VCO) showing various sources of control voltages and various output waveforms.

the keyboard, (c) a control voltage from another circuit, (d) an external control voltage, or (e) any combination of these. Nearly all VCOs produce sawtooth and pulse waveforms; many also provide sine waves (pure tones), triangular waves, square waves, and trapezoidal waves (see Fig. 27.3).

An interconnection between two synthesizer circuits is commonly called a "patch," regardless of whether the connections are made externally with patch cords or internally with switches. One of the simplest patches (and a fairly useful one) is shown in Fig. 27.4. The output of one VCO is used to control the frequency of another VCO, resulting in frequency modulation. In practice, the keyboard would probably be patched to VCO 2 and the amplitude of VCO 1 would be set manually to obtain a vibrato with the desired rate and depth (see Section 7.15).

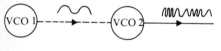

FIG. 27.4
Simple patch for generating a frequency-modulated (FM) signal: VCO 1 controls the frequency of VCO 2.

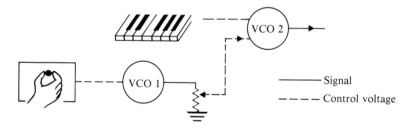

FIG. 27.5
Patch for obtaining a tone with frequency vibrato.

Using a low-frequency sawtooth as a control voltage provides a steadily increasing (or decreasing) frequency or glissando, as shown in Fig. 27.6. The patch is the same as that shown in Fig. 27.5, and the square-wave output from VCO 2 is shown.

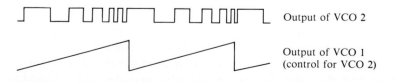

FIG. 27.6
A sawtooth control voltage to VCO 2 provides a square wave that steadily increases in frequency.

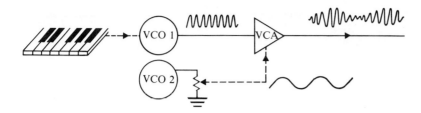

FIG. 27.7
Patch for obtaining amplitude
modulation (AM) by using a VCO to
control a VCA.

A *voltage-controlled amplifier (VCA)* has a gain that is proportional to the control voltage, which can come from any combination of the same sources for VCOs, shown in Fig. 27.3. A voltage-controlled amplifier modulates the amplitude of the input signal. Thus it can be used to provide an amplitude-modulated waveform, as shown in Fig. 27.7.

> *Note:* In all the schematic diagrams a dashed line (- - -) carries a control voltage, whereas a solid line (——) is a signal path.

Most musical instruments (and the singing voice as well) have a vibrato in which both the frequency and amplitude vary together. A signal of this type can be generated by using the low-frequency oscillator (LFO) with a limited frequency range, which serves as a source of modulation voltage. In Fig. 27.8, an LFO is used to control both a VCO and a VCA. Proper choice of frequency and amplitude will generate a pleasing vibrato that includes both AM and FM.

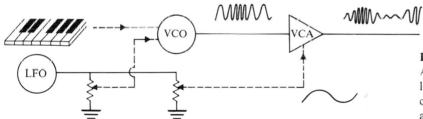

FIG. 27.8
A musical vibrato generated by using a low-frequency oscillator (LFO) to control both the frequency of a VCO and the gain of a VCA.

One of the most common functions of a voltage-controlled amplifier is to control the *envelope* of the tone, which includes the attack and decay plus variation in amplitude during its duration. For this purpose, the control voltage to the VCA is often supplied by an *envelope generator*. Envelope generators are often called attack-release (AR) generators or attack-decay-sustain-release (ADSR) generators.

An AR-type envelope generator is shown in Fig. 27.9(a). The attack rate and release rate can be controlled separately. An ADSR-type

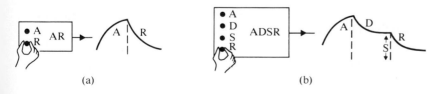

(a)   (b)

FIG. 27.9
Envelope generators: (a) attack-release (AR) generator; (b) attack-delay-sustain-release (ADSR) generator.

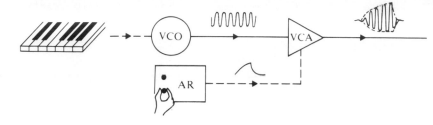

FIG. 27.10
Control of tone envelope by means of an envelope generator (AR) and a voltage-controlled amplifier (VCA).

envelope generator, with four selectable parameters, is shown in Fig. 27.9(b). "Decay" (D) refers to the initial decay to a sustain level (S), whereas "release" (R) refers to the final decay. A common patch that uses a VCA and an AR-type envelope generator to control the amplitude of a tone generated by a VCO is shown in Fig. 27.10.

27.3 VOLTAGE-CONTROLLED FILTERS

A *voltage-controlled filter (VCF)* may be used to vary the spectrum of a tone and thus its timbre. In Chapter 19, we discussed high-pass, low-pass, band-pass, and band-reject filters. Nearly all synthesizers have a voltage-controlled low-pass filter whose cutoff frequency f_c is determined by the control voltage. Larger synthesizers include multimode filters, which can perform any of the four basic filter functions mentioned above.

The way in which a voltage-controlled low-pass filter controls timbre is illustrated in Fig. 27.11 using the overtones of C_2 ($f = 131$ Hz). A sawtooth waveform used as a control voltage raises the cutoff frequency f_c so that the overtones appear one by one in the output. The fundamental ($f = 131$ Hz) always appears in the output, because it is below f_c.

FIG. 27.11
Change in timbre of a note by means of a voltage-controlled low-pass filter. The tone illustrated consists of seven harmonics of C_3 ($f = 131$ Hz). Note that each time the cutoff frequency F_c increases by 130 Hz, a new harmonic is added (the fundamental, present each time, is not shown).

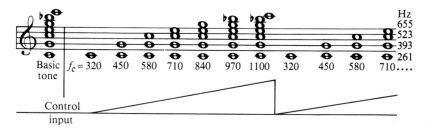

The control voltage to a VCF is often supplied by an envelope generator. In most musical instruments, the spectrum of the sound changes markedly during the attack. The patch shown in Fig. 27.12 uses one envelope generator (ADSR) to control the amplitude of the tone and a second one (AR) to control the spectrum.

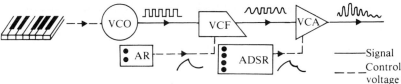

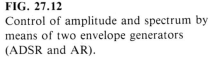

FIG. 27.12
Control of amplitude and spectrum by means of two envelope generators (ADSR and AR).

Voltage-controlled filters frequently incorporate a *resonator* circuit, which emphasizes components of the tone at or near the cutoff frequency f_c (see Fig. 27.13). The sharpness of the resonance is denoted by the parameter Q (Q is f_c divided by the frequency width of the resonant band). In some voltage-controlled filters, the gain at f_c is increased as Q is increased; in others, the gain at f_c remains the same, but the gain at other frequencies is reduced (this latter arrangement is called the *limit mode*).

The resonance feature of low-pass filters provides a very useful means of modifying sound waveforms. In the normal mode, if Q is made large enough, the filter will oscillate at the frequency f_c even without an input. In many synthesizers, a purer sine-wave tone can be generated this way than is possible with the voltage-controlled oscillators. Some synthesizers allow voltage control of both f_c and Q; in others, Q is controlled manually only. A control voltage to the Q input, if provided, allows the composer–performer to program the filter in and out of oscillation electrically.

Even more interesting are the effects obtained when Q is set just below the point at which self-sustaining oscillation begins. At this setting, any sharp pulse will cause the filter-resonator to "ring" at its resonance frequency f_c (that is, it begins to oscillate, but the oscillations die out rapidly). The duration of the ringing is controlled by Q; the higher the Q setting, the longer it lasts. The effects of various settings of Q are illustrated in Fig. 27.14 for a VCF with a sharply rising rectangular pulse applied.

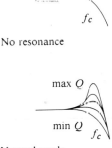

No resonance

Normal mode

Limit mode

FIG. 27.13
Characteristics of voltage-controlled low-pass filter with and without resonance.

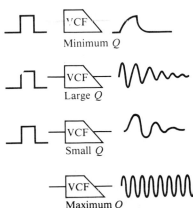

Minimum Q

Large Q

Small Q

Maximum Q

FIG. 27.14
Response of a voltage-controlled low-pass filter (VCF) for various Q settings. When Q is at minimum, frequencies above f_c are filtered out. As Q increases, the filter begins to ring, and finally goes into spontaneous oscillation.

27.4   CONTROL CHARACTERISTICS AND SOURCES

Control characteristics of synthesizer modules may be either *linear* or *exponential* (inverse-logarithmic). This means that as the control voltage changes on a linear scale, the parameter being controlled (frequency, amplitude, or whatever) will vary on a linear or logarithmic scale, as shown in Fig. 5.12.

The two different characteristics can be illustrated by first considering a voltage-controlled amplifier. If the control characteristic is linear, doubling the control voltage will double the amplifier gain; if it is exponential, increasing the control voltage by one volt will increase the gain by a factor of 3.2 (or, in other words, by 10 dB). This is shown in Table 27.1. Voltage-controlled amplifiers may have either linear or exponential control characteristics or both.

Figure 27.15(a) shows how linear and exponential control characteristics for a VCA appear when drawn on ordinary linear graph paper; Fig. 27.15(b) shows the same two characteristics drawn on semilogarithmic graph paper, similar to that in Fig. 5.13. Note that in Fig. 27.15(a), the linear characteristic is a straight line, whereas in Fig. 27.15(b), the exponential characteristic is a straight line.

In voltage-controlled oscillators, the exponential characteristic is nearly always used. A typical VCO control characteristic is one octave change in pitch per one volt change in the control voltage. A graph of control characteristics for a VCO, drawn as in Fig. 27.15, would have frequency on the vertical axis.

FIG. 27.15
Linear and exponential control characteristics for a voltage-controlled amplifier shown on (a) a linear graph; (b) a semilogarithmic graph.

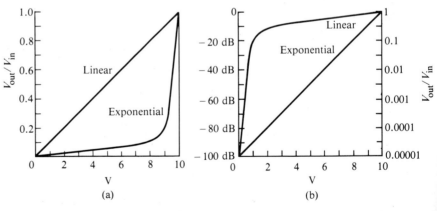

27.5   THE KEYBOARD AND OTHER CONTROL VOLTAGE SOURCES

One of the most important sources of control voltage for synthesizer modules is the *keyboard controller.* Nearly all synthesizers have a keyboard, and it provides a familiar and convenient way for musicians

TABLE 27.1   Control characteristics

Control voltage	Amplifier gain if control characteristic is		
	Linear	Exponential	
10	1.0	1.0	= 0 dB
9	0.9	0.3	= −10 dB
8	0.8	0.1	= −20 dB
7	0.7	0.03	= −30 dB
6	0.6	0.01	= −40 dB
5	0.5	0.003	= −50 dB
4	0.4	0.001	= −60 dB
3	0.3	0.0003	= −70 dB
2	0.2	0.0001	= −80 dB
1	0.1	0.00003	= −90 dB
0	0	0.00001	= −100 dB

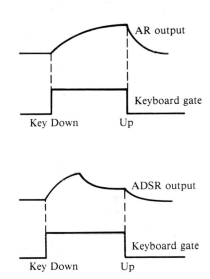

FIG. 27.16
Timing of envelope generators by keyboard gate signal.

to control various synthesizer parameters, especially the frequency of the voltage-controlled oscillators or the cutoff frequency of the voltage-controlled filters. The keyboard control voltage is usually available for patching to other modules as well.

In addition to supplying a control voltage, the keyboard controller usually supplies gate signals and trigger signals as well. A *gate* signal is a change in voltage that lasts as long as any key on the keyboard is depressed, then returns to the original voltage when all keys are released. A *trigger* signal, on the other hand, is a short voltage pulse that occurs when any key is depressed. Gate and trigger signals can perform many functions, one of which is to control the timing of events in the envelope generators, as shown in Fig. 27.16.

Most keyboards have a *portamento* or glissando control, which determines the rate at which the keyboard control voltage can change. With maximum portamento, the pitch of the oscillators controlled by the keyboard will "slide" from one note to another. A *pitch-bend* or *tuning* control allows one to increase or decrease the keyboard voltage a small amount (which will raise or lower the pitch of a VCO).

The keyboard control voltage is determined, of course, by what key is depressed. If the control characteristic of the VCO is one volt per octave, then each semitone step on the keyboard will normally change the keyboard control voltage by one-twelfth volt. Some keyboards, however, have provision for generating "microtones" by varying the size of the keyboard steps above and below normal. Keyboard control, gate, and trigger signals are shown in Fig. 27.17.

A *slidewire* or *ribbon controller* performs the same function as a keyboard controller, but allows the control voltage to be varied continuously rather than in steps. A *joystick controller* simultaneously generates two different control voltages as it is moved in two perpendicular directions, as shown in Fig. 27.18. The two control voltages

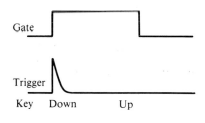

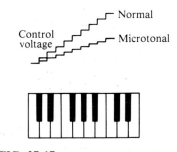

FIG. 27.17
Keyboard control, gate, and trigger signals.

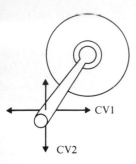

FIG. 27.18
Joystick controller. Moving the joystick in one direction varies one control voltage (CV 1), and moving it in a perpendicular direction controls CV 2.

CV 1 and CV 2 can be used to control two different modules. One way to use a joystick would be to control the pitch of a tone with CV 1 and the timbre with CV 2.

The *sample and hold* circuit generates voltage steps by sampling any signal fed into it each time a trigger is applied. The triggers may be supplied at a regular rate by an oscillator or on command from the keyboard. Sampling a sawtooth produces a voltage that ascends in steps, as shown in the second example in Fig. 27.19.

A *sequencer* supplies a repeating pattern of control voltages and triggers that can be used to generate a sequence of tones at a selected rate.

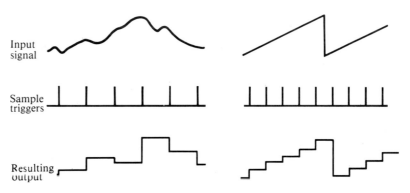

FIG. 27.19
Examples of control voltages produced by a sample and hold circuit.

27.6   RING MODULATORS

When a pure tone with a frequency f_1 is used to modulate another pure tone with a frequency f_2, the spectrum of the resulting tone shows sidebands at the sum and difference frequencies $f_2 + f_1$ and $f_2 - f_1$, as shown in Fig. 8.14. In a *ring modulator* or balanced modulator, the original frequencies f_1 and f_2 are suppressed, so only the sidebands appear in the output.

If the inputs to the ring modulator are complex tones with many harmonics, a rich mixture of inharmonic sum and difference tones results, and a large variety of sounds is possible. Another way of describing a ring modulator is that its output $(A \times B)/N$ is the product of two inputs A and B divided by some number N.

Some modulators can be operated either as a ring (balanced) modulator or as an unbalanced modulator with some of the original signals of frequencies f_2 and f_1 (as well as the sidebands $f_1 \pm f_2$) appearing in the output.

At one time balanced modulators were mainly constructed using a "ring" of four diodes, and that is why they are often called ring modulators. In most synthesizers in current use, however, balanced modulators employ integrated circuits called *multipliers,* either of analog or digital type.

Analog Multipliers

Like operational amplifiers (Section 19.5), *analog multipliers* are a useful product of integrated circuit technology. An ideal multiplier gives an output signal C which is proportional to the product of two input signals A and B (see Fig. 27.20):

$$C = k\, A \times B$$

The proportionality constant k is usually fixed at the time of manufacture, but in some multipliers it can be varied by supplying a control voltage at one of its terminals.

Multipliers are often classified according to the allowed polarities of the input and output signals. A multiplier that handles only positive inputs is called a *one-quadrant* multiplier, because the output would always lie in quadrant I in Fig. 27.21. In a *two-quadrant* multiplier, one of the input terminals accepts either positive or negative signals. A *four-quadrant* multiplier accepts bipolar (+ or −) signals on both inputs, and generates a bipolar output with the correct sign (positive if both inputs are positive or if both inputs are negative).

A four-quadrant multiplier acts as a balanced modulator, since the output is proportional to the product of the two inputs, as in Fig. 27.20. On the other hand, amplitude modulation (see Section 22.5) requires only a two-quadrant multiplier. If signal A (the "carrier") is to be modulated by signal B (the "modulator"), signal B would normally be combined with a DC bias voltage, as shown in Fig. 27.22.

A two-quadrant multiplier can also serve as a voltage-controlled amplifier (VCA). The proportionality constant k is adjusted so that the amplifier gain kB equals one when B reaches its maximum value.

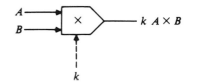

FIG. 27.20
A multiplier gives an output signal C that is proportional to the product of the input signals A and B. In some circuits, the proportionality constant k is controlled by an input voltage at one of the terminals.

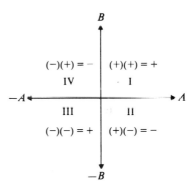

FIG. 27.21
Graph showing the polarity of the output signal of a multiplier circuit in each quadrant when the input signals A and B take on + or − values.

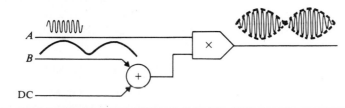

FIG. 27.22
Amplitude modulation with a two-quadrant multiplier ($\times$) plus an adder (+).

27.7   OTHER MODULES

Nearly all synthesizers have a *noise generator,* which produces a random pattern of voltage. If all frequencies are equally probable, the pattern is called *white noise,* an analogy with white light, which is a mixture of all the colors. If white noise is partially filtered so as to stress the low frequencies, the result is *pink noise.* White noise sounds like steam escaping from a radiator; pink noise sounds more like the roar of a waterfall.

The patch shown in Fig. 27.23 will generate a completely random melody. The sample-and-hold (SH) circuit samples white noise at a rate determined by the low-frequency oscillator (LFO). The resulting random assortment of voltages is used to control the frequency of a voltage-controlled oscillator (VCO).

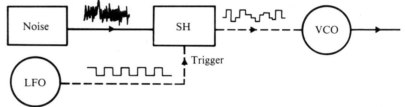

FIG. 27.23
Sampling white noise at regular intervals to generate a random melody.

Reverberation may be added with a digital delay or with a spring reverberation unit, as described in Section 26.7.

Most synthesizers have a *microphone input,* so that sound from musical instruments and other external signals can be processed in the synthesizer, modified, and perhaps combined with synthesizer-generated waveforms. For example, a signal from a musical instrument mixed with the output of a VCO in the ring modulator can produce a variety of unusual effects.

An *envelope follower* generates a control voltage proportional to the average signal input. It can be used, for example, to generate an envelope for use inside the synthesizer that duplicates the envelope of an external musical instrument. This envelope can be used internally to control a VCO, VCA, VCF, or other module.

We have by no means mentioned all the modules that are used in electronic music synthesizers. Besides the basic ones that are common to most synthesizers, each manufacturer adds some unusual features. New innovations are regularly appearing. It is a fertile field for the inventor as well as for the composer and performer.

27.8   THE ORGANIZATION OF A SYNTHESIZER

Although each manufacturer of synthesizers uses some unique circuits, there is a great deal of similarity in the characteristics and design of the basic modules or building blocks. This is fortunate, for it allows the composer or performer to develop a style on one synthesizer and adapt it quite easily to another.

The various building blocks fulfill the following functions:

1. Input and control,

2. Wave sources,

3. Wave modifiers, and

4. Mixing and output.

Together they constitute a sound "assembly line" somewhat as shown in Fig. 27.24.

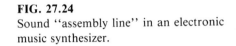

FIG. 27.24
Sound "assembly line" in an electronic music synthesizer.

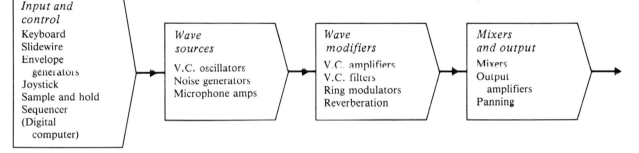

Interconnecting the modules, or patching, may be done either internally or externally. External patching is done with patch cords, which are plugged into jacks much in the manner of an old-fashioned telephone switchboard. External patching offers the composer maximum flexibility, because it allows virtually any input to be connected to any output. On the other hand, programming with patch cords is relatively slow and often leads to a maze of wires. To avoid confusion and possible error, most composers use patch cords of different colors or attach colored tapes to the plugs at each end of a patch cord. Figure 27.25 shows a synthesizer programmed with patch cords.

Internal programming or patching may be done with switches, slider pots, matrix pins, or a combination of these. Internal programming restricts the number of patches that can be made, but it greatly reduces the programming time and makes the use of synthesizers in live performance practical.

Figure 27.26 shows a portable synthesizer, which is convenient for live performances. The instrument is programmed by means of 22 switches and 34 slider potentiometers. Figure 27.27 shows the settings for a patch that will produce a saxophonelike sound. In this patch, the ADSR envelope generator is used to vary the width of pulses from VCO 2 (to give a "reedy" sound) as well as to control the VCF. The LFO (low-frequency oscillator) varies the frequency of VCO 2 to pro-

FIG. 27.25
A synthesizer programmed externally with patch cords. (Courtesy of Moog Music, Inc.)

FIG. 27.26
A portable synthesizer that is
programmed by switches and slider
potentiometers. (Courtesy of ARP
Instruments, Inc.)

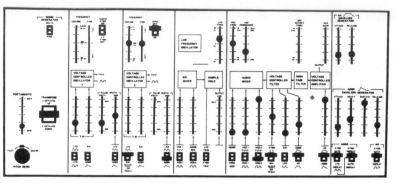

FIG. 27.27
A patch that will produce a saxophonelike sound on the synthesizer in
Fig. 27.26 (see the text for an explanation of the settings).

duce vibrato, and the AR envelope generator controls the VCA. A
small amount of resonance is added to the VCF.

27.9   COMPUTER-CONTROLLED SYNTHESIZERS

An interesting alternative to programming the synthesizer by patching
and by input from the keyboard or a slide wire is the use of a digital
computer to generate the desired control voltages. Such an arrangement is termed a *hybrid* synthesizer, in that it combines some desirable
features of both digital and analog systems. With a hybrid synthesizer,
the composer can write music in the form of a series of commands;
these are processed by the digital computer and converted into the
various control voltages needed to operate the voltage-controlled
modules that generate and process the desired waveforms. Digital control of synthesizers will be discussed further in Chapter 29.

Figure 27.28 illustrates two early examples of computer-controlled synthesizers. Figure 27.28(a) is a large studio synthesizer con-

FIG. 27.28
Computer-controlled or "hybrid"
synthesizer: (a) a large studio
synthesizer controlled by a small digital
computer; (b) an inexpensive portable
synthesizer controlled by a
microprocessor. (Part (a) courtesy of
Electronic Music Studio; part (b)
courtesy of PAIA Electronics.)

(a)

(b)

trolled by a small digital computer. Figure 27.28(b) is a small inexpensive synthesizer, sold as a kit, which is controlled by a microprocessor.

27.10   FM MUSIC SYNTHESIS

An entirely different type of music synthesis, using frequency modulation (FM), was invented by John Chowning and his colleagues at Stanford University (Chowning, 1973). FM synthesis is based on a simple principle: frequency modulation produces sidebands whose frequencies are spaced in integral multiples of the modulation frequency (see Section 22.5). With relatively simple circuits, it is possible to produce a large number of tones by FM synthesis.

Mathematically, we can describe frequency modulation as follows: When a carrier signal with frequency f_c and amplitude A is frequency modulated by a signal with frequency f_m, the resulting signal is described by:

$$y = A\pi\sin(2\pi f_c + I\sin 2\pi f_m t) \qquad (27.1)$$

$I = \dfrac{\Delta f}{f_m}$ is called the *modulation index*.

The spectrum of the modulated signal consists of the *carrier* with frequency f_c and *sidebands* with frequencies $f_c \pm n\Delta f$ ($\ldots f_c - 2\Delta f,\ f_c - \Delta f,\ f_c,\ f_c + \Delta f,\ f_c + 2\Delta f,\ \ldots$). The amplitudes of the sidebands depend on the modulation index I. As I increases, more energy is "stolen" from the carrier and distributed among an increasing number of sidebands.

The amplitudes of the sidebands are proportional to mathematical functions called Bessel functions $J_n(I)$ of order n.

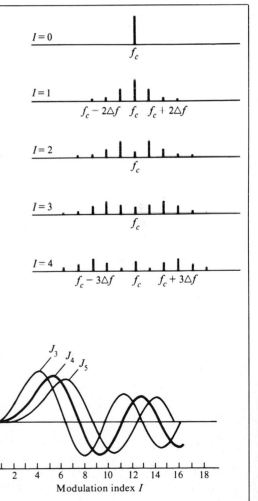

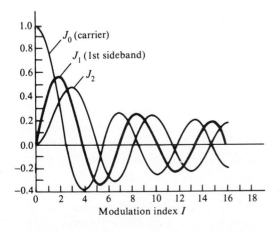

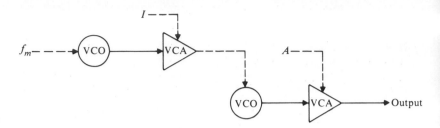

FIG. 27.29
Simple arrangement for FM synthesis.

Realization of FM synthesis can be accomplished with two VCOs and two VCAs, as shown in Fig. 27.29. The modulation frequency f_c is generated by the first VCO, whose output signal, scaled by the modulation index I, is used to modulate the frequency of the second VCO. As the modulation index I increases, more and more energy appears in an increasing number of sidebands. FM synthesis is frequently incorporated into digital synthesizers (see Section 28.5).

When the carrier frequency and the modulation frequency are the same, the sidebands will be a series of harmonics of this frequency. Tones with inharmonic partials, on the other hand, result from making the ratio of the modulation frequency to the carrier frequency an irrational number.

Although a reasonable assortment of sidebands can be generated by modulating a sinewave (single frequency) carrier by another sinewave, the use of complex waveforms for the carrier, the modulator, or both, could increase the possibilities for tone generation. Examples of *complex* FM synthesis include (Chowning and Bristow, 1986):

1. One modulator operating on parallel carriers;

2. One carrier with parallel modulators;

3. One carrier with cascaded modulators;

4. Multiple modulators, multiple carriers;

5. Feedback FM.

27.11 SUMMARY

Electronic music synthesizers create complex sounds by generating, altering, and mixing various electrical waveforms. They incorporate voltage-controlled modules (oscillators, amplifiers, filters, etc.), which can control or modify each other according to the way in which they are interconnected, or patched. Patching may be done externally with patch cords or internally with switches, slider pots, or pins in a matrix board. In hybrid computers, a small digital computer or microprocessor supplies control voltages for the various modules.

References

Chowning, J. M. (1973). "The Synthesis of Complex Audio Spectra by Means of Frequency Modulation," *J. Audio Eng. Soc.* **21**: 526.

Chowning, J., and D. Bristow (1986). *FM Theory and Applications.* Tokyo: Yamaha Music Foundation.

Hartmann, W. M. (1975). "The Electronic Music Synthesizer and the Physics of Music," *Am. J. Physics* **43**: 755.

Howe, H., Jr. (1975). *Electronic Music Synthesis.* New York: Norton.

Hutchins, B. (1975). *Musical Engineer's Handbook.* Ithaca, N.Y.: Electronotes.

Olson, H. F. (1967). *Music, Physics and Engineering,* 2nd ed. New York: Dover.

Olson, H. F., and H. Belar (1955). "Electronic Music Synthesizer," *J. Acoust. Soc. Am.* **27**: 595.

Towers, T. D. (1976). *Electronics in Music.* London: Butterworth.

Trythall, G. (1973). *Principles and Practice of Electronic Music.* New York: Grosset and Dunlap.

Glossary

ADSR, AR (envelope) generators A, D, S, and R refer to the parameters of an envelope: A = attack time; D = initial decay time; S = sustain level; and R = final release time.

envelope The manner in which amplitude varies with time; the envelope determines the attack and decay of a tone, among other things.

envelope follower A circuit that generates a control voltage to duplicate the envelope of an acoustical musical instrument or real sound.

exponential A function or characteristic that varies as e^y. In a typical exponential control characteristic, a change of one volt in the control voltage doubles some parameter such as frequency.

FM synthesis Generation of musical sound by frequency modulation of one signal by another.

hybrid synthesizer Use of a digital computer to control the analog circuitry in a synthesizer; also a synthesizer that combines digital and analog generators.

joystick controller Two potentiometers linked to a handle (resembling an airplane joystick) in such a way that horizontal motion varies one control voltage and vertical motion the other.

LFO (low-frequency oscillator) An oscillator used primarily to supply a time-varying control voltage.

multiplier An electronic circuit whose output is the product of two input signals.

musique concrète A term applied to music created by altering real sounds electronically.

operational amplifier ("op-amp") A high-gain amplifier generally used with a very large amount of negative feedback.

panning Applying a set of output signals to a series of speakers sequentially so that the sound appears to move in space.

patch (Verb): To interconnect; or (noun): a set of interconnections that programs a synthesizer to produce a certain sound.

pink noise Random noise that has been filtered so that the energy contained in each octave band is the same (it has more low-frequency strength than does "white" noise).

pitch bend A control that raises or lowers the keyboard output voltage in order to tune the voltage-controlled oscillators.

portamento Sliding from one note to another rather than changing the pitch abruptly.

Q A parameter that specifies the sharpness of a resonance (specifically, it is a ratio of energy stored to energy dissipated per cycle).

ring modulator A device that generates sum (f_2+f_1) and difference (f_2-f_1) frequencies from two signals of frequency f_1 and f_2 (similar to the balanced modulators used in communications circuits).

sample-and-hold circuit A circuit that, on command, "samples" a waveform and "holds" that voltage until it receives another command.

sequencer A circuit that switches in a series of predetermined voltages in sequence.

sidebands Sum and difference frequencies created in a modulator.

synthesizer An instrument that creates complex sounds by generating, altering, and combining various electrical waveforms, generally by means of voltage-controlled modules.

VCA (voltage-controlled amplifier) An amplifier whose gain varies with the control voltage.

VCF (voltage-controlled filter) A filter (usually low-pass) whose cutoff frequency varies with the control voltage.

VCO (voltage-coupled oscillator) An oscillator whose frequency varies with the control voltage.

Questions for Discussion

1. A clarinet tone has strong odd-numbered harmonics and weak even-numbered ones (see Chapter 12). To synthesize a clarinet sound, would you be more likely to start with a sine wave, a square wave, or a sawtooth wave?

2. What is the difference between white noise and pink noise?

3. What element in the oscillator shown in Fig. 27.29 corresponds to the bucket shown in Fig. 27.30?

4. What is the difference in the outputs of a ring (balanced) modulator and an unbalanced modulator?

Problems

1. The control characteristic of a certain VCO is given as one volt per octave.

   a) Is this a linear or exponential characteristic?

   b) If a control voltage of three volts causes it to oscillate at the frequency of C_3 ($f = 131$ Hz), what voltage will be required to give G_3 ($f = 196$ Hz), a fifth higher?

2. Show by a sketch similar to that shown in Fig. 27.12 what would happen if the control input were a square wave of sufficient amplitude to change f_c from 320 to 640 Hz. If the control characteristic were one volt/octave, what peak-to-peak amplitude of square wave would be required to do this?

3. If two waveforms with frequencies $f_1 = 200$ Hz and $f_2 = 350$ Hz are applied to a ring modulator, calculate all the new frequencies that can be constructed from these two tones and their second harmonics.

4. The portable synthesizer shown in Fig. 27.26 has a three-octave keyboard and, in addition, a switch to transpose two octaves up or two octaves down. What is the total playing range of the keyboard? If the control characteristic is one volt per octave, what is the maximum control voltage that must be supplied by the keyboard?

5. Make a drawing (to scale) of an envelope with a maximum amplitude of two volts and the following parameters: A = 50 ms, D = 20 ms, S = 1.5 V, R = 100 ms.

6. Explain how the patch shown in Fig. 27.13 could be used to generate a sound in which the rise time of the fundamental is shorter than that of the second harmonic. Which envelope generator should have the shortest attack time?

CHAPTER **28**

Digital Techniques for Generating and Recording Sound

In Chapter 20 we directed our attention to electroacoustic *transducers:* devices that convert sound signals to electrical signals (microphones) and back to sound signals (loudspeakers). Representing a sound by its electrical *analog* allows us to record, reproduce, and amplify it. In Chapters 26 and 27 we discussed some of the techniques for generating electrical signals that are the analog representation of musical sound.

The electrical analog of a musical sound is a complex voltage that varies with time. There are other ways to represent a musical sound. The undulating grooves on a phonograph record contain a mechanical analog of musical sound; the varying magnetization on recording tape is a magnetic analog representation of the sound. In this chapter we will discuss another very useful way to represent musical sound: by a sequence of numbers. This is called a *digital* representation.

Digital representation of sound offers several advantages over analog representation: Great accuracy is possible by using numbers with many digits; digital information is much less susceptible to distortion and interference from extraneous noise; it is easily processed by digital computers; and it can be efficiently recorded and retrieved. Rapid development of digital technology has made digital generation, processing, recording, and reproduction of sound cost competitive with analog methods, particularly for high-quality systems.

555

28.1   MUSIC BY NUMBERS: DIGITAL REPRESENTATION OF SOUND

The digital representation of a sound consists of a sequence of numbers that specify the sound pressure at regular time intervals. Although in principle it is possible to have a transducer that samples a sound field and generates the desired sequence of numbers directly, in practice it is more practical to convert the electrical (analog) signal from a microphone to a digital representation by means of an analog-to-digital converter, as described in Section 28.2. Time sampling is an essential part of the digital representation of sound.

Digital representation of sound is somewhat analogous to representing a changing visual scene by a sequence of still photographs in a movie camera or a sequence of images on a television screen. To represent rapid motion requires a more rapid sequence of images than are needed to represent slower motion. If the sequence of images is too slow, rapid motion appears "jumpy."

28.2   SAMPLING AND DIGITAL CONVERSION

In order to represent a continuous sound waveform as a sequence of numbers, one must (1) sample the waveform at regular intervals, and (2) express these samples as numbers. The sampler is very similar to the sample-and-hold circuit described in Section 27.5 (see Fig. 27.20). The samples, which represent values of the waveform at equal time intervals, are converted to numbers in an *analog-to-digital converter (ADC).*

According to the Nyquist sampling theorem, it can be shown mathematically that R samples per second are needed to represent a waveform with a maximum frequency $R/2$. Thus, to represent sound up to 15,000 Hz, the sampling rate must be 30,000 samples per second. Figure 28.1 is a schematic of sampling and digital conversion.

The precision with which each sample can be expressed depends on the number of digits used (or the "word length," in computer terminology). For example, if the numbers are 16 bits or binary digits in length, numbers from 0 to 65,536 can be expressed ($2^{16} = 65,536$). This allows for a theoretical dynamic range of about 96 dB (since $20 \log 65,536 = 96.3$).

Construction of a waveform from a sequence of numbers is done with a *digital-to-analog converter (DAC).* The output from the DAC

FIG. 28.1
A schematic of sampling and digital conversion.

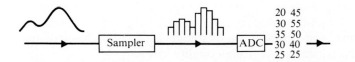

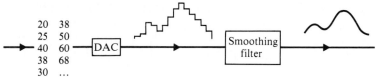

FIG. 28.2
A schematic of analog conversion and smoothing.

will be a stepped waveform, which is made smooth by passing it through a low-pass filter, as shown in Fig. 28.2.

28.3   ALIASING

Although waveform sampling, digital representation, and subsequent reconstruction of the waveform is an accurate process, errors and extraneous signals can occur, and special care is needed to minimize them. One such error is called *aliasing* or *foldover error*.

As mentioned in Section 28.2, a sampling rate R allows the accurate representation of frequencies up to $R/2$. If frequencies above $R/2$ are present, however, they will generate foldover frequencies f_f given by $f_f = R - f$. Thus, if frequencies above $R/2$ may be present in the original waveform, it is advisable to remove them with a low-pass filter before sampling.

Some examples of aliasing are illustrated in Fig. 28.3. Using a 44-kHz sampling frequency allows accurate representation of a 22-kHz

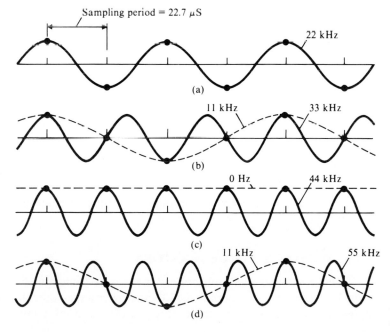

FIG. 28.3
Alias signals introduced by exceeding the Nyquist frequency (22-kHz for a sampling rate $R = 44$ kHz). (a) Signal frequency $f = 22$ kHz; no alias. (b) Signal frequency $f = 33$ kHz; foldover frequency $f_f = 11$ kHz. (c) Signal frequency $f = 44$ kHz; $f_f = 0$. (d) Signal frequency $f = 55$ kHz; $f_f = 11$ kHz.

signal, as shown in Fig. 28.3(a). However, a 33-kHz signal is sampled at intervals of $\frac{3}{4}$ of its period, and the samples themselves have a periodicity of 11 kHz, as shown in Fig. 28.3(b). Thus, by exceeding the Nyquist frequency $R/2$ (half of the sampling frequency R), we have introduced a signal with the foldover frequency $f_f = R - f = 44 - 33 = 11$ kHz. Sampling a 44-kHz signal (Fig. 28.3(c)) would give a foldover frequency of zero, and a 55-kHz signal (Fig. 28.3(d)) would again yield a 11-kHz signal ($R - f = -11$ kHz, but a negative foldover frequency is equivalent to a positive one).

Aliasing is analogous to the "wagon wheel effect" in motion pictures. As long as the spoke-passage frequency (wheel rotation frequency divided by the number of spokes) is less than the picture rate (24 frames per second for most professional motion pictures), the wagon wheel appears to rotate in the normal direction. When the spoke passage frequency exceeds the picture frequency, however, the wagon wheel appears to rotate backwards. (Ardent fans of Western movies will know that when the spoke passage frequency exceeds twice the picture frequency, the wagon wheel once again appears to rotate in the normal direction.)

28.4   QUANTIZATION ERROR AND DITHER

If our digital system handles 16-bit numbers, we will most likely divide the maximum voltage range into $2^{16} = 65,536$ equal levels. An ideal analog-to-digital converter (ADC) will express each voltage sample by the number of the closest discrete voltage level. This process is called *quantization*. The difference between the exact voltage and the nearest voltage level is called the *quantization error*. The maximum quantization error is one-half the spacing between levels, which in the digital representation corresponds to one-half the least significant bit.

With signals of large amplitude, there is little correlation between the signal and the extraneous signal due to quantization error; the latter is heard as random noise at a very low level. With low-level signals, however, noise due to quantization error has some correlation to the signal, and thus it results in signal distortion. To remove this distortion, a small analog noise signal (or *dither*) is sometimes added to the audio signal prior to sampling to randomize the effects of quantization error (see Pohlmann, 1985).

28.5   DIGITAL TONE GENERATORS

1. *Direct waveform computation*. One way to generate tones digitally is to simulate the operation of an analog voltage-controlled oscillator (see Section 27.2) by computing waveforms upon command. A

sawtooth waveform, for example, could be synthesized by increment-ing an *n*-bit register at regular intervals until it overflows, which is equivalent to changing the content from 2^{n-1} to zero. If the register is incremented in steps of two rather than in steps of one, it will fill twice as fast, and the sawtooth frequency will be twice as great. Using a 16-bit register, it is possible to adjust the oscillator frequency in steps of 0.61 Hz up to 20 kHz, while a 24-bit register makes steps of 0.0024 Hz possible (Chamberlain, 1987).

Just as the sawtooth generated by an analog VCO can be con-verted into other waveforms (see Section 27.2), a digital sawtooth can be converted to square, triangular, or sinusoidal waveforms by simple computation. A square wave can be computed by noting whether the sawtooth samples are positive or negative and accordingly setting a register to one of two values. A triangular waveform can be computed by taking the absolute value of the sawtooth samples, which is the digital equivalent of full-wave rectification in the analog domain.

2. *Wavetable lookup.* Most digital techniques for generating sound rely on stored waveforms rather than direct waveform computations from algorithms. The stored waveforms may be obtained by computa-tion, by sampling the sounds of musical instruments, or by a combina-tion of both. A wavetable lookup generator usually has an assortment of simple and complex waveforms stored in a read-only memory (ROM), which can be read out at different rates to generate sounds with different frequencies. The hardware used to accomplish this task is often called a *digital oscillator.*

There are two basic ways to scan a wavetable with a digital oscilla-tor: with a variable sample rate and with a fixed sample rate. The variable sample rate approach is a little like using a variable speed tape player. Slowing the tape speed causes the pitch to drop. In the case of a digital oscillator the variable sample rate is usually accom-plished by dividing a fixed clock rate (typically 10 MHz) by a select-able number N to determine the sample rate, as shown in Fig. 28.4.

The system illustrated in Fig. 28.4 works well for generating a single voice, but it becomes cumbersome for a polyphonic (multi-

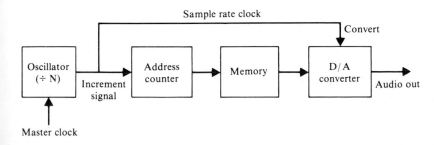

FIG. 28.4
A single voice, variable sample-rate oscillator.

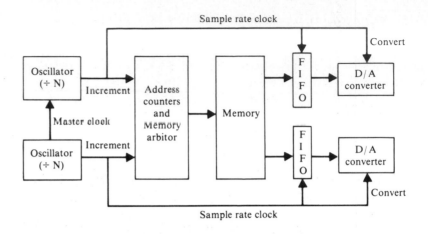

FIG. 28.5

Two oscillator variable sample-rate system with shared memory (after Mauchly, 1987).

voice) system, since each oscillator has its own memory. The shared-memory system, shown in Fig. 28.5, includes a "memory arbitration" scheme that prevents conflicts between two oscillators sampling the shared memory at different rates. Each voice includes a one-sample first-in-first-out buffer (FIFO).

In the variable-rate systems just described, the address sent to the waveform table was incremented by *one* for each sample. In a fixed sample-rate oscillator, on the other hand, the size of each step is determined by the desired frequency (Fig. 28.6). To achieve a reasonable frequency resolution, the step must have a fractional part of eight to ten bits. This requires the use of an *interpolator* to determine the sample that would have fallen between the two actual samples (Mauchly, 1987). Fixed-rate oscillators are used in popular synthesizers by Ensonic, Roland, and other manufacturers, as well as in an Ensonic digital oscillator chip included in the Apple IIGS microcomputer.

3. *Sampling.* Digital sound generators that use digitally stored waveforms from actual musical instruments are becoming very popular. Functionally, this technique is quite similar to the wavetable lookup technique described in the previous subsection. The waveforms may be stored in a read-only memory (ROM) or in a random access mem-

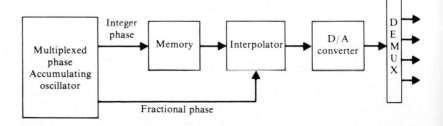

FIG. 28.6

Fixed sample-rate digital oscillator (after Mauchly, 1987).

ory (RAM), which allows the composer/performer to work with sounds of his or her own choosing. Smaller sampling synthesizers may store waveforms on a disc for transfer to a small RAM as needed.

One of the first musical instruments to generate sounds by reading out stored waveforms from real sources was the Allen Digital Computer organ (see Section 29.11). Among the manufacturers of synthesizers employing sampling techniques are Kurzweil, Fairlight, New England Digital (Synclavier), Ensonic, Casio, Korg, Oberheim, Roland, etc.

Synthesizers employing the sampling technique are particularly noted for their ability to reproduce the sounds of real instruments such as the piano, which are difficult to synthesize by other techniques. A popular technique is "splicing" together the sounds of two musical instruments, so that a tone changes its timbre from "violin-like" to "trumpet-like," for example.

4. *Frequency modulation.* FM synthesis, described in Section 27.10, is more easily implemented with digital circuits than with analog ones. Since it requires only two synthesis parameters (modulation index and ratio of modulating to carrier frequency) to produce a wide variety of timbres, it requires less memory space and computing speed than other digital methods. The basic FM system illustrated in Fig. 28.7 requires only two sinewave tables and a single multiplier.

As in analog FM synthesis, either the carrier signal or the modulating signal can be complex. This can be accomplished either by storing complex waveforms in the lookup tables or by combining the outputs from several sinewave tables. The advantage of the latter scheme is that the modulation indices as well as the frequencies of each component can be controlled independently.

The best known examples of digital synthesizers employing FM synthesis are probably the DX7 and other members of the Yamaha DX

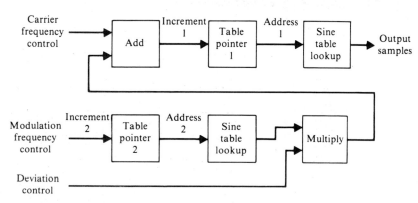

FIG. 28.7
Calculations required for FM synthesis (Reproduced with permission of the publisher, Howard W. Sams & Co., Indianapolis, *Musical Applications of Microprocessors* by Hal Chamberlain, copyright 1980.)

and TX families. Several large synthesizers, such as the Synclavier, combine FM synthesis with other techniques of tone generation.

5. *Other methods.* Other methods for digital sound generation go by such names as phase distortion, phase modulation, wave shaping, Fourier transformation, and vocal simulation (VOSIM). VOSIM employs a simple waveform described by only three parameters: a *width* parameter, specifying the widths of one or more pulses in the waveform; a *number* parameter, specifying how many pulses there are in each period; and a *decay* parameter. This waveform is repeated at the desired frequency (Chamberlain, 1987). The Fourier transformation technique uses an algorithm for fast Fourier transform (FFT) analysis to synthesize waveforms.

Digital synthesis of sound is a rapidly developing field, and it is almost certain that several useful new techniques for sound generation will be developed in the near future.

28.6 DIGITAL FILTERING

For any analog process there is an equivalent process in the digital domain. However, the two domains are very different, and a process that is easy in one domain may be difficult in the other. An analog VCO, for example, is quite difficult to manufacture and many designs tend to suffer from instability and inaccurate tuning at the high and low ends of the tuning range, whereas a digital oscillator is simple and stable. On the other hand, analog filters are relatively simple, whereas digital filters tend to be more complex and more expensive than digital oscillators.

Filtering in the digital domain amounts to calculating the results of difference equations involving consecutive samples of the waveform. The arithmetic operations prescribed by these equations are performed periodically at a rate equal to the sampling rate. The "circuit" elements in a digital filter include adders, multipliers, and delays.

Two examples of filter structures are shown in Fig. 28.8. The first one is known as a nonrecursive or finite impulse response filter. The

FIG. 28.8
Block diagrams of two filter structures: (a) nonrecursive or finite impulse responses; (b) recursive or infinite impulse responses (From Alles, "Music Synthesis Using Real Time Digital Techniques," 68: 436 © 1980 by IEEE.)

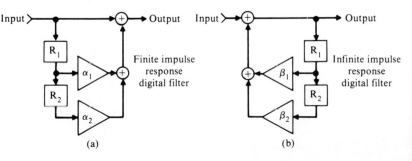
(a)   (b)

two storage registers R_1 and R_2 are loaded each sample period; α_1 and α_2 are two multipliers and the circles represent adders. Values of three consecutive samples contribute to the output, so first and second differences are calculated. Nonrecursive filter structures of this type are especially stable, and they are well suited to use in band-reject filtering.

The filter structure in Fig. 28.8(b) is known as a recursive or infinite impulse response filter. Again registers R_1 and R_2 are loaded at each sample period; β_1 and β_2 are multipliers. The feedback in this structure can result in gain or amplification at a particular frequency. It is particularly useful as a bandpass filter in which the coefficient β_1 controls the center frequency and β_2 controls the filter Q-value. (If $\beta_2 \rightarrow 1$, the filter can be used as an oscillator.) Recursive filters are generally more economical in execution time and storage requirements than nonrecursive filters, but they are more susceptible to distortion due to their nonlinear phase characteristic.

By adding multiple feedback paths to a recursive filter structure, a wide variety of filter characteristics are possible. The structure in Fig. 28.9, which has two feedback and two feed-forward paths, is sometimes called a *canonical* second-order filter because of its generality. In general the feedback paths are responsible for peaks (poles) in the frequency response curve, while the feed-forward paths cause dips (zeroes)(Chamberlain, 1987).

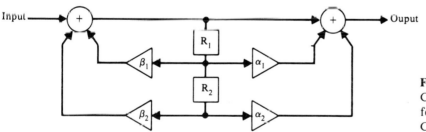

FIG. 28.9
Canonical second-order filter with both feedback and feed-forward paths (after Chamberlain, 1987).

28.7   DIGITAL RECORDING

Digital recording of data generally requires a medium that can exist in one of two states. The magnetic coating on recording tapes or a computer disc, for example, can be magnetized in one of two directions, to represent a 1 or 0; the surface of a compact disc can have a pattern of pits which represent binary 1's.

1. *Modulation and coding.* Rarely is digital information recorded on tape or disc in its raw state. The packing density and the reliability

can be improved markedly by using some type of *modulation*. Modulation is frequently used to encode information for transmission or storage. Amplitude modulation (AM) and frequency modulation (FM) are used to transmit audio information by radio. *Pulse code modulation* (PCM) is nearly always used for digital recording. A number of modulation codes or channel codes have been developed for computer peripherals; some work best when recording on magnetic tape, some on magnetic discs, and some are particularly well suited for optical discs.

One of the earliest channel codes was the NRZ (nonreturn to zero) code, in which ones and zeros were represented directly by two opposite states of the recording medium, as shown in Fig. 28.10(a). This system, which is used in video tape recorders, requires external synchronization to decode strings of ones or zeroes; therefore, for use in computer peripherals the NRZI (nonreturn to zero inverse) code, shown in Fig. 28.10(b), is preferred. In this code a one is represented by a transition, a zero by no transition (transitions usually occur in the middle of a bit cell). Other modulation or channel codes used in digital recording include PE (phase encoding), FM (frequency modulation), MFM (modified FM), Miller code, HDM (High-density modulation), and EFM (eight-to-fourteen modulation). The EFM code in Fig. 28.10(c), used in compact discs (Doi, 1983), will be discussed in Section 28.9.

FIG. 28.10
Examples of modulation or channel codes used in digital recording (after Doi, 1983).

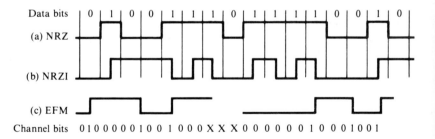

2. *Error detection*. With analog storage there is little or no opportunity to correct errors; information lost by damage is forever lost. One of the attractive features of digital recording, however, is that many types of errors can be detected and corrected or concealed by synthesizing new data.

Error detection techniques are generally based on the principle of redundancy. Redundancy is important in communication. Written language has a high level of redundancy (a message such as "AJL IS FIRGIVEN. PLEAOE COMW HOME" is readily understood), but numbers (such as 456,781) do not.

One way to assure redundancy would be to record every word twice. But if the two records are different, how do we decide which one is correct? By recording a third copy? We realize immediately that this would consume a great deal of recording space.

One fairly efficient technique for error detection is to record a *parity* bit. If the number of 1's in the data word is even (or zero) a 0 parity bit is recorded; if the number of 1's is odd, the parity bit is a 1. At playback, the parity bit is compared to the parity of the data word. If they are not in agreement, an error is present. If two data bits are in error, of course, the error will not be detected.

More sophisticated error detection codes have been developed for digital recording. One example is the cross-leaved Reed-Solomon code (CIRC) used in compact discs. Discussion of these codes would go beyond the scope of this book.

3. *Error correction.* Error correction codes tend to be highly mathematical. In general, they fall into two types: *block codes,* which use algebraic methods, and *convolutional codes,* which are statistical. A block-coded message generated by an encoder is formed from the data currently within a block, whereas a convolution-coded message includes data previously in the block as well.

Error correction codes attempt to reconstruct incorrect or lost data by making use of redundant data. This is analogous to supplying a missing word or correcting a misspelled word in a sentence by carefully noting its context. If an entire line of text or an entire sentence is missing, it becomes next to impossible to make the correction. This is analogous to a *burst error* in recorded digital data. Burst errors can be dealt with by *interleaving* the data: that is, scrambling the data in a carefully prescribed way so that a block of data, as recorded, originates from different blocks in the original data. This way, if a burst error occurs, the lost data will be distributed throughout several blocks upon de-interleaving, and error correction by computations may be possible. Most digital recording systems make use of interleaving (Nakajima, et al., 1983).

4. *Error concealment.* Even the best error correction code will be unable to deal with some errors. If an error cannot be corrected, the next best thing is to conceal it. Two useful error concealment techniques are interpolation and muting.

Interpolation is a way of estimating an unknown quantity that lies between two known quantities. Zeroth-order interpolation simply holds the previous data and repeats it to cover the missing or incorrect word. In first-order or linear interpolation, the replacement word is derived from the mean value of the previous and subsequent words. Higher-order interpolation techniques use more sophisticated mathematical algorithms to calculate substitute data (Pohlmann, 1985).

Muting is the process of substituting zeros for incorrect or missing words. The resulting silence is preferable to erratic sounds that might result from decoding erroneous data.

28.8 DIGITAL RECORDING ON MAGNETIC TAPE

For many years magnetic tape recording has been a standard method for recording analog data (see Section 21.6). Although the principles involved in recording digital and analog data on magnetic tape are the same, there are some important differences:

1. The bandwidth required for digital recording is at least 30 times greater than that required for analog recording;

2. Only two values (1 and 0) are used in digital recording, so linearity is not necessary;

3. Bits rather than waveforms are recorded, so the data bits in one channel can be recorded on several tracks, or several channels can be recorded on one track by means of time sharing (Nakajima, et al., 1983).

Analog tape recorders nearly always have stationary heads. In general, the maximum frequency that can be recorded depends on the tape speed and the head gap size (see Section 21.7). With tape speeds ranging from 4.8 to 38 cm/s ($1\frac{7}{8}$ to 15 in/s), it is possible to record frequencies across the entire range of audibility.

With a sampling rate of 44.1 kHz, recording 16-bit words requires a bit rate of $44,100 \times 16 = 705,600$ bits per second. Allowing for error detection, synchronization, etc., pulse frequencies of 1 MHz and higher will be recorded (this is reduced by one-half when a modulation code such as NRZ is used). Achieving the required high-frequency response is possible by using very high tape speeds or by using rotating heads.

1. *Rotating heads.* Rotating heads were developed for video tape recording where bandwidths of over 4 MHz are required. Two heads are attached to a cylindrical drum which rotates opposite to the direction of tape motion, as shown in Fig. 28.11. The tape is wrapped part

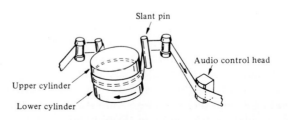

FIG. 28.11
Rotating head tape recorder for video or digital audio recording.

way around the drum at an angle, so that the tracks of recorded information run diagonally across the tape. The two heads record alternate tracks.

To use a video recorder for digital audio recording, the audio signal must be processed to conform to the video signal format. Synchronization pulses appropriate for the television format used in the video recorder (NTSC in the USA and Japan) are added. Video recorders use frequency modulation (FM), so changes in frequency represent ones and zeroes. An appropriate block of audio data goes into each video frame.

Using video tape recorders for digital audio offers an inexpensive way to achieve the high frequency response required. The additional electronic instrumentation required is fairly simple, since an FM modulator is already included in the VTR.

2. *Stationary heads.* Although the use of video tape recorders offers a low-cost method for digital audio tape recording, it makes tape editing and synchronous multitrack recording quite difficult. For this reason, most professional recording is done using recorders with stationary heads.

One of the first successful digital tape recorders, developed around 1976 by Thomas Stockham and colleagues at Soundstream, used a 42.5-kHz sampling rate and 16-bit quantization. It was used to master many early digital recordings. Around 1977, the 3M Company introduced a 32-track recorder with a sampling rate of 50-kHz. Sampling rates of 44.1 and 48 kHz are presently used in stationary head digital tape recorders. Tape speeds are generally 38 to 114 cm/s (15 to 45 in/s).

28.9   DIGITAL AUDIO TAPE RECORDERS FOR HOME USE

Digital audio tape recorders (DATs) appeared on the Japanese market early in 1987, and in Europe later that same year. As of this writing (1989), however, they are virtually unavailable in the USA. Under strong pressure from the music recording industry, the Congress has considered the pros and cons of restricting DATs unless they incorporate an anticopying safeguard to prevent direct digital copying of prerecorded digital tapes and compact discs. One bill required that all DATs incorporate a copy code scanner which would make it impossible to copy encoded source material. The "coding" described in the bill consisted of removing all frequencies between 3700 and 3900 Hz. Needless to say, this met with strong opposition from many audio engineers who objected to the resulting reduction in fidelity.

In 1985, representatives of a large number of manufacturers of DATs agreed on tentative specifications both for stationary-head recorders (S-DAT) and rotating-head recorders (R-DAT). A year later, a final technical standard for R-DAT was published (Feldman, 1987).

R-DAT uses a rotating head design and helical scan quite similar to that used in video tape recorders (see Section 28.8). The head rotates at 2000 rpm and the tape passes it at an angle of $6°23'$. Unlike videotape, however, the tape is wrapped around only $\frac{1}{4}$ of the head drum to minimize tape wear. Setting the two heads at slightly different angles reduces crosstalk between tracks and allows a denser track spacing than with standard video tape recorders. Tape width is 3.81 mm (0.15 in) and the tape speed is 8.15 mm/s (0.32 in/s). Cassettes of dimensions 73 $\times$ 54 $\times$ 10.5 mm ($2\frac{7}{8} \times 2\frac{1}{8} \times \frac{13}{32}$ in), slightly smaller than audio cassettes, allow a playing time of two hours.

DATs normally use a 48-kHz sampling rate, although a 32-kHz rate may be used for some home recordings. However, DATs will play back tapes recorded with sampling rates of 32, 44.1, or 48 kHz. Prerecorded tapes will probably use the same 44.1-kHz sampling rate used for CDs, so that tapes can be produced from the same digital masters. DATs are normally equipped with both digital and analog inputs and outputs. Some DATs incorporate circuitry to prevent direct (digital) duplication of copy-protected digital source material (from a CD or prerecorded digital tape with an appropriate anticopying "flag" coded in).

It is well to point out the difference between copying ("dubbing") a tape by connecting the analog output of a tape player to the analog input of a tape recorder, and copying by connecting the digital output of a tape player to the digital input of a DAT. The latter type of copying might be called "cloning," because it results in a nearly exact copy. A tape can be digitally copied (cloned) many times with little or no loss in fidelity. Large-scale digital copying or cloning is what the music recording industry wants to prevent.

28.10 COMPACT DISC DIGITAL AUDIO

Perhaps the most exciting development in high fidelity sound reproduction in recent years is the compact disc digital audio system. Only 12 cm in diameter, a compact disc can store more than 6 billion bits of binary data to be read out by a laser. This is equivalent to 782 megabytes, or more than the capacity of 1500 half-megabyte floppy discs. Over 275,000 pages of text, each holding 2000 characters, could be stored on a compact disc for display on a television monitor.

Used for digital audio, a compact disc stores 74 minutes of digitally encoded music. This music can be reproduced with very high

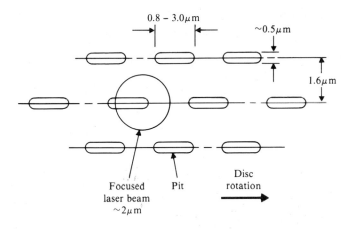

FIG. 28.12
Tracks of pits recorded on a compact disc. Also shown is the focused laser beam used to read the recorded information.

fidelity over the full audible range of 20 to 20,000 Hz. The dynamic range and the signal-to-noise ratios can both exceed 90 dB, and the sound is virtually unaffected by dust, scratches, and fingerprints on the disc. Unlike most other digital recording media, compact discs can be replicated in large quantities from a master disc.

Recorded information is stored in pits impressed into the plastic surface of the disc which is then coated with a thin layer of aluminum to reflect the laser beam (Fig. 28.12). Pits are about 0.5 μm wide and 0.11 μm deep, arranged in a spiral track similar to the spiral groove in a phonograph record, but much narrower. The track spacing on a compact disc is about 1.6 μm, compared to about 0.1 mm (100 μm) for the groove of a long-play phonograph record.

The track on a compact disc, which spirals from the inside out, is about three miles in length. The track of pits is recorded and read at a constant 1.25 m/s, so the rotation rate of the disc must change from about 8 to 3.5 revolutions per second as the spiral diameter changes. Each pit edge represents a binary 1, whereas flat areas within or between the pits are read as binary 0's.

The laser beam, applied from below the compact disc, passes through a transparent layer 1.2 mm thick and focuses on the aluminum coating, as shown in Fig. 28.13. The spot size of the laser on the transparent layer is 0.8 mm, but at the signal surface where the pits are recorded, its diameter is only 1.7 μm. Thus any dust or scratch smaller than 0.5 mm will not cause a readout error because it is out of focus. Larger blemishes are handled by error-correcting codes.

The optical pickup used to read a compact disc is shown in Fig. 28.14. A semiconductor laser emits a beam of infrared light (790-nm wavelength) that is eventually focused to a tiny spot 1.7 μm in diameter. The reflected beam is directed to a photodiode that generates an electrical signal to be amplified and decoded.

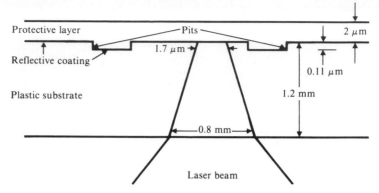

FIG. 28.13

Cross section of a compact disc (not to scale).

Included in the sophisticated optical pickup are a diffraction grating, a polarization beam splitter, a quarter-wavelength plate, and several lenses. A semiconductor laser employs an aluminum-gallium-arsenide (AlGaAs) pn junction, similar to that in a light-emitting diode (LED). The diffraction grating creates two secondary beams that are used for tracking the primary beam.

When the laser beam strikes a land area between two pits, it is almost totally reflected. When it strikes a pit (which appears like a

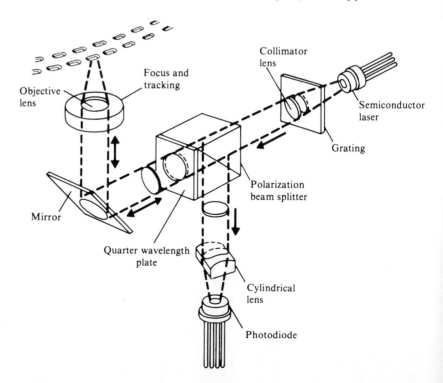

FIG. 28.14

Optical pickup in a compact disc player. Coherent light from a semiconductor laser is focused on one recorded track on the disc. The reflected light is directed to a photodiode.

bump from the reading side), whose 0.11-μm height is roughly one-quarter wavelength of the laser light, the part of the beam reflected from the pit cancels the part reflected from the land, so little or no light returns. The reflected light of varying intensity is directed on the photodiode; a change in intensity will eventually be interpreted as a binary 1 and unchanged intensity as a 0.

The compact disc digital audio system, a joint development of Philips (in The Netherlands) and Sony (in Japan), is a very sophisticated system. An elaborate error correction system, called "cross-leave Reed-Solomon code" (CIRC) was developed to deal effectively with both random errors, caused by inaccurate cutting, and burst errors, due to dirt and scratches on the disc (see Section 28.7). Packing density is maximized by rearranging the data bits, using a channel code called eight-to-fourteen modulation (EFM).

Compact disc players also include sophisticated servo systems for keeping the laser beam centered on the track and for keeping it focused exactly on the reflecting surface within the disc. Decoding the recorded information and digital-to-analog conversion in most players incorporates digital filtering plus a technique called oversampling (see box).

Although many of the technical details of the compact disc digital audio system are beyond the scope of this book, we briefly describe three innovative features: autofocusing, EFM coding, and oversampling.

1. *Autofocus.* To distinguish between pits and land areas, the laser beam must stay focused within about 0.5 μm of the reflecting surface. The flat surface of the disc, however, may have deviations as large as 0.5 mm, one thousand times greater. Thus the objective lens must refocus rapidly as the surface deviates during the rotation of the disc. A servo-driven autofocus system that uses a four-quadrant photodiode and a differential amplifier to control the servo disc makes the rapid refocus possible.

The cylindrical lens, shown just above the photodiode in Fig. 28.14, projects a circular laser spot on the photodiode if the reflecting surface of the disc is exactly at the focus of the objective lens, but an elliptical spot if it is above or below the focus, as shown in Fig. 28.15. The four-quadrant diode provides electrical signals proportional to the light intensity on each of its four quadrants, and these are combined to form the error focus signal, as shown in Fig. 28.15. The objective

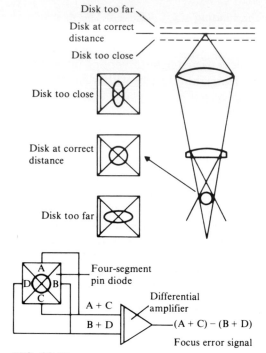

FIG. 28.15
A square pin diode photodetector, divided into
four identical triangular segments, and a
difference amplifier control a focus servo that
moves a cylindrical lens up and down to correct
for disc surface irregularities. (Miyaoka, © 1984
by IEEE.)

lens is attached to a movable coil in a magnetic field, similar
to that which drives a loudspeaker cone.

2. *EFM coding.* Compact discs use a modulation or channel
code called eight-to-fourteen modulation (EFM). The EFM is
a kind of NRZI code in which eight bits of data are represented by fourteen channel bits (see Fig. 28.10(c)). This may
seem wasteful, but by requiring that every one in channel bits
(corresponding to a transition) be separated by two to ten
channel zeros, EFM allows bit sizes to be less than the diameter of the laser beam that reads them. Thus EFM actually
results in a greater bit density. In the compact disc system,
a 1.7 μm diameter laser spot reads pits whose length varies
incrementally from 0.833 to 3.054 μm.

It turns out that of the 16,384 (2^{14}) possible 14-bit patterns, only 267 of them satisfy the specified requirement of

having two to 10 binary 0's in succession, but this is more than enough to code the 256 (2^8) possible 8-bit blocks. Blocks of 8 bits are translated into blocks of 14 bits using a conversion table (Miyaoka, 1984).

3. *Oversampling.* After decoding and error correction, the digital signal exists as a series of 16-bit words. Each word represents the numerical value of the acoustical signal, as sampled 44,100 times per second. The digital-to-analog converter (DAC) generates an electrical voltage of the appropriate magnitude for each word and holds it constant until the next word arrives (see Section 28.2). The resulting "staircase" curve resembles the original signal waveform plus a high-frequency signal with spectral components at 44.1 kHz and its harmonics, as shown in Fig. 28.16(a). These high-frequency components have to be removed by a low-pass filter.

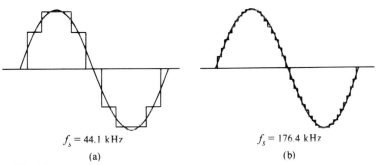

$f_s = 44.1$ kHz

(a)

$f_s = 176.4$ kHz

(b)

FIG. 28.16
A sinusoidal signal at 4.41 kHz sampled with a sampling rate f_s of 44.1 kHz (a) and with a frequency four times higher (b). In (b) the "staircase" curve approximates more closely to the analog waveform, and the high frequencies present in the staircase signal are more easily filtered out. (From D. Goedhard, R. J. van de Plassche and E. F. Stikvoort. Reprinted by permission.)

To avoid the use of sharp cutoff analog filters and to improve the signal-to-noise ratio, most manufacturers have introduced digital filters with oversampling, as shown in Fig. 28.17. Rather than suppress high-frequency components after the signal has been converted back to analog form, a digital oversampling filter is added before the DAC. In this filter, each 16-bit word is multiplied four times by different 12-bit coefficients, the resulting 28-bit words are averaged together, and outputed at a rate either twice (88.2 kHz) or four times (176.4 kHz) that of the original sampling frequency. This has the net effect of raising the high-frequency component in the staircase to 88.2 kHz or 176.4 kHz (as shown in Fig. 28.16(b)), which is more easily removed by filtering.

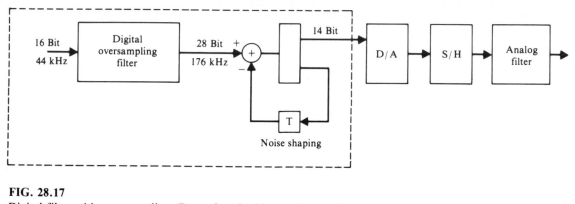

FIG. 28.17
Digital filter with oversampling (Reproduced with permission of the publisher, Howard W. Sams & Co., Indianapolis, *Principles of Digital Audio* by Kenneth C. Pohlmann, copyright 1985.)

References and Suggested Readings

Alles, H. G. (1980). "Music Synthesis Using Real Time Digital Techniques," *Proc. IEEE* **68**: 436.

Chamberlain, H. (1987). *Musical Applications of Microprocessors,* 2nd ed. Indianapolis: Hayden.

Chowning, J. N. (1973). "The Synthesis of Complex Audio Spectra by Means of Frequency Modulation," *J. Audio Eng. Soc.* **21**: 526.

Doi, T. (1983). "Channel Codings for Digital Audio Recordings," *J. Audio Eng. Soc.* **31**: 224.

Feldman, L. (1987). "Four DAT Recorders," *Audio* **71**(7): 36.

Mauchly, J. W. (1987). "Practical Considerations in the Design of Music Systems using VLSI," in *Music and Digital Technology,* ed. J. Strawn. New York: Audio Eng. Soc.

Miyaoka, S. (1984). "Digital Audio is Compact and Rugged," *IEEE Spectrum* **21**(3): 35.

Moorer, J. A. (1977). "Signal Processing Aspects of Computer Music," *Proc. IEEE* **65**: 1108.

Nakajima, H., T. Doi, J. Fukuda, and A. Iga (1983). *The Sony Book of Digital Audio Technology.* Blue Ridge Summit, PA: Tab Books.

Pohlmann, K. C. (1985). *Principles of Digital Audio.* Indianapolis: H. W. Sams.

Rossing, T. D. (1987). "The Compact Disc Digital Audio System," *Phys. Teach.* **25**: 556.

Glossary

algorithm Step-by-step directions for solving a problem (multiplying two numbers, for example).

analog-to-digital converter (ADC) A circuit that converts numbers from an analog to a digital representation.

binary number A number in base-two representation; a 1 or a 0.

bit A single binary digit.

byte A group of bits, usually eight.

Compact Disc Digital Audio (CD) An efficient means for digital sound recording that uses an optical pickup to read the recorded information.

digital-to-analog converter (DAC) A circuit that converts numbers from a digital to an analog representation.

digital oscillator A circuit that assembles a sequence of numbers to represent the desired waveform.

dither Low-level noise added to the signal to reduce the effect of quantization error.

eight-to-fourteen modulation (DFM) A code used to represent blocks of eight numbers on a compact disc.

foldover A spurious signal introduced into a digital sound by components with frequencies greater than half the sampling rate.

interleaving A means for rearranging recorded data to minimize the loss of information due to playback errors.

oversampling A method for increasing the rate of digital samples to the DAC in order to avoid the need for an analog filter with a sharp cutoff.

pulse code modulation A means for representing a sequence of binary numbers by a series of electrical pulses.

quantization error Difference between the exact voltage in a signal and the nearest voltage level selected by an analog-to-digital converter (ADC).

R-DAT Rotary head digital audio tape recorder.

S-DAT Stationary head digital audio tape recorder.

Questions for Discussion

1. Why are digital tape recordings virtually free from wow and flutter?

2. What is the purpose of the low-pass filter used in the playback of digital recordings?

3. A pocket calculator, like a large computer, uses binary numbers internally. Why is it unnecessary to understand binary numbers in order to use a pocket calculator?

Problems

1. Write the following decimal numbers in binary representation:
   a) 12;
   b) 22;
   c) 36.

2. Write the following binary numbers in decimal representation:
   a) 1101001;
   b) 100110;
   c) 1010101.

3. Try to multiply binary numbers 101 times 110, using the same procedure that you use to multiply ordinary decimal numbers. Check your answer by multiplying the corresponding decimal numbers.

4. If the sampling rate is 30 kHz, what is the highest frequency of audio signal that can be successfully coded in a digital representation? A frequency of 25 kHz in the original signal will give rise to what foldover frequency in the coded signal? How can foldover be prevented?

5. How many different levels can be represented by a 13-bit binary code? Show that this makes possible a dynamic range of 78 dB.

6. Show that a 16-bit representation of signals allows a signal-to-noise ratio of 96 dB.

7. Express each of the following dimensions in a CD as numbers of wavelengths of the laser light: pit width, pit depth, diameter of the focused laser spot.

8. Up to 6 billion bits of binary data stored on a compact disc are read out in 74 minutes. What is the maximum bit rate per second?

CHAPTER 29

Digital Computers and Musical Sound

29.1 DIGITAL COMPUTERS: ELECTRONIC BRAINS OR HIGH-SPEED MORONS?

Probably no technological development in recent history has done more to change our lives than has the digital computer. The invention of labor-saving machines brought about the industrial revolution of the nineteenth century; the digital computer is causing a second industrial revolution, due to its ability to make calculations with speed and accuracy, control machines and industrial processes, store and process vast files of information, and so on.

There are several ways in which computers can contribute to music. Music may be *composed* using a computer, *played* by a computer, or *both*. *Computer-composed music* is music in which the choice of each note is specified by a sequence of numbers or by a mathematically expressed rule. One procedure is to begin by generating a sequence of random numbers. These numbers are then screened through a series of arithmetic tests representing various styles of musical composition, and either accepted or rejected. The type of tests imposed by the programmer will tend to set the style of the music. Strict tests will tend toward a classical style; fewer rules will give the composition a randomness characteristic of more contemporary music. An early example of computer-composed music is the "ILLIAC Suite for

String Quartet'' composed by Hiller and Isaacson (1959) and performed by several well-known quartets.

Computer-synthesized music is music that consists of computer-generated sound. We have already shown in Chapter 28 how any sound can be described by a sequence of numbers. A composer could write down a long series of numbers, each representing a sample in time of a complex waveform; in practice, this would require a prodigious effort. Instead, the composer uses a code to describe the waveform. The computer reads this code and constructs the sequence of numbers needed to produce it. Programs for accomplishing this will be described in Section 29.5.

29.2 THE ORGANIZATION OF A DIGITAL COMPUTER

Every digital computer, small or large, has four major sections: *input/output, memory, arithmetic unit,* and *control.* Sometimes the arithmetic section and control section are together referred to as the *central processing unit,* or CPU.

The size of the memory and memory access time are important specifications that indicate the speed and flexibility of a computer. Memory size is specified as the number of storage locations or words (the term "word" is used to indicate a block of binary digits—usually 16 or 32—that may represent either a number or an alphabetic word). For example, "16 K of memory" usually means 16 units of 1024-word memory, or a total of 16,384 memory locations.

The memory access time means the length of time that it takes the computer to store or retrieve a word (number) stored in its memory. A typical memory access time of a medium-size computer is 10^{-7} second, or $\frac{1}{10}$ microsecond (0.1 μs). Some large computers have a memory access time of a few nanoseconds (10^{-9}s) or less.

Digital computers can operate with a wide variety of input and output devices, including magnetic disc files, line printers, keyboards, graphic displays, modems (connecting to telephone lines), work stations, time sharing terminals, etc. Since input and output operations are slow, in the computer's time scale at least, the input and output devices are generally connected to an input/output interface, to which the central processing unit need direct its attention only occasionally.

The central processing unit (CPU) is the heart and brain of the computer. It includes the arithmetic/logic unit (ALU) plus control and timing circuitry and registers. In a large mainframe computer, the CPU may include many circuit boards filled with components; in a microcomputer, on the other hand, the entire CPU will be contained in one integrated circuit called a *microprocessor.*

29.3 MEMORY

Computer memories can be classified in several ways. One is according to whether it can retain its contents when the power is turned off. A memory that can do so is called a *nonvolatile* memory; one that loses its contents is called *volatile*. Another way of classifying memories is according to addressability and read/write capabilities. A *random access* memory (RAM) is one in which data can be read from and written into any location with minimum delay. A *read only* memory (ROM) also offers random access to any location but only to read its contents. Data in a *serial* or *sequential access* memory, on the other hand, is available only at certain times and the access time may depend upon its storage location. Yet another way of classifying memory is according to the storage medium. Most commonly used are semiconductor, magnetic, and optical storage.

1. *Main memory.* The main memory in a computer is used to store the data and programs that are currently being executed by the CPU. The main memory must have a fast access time, and therefore it almost always is a semiconductor RAM.

2. *ROM.* Nearly every computer includes nonvolatile read-only memory (ROM). ROM is used to store computer languages (such as BASIC), tables of mathematical functions, programs that interface the computer with input/output devices, and a "bootstrap" memory that tells the computer what to do when it is first turned on ("booted").

Normally ROM is programmed when it is manufactured. However, programmable read-only memory (PROM) can be programmed by the user using a special device that "burns in" the program electrically. Less common is erasable/programmable memory (EPROM), that can be erased (usually by ultraviolet light) and reprogrammed.

3. *RAM.* Although older computers employed random access-magnetic core memories, random-access memory (RAM) in modern computers is nearly always a semiconductor integrated circuit (IC). RAM may be either static or dynamic. Static memory cells, which include several transistors, hold information as long as electrical power is supplied, whereas the simpler dynamic memory cells must be frequently "refreshed" to maintain their contents.

RAM is used for many different purposes in a computer in addition to its use in the main memory. It holds programs and subroutines transferred from mass storage and images displayed on the graphic display. An important section of RAM, called a *stack,* stores data temporarily in a location that is easily accessed by the CPU.

4. *Mass memory.* Mass memories are intended to store large amounts of data inexpensively. The most commonly used devices are magnetic hard discs and floppy discs. In both of these devices, the data is stored as the magnetic remnant state in a small spot on the magnetic oxide coating, as in magnetic tape recording (see Section 21.6).

Large computers employ magnetic disc drives whose removable magnetic storage discs are capable of storing 500 megabytes or more of information (a "byte" usually corresponds to 8 "bits" or binary numbers). In some computers, removable discs have replaced the larger magnetic tape storage units once a part of all computers; some computers include both disc and tape storage. Microcomputers often include hard discs or Winchester discs capable of storing up to 80 megabytes or more of data.

Flexible magnetic or "floppy" discs, used in most microcomputers, are 3½ to 8 inches in diameter (the smaller size is becoming standard) and store up to 10^6 bytes (1 MB) of data. These discs revolve rapidly, and the reading heads are moved mechanically to the desired track.

One of the simplest computer languages, which is easy to learn and adapts reasonably well to small computers, is BASIC. To illustrate the use of BASIC, we will consider a simple problem: the calculation of the mean (average) of a set of grades. The algorithm is simple: Sum the grades and divide by the number of grades summed. However, the program should be constructed so that it can be used for any number of grades, and this requires only a little additional care. We will designate the number of grades to be averaged as N, and call each grade X, where the index i goes from 1 to N. The mathematician would

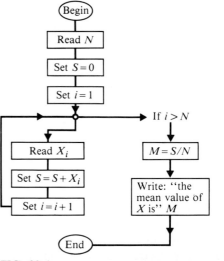

FIG. 29.1
Flowchart for a computer program that computes the mean of a set of numbers.

write the formula for the mean value of X as

$$\overline{X} = \frac{1}{N} \sum_i^N X_i,$$

but the computer does not understand this language.

Figure 29.1 shows a flowchart for the algorithm. Note the branch and the loop; the path leads N times around the loop until i exceeds N, at which point it branches to the right, and prints out the mean value.

A program in BASIC language designed to perform this operation might be as follows:

```
LIST
100   READ N
110   LET S = 0
120   LET I = 1
130   IF I < = N Then 150
140   GO TO 190
150   READ X
160   LET S = S + X
170   LET I = I + 1
180   GO TO 130
190   LET M = S/N
200   PRINT "THE MEAN VALUE OF X IS" M
800   DATA 6, 55.0, 85.0, 94.5, 76.0, 82.5, 88.7
999   END
```

When the computer comes to statement 100 READ N, it takes the first number in the data list (6) as the value for N. At statement 150 READ X, it takes 55.0 as X; each time it loops around to 150, it reads another value for X from the data list. Note that each time statements 160 and 170 are passed, the values of S and I change. Oh yes, zeros are written as 0 to distinguish them from the letter O.

A useful beginning text by Kemeny and Kurtz (1968) gives a simple program in BASIC language for harmonizing a melody. The serious reader should examine it carefully.

29.4 PROGRAMMING A DIGITAL COMPUTER

Problems for the computer must be presented in a precise and logical manner. Once the computer receives the problem, it is nearly impossible for the human operator to communicate with the computer or assist with "common sense." The problem must therefore be presented so that the computer can proceed to the desired result. If there is an ambiguous statement or a loophole in the logic, the computer will undoubtedly find it and hang up on it.

The steps usually followed in programming a computer are as follows:

1. Formulating the problem or defining the goal;

2. Constructing an *algorithm,* or step-by-step directions for solving the problem;

3. Diagramming the algorithm as a flowchart;

4. Writing the program in an appropriate language that can be understood by the computer;

5. Checking out ("debugging") the completed program.

Strictly speaking, the computer "understands" only commands written in machine language, which uses a suitable numerical code for every operation, including transfers from one register to another. However, machine language is unnecessarily detailed for most computer users, and a number of simpler languages have been developed, such as BASIC, FORTRAN, PASCAL, "C," etc. Before any of these user languages can be employed, a special program called a *compiler* or an *interpreter* must translate the program from user language into the correct machine language for the particular computer. One caveat: It is not always practical to employ user languages with a small computer, since translation by a compiler program might take too long and tie up too large a block of memory. Also, programs written in user language may not use the limited computing capability of a small machine to maximum efficiency.

The illustration of programming in the box below will be of interest to some readers, a bit too technical for some, and "old hat" to others. It is not essential to understanding the material in the remainder of this chapter.

29.5   COMPUTER SYNTHESIS OF MUSIC

Methods for using computers to synthesize music may be described as *off-line, interactive,* or *real time.* Most computer synthesis has been done off line; that is, the computation is done (most often on a large computer shared with other users) to create a passage or an entire work which is then converted by a digital-to-analog converter (DAC) in one batch. In an interactive mode, the composer computes a brief sound, listens to it, adjusts parameters, and then goes on to create another sound. Composition in the interactive mode generally requires the exclusive use of a computer or at least a time-sharing terminal or work station. Real-time computer synthesis, which requires exclusive use of a rather large computer or else special hardware to augment the computer, is a costly method unavailable to most composers.

Quite a number of programs have been written for off-line generation of musical sound with a digital computer. Most of them are descendants of programs developed by M. V. Mathews and his associates at the Bell Laboratories. A refined version, called MUSIC V, is described by Mathews (1969), while an adaptation MUSIC 4BF is described by Howe (1975). MUSIC V is written in FORTRAN, a language widely used in scientific programming. MUSIC 11 and MUSIC 360 are adaptations written in assembler language for the PDP 11 and IBM 360, respectively, by Barry Vercoe and colleagues at the Massachusetts Institute of Technology. Versions written in the popular "C" programming language include: C-MUSIC written by F. Richard Moore (University of California, San Diego); MUSIC 4C by Scott Aurenz (University of Illinois); and C-SOUND by Barry Vercoe.

The various MUSIC programs include a number of subroutines to which the composer adds two important components: an "orchestra" consisting of "instruments" of his or her own specification, and a "score" consisting of "notes" that are played by the instruments. The score may describe the notes in physical terms (seconds, decibels, and hertz) or in musical terms (beats, dynamic level (ppp to fff), and notes of the scale). Although individual notes may be described separately, programming can also be done at a higher compositional level with subroutines that handle phrases or melodies.

To describe the "instrument" on which the notes will be played, the composer may use building blocks such as digital oscillators, filters, envelope generators, multipliers, adders, etc. Block diagrams of two "instruments" are shown in Fig. 29.2. An important feature of the MUSIC V program is a catalog of musical sounds which enables the composer to listen to a large number of sounds that can be easily resynthesized (Mathews et al., 1974).

When the note and instrument specifications have been assembled, the computer processes them and computes the millions of numbers required to describe the sound. This computation, even with clever short cuts, tends to be time consuming and requires a large memory. Thus computer synthesis of music has tended to be restricted to composers with access to large, high-speed computers. This is changing, however, as microcomputers acquire greater speed and larger memories.

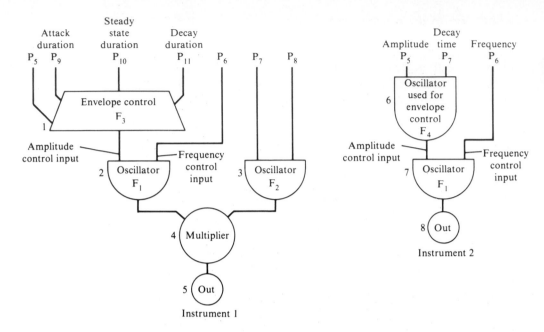

FIG. 29.2
Two "instruments" constructed from
unit generators of building blocks in
the MUSIC V computer synthesis
program (From Mathew, et al., 1974.
"Computers and Future Music,"
Science 183.263. Reprinted by
permission of AAS, copyright ©
1974.)

Music *work stations* have helped to make both interactive and
real-time methods of synthesis more convenient for the composer. A
work station provides the user with a set of task-related hardware and
software designed to optimize the interaction between user and machine for this task. An engineering work station, for example, may
be used by an aeronautical engineer to simulate airflow over a wing
pattern.

An example of a computer music work station is the CARL work
station developed at the Computer Audio Research Laboratory at the
University of California at San Diego. CARL is comprised of two
subsystems, one for general purpose (GP) computing and one for real-
time (RT) operations. The GP subsystem consists of a computer, associated disc drives, and displays; the RT subsystem includes interfaces
for sensing RT control inputs, a real-time synthesizer/processor, and
a computer. The two subsystems can work together or independently.
CARL includes several related programs designed to run in the time-
sharing UNIX operating system. Most of the programs, including C-
MUSIC, are written in the C language (Moore, 1987).

The Fairlight CMI (Computer Musical Instrument) synthesizer
might well be described as a computer music work station. This synthesizer, designed in Australia, has been used in the production of a
number of film sound tracks. The terminal part of the system includes
a 73-note musical keyboard, an alphanumeric computer keyboard,
and a graphics display screen with a light pen for drawing curves and

pointing out displayed items. The main computer includes two microprocessors, with a third one in the music keyboard.

The Fairlight, which has been termed a "toolbox" synthesizer because of its variety of input/output and storage options (Chamberlain, 1985), is designed for studio use. Some models include a 143-Mbyte hard disc and 14 Mbytes of waveform RAM, a storage capacity unmatched in other synthesizers. Up to eight synthesized or natural voices can be merged and played simultaneously when a key is depressed on the keyboard. A music composition language (MCL) and text editor allow complex scores to be typed in by the composer and compiled into the internal language of the synthesizer.

Another approach to real-time digital synthesis is to combine a general purpose computer with a fast signal processing computer of fixed architecture to create a "programmable signal processor" that can run synthesis calculations much faster than a general purpose computer of the same complexity and logic speed. An example of this approach is the DMX-1000 signal processing computer from Digital Music Systems, based on development work done at MIT (Chamberlain, 1985).

29.6 SMALLER AND SMALLER: THE DEVELOPMENT OF MICROELECTRONICS

In 1883, inventor Thomas Edison observed a flow of current between a hot filament and another conductor in an evacuated bulb. In 1903, Alexander Fleming made use of Edison's discovery and devised the first vacuum diode. In 1906, Lee DeForest invented the first three-element vacuum tube (triode) capable of amplification. Thus began the electronic age.

During World War II, intensive research was devoted to improving crystal detectors for microwave radar equipment. Partly as a result of this research, the *transistor* was invented in 1948 by William Shockley, John Bardeen, and Walter Brattain at the Bell Laboratories. This invention gave rise to a major revolution in electronics. Replacement of vacuum tubes with tiny transistors made it possible to reduce the size, the power requirements, and the cost of electronic equipment. More important, it led to the development of many new electronic devices, such as the digital computer, that would not have become practical if they had been constructed with vacuum tubes.

The development of microcircuits or integrated circuits (ICs) has been called the "second revolution" in electronics. Although it is difficult to point to a single invention or breakthrough, the technology of designing and fabricating smaller and smaller circuits has moved

ahead through research and development efforts at many labora
tories.

There are actually two classes of integrated circuits or microcircuits: digital and linear. *Digital* ICs are designed for on/off switching
or binary applications, and *linear* ICs are designed for linear amplifiers and other analog circuit applications. Both types became available
in the early to mid-1960s. By 1970, the microelectronics revolution
was in full swing.

One of the milestones in the microelectronics revolution was the
development of the *microprocessor,* which led to microcomputers or
personal computers. A microprocessor is an integrated circuit that
contains the central processing unit (CPU) for a microcomputer. Since
the first four-bit microprocessor was introduced by Intel in 1972, the
pace of development has been spectacular. Presently available 16-bit
and 32-bit microprocessors with over a million transistors exceed the
computing capabilities of mainframe computers of a decade or so ago.
Needless to say, microprocessors and microcomputers have had profound effects on computer music and digital synthesis of music.

29.7   DIGITAL SYNTHESIZERS

One of the results of the microelectronics revolution has been the
rapid development of digital synthesizers. In Chapter 28 we discussed
various digital circuits (tone generators, filters, etc.) that can serve as
building blocks in a digital synthesizer. A variety of architectures have
been employed in digital synthesizers; synthesis techniques have included additive, subtractive, FM, sampling, direct wave shaping, etc.
Thus far, at least, all-digital synthesizers have tended to be somewhat
more expensive than analog or hybrid synthesizers.

The largest-selling digital synthesizer to date has been the *Yamaha
DX-7,* shown in Fig. 29.3 (reportedly over 200,000 have been sold).
The DX-7 is the best known member of the DX family of synthesizers

FIG. 29.3
Yamaha DX-7 synthesizer (Courtesy of
Yamaha Corporation of America.)

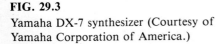

which feature FM-synthesis of sound (see Section 27.10). The DX-7, with up to 64 internal "voices," is designed for real-time performance; additional preset voices are available from a ROM cartridge or from floppy disks. Normally, the performer uses the *edit* mode to create or alter voices and performance parameters, and stores these in memory for use in the *performance* mode. Two voices can be played simultaneously, either in the *dual* mode (both voices combined over the entire keyboard) or the *split* mode (each voice is played using half the keyboard). Besides the keyboard, the performer controls the synthesizer by means of a modulation wheel (which controls the modulation index, see Section 27.10), a pitch-bend wheel (which "bends" the pitch up or down), a breath controller, several slider potentiometers, and two optional foot pedals.

Many digital synthesizers rely on (or at least incorporate) stored samples of real sounds to create new ones. Sampling and digital conversion by means of analog-to-digital converters (ADCs) was discussed in Section 28.2. Digital tone generators that use stored waveforms were described in Section 28.5.3.

If all the notes of an instrument exactly followed some scaling rule, it would be possible to sample and record a single note and play back the samples at different rates to synthesize other notes. In practice this is only partly satisfactory. The sound spectra of most musical instruments is substantially different at low, medium, and high frequencies (see, for example, the piano spectra in Fig. 14.5). This difference is illustrated in Demonstration 30 in Auditory Demonstrations (Houtsma, Rossing, and Wagenaars, 1987), in which the highest note of a bassoon is transposed downward in steps to create a scale. Such a transposed scale does not sound at all like a scale played on a bassoon.

In order for the synthesized sounds to closely resemble the natural sounds of an instrument, it would be necessary to record a large number of notes up and down the scale so that each recorded note would be transposed by only a few steps in either direction. This requires a great deal of memory, however, and is possible only in very large synthesizers or mainframe computers. Of course duplicating the natural sounds of instruments is only one function of a synthesizer, and probably a relatively minor one at that. Transposed sounds of instruments, although quite different in timbre from the original instruments, often serve very well as building blocks for creating new sounds.

The *Ensoniq Mirage,* shown in Fig. 29.4, is an example of a moderately priced digital sampling synthesizer. It is an 8-voice synthesizer with a variable sampling rate (8 to 33.3 kHz, expandable to 50 kHz) and sample times from two to eight seconds. The 61-note keyboard is

FIG. 29.4
Ensoniq Mirage synthesizer.

velocity sensitive, and three full-keyboard sounds, composed of 16 wave samples each, can be stored on a 3½-inch floppy disk. Several prerecorded sound diskettes are available. The Mirage includes 16 digital oscillators plus envelope generators, filters, etc. An optional visual editing system allows waveforms to be displayed on the screen of an Apple microcomputer (Fryer, 1986).

The *Kurzweil 250,* shown in Fig. 29.5, uses sampling rates ranging from 25 to 50 kHz, with corresponding sampling times of 20 to 10

FIG. 29.5
Kurzweil 250 synthesizer. (Courtesy of Kurzweil Music Systems, Inc.)

seconds. Coded representations of 40 musical sounds are stored in an extensive ROM in order to be on-line for immediate call. These representations include envelope information that describes how the spectra vary from lowest to highest notes and from soft to loud playing (Moog, 1986). Using this arrangement, the Kurzweil synthesizer has acquired an enviable reputation for its ability to synthesize realistic piano sounds.

A very successful digital synthesizer in the higher-priced studio synthesizer category has been the *Synclavier* by New England Digital,

which employs FM, additive and sampling techniques. This synthesizer, shown in Fig. 29.6, incorporates a spectrum analyzer to determine the harmonic content of real instrument sounds and resynthesizes up to 24 harmonics. Superimposing an FM-synthesized tone with a slightly different time envelope results in a more lifelike sound. Each pair of oscillators is called a "voice," and up to four voices or "partial timbres," can be controlled by each key (Lerner, 1983).

The Synclavier output is recorded on floppy discs. Up to 16 synchronized tracks can be recorded, as in the case of a multitrack tape recorder. A computer terminal can be used, in addition to the keyboard, for data entry.

The *Roland D-50* is a moderately priced all-digital synthesizer that combines subtractive synthesis (as in an analog synthesizer) with synthesis from sound samples, using a process called "linear arithmetic" synthesis. Each of two "partials" can be constructed by combining sounds synthesized by each of these two methods, and the two partials, in turn, can be mixed into a "tone." Two such tones can be synthesized and assigned to the upper and lower halves of the keyboard (split mode); a single tone can be assigned to the entire keyboard (whole mode); or both lower and upper tones can be played on the entire keyboard (dual mode). In the "whole" mode, 16 voices can be played, while the other two modes each allow eight-voice polyphony.

29.8 HYBRID DIGITAL/ANALOG SYNTHESIZERS

Hybrid synthesizers, which combine both digital and analog circuitry, have proven to be convenient and cost effective. A wide variety of arrangements are possible, ranging from an analog synthesizer controlled by a digital computer to a digital synthesizer which includes a few analog modules (such as voltage-controlled filters).

1. *Digital control of analog synthesizers.* Early in the development of electronic music, computers were called on to serve as large and

versatile *sequencers:* to supply repeating patterns of control voltages and triggers to several voltage-controlled synthesizer modules. Pioneering efforts by Donald Buchla and Peter Zinovieff advanced this technique (see Section 27.1; also Fig. 27.28(a)).

Meanwhile at Bell Laboratories, Max Mathews and F. Richard Moore developed the GROOVE system (generated real-time operation on voltage-controlled equipment), designed to add "performer nuance" to computer-generated music. All of the manual actions of the composer/performer (playing a keyboard, twisting knobs, etc.) are monitored and stored digitally in the computer as functions of time. These digital functions can be edited in any way the composer/conductor wishes: tempos and dynamic levels can be changed, particular voices can be emphasized, new time sequences can be added, etc. Thus edited, they are converted back to analog signals with a DAC and applied to voltage-controlled synthesizer modules. GROOVE allows a single musician to assume the roles of composer, performer, and conductor (Mathews and Moore, 1970).

Microprocessors, which made it possible to incorporate digital sequencers and control voltage generators within the synthesizer itself, have led to development of portable hybrid synthesizers. Nearly all analog synthesizers now incorporate digital control. Examples include the Oberheim *Xpander* and *Matrix 12* and the Rhodes *Chroma*.

The Rhodes Chroma, shown in Fig. 29.7, uses a 6809 microprocessor to control 16 analog oscillators (VCOs), VCFs, and VCAs. Envelope generators, however, are implemented by computer soft-

FIG. 29.7
Rhodes Chroma synthesizer. (Chroma is a product of Fender Musical Instruments.)

ware and applied via a DAC. A RAM stores 100 complete patches; the control program is stored in a ROM (Chamberlain, 1985).

The Chroma uses a separate 8039 microprocessor to scan the 61-key velocity-sensitive keyboard every millisecond. The control panel includes 50 numbered buttons which are used to specify 25 *control* parameters (voice board patches, ways of assigning keys to oscillators, output channel selects, envelope parameters, etc.) and 25 *audio* parameters (tuning offset, pitch control source, waveshape, pulse width, filter controls, etc.).

2. *Mixed analog and digital circuits.* A number of synthesizers mix analog and digital circuits. Since digital oscillators are relatively simple (Section 28.5), whereas digital filters tend to be more complex than analog filters (Section 28.6), some synthesizers (e.g., Roland JX-3P and JX-8P) combine digital oscillators with analog filters and amplifiers.

29.9   DRUM SYNTHESIZERS

A special class of synthesizers, often referred to as "drum machines," have become popular, especially in recording studios. Some drum machines produce synthesized or sampled real percussive sounds when a drummer strikes a contact pad. Others operate according to preprogrammed rhythms. Still others combine both modes of operation.

29.10   MIDI

Musical Instrument Digital Interface (MIDI) defines a standard interface and digital code for transmitting control and timing information in real time between synthesizers, sequencers, computers, drum machines, etc. On stage, performers use MIDI to network their synthesizers, sequencers, and keyboard controllers. In studios and research laboratories, MIDI provides an inexpensive, user-friendly link between large computers and tone-producing instruments.

The MIDI interface is *serial* and (to a certain extent) *bidirectional*. A MIDI instrument may have MIDI In, MIDI Thru, and MIDI Out connectors. MIDI Thru delivers an exact replica of the information received at MIDI In. The drive signal is a 5-mA current. MIDI has been described as a "16-channel party line."

MIDI serial data flows at a rate of 31.25 kbaud or 31,250 bits per second (which equals a 1 MHz clock rate divided by 32). A MIDI data stream is organized into 10-bit words; the first (start) bit and the last (stop) bit are always zeros, while the desired information is carried by the eight-bit byte in between.

There are two types of bytes: *status* and *data*. Status bytes always begin with a one and give the address for the data that follows or else define states or events with no numerical values. Data bytes begin with a zero and give the numerical value of some physical entity (e.g., the setting of some panel control).

Status bytes intended for all instruments in the system are called *system* bytes; *channel* bytes, on the other hand, contain a four-bit number that electronically addresses the message to one of 16 pre-assigned MIDI channels. Channel messages include: note on/note off, polyphonic key pressure, control change, program change, pitch bend change, voice mode (mono, poly, omni) on/off, etc. Good descriptions of MIDI protocol are given by Anderton (1986) and Moog (1986).

Considerable attention has been devoted to development of new control devices with MIDI outputs. Several stand-alone touch-sensitive MIDI keyboards have appeared, as have pitch trackers for guitars, voice, and various musical instruments. (One performer commented that he uses a MIDI guitar interface to "pad his rock-guitar solos with sampled Japanese subway screeches"). Continued development of nonkeyboard controllers will open up expanded opportunities for composers and performers to control sound parameters and create new sounds.

29.11   DIGITAL ELECTRONIC ORGANS

The Allen digital computer organ, mentioned in Chapter 26, will now be described in greater detail. It is an example of a complete concert instrument employing a digital sampling technique. It stores 5500 bits (binary digits) of information in read-only memories, as well as 1600 bits in regular read-write memories. In every microsecond (one-millionth of a second), it can perform ten additions and four multiplications simultaneously. In other words, during one complete cycle of vibration at the frequency of middle C (3.8 milliseconds), the processor will have performed 38,200 additions and 15,280 multiplications.

29.12   DIGITAL GENERATION OF ORGAN TONE

The first step in designing a digital organ is to collect data about desired sounds, which can then be stored in one of the memories. The Allen organ uses 16 sample intervals distributed over the first half of a voice envelope only. This saving on the amount of data that must be stored makes use of a clever trick. In constructing the voice envelope to be sampled, the phases of the harmonics that make up the tone have been adjusted to produce a waveform whose first half cycle and second half cycle are mirror images of each other except for sign. Since the ear is insensitive to the phase relationship between harmonics, this process has little if any effect on the tone quality.

The organization of the Allen Digital Computer Organ is shown in Fig. 29.8. The clock provides timing pulses to the rest of the system. The basic clock rate is 4 MHz (four million pulses per second), and the clock pulses are subdivided into four different phases.

The memories are designated *registration* memory and *specification* memory. Actually, there are several registration memories, one handling swell flute voices, one pedal voices, etc. The function of these memories will be described shortly. The frequency generator sup-

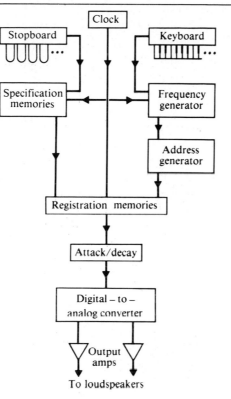

FIG. 29.8

Allen Digital Computer Organ System.

plies pulses to the address generator at the proper rate. Input to the processor is provided by the stop switches and the keyboard. In addition to the voices stored in the specification memory, the larger organs have several "alterable voices," stops that may be programmed by the organist by feeding punched cards into a card reader.

After all the processing has been completed, the digital information is ready to be converted to a conventional audio signal. This is done in a digital-to-analog converter (DAC) in a matter of microseconds. The audio signal then passes to conventional audio amplifiers and loudspeakers.

The microprocessor used in the Allen organ contains the equivalent of 48,000 transistors; twenty-two tiny MOS chips, each about one-tenth of an inch square, contain the computer circuitry. Larger organ models have two identical independent computers. The LSI approach, with its high reliability and low cost, has made the digital organ technically feasible.

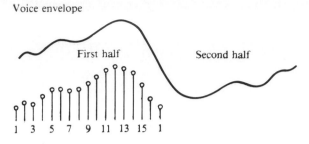

FIG. 29.9
Voice envelope and sample points used to obtain data for the digital generation of tone.

Figure 29.9 illustrates the voice envelope and the sampling process. Although it has the same harmonic structure as the original tone, the voice envelope will not look the same as the original tone viewed on an oscilloscope, because the phases of the harmonics have been altered. The 16 samples constitute the desired voice information, which is then stored as 16 seven-bit numbers ("words") in the specification or voice memory. The *specification memory* is a read-only memory (ROM), the voice information being built into it during its manufacture.

Voice reconstruction begins by the transfer of voice information from the specification memory to the registration memories. Each *registration* memory is a read-write or random-access memory (RAM), which means that information can be written in as well as read out. The function of the registration memory is to allow several voices to be combined into a single composite voice. There are several registration memories, corresponding to different divisions of the organ.

Once the registration memory is loaded with the proper voice information, it can be read out at a rate that is determined by which key on the keyboard is depressed. Circuits that participate in the readout process are shown in Fig. 29.10.

FIG. 29.10
Tone generation in a digital organ. Voice information is transferred from specification memory to registration memory and readout at a specified rate to reconstruct the tone.

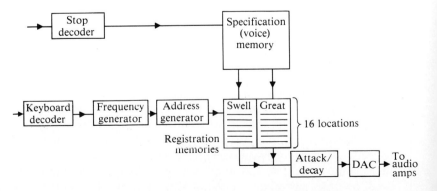

Pulses from the keyboard control the frequency generator, which in turn sends a string of pulses to the address generator at 32 times the desired audio frequency. Each time the address generator receives a pulse, it advances to the next higher address until it reaches 16, at which point it begins counting down, and the data are multiplied by −1 to reconstruct the second half of the tone cycle.

The registration memory readout is time-shared (by multiplexing) by the number of notes in the chord being played, up to a limit of 12. These (digitally coded) waveforms are then multiplied by a series of numbers that describe the *attack* and *decay* envelopes, and are then combined to produce a composite signal that is sent to the digital-to-analog converter for conversion to an audio tone. The entire complex process consumes only a small fraction of a second.

The digital generation of tone has several inherent advantages. Much more control of transient sound during attack and decay is possible. A slight lowering of pitch and change in harmonic content (in the manner of pipe organs) is possible. Chiff and wind noise can be added quite easily. The decay time on each manual can be made variable to compensate for the lack of reverberation in many smaller churches. Another interesting feature on digital organs is the transposer, which, by merely changing the position of control pulses, can accurately transpose the pitch of the entire organ up or down.

29.13   SUMMARY

Digital computers, which have had a great impact on science and commerce, are beginning to play a significant role in music. Digital computers use binary numbers and binary logic circuits. A list of instructions, called a program, directs the computer, and operations are synchronized by a clock. Compilers can be used to translate programs from algebraic to machine language. A number of programs have been written for generating music on digital computers; most of them are descendants of the MUSIC programs developed at the Bell Telephone Laboratories.

The microelectronics revolution has led to the development of a new generation of digital synthesizers controlled by microprocessors. Some of these synthesize new sounds from stored samples of real instruments; others make use of FM synthesis; some combine both digital and analog circuitry. Widespread use of MIDI interfacing has made it possible to use networks of computers, synthesizers, and other instruments by different manufacturers. An early example of a concert instrument using digital sampling techniques is the Allen digital computer organ.

References and Suggested Readings

Anderton, C. (1986). *MIDI for Musicians.* New York: Amsco.

Chamberlain, H. (1985). *Musical Applications of Microprocessors.* Indianapolis: Hayden.

Fryer, T. (1986). "Ensoniq Mirage DSK-8 Digitally Sampling Keyboard," *Recording Eng./Producer* **17**(2): 122.

Houtsma, A. J. M., T. D. Rossing, and W. M. Wagenaars (1987). "Auditory Demonstrations," Philips Compact Disc No. 1126–061, distributed by *Acoust. Soc. Am.*

Howe, H. Jr. (1975). *Electronic Music Synthesis.* New York: Norton.

Kemeny, J. G., and T. E. Kurtz (1968). *Basic Programming.* New York: Wiley.

Lerner, E. J. (1983). "Electronically synthesized music," *IEEE Spectrum* **20**(6): 46.

Mathews, M. V. (1969). *The Technology of Computer Music.* Cambridge, Mass: M.I.T. Press.

Mathews, M. V., and F. R. Moore (1970). "GROOVE-A Program to Compose, Store, and Edit Functions of Time," *Commun. ACM* **13**: 715.

Mathews, M. V., F. R. Moore, and J. C. Risset (1974). "Computers and Future Music," *Science* **183**: 263.

Moog, R. (1986). "MIDI: Musical Instrument Digital Interface," *J. Audio Eng. Soc.* **34**: 394.

Moore, F. R. (1987). "What is a Computer Music Workstation?" in *Music and Digital Technology,* ed. J. Strawn. New York: Audio Engineering Society.

Glossary

algorithm Step-by-step directions for solving a problem (multiplying two numbers, for example).

analog computer A computer that represents numbers as electrical voltages.

analog-to-digital converter (ADC) A circuit that converts numbers from an analog to a digital representation.

binary number A number in base-two representation; a 1 or a 0.

bit A single binary number.

byte A group of bits, usually eight.

central processing unit (CPU) The arithmetic and control sections of a computer.

clock pulse A timing pulse distributed throughout a computer to synchronize its various operations.

compiler A program used to translate another program into the machine language of a digital computer.

digital computer A computer that represents numbers as binary digits.

digital-to-analog converter (DAC) A circuit that converts numbers from a digital to an analog representation.

drum synthesizer A synthesizer designed to produce percussion sounds.

FM synthesis Synthesis that relies on frequency modulation of one signal by another (see Section 27.10).

hybrid synthesizer A synthesizer that combines analog and digital circuitry.

microprocessor The central processing unit of a computer constructed on a single integrated circuit chip.

MIDI A musical instrument digital interface that allows computers, synthesizers, and other instruments to communicate.

MUSIC V A widely used computer language, developed at the Bell Laboratories, for generating musical sound.

random-access memory (RAM) A computer memory in which any storage register can be addressed for read or write operations.

read-only memory (ROM) A computer memory that stores information permanently.

sequencer A device that supplies the control signals needed to generate repeating patterns of musical sound.

word A binary number or binary-coded alphabetic word.

*Questions for Discussion*_____

1. A pocket calculator, like a large computer, uses binary numbers internally. Why is it unnecessary to understand binary numbers in order to use a pocket calculator?

2. What are some advantages of using computer-controlled synthesizers rather than digital sound synthesis?

3. What are the advantages of programming in a "high level" language such as FORTRAN, BASIC, or COBOL rather than in "machine" language?

4. What is the main advantage of FM synthesis over additive or subtractive synthesis?

*Problems*_____

1. How many bits of information can be stored in:

   a) a 64 K memory of 16-bit words?

   b) an 80-megabyte hard disc?

2. a) How far does a beam of light travel in 10 ns?

   b) How far do radio waves travel in 10 ns?

   c) How far does a sound wave travel in 10 ns?

   (10 ns is the memory access time in a fast computer.)

3. A $3\frac{1}{2}$-in diameter floppy disc stores 0.5 MB. Assuming that a 1-in diameter circle in the center is not used, what is the average area of each storage cell? Estimate the gap width in the recording head (see Chap. 21).

4. Some microprocessors have the equivalent of a million transistors. Estimate the dimensions of a circuit board with 10^6 separate transistors.

Noise has been receiving increasing recognition as one of our critical environmental pollution problems. Like air and water pollution, noise pollution increases with population density; in our urban areas, it is a serious threat to our quality of life. Noise-induced hearing loss is a major health problem for millions of people employed in noisy environments. Besides actual hearing loss, however, human beings are affected in many other ways by high levels of noise. Interference with speech, interruption of sleep, and other physiological and psychological effects of noise have been the subject of considerable study recently.

Finding the technical solutions to many of our environmental noise problems requires the work of scientists and engineers with considerable knowledge of acoustics. On the other hand, many problems require social and political action rather than technical solutions. Thus it is important that the public be well informed about the basic principles of acoustics and noise control.

Chapters 30, 31, and 32 are intended as an introduction to environmental noise for the intelligent layperson, perhaps for future legislators or other governmental officials. Chapter 30 provides an introduction to the subject, Chapter 31 describes the way in which noise affects people, and Chapter 32 deals with the control of noise through both regulations and technical solutions.

PART EIGHT

Environmental Noise

CHAPTER 30

Noise in the Environment

Music and speech are not the only contributions humans have made to the world of sound. With few exceptions, advances in technology, such as the development of labor-saving machines, have resulted in a steady increase in the amount of unwanted sound, which we call *noise*.

If we try to think of all possible sound generators, we will probably find that they fall into one of the following categories:

1. Vibrating solid bodies, such as bars, membranes, plates, loudspeaker cones, etc.;

2. Vibrating air columns, such as those in musical instruments;

3. Flow noise in fluids due to turbulence, such as that which occurs in jet engines or from air leaking out of an air hose;

4. Interaction of a moving solid (such as a rotating propeller blade) with a fluid, or a moving fluid with a solid (such as air flowing in a duct or through a grill);

5. Rapid changes in temperature or pressure, such as the thunder caused by a lightning discharge or a chemical explosion;

6. Shock waves caused by motion or flow at supersonic speed.

In this chapter we will discuss some common noise sources that incorporate sound generators in one or more of these categories.

30.1   SOUND POWER AND MECHANICAL POWER

Fortunately for the environment, even the noisiest machines convert only a small part of their total energy into sound. A modern jet aircraft, for example, may produce a kilowatt or more of acoustic power, but this is less than 0.01 percent of its mechanical output of 55,000 kilowatts (Shaw, 1975). Automobiles emit approximately 0.001 percent (one-thousandth) of their power as sound, and household appliances are comparable to this in "efficiency" as noise sources.

Figure 30.1 compares the A-weighted sound power (see Section 6.2) of a number of large and small noise sources to their mechanical power. The appropriate values for the sound power level L_w are also given. The diagonal lines represent different percentages of mechanical power converted to sound power. The line labeled "FAA Rule 36" refers to the Federal Aviation Administration regulation on noise for new aircraft certified after 1969.

FIG. 30.1
Sound power compared to mechanical power for various machines. Also shown are the corresponding A-weighted sound power levels L_w (*A*). (After Shaw, 1975).

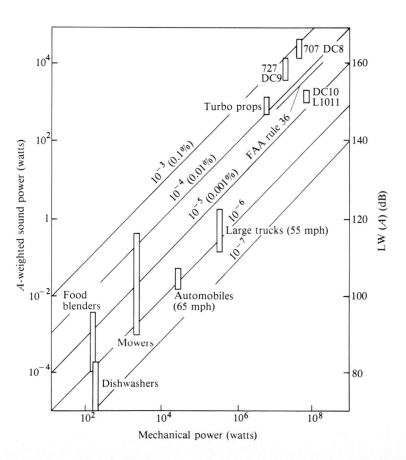

30.2 NOISE LEVELS

The sound pressure level L_p decreases as we move away from the sound source. In a free field (which may exist outdoors or in an anechoic room), the sound pressure level decreases by 6 dB each time the distance from the source doubles (see Section 6.2). Thus, the sound pressure level one or two meters from a small appliance may exceed that due to jet aircraft at a distance of ten kilometers.

The most commonly used measure of noise is the A-weighted sound pressure level $L_p(A)$, which de-emphasizes sounds of low frequency (see Fig. 6.5). Sound level meters are nearly always equipped with a weighting network, so that the value of $L_p(A)$ can be read directly. A-weighted levels are frequently expressed as dB(A) or dBA (see Section 6.4).

Since the sound pressure level at a given location may fluctuate at various times, the use of some averaging procedure is desirable. One way to describe noise is to express L_x, the level exceeded x percent of the time, for different values of x. Thus, L_{10} is the A-weighted sound level exceeded 10 percent of the time; L_{50} is the median level, since it is exceeded 50 percent of the time, and so on.

Another useful measure of noise is the *energy-equivalent level*, or *equivalent level* L_{eq}. It is defined as the decibel level of the steady noise that would give the same total energy over the same time period. For example, a single peak at 100 dB lasting 3.6 s is equivalent in energy to a whole hour of steady noise at 70 dB. The advantage of using L_{eq} rather than L_{50} or L_{90}, for example, in the description of noise is its sensitivity to short peaks of noise, which can be very annoying. Because noise that occurs at night is usually more annoying, a *day-night equivalent level* L_{dn} is frequently used to describe noise. In arriving at L_{dn}, one adds 10 dB to the sound level between 10:00 P.M. and 7:00 A.M. in the averaging process.

TABLE 30.1   Noise levels used to rate noise.

$L_p(A)$	A-weighted sound pressure level
L_{10}	A-weighted sound level exceeded 10 percent of the time
L_{50}	A-weighted sound level exceeded 50 percent of the time
L_{90}	A-weighted sound level exceeded 90 percent of the time
L_{eq}	Sound pressure level of steady noise that would give the same total energy as the noise being rated
L_{dn}	The same as L_{eq}, with 10 dB added to noise measured between 10:00 P.M. and 7:00 A.M.
PNL	Perceived noise level (used to rate aircraft noise)
EPNL	Effective perceived noise level (used to rate aircraft noise) takes into account both maximum loudness and duration of noise events

Still other noise levels may be found in the literature on noise control. Perceived noise level PNL (expressed in PNdB) and effective perceived noise level EPNL (expressed in EPNdB) are two criteria frequently used to describe aircraft noise. EPNL takes into account both the maximum loudness level and the duration of an event, such as an aircraft flyover. The noise-exposure forecast NEF, derived from EPNL, also takes into account the frequency of the events in a given neighborhood. Various noise levels are summarized in Table 30.1.

30.3 SOUND PROPAGATION OUTDOORS

Although sound propagation outdoors may approach that of a free field when conditions are ideal, this rarely takes place in actual practice. Atmospheric turbulence, temperature and wind gradients, molecular absorption in the atmosphere, and reflection from the earth's surface all affect propagation and cause fluctuations in the sound intensity level at the receiver.

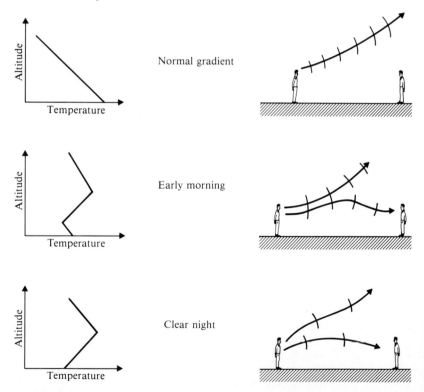

FIG. 30.2
Refraction of sound under different conditions.

Attenuation is strongly influenced by the type of ground cover present. The attenuation of noise through a dense forest may be as great as 20 dB per 100 meters. The attenuation through thick grass and shrubbery may be even greater. Mounds of earth placed beside highways are very effective noise barriers. Depressing a highway by about 4 meters typically reduces the sound level by 7 to 10 dBA.

Refraction, or the bending of sound by temperature and wind gradients, was described in Section 3.9. There it was shown that sound could be bent up away from the receiver (leading to attenuation in excess of the free field condition—a 6-dB loss when the distance from the source is doubled), or sound can be bent down toward the receiver, resulting in attenuation that is less than the free-field condition. Normally, the temperature decreases with altitude; thus there is an upward refraction, since sound travels faster in the warm air near the surface of the earth. Two examples of temperature inversion that will cause downward refraction are illustrated in Fig. 30.2. Mining companies, for example, carefully monitor the weather conditions to decide on the best time for blasting operations in order to minimize noise levels in surrounding communities.

Sound waves are weakly absorbed by the atmosphere itself. Atmospheric attenuation occurs mainly at high frequency, and depends rather strongly on humidity and temperature. Atmospheric attenuation in three different octave bands is shown in Fig. 30.3.

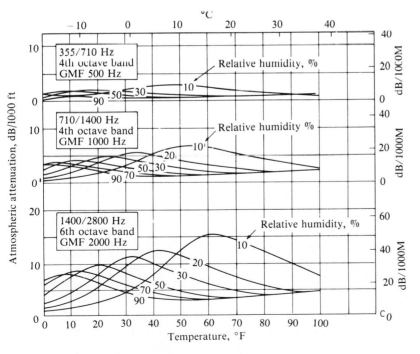

FIG. 30.3

Atmospheric attenuation of sound at 50, 1000, and 2000 Hz. (From *Noise and Vibration Control,* revised. © 1988 Leo Beranek. Used with author's permission.)

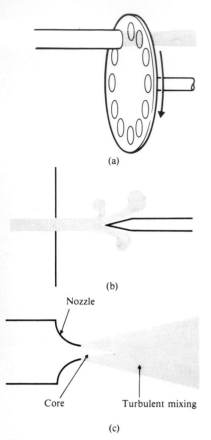

(a)

(b)

Nozzle

Core   Turbulent mixing

(c)

FIG. 30.4
Examples of three types of gas flow
noise: (a) siren (monopole source); (b)
edgetone generator (dipole source);
(c) gas jet (quadrupole source).

30.4 FLOW NOISE

Noise associated with gas flow may be of three general types; an example of each type is shown in Fig. 30.4. The first type occurs when the flow of air is interrupted at a regular rate, as in a siren, giving rise to what is called a *monopole* source. The sound power radiated from such a source varies as the fourth power of the flow velocity. A second type of noise is generated when a moving stream of air strikes a solid object, forming a *dipole* source. The noise of air blowing through a grill, the sound generated by a moving fan blade, and the aerodynamic whistles described in Section 12.8 are examples of this type of noise. Dipole radiation is even more dependent on flow velocity, the power varying as the sixth power of velocity (v^6). Finally there is *quadrupole* radiation, which is weak at ordinary flow velocities but increases with v^8, so that at high flow velocities, it becomes appreciable. The noise from a gas jet (as in a jet airplane engine) is of this type.

From the strong dependence of flow noise on velocity, we can see that the best way to minimize flow noise is to utilize high-volume, low-velocity flow whenever possible. Doubling the flow velocity increases the strength of monopole sources 16 times, that of dipole sources 64 times, and that of quadrupole sources 256 times.

The flow of air is often maintained by fans, which are common sources of noise. Fan noise is of two types: rotation noise and aerodynamic noise. Rotation noise, common to all rotating machinery, will be discussed in Section 30.5. Aerodynamic noise is generated by swirls of air, or *vortices*, caused by the blades moving through the air. Aerodynamic noise is amplified if there are stationary objects or guide vanes nearby with which the vortices interact.

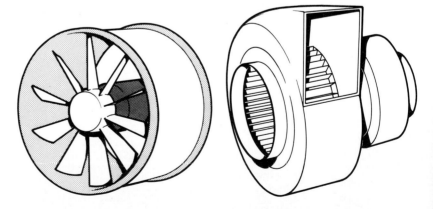

FIG. 30.5
Axial fan and centrifugal fan.

Fans are of two types: axial and centrifugal (see Fig. 30.5). Axial fans have propeller blades that push air in the direction determined by their pitch. Centrifugal fans spin the air, and centrifugal force causes an outward flow of air that is funnelled into the discharge duct. Centrifugal fans are noisiest at low frequency, which is less annoying. Axial fans, on the other hand, produce a more prominent noise, because the aerodynamic noise covers a wider range of frequency.

30.5   MACHINERY NOISE

In Section 30.1, we pointed out that the sound power output of various machines is typically from 10^{-7} to 10^{-3} of the corresponding mechanical power output. Although these are small fractions the sound power from the large number of machines found in a factory (or even a home) can create a noisy environment. In order to control environmental noise, we must understand the main noise sources in machines.

Possible noise sources in machines include unbalance in rotating parts, friction, bearing noise, gear noise, hum induced by magnetic fields, fan noise, impact noise, and turbulence. The noise source itself may not emit much airborne sound, but it may act as an energy source for vibrations that are transmitted through the structure to be emitted as sound by some other part of the machine. This is especially noticeable if the driving source vibrates at the natural resonance frequency of a large radiating member, such as a metal cover panel. Airborne noise is usually caused by a vibrating surface or by turbulent air.

Noise reduction in machines may be reduced by one or more of the following courses of action: reducing the noise generated by the source; decoupling the source from its resonators; damping the resonators; reducing the noise radiation from a radiating source or resonator. The best course of action frequently cannot be determined until the main noise sources have been located through acoustical testing.

Often the frequency of a prominent noise gives away its origin. Magnetic hum from a motor or transformer will usually occur at the power-line frequency (60 Hz) or its harmonics. Fan noise will have a strong component at the blade-passage frequency. Vibrations generated by rotating parts will be related to the frequency of rotation.

Inspection often indicates the radiating members. Sometimes they can only be determined by attaching *accelerometers*, which generate electrical signals proportional to acceleration when they vibrate. Using an accelerometer enables one to determine what surfaces are vibrating strongly at the frequency of the most prominent components in the noise spectrum.

A rectangular panel or thin plate has a large number of resonance frequencies. If it is supported (but not rigidly clamped) at its boundaries, its resonance frequencies are given by

$$f_{mn} = 0.48\, v_L h \left[\left(\frac{m}{l} \right)^2 + \left(\frac{n}{w} \right)^2 \right];$$

v_L = speed of longitudinal (sound) waves in a bar (m/s) (see Table 13.1);

h = thickness of plate (meters);

l, w = length and width of plate (meters);

m, n = integers (1, 2, 3 . . .).

If the plate is clamped at its boundaries, the lowest resonance frequency is nearly doubled, but the others are raised by a smaller factor.

Noise reduction in machines is most effective when applied at the source. This may involve balancing rotating parts, repairing or redesigning bearings, replacing gears, etc. Isolation of the source from resonators may be accomplished by installing resilient mountings or even by enclosing a noisy part in a housing. Panels that radiate noise can be damped by covering them with special materials having a high internal friction.

30.6   INDOOR NOISE

It is appropriate at this time to review Chapters 23 and 24, in which we discussed the acoustics of rooms. Let us for the moment look at Section 24.2, in which we focused on direct and reverberant sound fields, and in particular on Fig. 24.2, which shows the relationship between sound power level L_w and sound pressure level L_p in different types of rooms. This chart is useful in determining the amount of sound reinforcement needed in an auditorium, but it is also useful in determining the sound level to be expected from a noise source indoors.

The direct sound field for a nondirectional source ($Q = 1$) follows the $1/r$ relationship characteristic of a free field (see Section 6.2), and this is represented by the $R = \infty$ line in Fig. 24.2. The level of the direct sound decreases by 6 dB for each doubling of the distance, once we have moved out of the near field of the source. The sound level in a room is found to level off at a constant value as we move away from the source, and this is called the reverberant level. We have assumed sufficient reflecting surfaces in the room to distribute the reverberant sound evenly throughout the room; in many rooms, this will not be the case. In furnished residential rooms, for example, the rate of decrease appears to be 3 to 4 dB per doubling of distance throughout a large part of the room; in other words, there is a blending of the direct and reverberant fields. Similar behavior is noted in large, low rooms with distant sidewalls (Schultz, 1980).

In order to describe completely the strength of a noise source, both the *sound power level* (or sound power) and the *directivity* must be known at all frequencies. In Equations 24.1 and 24.2, the directivity is expressed by a single value Q, which is the ratio of the sound intensity in front of a source to that averaged in all directions. In the case of a complex source such as a large machine, this is insufficient.

Ideally, the sound power level of a source should be measured in a reverberation room and the directivity in an anechoic room. Often this is not practical. Many sophisticated methods have been devised for determining both sound power level and directivity in real rooms. They usually involve considerable computation based on a number of samples of the sound field (Beranek, 1971).

30.7   MOTOR VEHICLES

With more than 100 million passenger cars in the United States traveling over 10^{12} miles annually, automobiles generate megawatts of acoustic power. Fortunately, much of this power is radiated to areas with low population density, but an appreciable portion is generated in urban areas. Figure 30.6 shows typical noise levels for automobiles, trucks, and buses measured 50 feet from the highway as well as inside the vehicle itself. Sources of noise in an automobile or truck include the engine, the cooling fan, the drive train, the tires, aerodynamic turbulence, body vibrations, and the intake and exhaust systems.

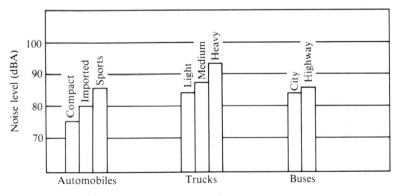

FIG. 30.6
Typical noise level for motor vehicles measured at 50 feet from the highway.

The modern automobile, when kept in good repair, is a relatively quiet vehicle at low speeds. Engine noise and aerodynamic noise are low at these speeds, and most of the noise is radiated from the exhaust system. Engine noise and aerodynamic noise increase with speed, however, and at high speeds tire noise also becomes an important factor. The average sound level, measured 50 feet from the center of the roadway, rises about 10 dB for each doubling of speed.

Although some work is being done on higher-performance mufflers, quieter tire treads, quieter fans, etc., for automobiles, it is unlikely that any great reduction in noise emission can be expected. The greatest problem with respect to automobile noise appears to be poor maintenance of equipment, especially muffler and exhaust systems.

Trucks generate about 10 times as much mechanical power as do automobiles, but they emit from 10 to 100 times as much acoustic power. Thus, they present a much more serious problem so far as road noise is concerned. Sources of truck noise are roughly the same as those of automobile noise, but the proportions assigned to each source are different. At low speeds, exhaust and fan noise are most important. At high speed, however, tire noise takes over completely. Tire noise at 55 mph for a single-chassis truck ranges from 75 to 95 dBA, depending on the design of the tread, as shown in Fig. 30.7.

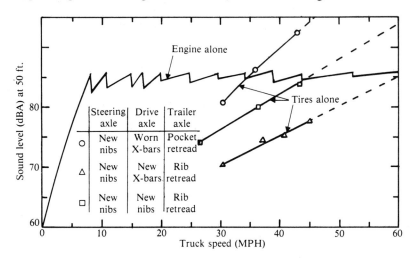

FIG. 30.7
Engine and tire noise at various speeds for an 18-wheel tractor-trailer. Note that the truck has shifted gears 11 times. (From Close and Atkinson, 1973. Reprinted by permission of *Sound and Vibration*.)

The main noise source in motor vehicles is the combustion or explosion of the fuel-air mixture inside the cylinders. Fortunately, this very powerful noise source is buried deep inside the massive engine, and therefore is well attenuated. Some of the energy of combustion does appear as noise, however, due to vibration of the entire engine as well as individual parts. Furthermore, when the exhaust and intake valves open, loud sounds of short duration are emitted, especially in the exhaust system, since the exhaust valve opens when the cylinder pressure is still quite high.

Diesel engines are noisier than gasoline engines, because the combustion is more sudden so that cylinder pressure rises more abruptly.

Furthermore, a diesel engine produces nearly as much noise under no load as it does under full load.

Engine cooling fans produce a substantial amount of noise unnecessarily. Fans must be large enough to cool the engine in a stationary vehicle with the engine running at low speed. When the vehicle is moving on the highway, there is little need for forced air flow. Furthermore, fan noise increases rapidly with speed, and at high speeds it can be as noisy as the engine itself. Substitution of an electrically driven fan or use of a magnetic clutch so that the fan operates only when needed not only reduces noise but results in a saving of fuel.

Using the authority granted by the Noise Control Act of 1972, the Environmental Protection Agency set the following noise limits for interstate motor carriers having gross weights of over 10,000 lb:

1. 86 dBA on highways with speed limits of 35 mph or less;

2. 90 dBA on highways with speed limits over 35 mph;

3. 88 dBA at full throttle with transmission in neutral.

These levels are to be measured 50 ft from the lane of travel, and apply to any grade. In addition, the rules provide for visual inspection of exhaust systems and tires, and forbid the use of tire treads "composed primarily of cavities in the tread which are not vented" (unless these tires can be shown not to exceed the 90-dBA limit at the legal speed limit).

The United States Environmental Protection Agency has estimated that from 5 to 55 percent of interstate trucks in different areas exceeded the standard in 1974, and that the cost of retrofitting them to conform to this standard would average $135 per truck. One of the most common modifications is the installation of a cooling fan with a thermostatically controlled clutch, which would be engaged only when needed for radiator cooling.

30.8   RAILROADS AND AIRCRAFT

Noise from railroad operations is not as widespread as noise from highways. Nevertheless, for persons living near a railroad, it can be an annoyance. Major sources of noise are locomotives, rail-wheel interaction, whistles and horns, yard retarders, refrigerator cars, maintenance operations, and loading equipment. A recording of a freight train passby is shown in Fig. 30.8.

The Environmental Protection Agency has proposed the following limits on locomotive noise measured 100 feet from the track:

1. 96 dBA for locomotives manufactured before 1980;

2. 90 dBA for locomotives manufactured in 1980 or after;

3. 73 dBA when idling (70 dBA for manufacture after 1980).

Railcars (or combinations of them) may not emit more than:

1. 88 dBA at speeds of up to 45 mph;

2. 93 dBA at speeds above 45 mph.

FIG. 30.8
Sound level of a freight train passby.
(From Close and Atkinson, 1973.
Reprinted by permission of *Sound and Vibration*.)

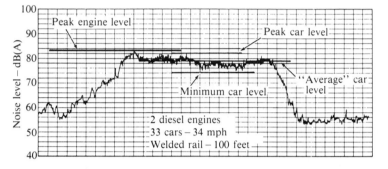

Some of the possible ways in which railroad noise can be reduced include equipping diesel locomotives with mufflers (an approximate 6-dB reduction), use of welded rails (an approximate 3-dB reduction), careful maintenance of rolling stock, and barrier walls around retarders in rail yards (up to a 20-dB reduction).

Aircraft noise is discussed in Sections 32.10 and 32.11.

30.9 SUMMARY

Noise has been receiving increasing recognition as one of our critical environmental problems. Various machines have sound power outputs that are only 10^{-7} to 10^{-3} of their mechanical power output. Nevertheless, this results in a substantial noise level. Various noise levels take into account the frequency of occurrence of annoying sounds as well as their levels. Atmospheric turbulence, temperature and wind gradients, atmospheric absorption, and reflections all affect the propagation of sound outdoors.

Noise associated with gas flow may be described as originating from monopole, diplole, or quadrupole sources. Machinery noise can originate in a number of places, the airborne noise often being radiated by panels or other surfaces at some distance from the source. Sources of noise in motor vehicles include the engine, the cooling fan, the drive train, tires, aerodynamic turbulence, intake, exhaust, and body vibrations. At high speeds, tire noises dominate, especially in the case of trucks. The EPA has written noise standards for both trucks and railroads.

References and Suggested Readings

Beranek, L. L., ed. (1971). *Noise and Vibration Control.* New York: McGraw-Hill.

Close, W. H., and T. Atkinson (1973). "Technical Basis for Motor Carrier and Railroad Noise Regulations," *Sound and Vibration* 7(10): 28.

Ingard, U. (1953). "The Physics of Outdoor Sound," *Proc. 4th National Noise Abatement Symposium.* (Reprinted in *Environmental Noise Control: Selected Reprints*, ed., T. D. Rossing, Am. Assn. Phys. Teachers, Stony Brook, N.Y., 1979).

Miller, T. D. (1976). "Machine Noise Analysis and Reduc-

tion," *Sound and Vibration* 1(3): 8.

Rossing, T. D. (1978). "Resource Letter ENC-1: Environmental Noise Control," *Am. J. Phys.* **46**: 444.

Schultz, T. J. (1980). "The Relation Between Sound Power Level and Sound Pressure Level Outside of Laboratories," *Noise Control Eng.* **14**: 24.

Shaw, E. A. G. (1975). "Noise Pollution—What Can Be Done?" *Physics Today* **28**(1): 46.

Stevens, S. S. (1961). "Procedure for Calculating Loudness: Mark VI," *J. Acoust. Soc. Am.* **33**: 1577.

Glossary

accelerometer A device used to measure vibration; its electrical output indicates its acceleration (see Section 1.4).

aerodynamic noise Noise generated by moving air or by the flow of air around a moving body (such as an automobile or truck).

day-night level L_{dn} Equivalent sound level that adds 10 dB to sounds that occur during the night-time hours of 10:00 P.M. to 7:00 A.M.

dipole source A noise source in which two halves are vibrating in opposite phase (e.g., two loudspeakers connected in opposite polarity, or the air at the leading and trailing edges of a fan blade).

equivalent level L_{eq} Sound pressure level that would give the same total energy as the noise being described.

monopole source A noise source in which the entire radiating surface vibrates in phase.

quadrupole source A noise source in which four parts vibrate alternately in phase.

refraction The bending of waves when the velocity changes (due to temperature and wind gradients, for example: see Section 3.9).

Questions for Discussion

1. Far more people are affected by truck noise than by airplane noise. Why are there fewer complaints, citizens' protest groups, etc., directed against highway noise?

2. Try to think of other examples of monopole, dipole, and quadrupole sound sources.

3. Describe the noise of a vacuum cleaner, and try to identify the two chief noise sources.

Problems

1. How many watts of sound power and mechanical power are delivered by a typical automobile?

2. Compare the attenuation at 500 Hz and 2000 Hz for the following atmospheric conditions.

   a) 70°F, 10 percent relative humidity

   b) 0°C, 30 percent relative humidity

3. Find the lowest resonance frequency of a steel panel ($v = 4905$ m/s) with dimensions $l = 1.5$ m, $w = 1.0$ m, and $h = 10^{-3}$ m.

4. Suppose that the velocity of air flow through a grill is cut in half but the grill area is doubled in order to maintain the same volume flow.

   a) By what factor will the sound power be reduced?

   b) By how many dB will the sound power level of the source decrease?

   c) Will the sound level in the room change by the same amount?

5. a) If a jet aircraft flying at an altitude of 10 km emits 1 kW of sound power, what will the sound level be on the ground? (Assume a free field; see Sections 6.1 and 24.2 and Equation 24.1.)

   b) If we add atmospheric attenuation of 30 dB/1000 m (see Fig. 30.3), what will the level be at the ground?

6. a) If an automobile traveling at 60 mph emits 0.01 W of acoustical power (Fig. 30.1), estimate the average continuous power from 100 million automobiles, each of which travels 10,000 miles per year.

   b) Make an estimate of the peak acoustical power (how many automobiles might be traveling at a peak hour of the day?).

CHAPTER 31

The Effects of Noise on People

Noise affects people in many ways. In addition to causing temporary and permanent hearing loss, noise interferes with speech communication, interrupts sleep, reduces human efficiency, and is believed to produce other physiological and psychological effects.

It is interesting to note that many of these important effects on community mental health have only recently received careful study. Thus it is understandable that there are differences of opinion about the seriousness of some of the more subtle physiological and psychological effects. Differing opinions have also been expressed about the levels of noise that can be tolerated without adverse effect on community health and welfare.

31.1 TEMPORARY HEARING LOSS

The ear has a remarkable ability to recover from conditions of overload if they are of short duration. Thus it is important to distinguish between noise-induced temporary threshold shift (NITTS or TTS) and noise-induced permanent threshold shift (NIPTS). Exposure to loud noise for a comparatively short time causes the ear to desensitize, especially at frequencies around 4000 Hz. However, if the overload is removed soon enough, the threshold will shift back almost to its normal level. If the stimulus continues or is repeated too frequently, however, a permanent shift can occur.

The primary measure of hearing loss is the hearing threshold level, the sound level of a tone that can just be detected. The threshold level at various test frequencies is usually measured with an *audiometer*. A pure-tone audiometer, shown in Fig. 31.1, allows tones of several frequencies to be applied at selected levels to one ear at a time. A plot of the threshold level at each frequency, compared to normal hearing, is called an *audiogram*. The reference threshold levels established for normal hearing with one type of earphone are also given in Fig. 31.1. Some audiometers automatically present the test tones and record the subject's responses on an audiogram.

Temporary threshold shifts can vary in magnitude from a change in hearing sensitivity of a few decibels in a narrow band of frequencies to shifts so large that the ear is temporarily deaf. The time required for hearing sensitivity to return to near-normal levels can vary from a few hours to two or three weeks. In spite of considerable research on the subject, the laws describing temporary threshold shift have not been completely determined. Figure 31.2 shows the hypothetical growth of threshold shift after exposure to noise of varying levels and duration. These curves are based on data from several laboratories, and on extrapolations made on the basis of data from animals.

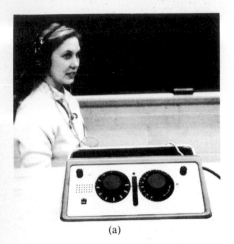

(a)

Frequency (Hz)	dB (re 20 μN/m^2)
125	45
250	25.5
500	11.5
1000	7
1500	6.5
2000	9
3000	10
4000	9.5
6000	15.5
8000	13

(b)

FIG. 31.1

(a) A pure-tone audiometer. (b) Pure-tone reference threshold levels for "normal" hearing.

FIG. 31.2

Hypothetical growth of temporary threshold shift due to exposure to noise of varying level and duration. (From J. D. Miller, (1974). "Effects of Noise on People," *J. Acoust. Soc. Am.* **56**: 729. Reprinted by permission from the Amer. Inst. of Physics.)

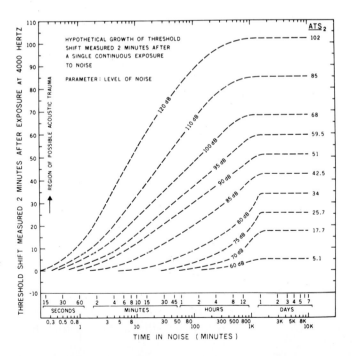

Although at this stage, curves of the type shown in Fig. 31.2 are hypothetical, they do indicate the way in which threshold shift depends on noise level and duration. For example, a shift of 34 dB, which is reached in one day at 80 dB, can be reached in about ten minutes at 110 dB. These data represent a "worst case," since the noise, centered at 4000 Hz, is in the range to which the ear is most susceptible to damage; and the threshold is measured at 4000 Hz, where threshold shifts are often the greatest.

Hypothetical curves for recovery from noise-induced threshold shifts are shown in Fig. 31.3. These curves are for seven-day exposure to noise at varying levels, and hence they could be considered continuations of the curves shown in Fig. 31.2. Recovery from a given shift is more rapid if the exposure time is less than eight hours, as can be seen by the single dotted curve, which presents recovery from a 102-min exposure to noise at a level of 95 dB. For shifts above 40 to 50 dB, recovery is not always complete.

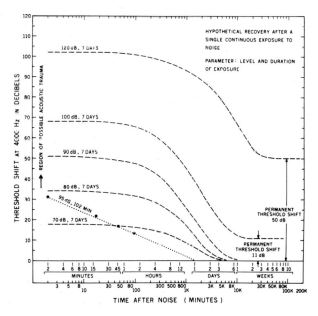

FIG. 31.3
Hypothetical recovery from various threshold shifts. (From Miller, 1974.)

Although the physiological basis of temporary hearing loss is not yet understood completely, the following statements summarize what is known about the relationship between noise and TTS (see Ward, 1969):

1. The growth of TTS (in dB) is a nearly linear function of the logarithm of time. That is, doubling the exposure time tends to double the threshold shift, as can be seen in Fig. 31.2.

2. Moderate TTS recovers in time, recovery usually being complete in 16 hours. However, when TTS reaches 35 or 40 dB, recovery takes days or weeks, and above 50 dB recovery may never be complete.

3. Low-frequency noise produces less TTS than does high-frequency noise.

4. Narrowband noise produces a maximum TTS one-half to one octave above the noise band rather than in it.

5. An intermittent noise produces much less TTS than does a steady one. In fact, TTS is nearly proportional to the fraction of the time that the noise is present.

6. Neither growth nor recovery of TTS appears to be influenced by drugs, medications, time of day, hypnosis, or state of mind. Thus it appears to be entirely a physiological effect.

31.2 PERMANENT HEARING LOSS

Noise-induced permanent threshold shift (NIPTS), like noise-induced temporary threshold shift (TTS), is usually greatest at frequencies around 4000 Hz. Figure 31.4(a) combines three audiograms to illustrate the progressive loss of hearing that might be found in a person who works in a noisy factory. In the early stages, hearing loss occurs mainly at frequencies between 2000 and 8000 Hz, and does not interfere very much with ordinary speech (which is carried mainly by sounds in the "speech band," 300–3000 Hz). As the exposure to noise continues, loss of hearing of speech becomes more apparent. The fact that noise-induced hearing loss generally occurs first at frequencies above the speech band provides a means of detecting a dangerous condition in order to take preventive action before the ability to hear and understand speech is lost.

In analyzing noise-induced hearing loss, it is important to note that hearing becomes less sensitive with advancing age, even in the absence of damaging noise exposure. This effect, called *presbycusis*, is most prominent at the higher frequencies, as indicated in Fig. 31.4(b). Actually, this process begins early in life; small children can usually hear up to 20,000 Hz, but adults can hear up to only 14,000 or 12,000 or 10,000 Hz (depending on age). By age 70, most people do not hear much above 8000 Hz, and the loss of hearing has a noticeable effect on frequencies within the speech band. Some television sets emit a fairly loud noise at 15,750 Hz, the frequency of the horizontal sweep oscillator. This noise is disturbing to young ears, but goes completely unnoticed by people of middle age.

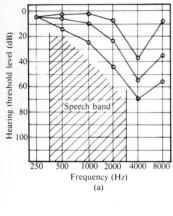

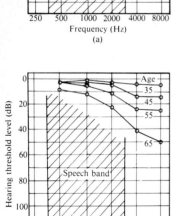

FIG. 31.4

(a) Audiogram showing progressive loss of hearing for typical workers in a noisy factory. (b) Loss of hearing with age (presbycusis) based on the data of Schneider et al. (1970).

Two interesting hypotheses have been suggested as bases for predicting noise-induced hearing loss. The *equal-energy hypothesis* states that equal amounts of sound energy will cause equal amounts of NIPTS regardless of the distribution of the energy with time. The *equal temporary effect*, or TTS, *hypothesis* states that the temporary threshold shift measured two minutes after cessation of an eight-hour exposure closely approximates the NIPTS that would result from a 10-to 20-year exposure to that same level. Neither of these hypotheses has been completely established, although there is a certain amount of evidence for each of them.

31.3   EAR DAMAGE

Although the eardrum can be ruptured by extremely large sound pressures, such as those that occur during blasts, the outer ear, eardrum, and middle ear are not ordinarily damaged by exposure to noise. The primary site of ear damage due to noise is the receptor organ of the inner ear, the organ of Corti (see Section 5.2). From extensive examination of animal ears, and occasional post-mortem examinations of noise-damaged human ears, it has been fairly well established that excessive exposure to noise destroys the delicate hair cells in the organ of Corti and eventually the organ itself (Miller, 1974). Figure 31.5 is a series of drawings, based on photographs of damaged organs, that illustrates increasing degrees of injury.

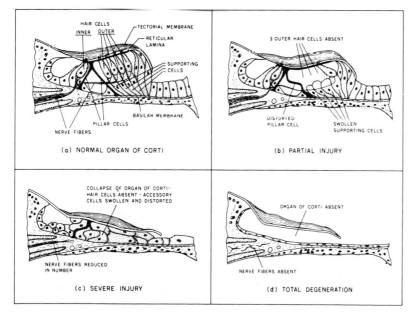

FIG. 31.5
Drawings of the human organ of Corti that illustrate increasing degrees of noise-induced permanent damage. (From Miller, 1974).

The mechanism by which the hair cells are destroyed is not completely clear to us. During the reception of sound, movement of the basilar membrane bends the hairs at the top of the hair cells, and causes the cells to stimulate the auditory nerve fibers. One theory suggests that constant overexposure to sound forces the cells to work at too high a metabolic rate for too long a time, leading to the death of the cells. These delicate receptor cells do not regenerate, and if they die of overwork, they are lost for life.

The ear is an exceedingly well-designed organ, except that it lacks protective devices. The eye, which, like the ear, is sensitive to exceedingly small stimuli, has both an adjustable iris and an eyelid to protect it from damage due to overload. This may be due to the fact that through millions of years of evolution, humans have been exposed to the intense light of the sun. However, frequent exposure to loud sounds largely began with the industrial revolution and the invention of gunpowder. In another million years or so, human ears may very well be equipped with "earlids" or some other similar protective mechanism.

One protective mechanism the ear does have, however, is the *stapedius reflex* or acoustic reflex, which densitizes the ear by as much as 20 dB. Loud noise triggers two sets of muscles that act in the middle ear. One set tightens the eardrum, and the other set draws the stirrup away from the oval window of the inner ear (see Section 5.2). The stapedius reflex may be triggered at sound levels between 80 and 95 dB in different individuals, depending on ear sensitivity. Unfortunately, it does not adequately protect the ear from impulsive or explosive sounds, because the reflex takes time. Although hearing sensitivity is measurably decreased in 30 or 40 ms, full protection may take as much as 200 ms. Thus in the case of a gunshot noise, the muscles "lock the barn door after the horse has already been stolen." It has been suggested that gun crews could be partly protected by stimulating their acoustic reflex with a loud tone a second or two before a gun is fired (Ward, 1962).

An intense sound impulse, especially if it is too rapid for the stapedius reflex to act, can rupture the eardrum, causing a temporary loss of hearing and the hazard of middle-ear infection. The membrane will nearly always heal, but the heavier scar tissue may lower the sensitivity slightly for sounds of high frequency.

31.4 HEARING PROTECTORS

Although it is much better to control noise at its source, people who are obliged to work in noisy environments can protect their hearing by the proper use of hearing protectors. Figure 31.6 shows the sound attenuation characteristics of various types of hearing protectors.

Cotton plugs are not very effective; cotton soaked in vaseline or soft wax is much better. Commercial ear plugs of the soft deformable plastic type, which make a good seal, can provide 20 to 30 dB of protection. Effective protection is offered by over-the-ear muffs, which are worn by personnel working at airports. For maximum protection, both muffs and earplugs can be used (see Fig. 31.6).

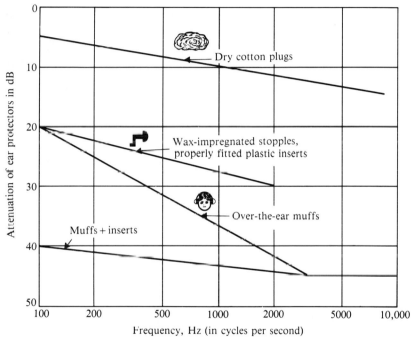

FIG. 31.6
Sound attenuation characteristics of various types of ear protectors. (From Berendt, Corliss, and Ojalvo, 1976).

31.5 SPEECH INTERFERENCE

Speech communication is extremely important to human society. Therefore, interference with speech is a particularly disturbing effect of environmental noise pollution. Even though it is possible to communicate, in some noisy environments, by speaking louder and slower or by decreasing the distance from talker to listener, the need to do so can be quite annoying.

Accuracy of speech depends on many factors, such as noise level, vocal effort, distance between talker and listener, acoustical character of the room, speech clarity, message familiarity, and so on. Thus it is not easy to predict the dependence on any one of these factors, such as noise level. Nevertheless, considerable research on the problem has led to graphs of the type shown in Fig. 31.7, which illustrates the maximum distances over which conversation is considered satisfactory. Figure 31.8 shows the conditions under which speech communication is possible at all. These graphs consider only direct sound, so they describe a free-field (outdoor) environment (see Section 6.2). In a room, reflected sound will reinforce the direct sound, and make communication possible over a greater distance.

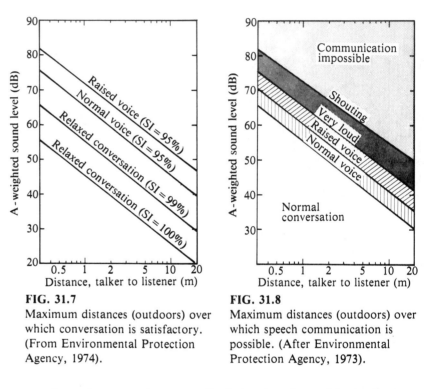

FIG. 31.7
Maximum distances (outdoors) over which conversation is satisfactory. (From Environmental Protection Agency, 1974).

FIG. 31.8
Maximum distances (outdoors) over which speech communication is possible. (After Environmental Protection Agency, 1973).

A speaker generates a complicated series of sound waves that contain many acoustical cues to the sounds of spoken language, as we described in Chapter 16. The listener sorts out these cues, combines them with other available information (such as context, knowledge of the speaker, etc.), and translates them into an intelligible message, if possible. Most speech cues are carried by sounds with frequencies in the range of 300 to 3000 Hz, although a few cues occur as low as 100 Hz and as high as 8000 Hz. Speech ordinarily carries an abundance of extra information, so that it is still understandable even when some cues are lost. Lower noise levels are necessary if the talker has poor articulation or speaks an unfamiliar dialect; young children, for example, have less precise speech than do adults. Also, the listener's ability to understand masked or distorted speech deteriorates with age (Palva and Jodinen, 1970).

A number of ways to rate speech interference by noise have been suggested. Perhaps the most accurate is the *articulation index* (AI), which considers the average speech level and average noise level in each of 20 frequency bands over the range of 250 to 7000 Hz. Somewhat simpler to measure is the *speech interference level* (SIL), which is the average of the sound pressure levels in the octave bands centered at 500, 1000, 2000, and 4000 Hz. (Older versions used octave bands centered at 850, 1700, and 3400 Hz, or at 500, 1000, and 2000 Hz, and there is some sentiment voiced for averaging in a fourth octave band centered at 4000 Hz.) The simplest measure is the A-weighted sound level $L_p(A)$, and this is sufficiently precise for most purposes.

Thus far we have discussed only the effects of rather broad bands of steady noise. We will now briefly consider the speech interference due to narrow-band noise and fluctuating noise level. It is well known that tones of lower frequency can mask tones of higher frequency more effectively than the converse (see Section 6.9). Thus, narrow-band noise interferes with speech most effectively if its frequency is near the low end of the speech band. Certain bands of noise interfere with certain phonemes more than others.

The speech interference due to fluctuating noise levels depends on the nature of the fluctuations as well as the equivalent sound level. For a given equivalent level L_{eq}, intermittent noise will nearly always cause less interference with speech than will steady noise. In Section 6.9, we pointed out that loud intermittent noise can mask speech sounds that occur 20 ms or so after the noise ceases (forward masking) and can even mask speech sounds that occur 5 or 10 ms before the onset of the noise (backward masking).

31.6   INTERFERENCE WITH SLEEP

From our everyday experience, we know that noise can interfere with sleep. Almost all of us have been awakened or kept from falling asleep by loud, strange, sudden sounds, and it is common to be awakened by an alarm clock. But it is also possible to become accustomed to sounds and sleep through them (even alarm clocks!). Apparently, unusual or unfamiliar sounds are most apt to disturb sleep.

The interference with sleep due to noise is not well understood at the present. Much of what is known is based on laboratory studies conducted on relatively few subjects. Field studies involving large numbers of persons are difficult to conduct. Among factors that appear to be important are the nature of the noise stimulus; the physical and emotional state of the subject; physiological differences such as sex and age; and the stage of sleep during which the noise occurs (see the box below).

There are several different way of classifying the stages of sleep, but one way simply numbers them from 1 to 5. As one relaxes and enters a stage of drowsiness, the electroencephalogram (EEG) pattern changes from rapid, irregular waves to regular variations at 9 to 12 Hz, known as the *alpha rhythm*. The person is relaxed, but not yet asleep. In sleep stage 1, the alpha waves diminish in amplitude and frequency. In stage 2, the alpha rhythm gives way to bursts of waves (*spindle* waves) mixed with single slow waves. In stage 3, 30 to 45 minutes later, bursts of large slow waves (*delta* waves) appear. In stage 4, the deepest sleep, delta waves occur over 50 percent of the time. An additional stage, called the *rapid eye movement* (REM) stage, exhibits many characteristics of stage 1. During a night's sleep, a person typically enters cycles from stage 4 to the REM-stage several times, and may go through the entire cycle of stages several times. The time spent awake generally increases with age (EPA, 1973).

A person's threshold for being awakened by noise appears to be lowest during the REM-stage and highest during stages 3 and 4. The amount of accumulated sleep time increases the probability of arousal, no matter what the stage of sleep. Stimuli of 50 dBA have been found to invoke some sort of response, either a change in sleep stage or an awakening, about half the time. When stimuli reach 70 dBA, awakening is the most likely response. In noise levels of 40 to 50 dBA, many subjects experience difficulty falling asleep and awaken rather easily.

Although further studies are needed to establish with any certainty the noise levels that interfere with sleep, it appears that indoor noise levels should not exceed 35 dBA in order to protect most of us from sleep interference.

31.7   OTHER PSYCHOLOGICAL EFFECTS

The effect of noise on the performance of various tasks has been the subject of several investigations in the laboratory and in actual work situations. When mental or motor tasks do not involve auditory signals, the effects of noise on human performance have been difficult to assess.

Some general conclusions can be drawn about the effects of noise on performance. Steady noises below about 90 dBA do not seem to affect performance, but intermittent noise can be disruptive. Noise with appreciable strength around 1000 to 2000 Hz is more disruptive than is low-frequency noise. Noise is more likely to reduce the accuracy of work than to reduce the total quantity of work (Miller, 1974).

Noise appears to interfere with the ability to judge the passage of time. Subjects in various experiments have judged time as passing too rapidly or too slowly, depending perhaps on the level of the noise they were hearing.

There is a general feeling that nervousness and anxiety are caused by exposure to noise or at least are intensified by it. Whether noise by itself causes a significant amount of stress is difficult to determine, because noise is often closely associated with events that involve fear and anxiety.

31.8   THE PHYSIOLOGICAL EFFECTS OF NOISE

Sudden noises are startling. They trigger a muscular reflex that may include an eyeblink, a facial grimace, inward bending of arms and knees, etc. These reflexes prepare the body for defensive action against the source of the noise. Sometimes these reflexive actions interfere with other tasks; sometimes they even cause accidents.

Constriction of blood vessels, reduction of skin resistance, changes in heartbeat, changes in breathing rate, dilation of pupils, and secretion of saliva have been observed in human response to brief sounds (Davis, Buchwald, and Frankmann, 1955). There is evidence that workers exposed to high levels of noise have a higher incidence of cardiovascular disorders; ear, nose, and throat problems; and equilibrium problems than do workers at lower levels of noise (Miller, 1974).

Laboratory experiments with animals have shown that intense noise can cause pathological effects such as hypertrophy of the adrenal glands, developmental abnormalities of the fetus, brain damage, and sexual misfunction. These often occur at levels above those that humans normally encounter, however.

31.9   SUMMARY

Noise affects people in many ways. One of the most serious effects is hearing loss, both temporary and permanent, due to noise exposure. Hearing threshold shifts usually occur first around 4000 Hz in amounts that depend both on noise level and duration. Ear damage centers in the organ of Corti, where prolonged exposure to noise results in the destruction of the hair cell sensors.

Because of the importance of speech communication, speech interference is a particularly annoying effect of noise. Most speech cues are carried by sounds in the range of 300 to 3000 Hz; thus, a speech interference level (SIL) is defined as the average of noise levels in the 500-, 1000-, and 2000-Hz octave bands. Noise can cause changes in a person's sleep stages or even awakening. It can also affect performance of mental and physical tasks, interfere with the judgment of time, intensify stress and anxiety, and trigger muscular reflexes.

References and Suggested Readings

American National Standards Institute (1969). "Specifications for Audiometers," ANSI S.3-1969. New York: American National Standards Institute (similar to International Standards Organization, ISO R839, 1964).

Berendt, R. D., E. L. R. Corliss, and M. S. Ojalvo (1976). *Quieting: A Practical Guide to Noise Control.* Washington, D.C.: National Bureau of Standards Handbook 119.

Davis, R. C., A. M. Buchwald, and R. W. Frankmann (1955). "Autonomic and Muscular Responses and Their Relation to Simple Stimuli," *Psychol. Monogr.* **69**(20): 1.

Environmental Protection Agency (1973). *Public Health and Welfare Criteria for Noise*, Report 550/9-73-003, U.S. Environmental Protection Agency, Office of Noise Abatement and Control, Washington, D.C.

Environmental Protection Agency (1974). *Information of Levels of Environmental Noise Requisite to Protect Public Health and Welfare with an Adequate Margin of Safety*, Report 550/9-74-004, U.S. Environmental Protection Agency, Office of Noise Abatement and Control, Washington, D.C.

Kryter, K. D. (1970). *Noise and Man*. New York: Academic Press.

Kryter, K. D. (1973). "Impairment to Hearing from Noise," *J. Acoust. Soc. Am.* **53**: 1211. (Several critiques of this paper also follow it.)

Lucas, J. S. (1975). "Noise and Sleep: A Literature Survey and a Proposed Criterion for Assessing Effect," *J. Acoust. Soc. Am.* **58**: 1232.

Miller, J. D. (1974). "Effects of Noise on People," *J. Acoust. Soc. Am.* **56**: 729.

Palva, A., and K. Jodinen (1970). "Presbycusis. V. Filtered Speech Tests," *Acta Oto-Laryngol.* **70**: 232.

Schneider, E. J., J. E. Mutchler, H. R. Hoyle, E. H. Ode, and B. B. Holder (1970). "The Progression of Hearing Loss from Industrial Noise Exposures," *Am. Ind. Hygiene Assoc. Jour.* **31**: 368.

Ward. W. D. (1962). "Studies on the Aural Reflex. III. Reflex Latency as Inferred from Reduction of Temporary Threshold Shift from Impulses," *J. Acoust. Soc. Am.* **34**: 1132.

Ward, W. D. (1969). "Effects of Noise on Hearing Thresholds," *Proc. of Conf. on Noise as a Public Health Hazard*, eds., W. D. Ward and J. E. Fricke. Washington: American Speech and Hearing Association.

Glossary

articulation index (AI) A means of rating speech interference by noise that considers the average speech level and average noise in each of 20 bands over the frequency range 250–7000 Hz.

audiogram A graph of a person's hearing threshold at several frequencies compared to thresholds for normal hearing.

audiometer An instrument that measures hearing thresholds of an individual at several audible frequencies.

electroencephalogram (EEG) A record of electrical potentials at several points in the brain (sometimes referred to as "brain waves").

equal energy hypothesis Postulates that risk of hearing loss is determined by the total amount of noise energy to which the ear is exposed each day, irrespective of its distribution in time.

equal temporary effect (TTS) hypothesis Postulates that risk of permanent hearing loss increases with average temporary loss (TTS).

hair cells Delicate sound-sensing cells in the organ of Corti that are destroyed by overexposure to noise.

permanent threshold shift (NIPTS) The amount that the threshold of hearing is raised irreversibly by exposure to noise.

presbycusis Gradual loss of hearing with age, especially at high frequency.

rapid eye movement (REM) state A stage of sleep during which a person can be awakened quite easily by noise (also the stage in which dreaming occurs).

speech cues Particular combinations of sounds or dynamic changes in sound by which a listener identifies phonemes (speech sounds).

speech interference level (SIL) The average of sound levels in the 500-, 1000-, and 2000-Hz octave bands.

temporary threshold shift (TTS) A reversible increase in the threshold of hearing that disappears in hours, days, or weeks depending on its severity (also called "auditory fatigue").

Home Experiments

1. Make your own test of the effect of noise on mental task. A good source of broadband noise is an FM radio tuned between stations; another would be a local discotheque. A mental task might be the solution of a set of arithmetic problems or copying names and numbers from the telephone directory. Try it at medium and loud noise levels (90 dBA and 100 dBA, if you have a sound level meter). If you do not have a sound level meter, be careful that the noise does not become uncomfortably loud; do not exceed the level you would expect at a loud concert.

2. Make a survey of acquaintances of varying ages to determine the average age at which people no longer hear the 15,750 Hz sound emitted by many television sets.

Problems

1. Sound levels of 67 dB, 71 dB, and 68 dB are measured in the octave bands centered at 500, 1000, and 2000 Hz, respectively. What is the speech interference level?

2. The sound energy that the ear receives depends on the average intensity multiplied by the time.

   a) Compare the intensities of sounds having levels of 85 dB and 110 dB.

   b) Compare the energy received by the ear during exposures for ten minutes at 110 dB and for eight hours at 85 dB (both of these exposures cause TTS of about 34 dB, according to Fig. 31.2).

3. If the A-weighted sound level outdoors is 50 dB, what is the maximum distance from talker to listener in order to carry on conversation with

   a) normal voice;

   b) very loud voice.

4. All the curves in Figs. 31.7 and 31.8 have essentially the same slope. How much does the allowed sound level decrease when the talker-to-listener distance doubles? Explain this result by referring to Section 6.2.

CHAPTER 32
The Control of Noise

The words "silencing" and "soundproof" are two very common, yet misleading terms. Both suggest the elimination of noise, which generally is not possible (or even desirable). What is desirable is the reduction of noise to levels that are not injurious to the health and well-being of people who live and work in that particular environment.

32.1 ANALYZING A NOISE PROBLEM: SOURCE-PATH-RECEIVER

A straightforward approach to solving a noise control problem is to divide it into its three basic elements: source, path, and receiver. The *source* may be a noisy machine or appliance, a jet aircraft, a noisy highway, a neighbor's loudspeaker or lawnmower, or any of a large number of mechanical noisemakers that are common in our society. The *path* may be a direct line-of-sight air path, a structural path through a building, or a complex path that includes propagation through ducts, windows, wall panels, etc. The *receiver* in which we are interested is a person or a group of persons receiving unwanted sound.

There is little doubt that the most effective place to control noise is at its source. The best policy is to select quiet machines or appliances initially. This would become easier if product noise labels were to become mandatory, as proposed by the Environmental Protection Agency at one time. Noisy machines can often be quieted by a few simple modifications such as the installation of resilient mounts, the damping of vibrating panels, the slowing down of air velocities, and so on. In other cases, major re-engineering is required. Sound-attenuating housings around the machines are frequently helpful.

After careful attention has been paid to controlling noise at the source, the next line of defense is to cause the sound to be attenuated as much as possible along its path from source to receiver. This can be done by

1. Absorbing sound energy along the path;

2. Reflecting sound back toward the source by means of barriers;

3. Eliminating alternative paths for sound transmission.

It is usually more difficult to attenuate sound outdoors than indoors.

When efforts to reduce noise at its source or along the transmission path prove insufficient, one or more of the following measures may be taken to protect the receiver:

1. Using ear protectors (see Section 31.4);

2. Limiting the time of exposure to noise;

3. Operating noisy machines by remote control from a sound-attenuating enclosure; and

4. Making use of masking noise to reduce the annoyance due to pure-tone noise or nearby conversations.

32.2 NOISE REGULATIONS

Like most aspects of environmental protection in the United States, noise regulation is shared by federal, state, and local governments. The federal government sets general noise standards, especially in areas where they will affect interstate commerce, including the health and hearing of workers employed by companies that engage in interstate commerce. State noise control measures have typically dealt with motor vehicle noise, airport noise, land use, and related aspects of noise control. Local governments regulate the use of noisy machines, the noise of vehicles (including recreational vehicles) that operate within their boundaries, nuisance noise, noise transmission within apartment buildings, and so forth.

The *Noise Control Act of 1972* was the first major piece of federal legislation in the field of noise control. This act directed the United States Environmental Protection Agency (EPA) to develop and publish information on hazardous noise levels, to identify major noise sources, and to define permissible noise levels for them. The EPA was also directed to coordinate all federal noise research and control programs and to provide technical assistance to state and local governments.

Among the major sources of noise that were identified by the EPA are portable air compressors, medium and heavy trucks, motorcycles, buses, garbage trucks, jackhammers, railroad cars, snowmobiles, and lawnmowers. However, the plans that the EPA developed in the 1970s were effectively scrapped in 1980, and the EPA has had no national noise program since 1981.

Other U.S. government agencies are also concerned with special areas of noise control.

The *Federal Aviation Administration* (FAA) sets criteria and standards for aircraft noise.

The *Federal Highway Administration* has adopted noise control standards for motor vehicles, and the *Bureau of Motor Vehicle Safety* shares with state and local agencies the responsibility for enforcement. Both of these agencies, as well as the FAA, are in the Department of Transportation at this time.

The *Occupational Safety and Health Administration* (OSHA) sets regulations designed to protect the hearing of workers in companies engaged in interstate commerce. Popularly known as "OSHA standards," the current limits on noise exposure are given in Table 32.1.

The *Department of Housing and Urban Development* (HUD) has developed standards for sound-insulation characteristics of walls and floors in multifamily residences and other buildings that qualify for HUD mortgage insurance. It also sets guidelines for permissible noise levels at sites of housing developments.

TABLE 32.1   Permissible occupational noise exposure

Average sound level, dB (A-weighted, slow response)	Hours per day
90	8
92	6
95	4
97	3
100	2
102	$1\frac{1}{2}$
105	1
110	$\frac{1}{2}$
115	$\frac{1}{4}$

The *National Institute for Occupational Safety and Health* (NIOSH) sets standards to protect hearing in underground mines, and these standards are enforced by the *Mine Safety and Health Administration.*

The Noise Control Act of 1972 leaves the primary responsibility for controlling noise in the hands of state and local governments. Several states have adopted and are enforcing standards for trucks that are similar to the federal standards.

Many cities have adopted noise regulations, some of them based on the Model Community Ordinance published by the EPA and the National Institute of Municipal Law Officers in 1975. Noise control laws that are based on performance standards are generally superior to "general nuisance" laws, which are difficult to enforce. A number of cities have included noise standards in their building codes, although cities in the United States are in general far behind European cities in this regard.

32.3 EXPOSURE TO OCCUPATIONAL NOISE

Some of the noisiest environments in the world are found within factories in which thousands of workers earn their living. Although hearing loss among factory workers has been common, only recently has occupational noise been regulated.

The Occupational Safety and Health Act of 1970 set limits on permissible noise exposure for workers based on the amount of time they are exposed to these levels. These standards have precipitated some rather heated discussion, and there is considerable feeling that the limits should be lowered.

The combinations of exposure times and sound levels given in Table 32.1 are considered the limit of daily dose that will *not* produce disabling loss of hearing in more than 20 percent of a population exposed throughout a working lifetime of 35 years. For exposure at two or more different levels, effects are combined according to the relationship

Fraction of allowed dose $= C_1/T_1 + C_2/T_2 + \cdots + C_n/T_n,$

where C_n represents the actual exposure time at a given sound level, and T_n is the time permitted if exposure were all at that level.

When the sound level exceeds the maximum permissible combination of dose and time given in Table 32.1, the company should

1. Take steps to reduce the noise at its source;

2. Provide hearing protection; and

3. Carry out a program to test and conserve hearing.

32.4 PRODUCT LABELING

There are indications that some people prefer to purchase a quiet product even if it costs slightly more than a noisier one. Phrases such as "whisper quiet" often appear in product advertising. Nevertheless, it is difficult for the prospective buyer to compare the noise output from different brands.

The American National Standards Institute published a method for rating noise emission by small stationary sources to form the basis for product noise labeling, but the proposed method was never adopted and has been replaced by a method based on A-weighted sound power level (see Section 6.2).

32.5 WALLS AND FLOORS

When an airborne sound wave strikes a solid wall, the largest part is reflected, whereas smaller portions are absorbed and transmitted through the wall. The coefficients of *reflection, absorption,* and *transmission* are determined by the physical properties of the wall and by the frequency of the sound and its angle of incidence to the wall.

The *transmission coefficient* τ is defined as the fraction of the acoustic power incident on the wall that is transmitted through the wall to the other side. The principle of conservation of energy tells us that the transmitted power is that part of the incident power that is neither reflected nor absorbed. This fraction is expressed as *transmission loss* (TL) in decibels:

$$\tau = \frac{W_{\text{transmitted}}}{W_{\text{incident}}} \ ;$$

$$\text{TL} = 10 \log \frac{1}{\tau}.$$

Sound waves striking a wall can bend it, or shake it, or both. (These motions may be described as flexural and compressional waves, respectively, in the wall.) At low frequency, the sound transmission loss in a solid wall follows a *mass law*; it increases with increasing frequency f and mass M of the wall. Formulas for calculating the transmission loss at low frequency are given in the box below.

For waves that approach a wall of large dimensions with normal (perpendicular) incidence, the transmission loss is

$$(\text{TL})_0 = 10 \log \left(1 + \frac{\pi M f}{400} \right),$$

where M is wall area mass (in kg/m²) and f is frequency (in Hz). In a room, it is a good approximation to assume the sound waves of low frequency to be randomly distributed over all angles from 0 to 80°. This decreases the transmission loss by about 5 dB:

$$TL = 10 \log \left(1 + \frac{\pi M f}{400}\right) - 5.$$

From the formulas above, two facts are clear:

1. To reduce transmission of sound between adjoining rooms, the common wall should be as heavy as possible; and

2. Low-frequency sounds are the most difficult to block (this should be clear if you recall that sound will "leak through" from a hi-fi system playing in the next room).

Transmission loss for a wall may fall considerably below that predicted by the mass law. This may be due to any of the following effects:

1. Wall resonances that occur at certain frequencies;

2. Excitation of bending waves at the "critical frequency" where they travel at the same speed as certain sound waves in air; and

3. Leakage of sound through holes and cracks.

The transmission losses for walls of several materials are shown in Fig. 32.1. Note the dip in TL at the critical frequency, which is different for each material.

Leakage of sound through small holes or cracks in walls tends to be underestimated all too often in building construction. Openings around pipes and ducts and cracks at the ceiling and floor edges of walls allow the leakage of airborne sound. Common causes of leakage in party walls separating apartments may include back-to-back electrical outlets or medicine cabinets. Cracks under doors are especially bad. Figure 32.2 illustrates the effect of holes of various sizes on the transmission loss of walls.

It is often desirable to provide a single-number rating of a wall for purposes of comparison. This is done by measuring the transmission loss of a wall sample at 16 different frequencies, and comparing these values to standard curves. The resulting single number is called the *sound transmission class* (STC) of the wall. Table 32.2 gives the sound transmission class for various wall structures, and also indicates the degree of privacy that they offer. The Department of Housing and

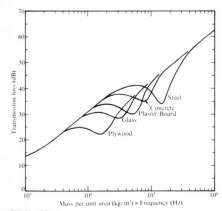

FIG. 32.1
Transmission loss (TL) of a wall as a function of mass and frequency. Note the drop in TL near the critical frequencies for exciting bending waves.

Urban Development recommends at least STC 55 for walls between two apartments (party walls) in nonurban areas.

In the case of floor-ceiling combinations, the transmission of impact noise is generally more important than the transmission of airborne sound. The impact-sound transmission level (ISTL) is measured with the help of a *standard tapping machine,* and the resulting data are used to determine the *impact isolation class* (IIC) of a floor-ceiling combination. One of the easiest ways to increase the impact-isolation of a floor-ceiling structure is to cover the upper surface with a thick carpet over a resilient pad.

32.6 BARRIERS—INDOORS AND OUTDOORS

Sound barriers, which block the direct sound path from source to receiver, can result in appreciable noise reduction, both indoors and out-

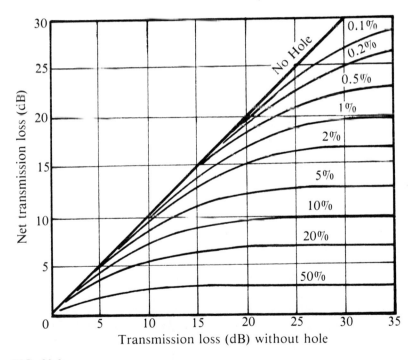

FIG. 32.2
The effect of a hole on transmission loss TL. The horizontal axis is the transmission loss without the hole; the vertical axis is the transmission loss with the hole. The numbers that label the individual curves give the hole area as a percentage of total wall area.

TABLE 32.2 Sound transmission class (STC) for various wall structures.

STC rating	Privacy afforded	Wall structure
25	Normal speech easily understood	¼" wood panels nailed on each side of 2×4 studs
30	Normal speech audible but not intelligible	⅜" gypsum wallboard nailed to one side of 2×4 studs
35	Loud speech audible and fairly under-standable	⅝" gypsum wallboard nailed to both sides of 2×4 studs
40	Loud speech audible but not intelligible	Two layers of ⅝" gypsum wallboard nailed to both sides of 2×4 studs
45	Loud speech barely audible	Two sets of 2×3 studs staggered 8" on centers fastened to 2×4 base and head plates with two layers of ⅝" gypsum wallboard nailed on the outer edge of each set of studs
50	Shouting barely audible	2×4 wood studs with resilient channels nailed horizontally to both sides with ⅝" gypsum wallboard screwed to channels on each side
55	Shouting not audible	3⅝" metal studs with 3" layer of glass fiber blanket stapled between studs. Two layers of ⅝" gypsum wallboard attached to each side of studs.

Source: Berendt, Corliss, and Oljavo (1976).

doors. Unfortunately, their potential is often overestimated, and the actual noise reduction achieved from barriers is disappointing.

Sound transmission *through* a barrier is generally less important than sound transmission *around* a barrier. In Fig. 32.2, for example, we can see that as the open area increases in percentage of the total area, the transmission loss becomes more or less independent of the TL of the wall itself. Thus, the primary attention should be directed at the "flanking" paths that allow sound transmission over or around a barrier.

A typical situation in an indoor office is shown in Fig. 32.3. There are three types of transmission paths to be considered: transmission through the barrier (path SCR), diffraction around the barrier (path SBR), and reflection from the ceiling (paths SAR, SDER, etc.).

Transmission through the barrier is similar to that through a full wall of the same construction; unless the barrier is flimsy, direct transmission will be much less than will transmission by diffraction and reflection.

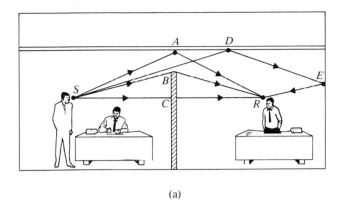

(a)

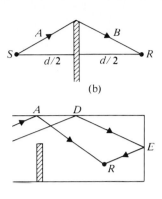

(b)

(c)

FIG. 32.3
(a) Transmission paths through and around a barrier. (b) Diffraction around the barrier. (c) Reflection paths around a barrier.

The diffraction of sound of a given frequency around a barrier depends on the Fresnel number N, which is expressed as

$$N = \frac{2}{\lambda}(A + B - d),$$

where λ is the wavelength of sound, lengths A, B, and d are as shown in Fig. 32.3(b). Figure 32.4 illustrates the attenuation that can be achieved with a barrier when diffraction is the only type of transmission. This curve assumes a fairly large open area above the barrier. If the barrier extends nearly to the ceiling, the attenuation will be increased.

Transmission by reflection depends on the acoustic properties of the ceiling, the size of the opening above the barrier, and the nature of the walls on the source and receiver sides of the barrier. A highly absorbent ceiling is essential in an open-plan office or school.

Even when carefully designed, barriers in open-plan offices and schoolrooms cannot provide anywhere near the privacy or noise reduction of full-size walls. It has frequently been found that to obtain any degree of privacy in an open-plan environment, work areas must be spread out more than is necessary for enclosed work spaces, thus offsetting the economy anticipated from open planning. Even so, the nonacoustical advantage of open planning may outweigh the acoustical disadvantages in some situations.

Barriers are frequently used outdoors to attenuate sound from a noisy highway or railroad, for example. The same physical laws that govern the performance of indoor barriers serve for outdoor barriers as well but the design considerations are different. Most of the sound transmission outdoors is due to diffraction over the barrier (see Fig. 32.4). Since source-receiver distances are greater outdoors, achieving a large Fresnel number ($A + B - d$) requires a very high barrier.

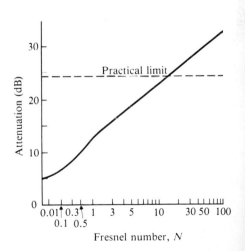

FIG. 32.4
Attenuation of sound by a barrier if diffraction were the only mode of transmission.

Also, refraction of sound caused by wind or temperature gradients may lead to additional transmission of sound over the barrier (see Fig. 3.18b). Sound barriers are much more effective in attenuating high-frequency noise than low-frequency noise.

32.7 ENCLOSURES

Consider the demonstration experiment shown in Fig. 32.5 (better yet, perform the experiment yourself). First, a small noise source, such as a battery-operated doorbell or buzzer, is allowed to operate on a wooden table. The sound level is measured at some nearby location. The source is now covered or enclosed with acoustic tile or other absorbing material (Fig. 32.5b); the reduction in sound level is minimal. Next, a small piece of the same material is placed under the source to decouple it from the table (c); a larger reduction in sound level results. Now a lightweight solid enclosure (e.g., a small wastebasket or cookie can) is placed over the source (d); the sound level is reduced by 6 dB (half the sound pressure); the can and pad together (e) give 11 dB. Finally, the absorbing material is placed *inside* the enclosure (f), and the sound level is further reduced by about 3 dB.

If the metal can in Fig. 32.5 makes a reasonably good seal, it reduces the sound to some 10 or 11 dB below its level inside the can. If there is insufficient absorption inside the can, however, the reverberant level builds up inside the can, so the net reduction is only 6 dB (Fig. 32.5d). Placing absorbing material on the inside will absorb the reverberant sound so that the sound attenuation by the can results in a greater reduction in the noise level outside. Absorbing material on the outside of the can would be of little use.*

FIG. 32.5

Demonstration of noise reduction with an enclosure; (a) noise source: (b) covered with absorbing material; (c) placed on a resilient pad; (d) covered with a metal can; (e) can and pad; (f) can lined with absorbing material.

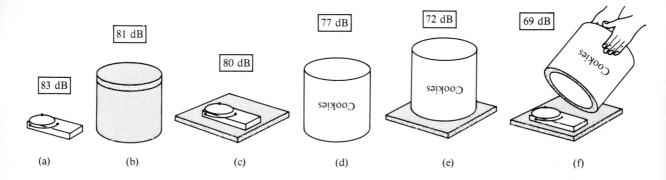

(a)   (b)   (c)   (d)   (e)   (f)

---

* To constantly remind me of this fact, I have a noisy duct from an exhaust fan passing through my office. Some well-meaning person carefully installed absorbing material on the outside, rather than lining the inside of the duct.

If the noise source needs ventilation, so that the enclosure cannot be made airtight, less noise reduction is possible. A muffler can be included in the air inlet and outlet ducts to reduce noise transmission, especially if the noise source emits sound at certain frequencies more than others. If mufflers are not practical, the air inlet and outlet ducts should be lined with absorbing material, and they should make as many bends as possible. Vented enclosures are shown in Fig. 32.6.

32.8 SOUND ABSORBERS

Absorption of sound requires the conversion of sound energy to heat. One of the most effective ways to absorb airborne sound is to use a porous material made of many small fibers or cells. The rush of air particles back and forth among the fibers, due to passage of a sound wave, generates heat and absorbs the sound wave. Because the energy of a sound wave is small, the temperature rise in the material is not normally measurable.

Porous absorbers are more effective at high frequencies than at low frequencies (see Table 20.1). Also, porous absorbers are more effective if mounted away from a reflecting surface where the air-flow velocity is greater. (Recall from Chapter 3 that at the hard surface of a reflector, the sound pressure is a maximum but the particle velocity is zero.)

A second type of absorber is a panel that is set into vibration by an airborne sound wave. Sound energy is converted into heat by inter-

FIG. 32.6
(a) A noise-reducing enclosure requiring air circulation. (From Berendt, Corliss and Oljavo, 1976). (b) A homemade air blower that is 10 to 15 dB quieter than vacuum cleaner blowers. (From "Acoustics of Percussion Instruments, Part I," *Phys. Teach.* **14**: 546. Reprinted by permission from the Amer. Assoc. of Physics Teachers.)

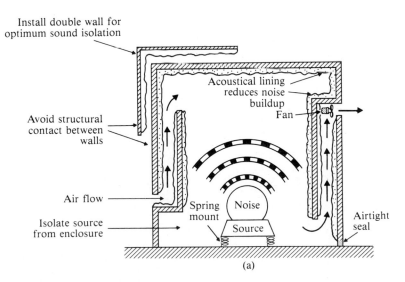

Install double wall for optimum sound isolation

Acoustical lining reduces noise buildup

Fan

Avoid structural contact between walls

Air flow

Spring mount

Noise Source

Isolate source from enclosure

Airtight seal

(a)

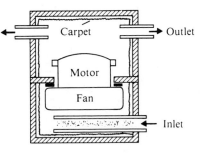

Carpet — Outlet

Motor

Fan

Inlet

(b)

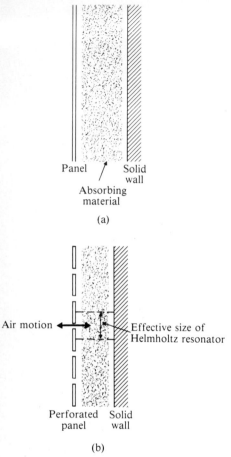

Panel | Solid wall
Absorbing material
(a)

Air motion — Effective size of Helmholtz resonator

Perforated panel | Solid wall
(b)

FIG. 32.7
(a) Panel absorber spaced away from a reflecting wall (the air space may contain a porous absorber).
(b) Perforated panel backed by porous material.

nal friction in the panel. A panel absorber is effective mainly at low frequencies (see Table 20.1). The frequency of maximum absorption can be controlled by choice of panel weight and by the depth of air space behind the panel, and the absorption can be increased by filling the air space behind the panel with a porous absorbing material.

An absorbing surface that combines the principles of both porous and panel absorbers is a perforated panel backed by porous material. The holes act as resonators, very similar to Helmholtz resonators, without cavity walls (see Section 2.3). The frequency of maximum absorption can be controlled by choice of panel thickness, the size and spacing of the holes, and the depth of the space behind the panels. Panel absorbers with and without holes are shown in Fig. 32.7.

Perforated tiles, or acoustical tiles, are commonly used as absorbers. The front surface is porous with openings that penetrate into the interior. If the openings are regular in size and spacing, the material will show an absorption maximum, typically near a frequency of 1000 Hz. If the holes are random, the absorption maximum will be broader. Acoustical tiles of several designs are shown in Fig. 32.8.

32.9 HEATING AND AIR CONDITIONING NOISE

Heating, ventilating, and air-conditioning systems are a most annoying source of noise in many homes, offices, classrooms, and even concert halls. In nearly all cases, the annoyance is completely unnecessary. In some cases, the level of the noise indicates a malfunction that may also reduce efficiency.

Several different types of noise may be encountered in air-handling systems. These include the following:

1. Noise from motors, blowers, and compressors that is transmitted through the air ducts. Unlined ducts act as "waveguides," delivering sound as well as fresh air efficiently throughout the building.

2. Noise from air flow and turbulence within the ducts, especially at inlet and outlet grills.

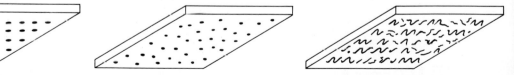

FIG. 32.8
"Acoustical" tiles with porous surfaces.

3. Mechanical noise, as in (1), transmitted through the building structure, which results in a low-pitched rumble.

4. Noise from fresh air intake and exhaust ducts and compressors outside the building that radiate noise throughout the neighborhood. Window-mounted air conditioners may be particularly annoying to neighbors who wish to sleep with their windows open (or even closed).

In planning new construction, residential or otherwise, it is most important to select quiet heating and ventilating equipment, and to isolate it as much as possible from the quietest parts of the building. Centrifugal or squirrel-cage fans are usually less noisy than are axial or propeller fans. Motors, blowers, and compressors should be mounted on resilient pads to isolate vibrations. Heavy equipment in large buildings should be bolted to a concrete slab that is isolated from the rest of the building by resting either on vibration isolators or on its own foundation. The blower should be mechanically isolated from supply and return ducts by a flexible "boot."

Noise transmitted through ducts can be reduced by installing duct liners. A one-inch-thick lining will reduce high-frequency noise by as much as 10 dB per meter, although low-frequency noise may require a thicker lining for effective absorption. Both supply and return ducts should be lined; lining is most effective when installed near the open or grill end of the duct. In difficult cases, expansion or plenum chambers may be necessary.

Aerodynamic noise generated in ducts and grills can be minimized by keeping air velocities low (see Section 30.4). Sharp corners, ragged joints, and dampers can cause noise-generating turbulence in air ducts, as shown in Fig. 32.9. If removing a grill results in a substantial reduction in noise, replacement with a quieter grill should be considered. Quiet grills or diffusers are made of heavy-gauge metal with widely spaced streamlined deflectors devoid of sharp corners and edges.

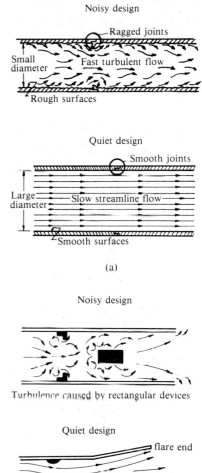

(a)

(b)

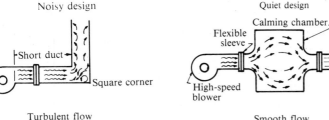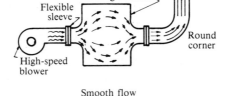

(c)

FIG. 32.9

Examples of duct designs that reduce noise from air turbulence. (From Berendt, Corliss, and Ojalvo, 1976).

Many practical suggestions for quieting home heating, air-conditioning, and plumbing systems are given in an inexpensive Bureau of Standards (now NIST) handbook (Berendt, Corliss, and Oljavo, 1976).

32.10   AIRCRAFT NOISE

Control of aircraft noise is one of the most challenging of urban environmental issues. Airlines transport approximately 80 percent of all intercity passenger traffic traveling by common carrier in the United States. In spite of efforts to develop high-speed ground transport, the number of aircraft takeoffs and landings near major cities continues to grow. Furthermore, as land prices rise, residential dwellings encroach on noise buffer zones near airports in increasing numbers.

The modern jet aircraft is actually a rather inefficient noise source, radiating less than 0.01 percent of its total power as sound. Nevertheless, this may exceed one kilowatt of sound power because of the prodigious amount of mechanical power generated by the engine. The development of the turbofan engine in 1960 led to greater efficiency and somewhat less total sound power, but it added a new source of annoyance: a sirenlike whine from the fan. Sound power radiation from early long-range jets, such as the Boeing 707 and the Douglas DC-8, exceeded 10 kilowatts, which meant that A-weighted sound levels at 1000 ft often exceeded 100 dB. In the mid-1950s, considerable work was begun on the quieting of large jet engines.

A turbofan jet engine produces two main types of noise. The first is due to the turbulence created when the high-velocity jet of gas reacts with the quiescent atmosphere. This noise, which has a considerable low-frequency component, dominates during takeoff and climb. The second type of noise is the high-pitched whine of the fan, which becomes dominant during a landing approach with reduced power. Two major engineering developments have spearheaded the attack on jet engine noise: the development of acoustic linings for engine nacelles and the "high bypass-ratio" engines, now used in the wide-body 747, DC-10, L-1011, 767, and A-300 aircraft.

A cutaway view of the high bypass-ratio engine used in the DC-10 is shown in Fig. 32.10. In this type of engine, a fan supplies intake air to the engine but also blows a stream of bypass air, which surrounds the primary jet and mixes with it. The new acoustical linings, some of them capable of functioning at temperatures of 700°F and sound levels up to 170 dB, are especially designed to attenuate fan whine and noise in the frequency range between 1000 and 5000 Hz. Duct linings developed for the DC-10 are shown in Fig. 32.11.

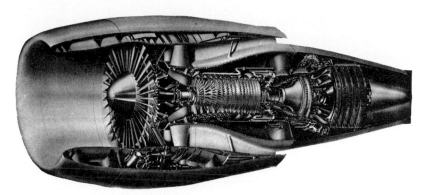

FIG. 32.10
Cutaway view of high bypass engine used in the DC-10. (From "Air Transport Noise Reduction" by Robert J. Koenig from *Noise Control Engineering Journal* (May/June 1977). Reprinted by permission.)

FIG. 32.11
Examples of sound-absorbing linings used in the DC 10. (From "Noise Control Features of the DC-10" by Alan H. Marsh from *Noise Control Engineering Journal* (May/June 1975). Reprinted by permission.)

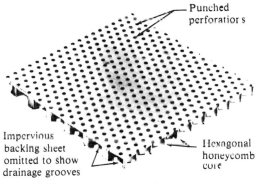

(a) Perforated aluminum sheet bonded to aluminum honeycomb core

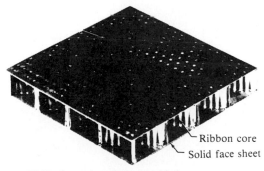

(b) Perforated fiberglass laminate bonded to slotted fiberglass laminate double-diamond core

(c) Perforated steel sheet brazed to corrugated steel core

(d) Perforated steel sheet welded to ribbed-ribbon steel core

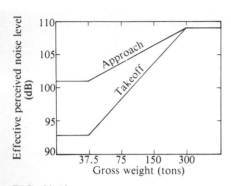

FIG. 32.12
EPNL limits during approach and takeoff vs. aircraft gross weight.

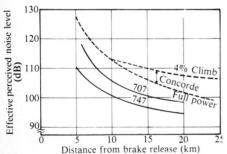

FIG. 32.13
Effective perceived noise levels of several aircraft. (From Wesler, 1975. Reprinted by permission of *Sound and Vibration*.)

A quantity called the *effective perceived noise level* (EPNL), expressed in dB, is used in Federal Aviation Regulation (FAR) Part 36 for certification of aircraft noise. FAR Part 36 imposes noise level limits on aircraft certified after 1969. The maximum EPNL's during takeoff and landing are specified as a function of the maximum gross takeoff weight of the aircraft. The noise limits imposed on large aircraft during approach and takeoff by FAR Part 36 are shown in Fig. 32.12.

It is possible to retrofit existing jet engines in order to meet the Part 36 standards. A quiet nacelle has been designed for the JT3D used in Boeing 707 aircraft. The FAA has estimated that replacing existing engine nacelles with quieter ones would add about one percent to direct operating costs over the remaining lifetime of the fleet.

Noise limits for small aircraft are also set by Part 36. The maximum A-weighted sound level during a flyover at 1000 ft with full throttle varies from 68 dB for aircraft weighing less than 1320 lb to 80 dB for aircraft of 3300 lb or more.

32.11   SUPERSONIC AIRCRAFT

Although the United States decided to terminate its program to develop a civilian SST in the mid-1960s, the Russian TU-144 and the Anglo-French Concorde are now operational. Supersonic transports present unique noise problems around those airports that they serve.

Both the TU-144 and the Concorde use afterburners for increased thrust during takeoff. Afterburners increase the noise emission, especially at low frequency where atmospheric absorption is very low. These low-frequency sounds are also more apt to excite building vibration and rattle. On the other hand, the afterburners can be cut out shortly after takeoff to reduce noise by sacrificing rate of climb. Effective perceived noise levels on takeoff of the Concorde, the 707-320B (not retrofitted), and the 747 are shown in Fig. 32.13.

The Concorde is only slightly noisier than a 707. However, if we compare it to a DC-10 with high bypass-ratio engines, we find that it emits substantially higher levels.

A *sonic boom* is a pressure transient of short duration that occurs during the flyover of an aircraft at a speed that exceeds the speed of sound (approximately 770 mph or 343 m/s at low altitude). A sonic shock wave is analogous to the bow wave produced by a boat moving through water. At ground level, a momentary overpressure of 10 to 100 N/m² occurs, followed a moment later by a similar underpressure as the pressure fronts (Mach cones) pass by (see Fig. 32.14). For the Concorde, flying at an altitude of 40,000 ft, the overpressure on the ground is about 20 N/m² (about 10^{-3} atm), and the time between overpressure and underpressure is approximately 0.2 to 0.3 second.

The "boom carpet," a term applied to the zone affected by the sonic boom, is roughly one mile wide for each thousand feet of altitude of the aircraft. Thus, a flyover at very high altitude produces a less intense sonic boom, but a wider boom carpet. Regulations of the FAA forbid sonic booms over land areas in United States territory by civilian aircraft, although military craft continue a limited number of supersonic operations over land.

The increasing use of helicopters, short-takeoff and landing (STOL) aircraft, and even vertical-takeoff and landing (VTOL) aircraft will present new noise problems, because these craft will operate closer to areas of high population density. Minimizing their noise emission will require careful engineering.

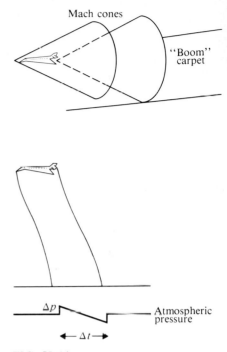

FIG. 32.14
Structure of the sonic boom generated by aircraft flying at supersonic speed.

32.12   ULTRASOUND AND INFRASOUND

Ultrasound is sound with frequencies higher than the audible range (greater than 20,000 Hz). *Infrasound* is sound with frequencies lower than the audible range (less than 16 Hz). There are indications that at high enough sound pressure levels, both types of inaudible sound can affect people adversely.

Ultrasonic waves are emitted by jet engines, high-speed drills, cleaning devices, etc. There have been reports of people experiencing nausea, headache, and changes in blood sugar, due to exposures to ultrasound of high intensity. Fortunately, ultrasonic waves are absorbed very strongly by air, so that ultrasound does not travel very far from the source. At sound pressure levels below 105dB, no adverse effects have been reported (U.S. Environmental Protection Agency, 1973).

Infrasound of moderate intensity is radiated by a number of natural sources, including waterfalls, volcanoes, ocean waves, earthquakes, wind, and thunder. There are also a number of sources, which radiate higher intensities, such as jet aircraft, air-conditioning systems, and other machinery. Headaches, coughing, blurred vision, and nausea are among symptoms reported from excess exposure to high levels of infrasound (U.S. Environmental Protection Agency, 1973).

32.13   SUMMARY

Noise control problems can be dealt with by considering the *source*, *path* and the *receiver*. The most effective approach is to quiet the source by damping of vibrating panels, reducing air velocity, etc. Regulation of noise is shared by federal, state, and local governments. Occupational noise should be limited under the provisions of the Occupational Safety and Health Act. Labeling of certain noisy appliances and hearing protectors has been proposed by the EPA.

Noise transmission through walls decreases as the frequency of the noise and the wall mass increase; attenuation of low-frequency noise requires heavy walls. Wall resonances, excitation of bending waves (at the "critical" frequency), and leakage through holes can cause excessive sound transmission through the walls. Sound barriers are used as substitutes for walls in open-plan rooms and as aids in the reduction of outdoor noise. Sound reflected and diffracted around a barrier reduces its attenuation substantially below that of a comparable airtight wall, however. Enclosures are frequently used to reduce the noise from machines. Porous surfaces absorb well at high frequency, whereas free panels absorb well at low frequency. Perforated panels and similar materials combine both principles of absorption.

Heating, ventilating, and air-conditioning systems distribute noise of aerodynamic and mechanical origin. Aircraft noise is being reduced by technological advances and by government regulation, but this noise reduction is offset to some extent by the increase in the number of flight operations per day at major airports. Thus, aircraft noise continues to be one of our major environmental noise problems. Sonic boom is a pressure transient that occurs during the flyover of an aircraft at supersonic speed. The terms ultrasonic and infrasonic refer to sound above and below the audible range of frequencies, respectively.

References

American National Standards Institute (1975). "Method for Rating the Sound Power Spectra of Small Stationary Noise Sources," ASA Std. 4-1975 (ANSI 3.17-1975), Acoustical Society of America, New York.

Bazley, E. N. (1966). *The Airborne Sound Insulation of Partitions.* London: Her Majesty's Stationary Office.

Berendt, R. D., E. L. R. Corliss, and M. S. Oljavo (1976). *Quieting: A Practical Guide to Noise Control,* Handbook 119. Washington, D.C. National Bureau of Standards.

Beranek, L. L. (1971). *Noise and Vibration Control.* New York: McGraw-Hill.

Magrab, E. B. (1975). *Environmental Noise Control.* New York: Wiley.

Marsh, A. H. (1975). "Noise Control Features of the DC-10," *Noise Control Eng.* **4**: 130.

Rossing, T. D. (1979). *Environmental Noise Control, Selected Reprints.* Stony Brook, N.Y.: American Association of Physics Teachers.

Taylor, R. (1970). *Noise.* Baltimore: Penguin.

U.S. Environmental Protection Agency (1973). *Public Health and Welfare Criterion for Noise,* Report 550/9-73-003. U.S. Environmental Protection Agency, Office of Noise Abatement and Control, Washington, D.C.

U.S. Gypsum (1972). *Sound Control Insulation.* Chicago: U.S. Gypsum Company.

Wesler, J. E. (1975). "Noise and Induced Vibration Levels from Concorde and Subsonic Aircraft," *Sound and Vibration* **9**: 10, 18.

Glossary

aerodynamic Having to do with the flow of air and its interaction with objects in its path.

critical frequency The frequency of bending (flexural) waves in a panel that can be excited by sound waves traveling at the same speed.

FAR Part 36 A regulation by the Federal Aviation Administration that sets standards for noise from new aircraft.

Fresnel number The parameter that determines the sound diffracted around a barrier.

IIC (impact isolation class) A number that describes the effectiveness of a ceiling-floor structure in attenuating impact sound.

infrasonic Having a frequency below the audible range.

OSHA (Occupational Safety and Health Agency) The agency that publishes industrial safety standards set by the Occupational Safety and Health Act.

sonic boom Pressure transient that occurs during the flyover of an aircraft faster than the speed of sound.

STC (sound transmission class) A number that describes the effectiveness of a wall structure in attenuating airborne noise.

supersonic Having a speed greater than that of sound (approximately 340 m/s or 770 mi/hr).

TL (transmission loss) A number that describes the reduction in the sound transmitted through a wall relative to the incident sound.

turbofan engine A type of jet aircraft engine that uses a large fan to drive air into or through the engine.

ultrasonic Having a frequency above the audible range.

waveguide A device that transmits waves (e.g., sound, light, or radio waves) over a particular path minimizing their tendency to propagate in all directions.

Questions for Discussion

1. Why is an enclosure made of a thick blanket of an absorbing material ineffective in attenuating noise?

2. Room A in the figure below contains a noisy machine and Room B is an office. You have insufficient absorbing material to cover the walls and ceiling of both rooms. Would you use the available material in Room A, in Room B, or part of it in each?

3. In response to your complaints about the noise from the apartment above you, the landlord offers to install acoustical tile on your ceiling. Would you expect this to be effective in reducing the noise? Why? What measures for reducing the noise would you suggest?

Problems

1. Sound pressure from a dipole source, such as a fan, is proportional to the sixth power of the air velocity (Section 30.4). If the velocity is doubled, how much will the sound pressure increase? How much will the sound pressure level (in dB) increase?

2. A common wall between two rooms has a transmission loss of 30 dB. If the sound level on the source side is 60 dB, how much sound power is actually transmitted through the wall, if it has an area of 20 m²?

3. Suppose the wall described in Problem 1 has an opening with an area of 0.2 m². What is the transmission loss of the wall with the hole? Is more sound power transmitted through the wall or through the hole?

4. Calculate the total force on one wall of a typical house due to an overpressure of 20 newtons/m² in a sonic boom.

5. If workers are exposed to an A-weighted sound level of 97 dB for one hour, how long can they then be exposed to 92 dB before they reach the maximum exposure allowed?

6. Using Fig. 32.1, estimate the critical frequency for
   a) A plywood wall with an area mass density of 10 kg/m²;
   b) A glass window with an area mass density of 5 kg/m²;
   c) A concrete wall with an area mass density of 500 kg/m².

7. Using Fig. 32.1, estimate the transmission loss (TL) at 100 Hz and at 1000 Hz for each of the three walls described in Problem 6.

CHAPTER 33
Measuring Instruments

In this chapter, we will describe a number of electronic test instruments. You have probably observed a number of them in use in classroom demonstrations, in a recording studio, in a television repair shop, or in a physics laboratory. It is not necessary to understand their construction to be able to use them; in some cases, however, it is very helpful. As with musical instruments, *practice leads to proficiency.*

33.1 THE CATHODE-RAY OSCILLOSCOPE

The cathode-ray oscilloscope is an instrument used to display electrical signals on a fluorescent screen. Its most important component is a cathode-ray tube, similar to the picture tube in a television receiver. The cathode-ray tube contains an electron source, accelerating and focusing electrodes for shaping the electrons into a beam, and two sets of deflection plates for deflecting the beam either horizontally or vertically. Electronic amplifiers amplify the input signals before applying them to the deflection plates. A simplified diagram of the cathode-ray oscilloscope is shown in Fig. 33.1.

The cathode-ray oscilloscope usually performs one of two operations:

1. It displays *one voltage as a function of another* by applying one voltage to the horizontal plates and one to the vertical plates.

2. It displays *a voltage as a function of time* by applying a steadily increasing voltage (sweep) to the horizontal plates, so that the electron beam moves across the screen at a constant speed while it is deflected up and down by a voltage from the vertical amplifier.

FIG. 33.1
A simplified diagram of a cathode-ray oscilloscope.

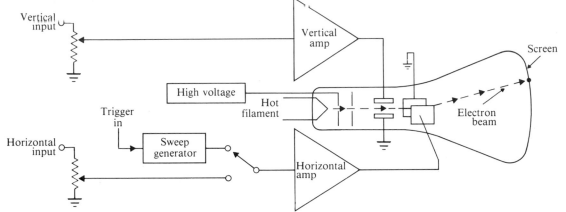

The most useful oscilloscope is one that has a calibrated sweep time, so that time can be obtained directly from the displayed pattern. The sweep time will usually read in seconds or milliseconds per centimeter of beam travel across the screen. By noting the length of one complete wave in centimeters and multiplying by the sweep time setting in seconds per centimeter, the period (in seconds), and therefore the frequency (in hertz), can be determined. The vertical deflection should also be calibrated in volts or millivolts per centimeter, so that the amplitude of the wave can also be determined from the oscilloscope picture. Another important feature is a trigger circuit, which automatically starts each sweep at the same point on the wave (similar to the horizontal hold on a television set).

FIG. 33.2
A laboratory oscilloscope. (Courtesy of B & K Precision.)

Figure 33.2 shows a basic laboratory oscilloscope with a calibrated sweep time and a calibrated vertical deflection. Figure 33.3 is a closeup photograph (called an oscillogram) of a display. The vertical calibrator was set at 2 V/cm and the time base at 5 ms/cm, so the total (peak-to-peak) amplitude is $4 \times 2 = 8$ V, and the period is $2.6 \times 5 = 13$ ms. The frequency is the reciprocal of the period ($1/13$ ms), or 77 Hz.

Oscilloscopes are available with some rather elaborate and useful features. Dual-trace oscilloscopes display two different waveforms at once so that they can be compared (the input and output signals of an amplifier, for example). Storage oscilloscopes can store a transient waveform for a long period of time (the sound of a drum beat or a plucked string, for example). Some oscilloscopes have a differential input; one waveform can be subtracted from another (to reduce noise, for example). Some oscilloscopes have an extremely fast response time, but this is unimportant in acoustical measurements.

33.2 THE AUDIO GENERATOR

Audio oscillators or generators provide a sine wave or pure tone over the entire audible frequency range. Some audio generators also provide a square-wave output, and others provide frequency modulation, although these are features more often found in function generators. Some generators have a low-impedance output of sufficient power to drive a small loudspeaker, but more often the output is at a high impedance so that an amplifier must be used. An audio oscillator is shown in Fig. 33.4.

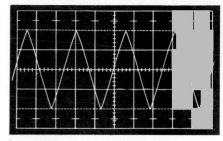

FIG. 33.3
An oscillogram of a triangle waveform (vertical calibration, 2 V/cm; time base, 5 ms/cm).

FIG. 33.4
An audio generator. (Courtesy of Hewlett-Packard.)

33.3 THE FUNCTION GENERATOR

The term function generator usually describes an audio generator that provides several different waveforms, including sine, square, and triangular (and sometimes sawtooth, pulse, and trapezoid). A function generator is shown in Fig. 33.5, and typical waveforms are shown in Fig. 33.6. A useful feature on most function generators is the ability

FIG. 33.5
A function generator. (Courtesy of Hewlett-Packard.)

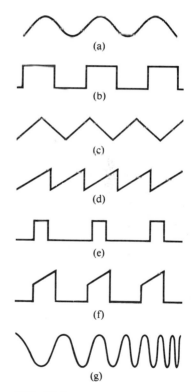

FIG. 33.6
Typical waveforms provided by a function generator: (a) sine; (b) square; (c) triangle; (d) sawtooth; (e) pulse; (f) trapezoid; (g) swept sine wave.

to sweep or modulate the frequency of the oscillator electronically, either by a built-in sweep circuit or by a signal from another generator. Thus, the function generator can operate in a manner similar to the voltage-controlled oscillators (VCO's) found in electronic music synthesizers (see Chapter 27). If the generator can be swept through the entire audio range, it can be used to display the frequency response of an amplifier or a filter on an oscilloscope screen.

Several companies offer inexpensive function generators and audio generators as kits. It is also possible to buy an integrated-circuit chip that contains virtually all the components of a function generator.

33.4   ELECTRONIC TUNERS

The Conn Strobotuner, described in Section 9.6, uses stroboscopic principles to measure sound frequency. A special disc (see Fig. 9.6) rotates at one of twelve selected speeds, each corresponding to a note on the musical scale. Neon lamps, driven by the sound from a microphone, flash on and off to indicate the frequencies of both the fundamental and prominent overtones in a musical tone.

Another type of tuner uses computer-type circuits to produce a standard frequency or frequencies. The Widener AccuTone tuner, for example, uses punched cards to select frequencies of the desired scale. Cards are available to generate the just, Pythagorean, tempered, and meantone scales. Another set of cards duplicates the stretched scale of a particular piano, and it is possible to punch cards to obtain a scale of one's own choosing. The Conn Strobotuner and the Widener AccuTone are shown in Fig. 33.7.

33.5   THE FREQUENCY COUNTER

Another instrument that measures frequency with great accuracy is the electronic frequency counter, which counts the number of vibrations that occur during a given time interval and displays this information on a digital readout. Usually the time interval is one second, but if a more accurate measurement of frequency is desired, the counting interval may be ten seconds (thus measuring frequency to 0.1 Hz).

Some frequency counters measure the period of a vibration (like a very fast stopwatch) instead of its frequency, if desired, in order to obtain greater accuracy at low frequency. The time base in most counters is obtained from a crystal-controlled oscillator, although some counters use the 60-Hz alternating current from the electrical power line. An electronic frequency counter is shown in Fig. 33.8.

(a)

(b)

FIG. 33.7
Two types of electronic tuner:
(a) Conn Strobotuner (Conn).
(b) Widener AccuTone (Courtesy of Widener Engineering, Austin, TX.)

33.6 THE SOUND LEVEL METER

A sound level meter consists of a microphone, an amplifier, some type of weighting network, and a meter calibrated in decibels. It is used primarily for measuring sound levels but also serves conveniently as a calibrated microphone-preamplifier combination in the laboratory. Sound level meters are described in Section 6.4.

Nearly all sound level meters have a frequency response that conforms to one or more of the weighting curves specified by the American National Standards Institute. Most sound level meters offer A-, B-, and C-weighting as shown in Fig. 6.5. Instruments that meet the ANSI standards are specified as Type 1 (precision), Type 2 (general purpose), type 3 (survey), or Type S (special purpose) meters. The C-weighting gives a nearly flat response over the audible range, whereas the A-weighting provides a much-reduced sensitivity at low frequencies in a manner similar to the ear (see Fig. 6.5). Thus, the C-weighting is used to measure the output of a loudspeaker, for example, whereas the A-weighting is used to measure environmental noise. Some special-purpose sound level meters have a single weighting network.

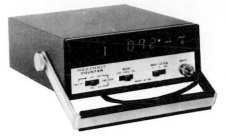

FIG. 33.8
An electronic frequency counter. (Courtesy of Heath/Zenith Co.)

33.7 THE SPECTRUM ANALYZER

A spectrum analyzer is a very useful instrument in any type of acoustical research, since it measures the relative strengths of the various partials (the scientist calls this Fourier analysis of a complex waveform). You were introduced to spectra of complex vibrations in Section 2.7, and to Fourier analysis of complex tones in Section 7.10. Numerous sound spectra have appeared throughout this book, especially in Part III, "Musical Instruments."

Spectral analysis may use either digital or analog techniques. Digital analysis is usually made with a computer, a microprocessor, a pocket calculator, or a specially designed fast Fourier transform (FFT) analyzer. Computer programs for FFT have appeared frequently in the literature, including some adapted for pocket calculators (Schmidt, 1977). Analog spectral analysis is accomplished with a spectrum analyzer, which may be one of several types.

Multiple Filters

One type of spectrum analyzer uses a series of filters, each of which responds to a band of frequency. The most common filter bandwidths are one octave and one-third octave. For octave-band filters covering the audible range, the standard center frequencies are 31.5, 63, 125, 250, 500, 1000, 2000, 4000, 8000, and 16,000 Hz. Thirty-one one-third

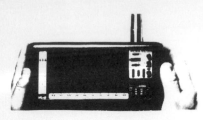

FIG. 33.9
A one-third octave spectrum analyzer that displays the output on an array of light-emitting diodes. (Courtesy of Ivie Electronics.)

FIG. 33.10
An octave-band analyzer. (Courtesy of GenRad, Inc.)

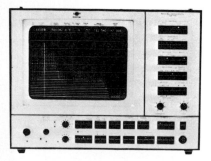

FIG. 33.11
A real-time spectrum analyzer. (Courtesy of B & K Instruments.)

octave filters cover the same range; every third filter has one of the center frequencies above. A one-third octave analyzer that displays the spectrum on an array of light-emitting diodes is shown in Fig. 33.9.

Tunable Filters

Another type of spectrum analyzer uses a single filter that can be tuned, either continuously or in steps. Filter bandwidths in different analyzers vary widely; some are as narrow as one-seventieth octave, some are as wide as one octave, some have a fixed number of hertz. A general rule is that the narrower the filter bandwidth, the more slowly the frequency must be changed in order to obtain an accurate spectrum. An octave-band analyzer is shown in Fig. 33.10.

Real-Time Analyzers

A real-time spectrum analyzer is one which computes the spectrum of a waveform in a very short time, so that rapid changes in the spectrum can be observed. Real-time spectrum analyzers are very useful in the analysis of sound and vibration; they are also quite expensive.

Some spectrum analyzers, such as the one-third octave analyzer shown in Fig. 33.9, display spectra in real time. However, most real-time spectrum analyzers have greater resolution than is practical with multiple filters. Most often, the waveform is rapidly sampled at regular intervals and from these samples a fast Fourier transform is computed. A real-time FFT analyzer is shown in Fig. 33.11.

The Sound Spectrograph

The sound spectrograph, originally developed at the Bell Telephone Laboratories, is widely used to study speech sounds. A current model is shown in Fig. 33.12; a schematic diagram was shown in Fig. 16.2, along with a description of its operation. Several sound spectrograms appeared in Chapters 16 and 17.

33.8 DISTORTION ANALYZERS

Special instruments that measure *harmonic* distortion and *inter-modulation* distortion in amplifiers and other audio components are useful (see Section 22.8). A harmonic distortion analyzer includes a tone generator plus a high-pass filter, which suppresses the fundamental frequency but allows the overtones to be measured. Harmonic distortion is expressed as a percentage:

$$\% \text{ harmonic distortion} = \frac{\text{sum of power in harmonics 2 and up}}{\text{total power}} \times 100\%$$

An intermodulation distortion analyzer supplies two frequencies and measures the extent to which combination tones are generated in the component under test. The testing of amplifiers is typically done with two widely different frequencies ($f_1 = 100$ Hz, $f_2 = 5000$ Hz, for example); the testing of loudspeakers for intermodulation distortion is done with two frequencies within the speaker's range of reproduction. Intermodulation distortion results in the production of components at frequencies $f_2 + f_1$, $f_2 - f_1$, $f_2 + 2f_1$, etc. The total intermodulation distortion is the variation in amplitude of f_2 divided by the average amplitude of f_2.

Examples of waveforms with harmonic and intermodulation distortion are shown in Fig. 22.10. A harmonic distortion analyzer and an intermodulation distortion analyzer are shown in Fig. 33.13.

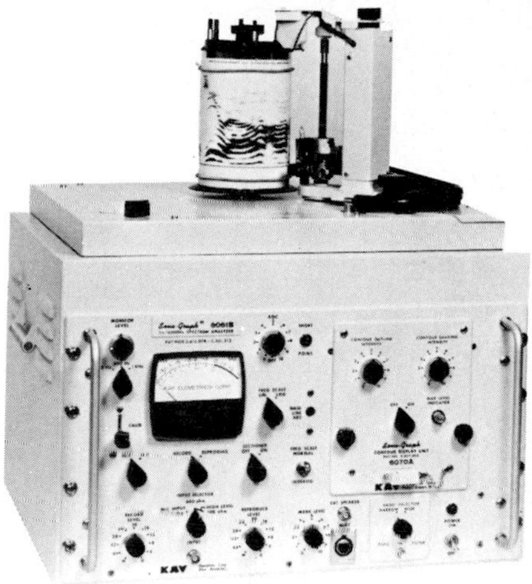

FIG. 33.12
A sound spectrograph. (Courtesy of Kay Elemetrics.)

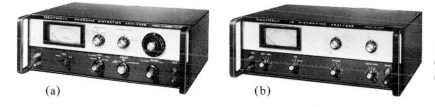

(a)   (b)

FIG. 33.13
(a) A harmonic distortion analyzer
(b) An intermodulation distortion analyzer. (Courtesy of Heath/Zenith Co.)

33.9   THE LEVEL RECORDER

A graphic level recorder, which makes a permanent record of time-varying data, is a very useful item of equipment in acoustics laboratories. When measuring reverberation time, for example, the sound level (usually within a single octave band) is recorded as a function of time, and from this recording the reverberation time can be determined (see Fig. 23.5). The decay time of sound from a percussion musical instrument (see Fig. 13.5) or a percussive noise source can be recorded similarly on a graphic level recorder with a fast response.

Graphic level recorders are often coupled to other instruments, such as spectrum analyzers or audio generators, so that recordings of level versus frequency, for example, can be made. In Fig. 33.14(b), a graphic level recorder is shown coupled to a spectrum analyzer by a chain drive, so that the frequency of the spectrum analyzer is tuned through its range as the paper chart advances. The result, when specially calibrated paper is used, is a graph of level versus frequency.

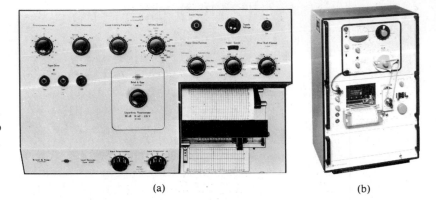

(a)   (b)

FIG. 33.14
(a) A graphic level recorder
(courtesy of B & K Instruments, Inc.).
(b) A graphic level recorder coupled to
a spectrum analyzer in order to record
the spectrum of a complex sound
(courtesy of GenRad, Inc.).

33.10 THE IMPEDANCE TUBE

An impedance tube consists of a sound source (usually a small loudspeaker) together with a probe microphone that can sample the sound pressure within the tube connecting the source to the item under test. One important use for an impedance tube is to measure the sound absorption coefficient of various types of building materials. Another is to study the acoustical behavior of mufflers and other types of acoustical filters. The apparatus shown in Fig. 11.3, used to graph the input impedance of musical wind instruments, is a special type of impedance tube with a fixed microphone. A laboratory type of impedance tube with a movable probe microphone is shown in Fig. 33.15.

FIG. 33.15
An impedance tube. (Courtesy of B & K Instruments, Inc.)

References

Schmidt, S. A. (1977). "Fourier Analysis and Synthesis with a Pocket Calculator," *Am. J. Physics* **45**: 79.

Appendix

━━━━━━━━━━━━━

A.1 UNITS: SI AND OTHER SYSTEMS

The preferred system of units, based on the meter, kilogram, and second, is called the *Système International*, or *SI*. It is also referred to as the *MKS system*. It is recommended for all scientific publications, and will become more familiar as the United States converts to metric measure. The SI system is used throughout this book, although occasionally the English system is used as well.

The *English system*, in use for many years in the United States, uses the foot as a unit of length, the pound as a unit of force, and the second as a unit of time. It has a number of disadvantages, the greatest of which is that it is not used in most countries of the world. It is a clumsy system, and is generally avoided in scientific work.

Another metric system that is in common use is the *cgs system*, based on the centimeter, the gram, and the second. Units in this system are considerably smaller than the corresponding SI units; conversions between the two are relatively simple.

Let us compare these three systems by referring to Newton's law: force = mass × acceleration.

System	Force unit	Mass unit	Acceleration unit
SI (MKS)	newton (N)	kilogram (kg)	meters/second^2 (m/s^2)
cgs	dyne (dyn)	gram (g)	centimeters/second^2 (cm/s^2)
English	pound (lb)	slug (slug)	feet/second^2 (ft/s^2)

Other physical quantities and units are given in Table A.1.

TABLE A.1 Physical quantities and units.

Quantity	Symbol	SI unit	Abbrev.	cgs unit	Abbrev.	English unit	Abbrev.	Formula
Distance, length	d, L	meter	m	centimeter	cm	foot	ft	
Mass	m	kilogram	kg	gram	g	slug	—	
Time	t	second	s	second	s	second	s	
Speed	v	(meters/sec)	m/s	(centimeters/ second)	cm/s	(feet/second)	ft/s	$v = d/t$
Force	F	newton	N	dyne	dyn	pound	lb	$F = ma$
Energy,	E	joule	J	erg	erg	(foot-pound)	ft-lb	
work	$\mathscr{W}$	joule	J	erg	erg	(foot-pound)	ft-lb	$\mathscr{W} = Fy$
Pressure	p	pascal (N/m²)	Pa	(dyne/cm²)	dyn/cm²	(pounds/inch²)	lb/in²	$p = F/A$
Power								
mechanical	$\mathscr{P}$	watt	W	(ergs/second)	erg/s	horsepower	hp	$\mathscr{P} = \mathscr{W}/t$
sound	W	watt	W					$W = pU$
electrical	P	watt	W					$P = I^2R = V^2/R$
Potential	V	volt	V					
Current	I	ampere	A					$I = V/R$
Resistance, impedance	R, Z	ohm	Ω					
Frequency	f	hertz	Hz					$f = 1/T$

One of the features of the metric system is that units are related by powers of ten. New units can be derived by using standard prefixes. For example, a kilogram is 1000 grams; a centimeter is $^1/_{100}$ meter, etc. Table A.2 gives the standard prefixes, Table A.3 gives factors for converting from one system of units to another.

TABLE A.2 Standard prefixes.

Prefix	Symbol	Factor
terra	T	10^{12}
giga	G	10^9
mega	M	10^6
kilo	k	10^3
milli	m	10^{-3}
micro	μ	10^{-6}
nano	n	10^{-9}
pico	p	10^{-12}

Other prefixes are peta (10^{15}), exa (10^{18}), deci (10^{-1}), centi (10^{-2}), femto (10^{-15}) and atto (10^{-18}).

TABLE A.3 Conversion of units.

1 meter = 39.37 in = 3.281 ft
1 kilogram = 0.0685 slug
1 newton = 10^5 dyne = 0.225 lb
1 joule = 10^7 ergs = 0.738 ft-lb
1 pascal (N/m^2) = 10 dyn/cm^2 = 1.45×10^{-4} lb/in^2 = 9.872×10^{-6} atm
1 foot = 0.3048 m
1 pound = 4.448 N
1 horsepower = 746 W

A.2   POWERS OF TEN AND LOGARITHMS

It is convenient to use *scientific notation* to express large and small numbers. The number 30,000 can be obtained by multiplying $3 \times 10,000$, or $3 \times 10 \times 10 \times 10 \times 10$. This is written 3×10^4 or 3.0×10^4; on some electronic calculators, this appears as "3.0 E4." Table A.4 further illustrates scientific notation.

TABLE A.4 Scientific notation

$$1,000,000 = 10^6 \text{ or E6}$$
$$100,000 = 10^5 \text{ or E5}$$
$$10,000 = 10^4 \text{ or E4}$$
$$1,000 = 10^3 \text{ or E3}$$
$$100 = 10^2 \text{ or E2}$$
$$10 = 10^1 \text{ or E1}$$
$$1 = 10^0 \text{ or E0}$$
$$.1 = 10^{-1} \text{ or E} - 1$$
$$.01 = 10^{-2} \text{ or E} - 2$$
$$.001 = 10^{-3} \text{ or E} - 3$$
$$.0001 = 10^{-4} \text{ or E} - 4$$
$$.00001 = 10^{-5} \text{ or E} - 5$$
$$.000001 = 10^{-6} \text{ or E} - 6$$

Examples:
$$23,500 = 2.35 \times 10^4 \text{ or } 2.35 \text{ E4};$$
$$0.084 = 8.4 \times 10^{-2} \text{ or } 8.4 \text{ E} - 2;$$
$$0.005307 = 5.307 \times 10^{-3} \text{ or } 5.307 \text{ E} - 3;$$
$$186,000 = 1.86 \times 10^5 \text{ or } 1.86 \text{ E5}.$$

Arithmetic in Scientific Notation

To *multiply* in scientific notation, we add the exponents; to *divide*, we subtract the exponents. For example:

$$(3.1 \times 10^4) \times (4.2 \times 10^5) = 13.02 \times 10^9 = 1.302 \times 10^{10};$$

$$(3.1 \times 10^4) \times (2.3 \times 10^{-2}) = 7.13 \times 10^2;$$
$$(6.4 \times 10^5)/(4.1 \times 10^3) = 1.56 \times 10^2;$$
$$(7.35 \times 10^3)/(3.2 \times 10^{-2}) = 2.30 \times 10^5.$$

Logarithms

The logarithm to the base 10 of a number x is the power to which 10 must be raised to equal x. That is, if $x = 10^y$, then $y = \log x$. Logarithms of some numbers are given in Table A.5, and several useful identities involving logarithms are listed in Table A.6.

TABLE A.5 Logarithms of some numbers.

x	$\log x$	x	$\log x$	x	$\log x$
1	0	10	1.000	$1/2$	-0.301
2	0.301	20	1.301	$1/3$	-0.477
3	0.477	30	1.477	$1/4$	-0.602
4	0.602	40	1.602	$1/5$	-0.699
5	0.699	50	1.699	0.1	-1.000
6	0.778	60	1.778	0.2	-0.699
7	0.845	70	1.845	0.3	-0.523
8	0.903	80	1.903	0.4	-0.398
9	0.954	90	1.954	0.5	-0.301
10	1.000	100	2.000		

TABLE A.6 Identities.

$$\log AB = \log A + \log B$$
$$\log A/B = \log A - \log B$$
$$\log A^n = n \log A$$
$$\log 1/A = - \log A$$

A.3 SOLVING EQUATIONS

To solve an equation for an unknown quantity, perform the same arithmetic operations to both sides of the equation, step by step, until the unknown quantity stands alone. Examples:

$6t = 30$	Divide by 6 to get $t = 5$;
$5x + 7 = 32$	Subtract 7: $5x = 25$
	Divide by 5: $x = 5$;
$4p^2 + 8 = 44$	Subtract 8: $4p^2 = 36$
	Divide by 4: $p^2 = 9$

Take the square root: $p = 3$ (strictly speaking, $p = \pm 3$, because $(-3)(-3) = 9$ also).

A.4 GRAPHS

Graphs are very useful in showing how one quantity relates to another. Often one of the quantities is time. A graph of pressure versus time, for example, which represents the waveform of a sound wave, conveys a great deal of information about the sound.

A Linear Graph

If two quantities are related by the equation $y = 2x + 3$, the graph of y versus x will be a straight line having a slope of 2, as shown in Fig. A.1.

$y = 2x + 3$

Figure A.1

A Logarithmic Graph

If two quantities are related by the equation $y = 10 \log x$, the graph of y versus x will not be a straight line, unless the scale for x is logarithmic, as shown in Fig. A.2.

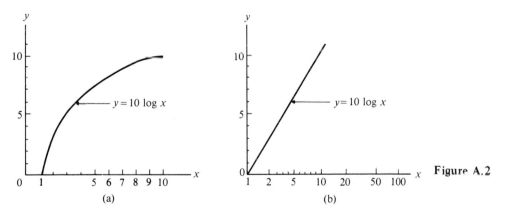

Figure A.2

Determining the Slope of a Curve

If a curve is a straight line, its slope is the number of units of "rise" for each unit of "run." The slope can be calculated by using any two points on the curve and determining Δy and Δx between those two points. If a curve is not a straight line, a tangent should be drawn to the curve at a particular point and Δy and Δx determined for two points on the tangent line, as shown in Fig. A.3(b).

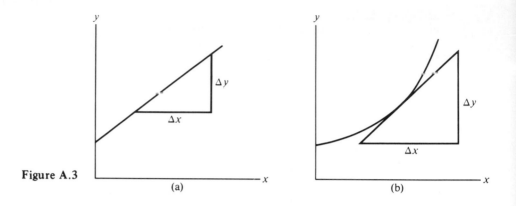

Figure A.3

(a)   (b)

A.5 DECIBELS

Sound pressure level L_p, sound intensity level L_I, and sound power level L_W are all measured in decibels. The formulas are as follows:

$L_p = 20 \log p/p_0,$ $p_0 = 2 \times 10^{-5} \, \text{N/m}^2 = 20 \, \mu\text{Pa}$ (20 micropascals);
$L_I = 10 \log I/I_0,$ $I_0 = 10^{-12} \, \text{W/m}^2;$
$L_W = 10 \log W/W_0,$ $W_0 = 10^{-12} \, \text{W}$ (watts).

Table A.7 lists the pressure ratios and power (or intensity) ratios for different decibels.

TABLE A.7 Conversion from decibels to pressure or power.

Pressure ratio	Power ratio	− dB +	Pressure ratio	Power ratio
1.0000	1.0000	0	1.000	1.000
.9886	.9772	.1	1.012	1.023
.9772	.9550	.2	1.023	1.047
.9661	.9333	.3	1.035	1.072
.9550	.9120	.4	1.047	1.096
.9441	.8913	.5	1.059	1.122
.9333	.8710	.6	1.072	1.148
.9226	.8511	.7	1.084	1.175
.9120	.8318	.8	1.096	1.202
.9016	.8128	.9	1.109	1.230
.8913	.7943	1.0	1.122	1.259
.8810	.7762	1.1	1.135	1.288
.8710	.7586	1.2	1.148	1.318
.8610	.7413	1.3	1.161	1.349
.8511	.7244	1.4	1.175	1.380

Pressure ratio	Power ratio	– dB +	Pressure ratio	Power ratio
.8414	.7079	1.5	1.189	1.413
.8318	.6918	1.6	1.202	1.445
.8222	.6761	1.7	1.216	1.479
.8128	.6607	1.8	1.230	1.514
.8035	.6457	1.9	1.245	1.549
.7943	.6310	2.0	1.259	1.585
.7852	.6166	2.1	1.274	1.622
.7762	.6026	2.2	1.288	1.660
.7674	.5888	2.3	1.303	1.698
.7586	.5754	2.4	1.318	1.738
.7499	.5623	2.5	1.334	1.778
.7413	.5495	2.6	1.349	1.820
.7328	.5370	2.7	1.365	1.862
.7244	.5248	2.8	1.380	1.905
.7161	.5129	2.9	1.396	1.950
.7079	.5012	3.0	1.413	1.995
.6998	.4898	3.1	1.429	2.042
.6918	.4786	3.2	1.445	2.089
.6839	.4677	3.3	1.462	2.138
.6761	.4571	3.4	1.479	2.188
.6683	.4467	3.5	1.496	2.239
.6607	.4365	3.6	1.514	2.291
.6531	.4266	3.7	1.531	2.344
.6457	.4169	3.8	1.549	2.399
.6383	.4074	3.9	1.567	2.455
.6310	.3981	4.0	1.585	2.512
.6237	.3890	4.1	1.603	2.570
.6166	.3802	4.2	1.622	2.630
.6095	.3715	4.3	1.641	2.692
.6026	.3631	4.4	1.660	2.754
.5957	.3548	4.5	1.679	2.818
.5888	.3467	4.6	1.698	2.884
.5821	.3388	4.7	1.718	2.951
.5754	.3311	4.8	1.738	3.020
.5689	.3236	4.9	1.758	3.090
.5623	.3162	5.0	1.778	3.162
.5559	.3090	5.1	1.799	3.236

Pressure ratio	Power ratio	− dB +	Pressure ratio	Power ratio
.5495	.3020	5.2	1.820	3.311
.5433	.2951	5.3	1.841	3.388
.5370	.2884	5.4	1.862	3.467
.5309	.2818	5.5	1.884	3.548
.5248	.2754	5.6	1.905	3.631
.5188	.2692	5.7	1.928	3.715
.5129	.2630	5.8	1.950	3.802
.5070	.2570	5.9	1.972	3.890
.5012	.2512	6.0	1.995	3.981
.4955	.2455	6.1	2.018	4.074
.4898	.2399	6.2	2.042	4.169
.4842	.2344	6.3	2.065	4.266
.4786	.2291	6.4	2.089	4.365
.4732	.2239	6.5	2.113	4.467
.4677	.2188	6.6	2.138	4.571
.4624	.2138	6.7	2.163	4.677
.4571	.2089	6.8	2.188	4.786
.4519	.2042	6.9	2.213	4.898
.4467	.1995	7.0	2.239	5.012
.4416	.1950	7.1	2.265	5.129
.4365	.1905	7.2	2.291	5.248
.4315	.1862	7.3	2.317	5.370
.4266	.1820	7.4	2.344	5.495
.4217	.1778	7.5	2.371	5.623
.4169	.1738	7.6	2.399	5.754
.4121	.1698	7.7	2.427	5.888
.4074	.1660	7.8	2.455	6.026
.4027	.1622	7.9	2.483	6.166
.3981	.1585	8.0	2.512	6.310
.3936	.1549	8.1	2.541	6.457
.3890	.1514	8.2	2.570	6.607
.3846	.1479	8.3	2.600	6.761
.3802	.1445	8.4	2.630	6.918
.3758	.1413	8.5	2.661	7.079
.3715	.1380	8.6	2.692	7.244
.3673	.1349	8.7	2.723	7.413
.3631	.1318	8.8	2.754	7.586
.3589	.1288	8.9	2.786	7.762

Pressure ratio	Power ratio	− dB +	Pressure ratio	Power ratio
.3548	.1259	9.0	2.818	7.943
.3508	.1230	9.1	2.851	8.128
.3467	.1202	9.2	2.884	8.318
.3428	.1175	9.3	2.917	8.511
.3388	.1148	9.4	2.951	8.710
.3350	.1122	9.5	2.985	8.913
.3311	.1096	9.6	3.020	9.120
.3273	.1072	9.7	3.055	9.333
.3236	.1047	9.8	3.090	9.550
.3199	.1023	9.9	3.126	9.772
.3162	.1000	10.0	3.162	10.000
.3126	.09772	10.1	3.199	10.23
.3090	.09550	10.2	3.236	10.47
.3055	.09333	10.3	3.273	10.72
.3020	.09120	10.4	3.311	10.96
.2985	.08913	10.5	3.350	11.22
.2951	.08710	10.6	3.388	11.48
.2917	.08511	10.7	3.428	11.75
.2884	.08318	10.8	3.467	12.02
.2851	.08128	10.9	3.508	12.30
.2818	.07943	11.0	3.548	12.59
.2786	.07762	11.1	3.589	12.88
.2754	.07586	11.2	3.631	13.18
.2723	.07413	11.3	3.673	13.49
.2692	.07244	11.4	3.715	13.80
.2661	.07079	11.5	3.758	14.13
.2630	.06918	11.6	3.802	14.45
.2600	.06761	11.7	3.846	14.79
.2570	.06607	11.8	3.890	15.14
.2541	.06457	11.9	3.936	15.49
.2512	.06310	12.0	3.981	15.85
.2483	.06166	12.1	4.027	16.22
.2455	.06026	12.2	4.074	16.60
.2427	.05888	12.3	4.121	16.98
.2399	.05754	12.4	4.169	17.38
.2371	.05623	12.5	4.217	17.78
.2344	.05495	12.6	4.266	18.20
.2317	.05370	12.7	4.315	18.62

Pressure ratio	Power ratio	− dB +	Pressure ratio	Power ratio
.2291	.05248	12.8	4.365	19.05
.2265	.05129	12.9	4.416	19.50
.2239	.05012	13.0	4.467	19.95
.2213	.04898	13.1	4.519	20.42
.2188	.04786	13.2	4.571	20.89
.2163	.04677	13.3	4.624	21.38
.2138	.04571	13.4	4.677	21.88
.2113	.04467	13.5	4.732	22.39
.2089	.04365	13.6	4.786	22.91
.2065	.04266	13.7	4.842	23.44
.2042	.04169	13.8	4.898	23.99
.2018	.04074	13.9	4.955	24.55
.1995	.03981	14.0	5.012	25.12
.1972	.03890	14.1	5.070	25.70
.1950	.03802	14.2	5.129	26.30
.1928	.03715	14.3	5.188	26.92
.1905	.03631	14.4	5.248	27.54
.1884	.03548	14.5	5.309	28.18
.1862	.03467	14.6	5.370	28.84
.1841	.03388	14.7	5.488	29.51
.1820	.03311	14.8	5.495	30.20
.1799	.03236	14.9	5.559	30.90
.1778	.03162	15.0	5.623	31.62
.1758	.03090	15.1	5.689	32.36
.1738	.03020	15.2	5.754	33.11
.1718	.02951	15.3	5.821	33.88
.1698	.02884	15.4	5.888	34.67
.1679	.02818	15.5	5.957	35.48
.1660	.02754	15.6	6.026	36.81
.1641	.02692	15.7	6.095	37.15
.1622	.02630	15.8	6.166	38.02
.1603	.02570	15.9	6.237	38.90
.1585	.02512	16.0	6.310	39.81
.1567	.02455	16.1	6.383	40.74
.1549	.02399	16.2	6.457	41.69
.1531	.02344	16.3	6.531	42.66
.1514	.02291	16.4	6.607	43.65

Pressure ratio	Power ratio	− dB +	Pressure ratio	Power ratio
.1496	.02239	16.5	6.683	44.67
.1479	.02188	16.6	6.761	45.71
.1462	.02138	16.7	6.839	46.77
.1445	.02089	16.8	6.918	47.86
.1429	.02042	16.9	6.998	48.98
.1413	.01995	17.0	7.079	50.12
.1396	.01950	17.1	7.161	51.29
.1380	.01905	17.2	7.244	52.48
.1365	.01862	17.2	7.328	53.70
.1349	.01820	17.4	7.413	54.95
.1334	.01778	17.5	7.499	56.23
.1318	.01738	17.6	7.586	57.54
.1303	.01698	17.7	7.674	58.88
.1288	.01660	17.8	7.762	60.26
.1274	.01622	17.9	7.852	61.66
.1259	.01585	18.0	7.943	63.10
.1245	.01549	18.1	8.035	64.57
.1230	.01514	18.2	8.128	66.07
.1216	.01479	18.3	8.222	67.61
.1202	.01445	18.4	8.318	69.18
.1189	.01413	18.5	8.414	70.79
.1175	.01380	18.6	8.511	72.44
.1161	.02349	18.7	8.610	74.13
.1148	.01318	18.8	8.710	75.86
.1135	.01288	18.9	8.811	77.62
.1122	.01259	19.0	8.913	79.43
.1109	.01230	19.1	9.016	81.28
.1096	.01202	19.2	9.120	83.18
.1084	.01173	19.3	9.226	85.11
.1072	.01148	19.4	9.333	87.10
.1059	.01122	19.5	9.441	89.13
.1047	.01096	19.6	9.550	91.20
.1035	.01072	19.7	9.661	93.33
.1023	.01047	19.8	9.772	95.50
.1012	.01023	19.9	9.886	97.72
.1000	.01000	20.0	10.000	100.00

Pressure ratio	Power ratio	$-$ dB $+$	Pressure ratio	Power ratio
3.162×10^{-1}	10^{-1}	10	3.162	10
10^{-1}	10^{-2}	20	10	10^{2}
3.162×10^{-2}	10^{-3}	30	3.162×10	10^{3}
10^{-2}	10^{-4}	40	10^{2}	10^{4}
3.162×10^{-3}	10^{-5}	50	3.162×10^{2}	10^{5}
10^{-3}	10^{-6}	60	10^{3}	10^{6}
3.162×10^{-4}	10^{-7}	70	3.162×10^{3}	10^{7}
10^{-4}	10^{-8}	80	10^{4}	10^{8}
3.162×10^{-5}	10^{-9}	90	3.162×10^{4}	10^{9}
10^{-5}	10^{-10}	100	10^{5}	10^{10}

A.6   A PROPOSAL FOR THREE NEW CLEFS

Music is most often written on two staffs, designated by treble and bass clefs. The staff with the bass clef extends from G_2 to A_3 (98 to 220 Hz), and the staff with the treble clef extends from E_4 to F_5 (330 to 698 Hz). However, the piano keyboard extends from A_0 to C_8 (27.5 to 4186 Hz), and the range of human hearing may extend to C_{10} (16,744 Hz) and beyond. Notes above and below the staff are written by using either ledger lines (extra lines added above or below the staff) or notations such as "8va," "15va," etc. Figure A.4 shows notations commonly used for the notes C_6 (1047 Hz) and C_7 (2093 Hz).

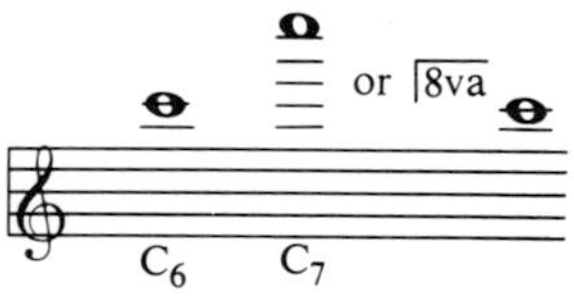

Figure A.4

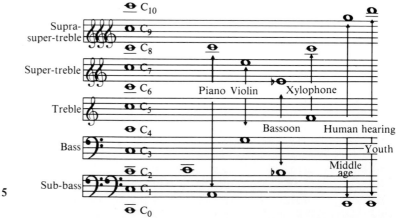

Figure A.5

Adding two new staffs above the treble clef and one new staff below the bass clef avoids the use of ledger lines and the various "-va" notations. The five staffs together allow notation of over nine octaves (F_0 to G_9) without ledger lines. The audible overtones of all musical instruments, as well as the fundamental notes, can be written on musical staffs. The entire piano range (A_0 to C_8) requires only the lower four staffs, the upper one being for overtones. The new clefs are shown in Fig. A.5.

I have named these clefs "sub-bass" (SB), "super-treble" (ST), and "supra-super-treble" (SST). The SB clef is two octaves below the bass clef, so its staff has the same note names; the ST and SST clefs are two and four octaves above the treble clef, respectively, so their staffs have the same note names as the treble. Note that whereas there is one line (middle C) between the treble and bass staffs, there are two lines between each of the others.

A.7   NOTATION USED FOR NOTES

Throughout this book we have used the U.S.A. standard note designation, which is widely used in world literature as well. However, many European musicians use a different notation for octaves, and designate sharps and flats by "is" and "es" respectively, following the note name. B_3 is denoted by "h," and $B_3^\flat$ by "b".

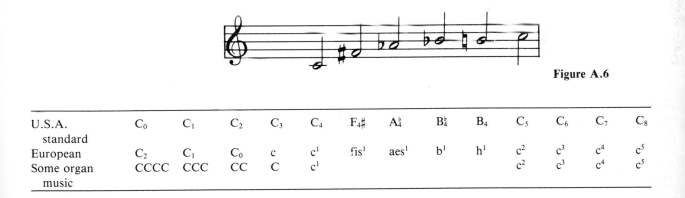

Figure A.6

	C_0	C_1	C_2	C_3	C_4	$F_4^\sharp$	$A_4^\flat$	$B_4^\flat$	B_4	C_5	C_6	C_7	C_8
U.S.A. standard	C_0	C_1	C_2	C_3	C_4	$F_4^\sharp$	$A_4^\flat$	$B_4^\flat$	B_4	C_5	C_6	C_7	C_8
European	C_2	C_1	C_0	c	c^1	fis^1	aes^1	b^1	h^1	c^2	c^3	c^4	c^5
Some organ music	CCCC	CCC	CC	C	c^1					c^2	c^3	c^4	c^5

Answers To Odd-Numbered Problems

Chapter 1

3. 1000 kg/m^3, 920 kg/m^3

5. 1.86 m/s^2 = 0.19 g

9. For cylinder 2 m long, 1 m in circumference, $F = 2 \times 10^5$ N

11. 83%, dissipated as heat (core loss, friction, etc.)

Chapter 2

1. (a) 49 N/m; (b) 1.11 Hz

3. 0.99 m

5. 0.061 J

7. 3.05 Hz and 1.56 Hz, compared to 3.15 Hz and 1.58 Hz

Chapter 3

1. (a) 3 × 10^6 Hz; (b) 4 × 10^{14} Hz; (c) 6 × 10^{14} Hz; (d) 10^{10} Hz

3. From $t = 23°$ to 37°, $\Delta v = 8.4$ m/s, $\Delta f/f = 2.4\%$

5. 260 m/s

7. $t = 2.42 \times 10^{-3}$ s

9. 663 m

11. 349.9 m/s and 349.3 m/s

Chapter 4

1. 15 Hz

3. (a) 35 Hz, 70 Hz; (b) 18 Hz, 53 Hz

5. 2.5 Hz

7. 2550 Hz, 5100 Hz, 7650 Hz

Chapter 5

1. 2942 Hz

3. 5.5 × 10^{-7} N

5. 8.7 × 10^{-4} s for $L_2 - L_1 = 30$ cm

7. (a) 8.0 × 10^{-5}; (b) 30; (c) 1.73 × 10^3; (d) 4.19946 × 10^2 ≃ 420

9. (a) $x = 2$; (b) $x = 1000$; (c) $x = 20$; (d) $x = \frac{1}{2}$

Chapter 6

1. 43 dB, 25 dB, 6 dB, 4 dB, -2 dB, 18 dB
3. 56 dB
5. $L_W = 117$ dB; $L_p = 106$ dB, 94 dB
7. 39 phons, 1 sone; 85 phons, 19 sones
9. 6.3×10^{-3} N/m^2 (or Pa); 10^{-7} W/m^2

Chapter 7

1. At 21.7 cm from either end
3. c.b./jnd = 30, 31, 50, 38
5. 100 s^{-1}, 200 s^{-1}, 300 s^{-1}; 100 Hz
9. 650 Hz, 950 Hz; about 160 Hz
11. 9% greater
13. about 700 Hz
15. at $t = 10$ ms: $x = 0.62$ (high-frequency partials); $y = 0.27$ (mid-frequency partials); $z = 0.11$ (fundamental)

Chapter 8

1.

5. 60*, 120, 180*, 480, 600, 720, 900

7.

Perfect fourth		Major third	
C_2:F_2 ($f_2/f_1 = 4/3$)		C_2:E_2 ($f_2/f_1 = 5/4$)	
mf_1	nf_2	mf_1	nf_2
130.8	174.4	130.8	163.5
261.6		261.6	R 327
392.4	R 3488	392.4	
523.2	= 5232	523.2	R 490.5
654.0	R 6976	654.0	= 654.0
784.8		784.8	R 817.5
915.6	R 8720		
1046.4	= 1046.4		

9. 799 kHz, 801 kHz

Chapter 9

3. B_0, B_1, B_2, B_3, B_4, B_5, B_6, B_7, B_8, B_9
5. ET $= 2^{7/12} \cdot 2^{5/12} = 2$; $2^{9/12} \cdot 2^{3/12} = 2$

Just: $\frac{3}{2} \cdot \frac{4}{3} = 2$; $\frac{5}{3} \cdot \frac{6}{5} = 2$
Pyth: $\frac{3}{2} \cdot \frac{4}{3} = 2$; (1.687) (1.184) $= 2$

7. $\frac{32}{27}$
9. 37.6¢ flat

Chapter 10

1. Turn the figure upside down and read from bottom to top.
3. 0.21 mm; 2.3 m/s
5. G_3 and D_4; viola resonances are somewhat higher (about $A_3^\sharp$ and F_4)
7. 373, 746 and 520, 1039 compared to 384, 795 and 574, 980
   Air in the guitar cavity has resonances like two closed pipes equal to length and width (i.e., nearly plane waves propagate back and forth between ribs).

Chapter 11

1. 8.1 cm
3. 100, 255, 380, 495, 610, 730, 840, 960 Hz
   Bugle notes: 233, 349, 466, 587, 698, 831, 932
   (Lower resonances are sharp by nearly a semitone)
5. 1.07, 1.18, 1.20, 1.38, 1.41

Chapter 12

1. (a) 128 Hz; (b) Frequency of D_3 is 15% greater; (c) Flare of bell shortens effective pipe length and reed appears capacitive (played below its resonance frequency)
3. 73%
5. (a) 408 Hz; (b) 14% lower; (c) end correction (very large at mouth end; see Sect. 14.9)

Chapter 13

1. 1 : 2.71 : 5.15 : 8.43 : 12.21 compared to 1 : 2.76 : 5.40 : 8.93 (thin bar)
   Glockenspiel bar shows thick bar behavior.
3.

$$\frac{f_t}{f_L} = \frac{9\pi t}{4L\sqrt{12}}; \ 0.0858 \text{ compared to } 0.0888$$

5. 1.77

7. 2440, 3180, 3980, 4810 Hz; 802 Hz, $G_5 + 39$¢

Chapter 14

1. 0.0156 m or 16 mm; not very practical

3. Piano is longer; 1.054

5. 1.0039; 6.7¢

7. 34.5 ft, 17.2 ft, 8.6 ft; compared to 32, 16, 8 ft.

Chapter 15

1. 780 Hz, 2239 Hz, 3898 Hz; compare 690, 2610, 3570 Hz

5. 536 Hz, 1608 Hz, 2680 IIz, 3752 Hz

7. 1426, 4279, 7132 Hz

9. 546 Hz

Chapter 16

1. 81 Hz male

3. di: 300 Hz in 0.05 s; da: −300 Hz in 0.05 s; du: −700 IIz in 0.1 s

Chapter 17

1. Male: 82–311 Hz, 330–440 Hz, 494 Hz up; Female: 175–370 Hz, 392–659 Hz, 698–1046 Hz

3. /u/ 300, 750, 2500 vs 350, 640, 2550
/ɑ/ 600, 1000, 2500 vs 700, 1200, 2600
/i/ 300, 1800, 2500 vs 300, 1950, 2750

5. (a) 0.01 W; (b) 0.1 W; (c) 1.6 W

Chapter 18

1. (a) 2.5 A; (b) 1 A; (c) 4.25 A; (d) 1 A

3. (a) 2906 Hz; (b) 3.2 Hz

5. (a) 2.31 Ω; (b) 4 Ω

7. 0.56 W; 0.63 W

Chapter 19

1. 7958 Hz (low pass); 318 Hz (low pass)

3. 5000, 37 dB

5. (a) Positive feedback at f_0; (b) 712 Hz

7. (b) 25, 35, 37, 37.5, 37.5, 37, 34, 25; (d) with: 20–20,000 Hz; without: 45–11,000 Hz

Chapter 20

1. 10 dB greater

3. (a) 20 cm; (b) 860 Hz

5. 200 μV

7. 1 mV, 94 dB

9. 1%

11. 20 Hz

Chapter 21

1. (a) 700; (b) 483 m; (c) 21 min.; (d) 0.38 m/s

3. 1.35; a perfect fourth (+20¢)

5. ⅛ inch, 1/16 inch, 1/32 inch

Chapter 22

1. 4 Ω

3. 1.94 A, 2 A

5. 980–2840 dynes or 0.01-0.03 N

7. 2251 Hz

Chapter 23

1. (a) 14 m^2 vs 8 m^2; (b) 4 m^2 vs 71 m^2

3. $t_1 = 23$ ms; side

5. 2.0 s, 1.7 s, 1.5 s, 1.5 s

Chapter 24

1. 900 Hz, 6482 Hz

3. 707 Hz, 354 Hz

5. 69 Hz; yes

Chapter 25

1. 17.2 Hz, 34.4 Hz, 34.4 Hz

Chapter 26

3. 2½ cycles, 7½ cycles, 12½ cycles

Chapter 27

1. (a) exponential; (b) 3.58 V
3. 50, 150, 300, 350, 500, 550, 600, 750, 900, 1050, 1100

Chapter 28

1. (a) 1100; (b) 10110; (c) 100100
5. 8192; 20 log 8192 = 78.3
7. 0.63, 0.14, 2.15

Chapter 29

1. (a) 1.05×10^6; (b) 6.40×10^8
3. 2.2×10^{-6} in^2 or 1.4×10^{-3} mm^2

Chapter 30

1. 15–30 mW of sound power and 25 kW of mechanical power
3. 1.1 Hz
5. (a) 59 dB; (b) −241 dB

Chapter 31

1. 68.7 dB
3. (a) 4 m; (b) 20 m

Chapter 32

1. 64 times, 18 dB
3. 19 dB; through the hole
5. 4 hr
7. (a) 14 dB, 24 dB; (b) 8 dB, 25 dB; (c) 37 dB, 56 dB

Index